Radiation Detection

Radiation Detection

W H Tait, PhD
Senior Lecturer in Physics
North East London Polytechnic

Butterworths
London   Boston
Sydney   Wellington   Toronto   Durban

The Butterworth Group

United Kingdom	**Butterworth & Co (Publishers) Ltd** London: 88 Kingsway, WC2B 6AB	
Australia	**Butterworths Pty Ltd** Sydney: 586 Pacific Highway, Chatswood, NSW 2067 Also at Melbourne, Brisbane, Adelaide and Perth	
Canada	**Butterworth & Co (Canada) Ltd** Toronto: 2265 Midland Avenue, Scarborough, Ontario, M1P 4S1	
New Zealand	**Butterworths of New Zealand Ltd** Wellington: T & W Young Building, 77–85 Customhouse Quay, 1, CPO Box 472	
South Africa	**Butterworth & Co (South Africa) (Pty) Ltd** Durban: 152–154 Gale Street	
USA	**Butterworth (Publishers) Inc** Boston: 10 Tower Office Park, Woburn, Massachusetts 01801	

All rights reserved. No part of this publication may be reproduced or transmitted in any form or by any means, including photocopying and recording, without the written permission of the copyright holder, application for which should be addressed to the Publishers. Such written permission must also be obtained before any part of this publication is stored in a retrieval system of any nature.

This book is sold subject to the Standard Conditions of Sale of Net Books and may not be re-sold in the UK below the net price given by the Publishers in their current price list.

First published 1980

ISBN 0 408 10645 X

© W. H. Tait, 1980

British Library Cataloguing in Publication Data

Tait, W. H.
 Radiation detection.
 1. Nuclear counters 2. Ionizing radiation – Measurement
 I. Title
 539.7'7   QC787.C6   80-40240

ISBN 0-408-10645-X

Typeset by Butterworths Litho Preparation Department
Printed and bound in Great Britain by Redwood Burn Limited, Trowbridge and Esher

Preface

This is an introductory treatment of radiation detection which, nevertheless, extends to a fairly detailed account of detector types, properties and functions. The hope is that it will interest readers of diverse academic and disciplinary backgrounds and that it will serve as the basis for relevant studies up to, including and, in some instances, beyond the level of a first degree. An ambitious target, perhaps, but one that is made possible by the specialised nature of the subject.

The subject is usually known as radiation detection, but it includes both the detection of ionising radiations and the measurement of their properties. It is largely concerned with hardware, that is, with the design and use of radiation detectors, but both of these operations require some knowledge of the properties of ionising radiations, on the one hand, and of the demands made by various applications, on the other. As a result, the book is about ionising radiations, their detection, and the detector functions required of their major applications. It is for the user of detection systems as well as the would-be designer.

The early chapters deal with some basic concepts and the properties of ionising radiations. Some readers will find this material quite elementary or, at least, of a revisionary nature, while others may regard it as a complete course on ionising radiations. In both cases, this part of the book serves an important purpose in defining concepts, terminology and formulae in relation to data. With the range of data supplied in the appendices, the book should be quite self-sufficient in this respect.

The next two chapters present the general properties and the major functions of detection systems without giving much consideration to their specific natures. In effect, this amounts to treating the detector as a 'black box'. It is an approach which has been developed over some years of teaching the subject to students of mixed abilities and disciplinary origins. It defines user and designer criteria, and constructs a framework for the systematic study of detection systems.

The remaining chapters provide detailed descriptions of the most widely used detectors, in turn, and of nuclear electronic systems. Readers will tend to concentrate on those systems which are of immediate, personal concern. This is a natural attitude, which the presentation is designed to accommodate. It should be pointed out, however, that this is a developing subject in which a selective education may soon become an obsolete one.

In conclusion, it is hoped that, with the help of this book, the reader will learn to use detection systems competently, to select and adjust them to best advantage, and, with some further, more specialised reading, design them.

W. H. Tait

Contents

1 Basic Concepts 1
1.1 Introduction 1
1.2 Classical Models 2
1.3 Relativity 8
1.4 Quantum Theory 10

2 The Structure of Matter 18
2.1 Atoms 18
2.2 Atomic Systems 25
2.3 Nuclei 29

3 Ionising Radiations 37
3.1 Introduction 37
3.2 X-radiation 38
3.3 Electrons 43
3.4 Positive Ions 44
3.5 Radioactive Decay Theory 45
3.6 Alpha Particles 47
3.7 Beta Particles 50
3.8 Gamma Radiation 55
3.9 Neutrons 57
3.10 Fundamental Particles 60
3.11 Radiation Sources 61

4 Interactions with Matter 66
4.1 General Theory 66
4.2 Attenuation 73
4.3 Charged Particles 80
4.4 Photons 89
4.5 Neutrons 98

5 Detector Properties 107
5.1 General Features 107
5.2 Detection 109
5.3 Energy Measurement 117
5.4 Position Measurement 122
5.5 Time Measurement 124
5.6 Discrimination 127
5.7 Signal Formation 136

6 Detector Functions 139
6.1 Counting 139
6.2 Pulse Height Spectrometry 144

6.3 Dosimetry 150
6.4 Imaging 157
6.5 Timing 172

7 Gas Counters 176
7.1 Introduction 176
7.2 Ionisation Chambers 177
7.3 Proportional Counters 186
7.4 Geiger-Müller Counters 191
7.5 Spark Chambers 193
7.6 Multiwire Proportional Counters 197
7.7 Functions 202

8 Solid-State Detectors 209
8.1 Introduction 209
8.2 Semiconductor Counters 210
8.3 Integrating Solid-State Devices 224
8.4 Functions 226

9 Scintillation Counters 237
9.1 Basic Systems 237
9.2 Inorganic Scintillators 245
9.3 Organic Scintillators 250
9.4 Functions 254

10 Visual Imaging Systems 267
10.1 Films 267
10.2 Properties of Films 271
10.3 Functions 279
10.4 Fluorescent Screens 282
10.5 Xerography 284
10.6 Ionography 287

11 High-Energy Particle Detectors 291
11.1 General Principles 291
11.2 Nuclear Emulsions 292
11.3 Cloud Chambers 294
11.4 Bubble Chambers 296
11.5 Čerenkov Counter 299
11.6 Imaging Gas Counters 304

12 Nuclear Electronics 305
12.1 Pulses 305
12.2 Nucleonic Units 316
12.3 Nucleonic Systems 332
12.4 Computerised Systems 342

Appendix 1 348
Appendix 2 350
Appendix 3 354
Appendix 4 386

Index 399

Chapter 1
Basic concepts

1.1 INTRODUCTION

Radiation detection involves the detection and investigation of ionising radiations. A major part of the subject relates to the study of radiation detectors and detection systems, but it is not solely concerned with hardware and includes all aspects of the detection process. It also overlaps with a number of peripheral subject areas which are conveniently referred to as 'technology', 'applications' and 'theory'. A complete treatment of the subject must include selected material from these three areas.

Technology represents manufacturing capability and derives, mainly, from materials science and electronics. It determines the limits of performance which can be achieved with various detection systems. The *applications* include a wide range of techniques employing ionising radiations and these define the desirable characteristics of detection systems, that is, the design objectives. The *theory* is the theory of *atomic and nuclear science* and provides a basis for the design of detection systems and for the interpretation of results. In the present treatment, technology and applications are discussed in later chapters, but only to the extent that their effects on the detection process may be appreciated. Further information can be obtained, if required, from other sources. These remarks also apply to the basic theory, but this is reviewed in the first few chapters in order to generate a coherent impression.

One of the difficulties encountered in the study of atomic systems is produced by the conflict between the psychological need to visualise these systems and the inherently unfamiliar and, therefore, abstract concepts required to do so. The problem reflects the divergence between classical and modern physics. Essentially, classical theory is a natural extension of everyday human experience and relies on concepts such as position, mass and velocity, which are easily visualised. Modern theory, on the other hand, has to deal with concepts, such as spinning massless particles and the creation of mass from pure energy, that are well outside the range of normal experience. For this reason, atomic and nuclear theory depends, more than most subjects, on the construction of *models*.

A model is used to *represent* a system and, conversely, a system cannot be represented, or described, without recourse to a model. It provides the means of visualising the system as well as a basis for the development of appropriate terminology and concepts. In many cases, it leads to the establishment of formulae which can be used to describe the system quantitatively and predict its behaviour. It is important to realise, however, that a model is an assumption and may be superseded by a better or more sophisticated model which represents a system more accurately or more extensively, in terms of its observed behaviour. Thus, the models of classical physics have been superseded by those of modern physics in order to 'explain', that is, to represent certain observed properties. The *theory of relativity* is required when very high velocities are involved and the *quantum theory* is necessary for very small systems of about the size of an atom. Both have to be considered when discussing ionising radiations and together they comprise modern physics.

Before we consider these theories, it is useful to make a note of two general properties of models. First of all, it must be realised that diagrams, graphs and even mathematical formulae may be regarded as models. They are more abstract than the simple physical analogues, which might be preferred, but they are often the most direct and least ambiguous means of representing complex systems. Secondly, it should be borne in mind that the best model for a particular purpose is not always the most sophisticated one available. On the contrary, it is the simplest one that is sufficient to do the job. For example, the quantum theory is able to describe all the properties of an electron but is unnecessarily complicated when applied to simple particle-type behaviour, for which classical theory is more suitable. Similarly, quantum theory can 'explain' the diffraction of electrons but the classical wave model is usually the best way of dealing with this phenomenon. Indeed, one of the main reasons for introducing relativistic and quantum models is to make it clear when they are not required.

1.2 CLASSICAL MODELS

Particles

In classical theory, a particle is represented as a small piece of matter with no internal structure. Its properties are modelled on those of fairly large objects, such as balls and missiles, which typically comprise more than 10^{24} component particles and never achieve really high velocities. On this basis, it is assumed that the position of any particle can be defined with absolute precision, in terms of spatial coordinates at any given time. The particle is also assumed to have properties (such as size, mass, velocity, momentum, energy and angular momentum) which are defined, measured and related to each other in the same ways as are those of much larger and slower-moving objects. These assumptions are quite valid in all circumstances except those in which relativistic or quantum effects become significant.

The behaviour of a particle is described by an *equation of motion* which is a relationship between two or more of its properties and is often expressed as a mathematical formula. Such equations may be constructed in terms of so-called natural laws, such as the *conservation laws*. It is found, for example, that the total energy of a system remains constant; that is, it is conserved. The same is true of momentum, angular momentum and electric charge.

There are two types of energy: kinetic energy, T, and potential energy, V. These are interconvertible and it is the total energy, $E = T + V$, which is conserved. Kinetic energy is due to movement and is equal to $\tfrac{1}{2}mv^2$, where m is the mass of the particle and v its velocity. Potential energy is produced by the action of a force. It is the work done in pushing the particle, to its present position, against a force, or the work which would be done by the particle if it were released and moved away by the force — converting its potential energy into kinetic energy in the process. A particle which is under the influence of a force is said to be in a *field*. If it is stationary, it has potential energy but no kinetic energy. A particle which is not under the influence of a force is said to be in *free space*. If it is moving, it has kinetic energy but no potential energy.

The SI unit of energy is the joule (J) but, in radiation science, this is an inconveniently large quantity and the preferred unit is the *electronvolt* (eV). This is defined by reference to the effect of an electric field on a charged particle, but it is applicable to all types of radiation, including uncharged particles and electromagnetic radiations. The electric potential at a particular point in an electric field is defined to be the potential energy of a unit charge at that point. Unfortunately, electric potential, like potential energy, is denoted by V and the full potential energy of a charge, Q, at the point is the product QV. In SI units, Q is measured in coulombs (C) and V in volts (V). An electronvolt is the potential energy of an electron at an electric potential of one volt and, since the electron charge is 1.6×10^{-19} C, the electronvolt and its standard multiples are given by

$$1 \text{ eV} = 1.6 \times 10^{-19} \text{ J} \tag{1.1}$$

1 kilo-electronvolt, i.e.   1 keV = 10^3 eV
1 mega-electronvolt, i.e.   1 MeV = 10^6 eV
1 giga-electronvolt, i.e.   1 GeV = 10^9 eV
(or 1 beva-electronvolt, i.e. 1 BeV = 10^9 eV)

The momentum, p, of a particle is given by the formula $p = mv$. If the particle has any circular movement, it is said to have angular momentum which is the moment of its momentum about some axis of rotation. As for energy, angular momentum has two components. One is *orbital angular momentum* due to the motion of a particle, in an orbit, around some external axis. The other is due to rotation about the particle's own internal axis and is known as *intrinsic angular momentum*, or *spin*. These are denoted by L and S respectively and they are *vectors*; that is, they have magnitude and direction. The direction of angular momentum is defined to be about the axis of rotation in such a way that the rotation is in the form of a right-handed screw in the direction of the axis.

Waves

In classical theory, a wave is quite different from a particle. It can be described as a disturbance, or displacement, which travels through a transmitting medium with a constant velocity v that is a *property of the medium*. A surface wave on water is a typical visual representation. A wave has no precisely defined position, size, mass, electric charge, momentum or discrete energy. It does transfer energy through the medium, but not in discrete lumps and not along with mass. Each piece of the medium can be imagined to bump against the next piece and transfer energy in that way.

Basically, there is only one equation of motion for waves and it is known as the *wave equation*. This is a differential equation, however, and it is solved under various 'boundary conditions' to produce a variety of functions which describe

4 Basic concepts

specific waves. These are known as *wave functions*. They may represent three-dimensional waves, such as that described by the surface of a vibrating ball, they may be two-dimensional waves on the surface of water or a drum, or they may be one-dimensional waves on a string, but they have certain properties in common. Specifically, they are *periodic functions*; that is, they repeat themselves as regular functions of time and space. This can be illustrated by reference to the simplest form of wave, namely, the one-dimensional sinusoidal wave. In this case, the wave function, Ψ (psi), is given by

$$\Psi = a \sin (kx - \omega t) \tag{1.2}$$

or, if preferred, by $a \cos (kx - \omega t)$, which is the same shape but displaced along the x axis, as shown in *Figure 1.1*. In this expression, a is the *amplitude* of the

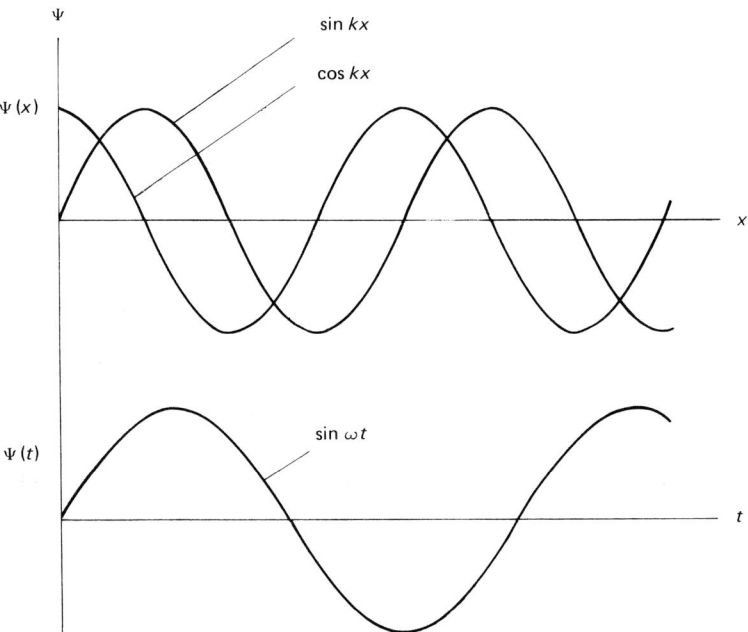

Figure 1.1 Sinusoidal waves in one dimension

wave, which is the maximum value of its displacement. The term ω (omega) is the *angular frequency* and k is known as the *reduced wavenumber*. ω and k are wave *parameters*, that is, they are constants for a given wave but vary from one wave to another. They are not easily visualised and are usually replaced by more useful parameters. These are the *wavelength*, λ, which is the distance between one wave peak and the next one, that is, the length of one complete cycle, and the *frequency*, ν (nu), which is the number of cycles completed in unit time. An alternative time parameter is the *period*, T, which is the time taken to complete one cycle, and, although it is seldom used, a corresponding alternative to the wavelength is the *wavenumber*, K, which is the number of wavelengths travelled in unit time. These parameters are related as follows:

$$k = 2\pi K = \frac{2\pi}{\lambda} \qquad K = \frac{1}{\lambda}$$

$$\omega = 2\pi\nu = \frac{2\pi}{T} \qquad \nu = \frac{1}{T} \tag{1.3}$$

so that the wave function can be expressed as

$$\Psi = a \sin 2\pi \left(\frac{x}{\lambda} - \frac{t}{T} \right) \tag{1.4}$$

Electromagnetic radiations

Electromagnetic radiations are waves produced by moving, or — more accurately — by *accelerating* electric charges. They are similar to other types of wave except that they do not require a transmitting medium. They are transmitted through a vacuum with a velocity c, equal to 2.998×10^8 m s^{-1}, which is a universal constant, that is, a property of free space. For propagation in one direction, along the x axis, they can be described by the wave function of equation 1.2, but the displacement, Ψ, now defines the magnitude of an electric field and a magnetic field perpendicular to each other and to the x axis. An electron in the path of such a wave behaves just like a piece of a vibrating string or a cork on a water surface — it oscillates laterally, across the x axis, in phase with the wave vibrations.

A visual model of an electromagnetic wave can be constructed by imagining a line of electrons displaced into a sinusoidal profile, like that of a water wave, by the passage of the radiation. This can be represented as a graph of Ψ against x or t as in *Figure 1.1*. The same graph can be used to show potential energy against x or t, since this is proportional to physical displacement, so it is a short step to represent an electromagnetic wave in a vacuum by the same graph, in which Ψ is associated with some type of electric potential. The graph is the model and equation 1.2 is its mathematical representation.

Electromagnetic radiations include gamma rays, X-rays, ultra-violet light, visible light, infra-red radiation and radio waves. As shown in *Figure 1.2*, these are identical except in wavelength and frequency. For the long, radio waves, wavelengths λ are measured in metres but for the much shorter light waves the ångstrom (Å) may be used, where

$$1 \text{ Å} = 10^{-10} \text{ m} \tag{1.5}$$

The frequency, ν, is measured in cycles per second (c s^{-1}) or hertz (Hz), where

$$1 \text{ Hz} = 1 \text{ c s}^{-1} \tag{1.6}$$

and, since λ is the distance travelled in the period T, the velocity c of an electromagnetic wave in a vacuum is given by

$$c = \lambda/T = \lambda\nu \tag{1.7}$$

6 Basic concepts

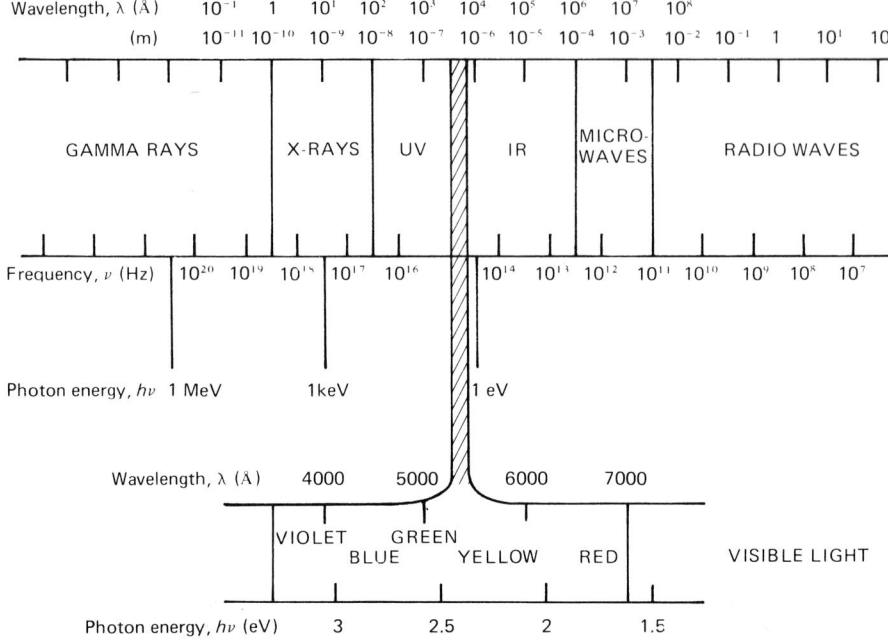

Figure 1.2 Spectrum of electromagnetic radiation

In a refracting medium, of course, the wave is slowed down to a velocity v equal to c/n, where n is the refractive index of the medium. The frequency remains the same, but the wavelength is reduced.

From an experimental point of view there is one very important difference between waves and particles: waves produce interference and diffraction effects. Two light rays, for example, may arrive at one point on a screen in such a way that they are always exactly out of phase so that they produce equal but opposite displacements. The net effect on, say, an electric charge, is zero and, together, they generate zero illumination. This effect is called *cancellation* and, in classical theory, it is quite impossible for two particles to do the same. A diffraction pattern consists of alternate light and dark regions, where the waves superimpose in phase (reinforcement) and out of phase, and is convincing evidence of the wave nature of the radiation involved.

Bound systems

A bound system is one in which wave motion is confined to a particular region. For example, this may be a vibrating string, fixed at each end, or a three-dimensionally vibrating ball. In such a system the wave is refracted back and forth within the system to form *stationary* or *standing waves* whose outer, or envelope, profiles do not change with time. Unlike the progressive waves, discussed above, stationary waves can have only certain *allowed* wavelengths. For a string of length L, for example, the displacement, ψ, must be zero at $x = 0$ and $x = L$, so that L must always be an integral number n of half wavelengths, as illustrated in Figure 1.3. Thus, the wave function is a solution of the wave equation for the boundary conditions $\psi = 0$ at $x = 0, x = L$. It has the general form

$$\psi_n(x) = a \sin\left[2\pi\left(\frac{x}{2L/n}\right)\right] = a \sin\left(\frac{n\pi x}{L}\right) \tag{1.8}$$

where the capital, Ψ, has been replaced by ψ, as is customary for the *time-independent* function. ψ describes the outer profile of the wave, but not its time-dependent movement between these extreme positions, which is usually of less interest.

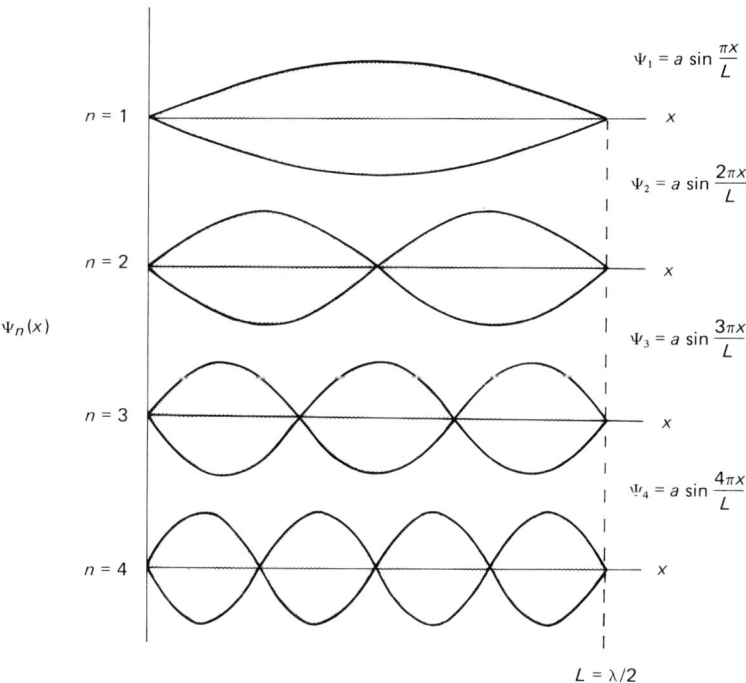

Figure 1.3 One-dimensional stationary waves in a bound system of length L

In equation 1.8, n can have the values $1, 2, 3, \ldots$, giving, for a vibrating string, the fundamental mode ($n = 1$) and the overtones, with shorter wavelengths. Since $v = \lambda \nu$ is a constant, these overtones have higher frequencies than the fundamental mode. The number n is the classical equivalent of a *quantum number*. It is an integer label, or index, which appears in the wave function and restricts the function to a finite set of allowed forms. Such a wave function may be called an *eigenfunction* and each vibrational mode is represented by one quantum number and its corresponding *eigenstate* function. Similar results are obtained for all bound systems but two-dimensional systems require two quantum numbers and three-dimensional systems need three to define all the allowed states, or eigenstates.

1.3 RELATIVITY

Einstein's special theory of relativity concerns the measurement of the properties of objects which are moving relative to the observer. To be exact, it applies to systems moving with uniform relative velocity but can be used, to a good approximation, for situations in which there is some relative acceleration. It becomes important only when these relative velocities are very large; that is, when they approach c, the velocity of light in a vacuum. This is usually the case in astronomy, when fast-moving stars or galaxies are being examined, and is often the case in radiation science, since ionising radiations may have high velocities relative to the observer.

The theory is well documented and a short bibliography is given at the end of this chapter. For present purposes, it is necessary to learn only how and when to use it, but some background information may serve to emphasise the distinction between classical and relativistic mechanics.

If a light-emitting object is receding away from the observer with a relative velocity v, then the measured velocity of light from the object should be $c - v$. Similarly, if the object and the observer are coming together with a relative velocity v, the velocity of light should be measured as $c + v$. In general, the velocity of light emitted from different moving sources, such as the stars, should depend on the relative velocities of these objects. In fact, however, it does not, and is always observed to be c. The velocity of light, and of all electromagnetic radiation, in a vacuum, is not affected by any relative motion between the source and the observer.

The observation is a dramatic departure from classical physics. It appears to be impossible but it is a fact. It must be accepted, along with a number of consequences. Basically, these are that the *measured* properties of an object which is moving, with respect to the observer, are all functions of the relative velocity of the object. For example, the *length* of an object that is at rest, as far as the observer is concerned, is measured as l_0, the *rest length*, but, if the object is moving past him with a relative velocity v, its length appears to *decrease* to l, given by the *Lorentz contraction* as

$$l = l_0 \left(1 - \frac{v^2}{c^2}\right)^{1/2} \tag{1.9}$$

Since the term v/c occurs frequently in relativistic formulae, it is often replaced by β and the term $1/(1 - v^2/c^2)^{1/2}$ may be replaced by γ.

Note that if v is small, that is, if $v \to 0$, this expression reduces to $l = l_0$, the classical formula. As an example, the length of an object moving at $v = 0.8c$ is measured as

$$l = l_0 \left(1 - \frac{0.64c^2}{c^2}\right)^{1/2} = 0.6 l_0$$

The *mass* of an object is also affected by relative velocity. If the object's *rest mass* is m_0, then its observed, and effective, mass m at a velocity v is *increased* to

$$m = m_0 \bigg/ \left(1 - \frac{v^2}{c^2}\right)^{1/2} = m\gamma \tag{1.10}$$

The rest mass of an electron is 9.108×10^{-31} kg, so the mass of an electron moving at a velocity of $0.8c$ is measured as 15.18×10^{-31} kg.

Time appears to pass more slowly for moving objects. If the time required for a certain process to take place in a stationary object is t_0, then the time t required for the same process to be completed in a moving object is given by the *time dilation* formula

$$t = t_0 \bigg/ \left(1 - \frac{v^2}{c^2}\right)^{1/2} = t\gamma \tag{1.11}$$

For example, the time taken for a muon to disintegrate into its decay products is measured by its half-life. The half-life of a stationary muon is 1.52 μs (microseconds, i.e. 10^{-6} second). Muons are found in cosmic radiation, moving at high velocities, and in such circumstances half-lives of the order of 2 μs are measured.

In classical theory, the *momentum*, p, of a moving particle is given by $p = mv$. The relativistic formula for momentum is obtained by inserting the relativistic expression for mass, m, to get

$$p = m_0 v \bigg/ \left(1 - \frac{v^2}{c^2}\right)^{1/2} = m_0 v \gamma \tag{1.12}$$

As for all the other relativistic formulae, this reduces to the classical one as v becomes small.

When the theory of relativity is applied to the concept of *energy* an even more dramatic result is obtained. In classical theory, kinetic energy is given by $T = \tfrac{1}{2}mv^2$, but this formula is not a fundamental relationship; it is derived by means of classical assumptions. When the derivation is repeated using fully relativistic mathematics, the formula obtained is

$$T = m_0 c^2 \left[\left(1 - \frac{v^2}{c^2}\right)^{-1/2} - 1\right] \tag{1.13}$$

or, using equation 1.10,

$$T = mc^2 - m_0 c^2 \tag{1.14}$$

Using the binomial theorem to expand the γ term, it can be shown that equation 1.13 reduces to the classical formula for small values of v, but the equivalence of the two formulae can be demonstrated more easily by using both of them to calculate, for example, the kinetic energy of a particle of rest mass m_0, moving at a velocity v equal to $0.1c$. Both formulations give the result $0.005 m_0 c^2$.

The implications of equation 1.14 can be made clearer by writing it as

$$mc^2 = T + m_0 c^2$$

Since m_0 and c^2 are constants, this shows a direct proportionality between mass, m, and kinetic energy. In effect, it shows that the term mc^2 has the dimensions of energy and can be defined as *total energy*, E, with the term $m_0 c^2$ representing a similar quantity for a rest mass and known as the *rest mass energy*. This conclusion can be formulated as

$$E = mc^2 \tag{1.15}$$

This is an important relationship. It states that the total energy of a particle is equal to its kinetic energy plus a rest mass energy, E_0, equal to $m_0 c^2$, and suggests that mass and energy are interconvertible. This is found to be the case, in practice. Nuclear fission and fusion depend upon the conversion of mass into energy and particles are bound in atoms and nuclei by energy converted from their mass. As an example of the reverse process, pure electromagnetic energy is converted into the mass of an electron and a positron in the pair production effect, described later. In these and many other processes, the conversion is described by equation 1.15 and the equivalence is so complete that the mass of a particle is frequently given in energy units. For example, the rest mass energy of an electron is 0.511 MeV. If it is moving, its energy is greater, of course, and if it is in a force field it also has potential energy.

In classical physics, the relationship between energy and momentum is a useful one. Since kinetic energy is given by $\frac{1}{2}mv^2$ and momentum by mv, these are related by the formula

$$T = p^2/2m \tag{1.16}$$

The corresponding relativistic relationship is derived by squaring both sides of equation 1.10, for relativistic mass, and inserting $E^2 = m^2 c^4$ and $p^2 = m^2 v^2$ to get

$$E^2 = p^2 c^2 + m_0^2 c^4 \tag{1.17}$$

Thus, massless particles and electromagnetic radiation can have momentum, given by $E = pc$.

In general, there is a gradual transition from conditions in which classical formulae apply to those in which relativistic formulae apply. The former become less accurate as the velocity of the observed particle increases. It is often difficult to decide when the more complicated relativistic theory is required. One way to resolve this problem is to calculate the property with both formulations and estimate the error incurred by using classical theory. If this is acceptable, subsequent calculations can be based on classical relationships. For example, for a particle moving at a velocity of $0.1c$, a non-relativistic estimation of mass gives m_0, which is $0.995m$ and, therefore, 0.5% too small. At a velocity of $0.2c$, the error is 2.5%, at $0.3c$ it is 4.6%, and so on.

1.4 QUANTUM THEORY

The first point to be made about the quantum theory is that there is a lot of it and readers are referred to the bibliography at the end of this chapter for a full treatment. The present discussion will be concerned with only two major features of the theory, both dealing with the fact that, for very small systems, the classical distinction between particles and waves becomes invalid. The first example of this effect is provided by particles and waves in the free state and is widely known as *wave–particle duality*. It is important in the present context because it applies to ionising radiations. The second example is found in *bound systems*, such as atoms, nuclei and molecules, which have been regarded, classically, as particulate, but are more successfully treated as stationary-wave systems.

Wave—particle duality

Classically, there are two distinct types of ionising radiation: electromagnetic radiation and particles. The former includes X-rays and gamma rays and is represented as a continuous wave. It does not have a well-defined position, along its direction of travel, but it does have well-defined wave parameters in the form of a frequency, ν, and a wavelength, λ. Particles, on the other hand, do have well-defined positions in space and they have discrete quantities of mass, m, energy, E, and momentum, p. Unlike waves, they cannot add together to produce a zero effect, as in a diffraction pattern.

There is, however, convincing experimental evidence that short-wavelength electromagnetic radiations behave like particles in that they are localised into discrete pulses, called *photons*. A photon has no mass but it does have a discrete *quantum* of energy, E, and a quantum of momentum, p, like a particle. These properties were originally detected by Planck, in 1901, and Einstein, in 1904. Later, in 1923, Compton showed that photons undergo collision processes with electrons (Compton scattering) in which they behave exactly like the billiard ball analogues of particles.

In 1924, de Broglie suggested that, conversely, a particle may behave like a wave and can be described by a wave function with wave parameters, ν and λ. Subsequently it was confirmed by Davisson and Germer in 1927 that electrons do produce diffraction patterns exactly as predicted by the de Broglie hypothesis. It is now well known that other particles diffract like waves, and electron diffraction is an established technique.

A particle of energy E has an associated wave frequency ν, and a photon of frequency ν has a quantum of energy, E, both given by

$$E = h\nu \tag{1.18}$$

where h is Planck's constant, equal to 6.625×10^{-34} J s, or 4.14×10^{-15} eV s.

Also, a particle of momentum p has an associated wavelength λ and a photon of wavelength λ has a quantum of momentum, p, both given by

$$p = h/\lambda \tag{1.19}$$

Consequently, a particle and a wave can both be represented by a wave function, of the form

$$\Psi = a \sin 2\pi \left(\frac{x}{\lambda} - \frac{t}{T}\right)$$

$$= a \sin 2\pi \left(\frac{px}{h} - \frac{Et}{h}\right)$$

and, using the term $\hbar$ ('h bar') to denote $h/2\pi$ (this ratio occurs frequently in quantum theory), the wave function reduces to

$$\Psi = a \sin [(px - Et)/\hbar] \tag{1.20}$$

For those who wish to pursue the matter, it may be noted that in quantum theory an exponential form of the wave function, $\exp [i(kx - \omega t)]$ or $\exp [i(px - Et)/\hbar]$, is preferred because it provides a more complete description and is more easily handled than the sine function.

Equations 1.18 and 1.19 must be applied to particles with some care. It should be noted that the energy E refers to the total energy as given by equation 1.17, and that the particle velocity v is not equal to $\nu\lambda$, as given by equation 1.7. The reason for this latter fact is that v is the velocity of the group of waves, or *wave packet*, representing the particle, and not that of a single wave of frequency ν and wavelength λ.

In practice, it is usually the photon energy or the particle's kinetic energy which is well known, rather than the momentum, and it is the wavelength which is required, rather than the frequency, so it is convenient to convert equations 1.18 and 1.19 to a single relationship between energy and wavelength. For a photon, this is best derived from equations 1.17 and 1.19:

$$\lambda = hc/E \tag{1.21a}$$

and for a particle, from equations 1.16 and 1.19,

$$\lambda = h/(2m_0 T)^{1/2} \tag{1.21b}$$

where T is the *kinetic energy* of the particle and, when the meaning is clear from the context, may be written as E.

These are useful relationships, but they can be simplified by inserting values of h, c and m_0 to obtain, for a photon,

$$\lambda = 12.4/E \tag{1.22a}$$

for an electron,

$$\lambda = 0.388/T^{1/2} \tag{1.22b}$$

and for a neutron,

$$\lambda = 0.009\,05/T^{1/2} \tag{1.22c}$$

where λ is in ångstroms and T (the kinetic energy of the particle) is in kiloelectronvolts.

These equations, 1.18–1.22, define a useful quantum model of a particle and a photon but they fail to provide a visual model and they raise a number of conceptual problems. For example, it is not clear how a particle or photon can be localised to a point in space and, at the same time, be described by equation 1.20 as a continuous wave. Nor is it easy to imagine what is the physical interpretation of the displacement, ψ, of a particle wave.

Briefly, the answer to both problems lies in the fact that for a single particle or photon, *position is not well-defined*. In other words, there is no such property as exact position. Indeed, the position of a particle is described by a function Ψ of space and time coordinates. Ψ is related to the *probability* that a particle will be found at a given position at any particular time. To be more precise, the modulus squared of Ψ, that is, $|\Psi|^2$, is equal to this probability. If Ψ is a real function then this is the same as Ψ^2, but in quantum theory most wave functions are complex, that is, they contain i, the square root of -1, and it is necessary to take the square of the modulus in order to produce a real physical value. Ψ is called the *probability amplitude* and $|\Psi|^2$ is the *probability density*. $|\Psi|^2$ is analogous to the probability distribution functions of classical mechanics, such as the Gaussian function.

If this idea is applied to the wave function given in equation 1.20 or, better, to the exponential form of this function, it is seen that there is a uniform probability of finding the particle or photon at any point in its path. It is not localised. To improve matters, the model must be refined slightly. This is done by representing the particle or photon not by one wave but by the sum of a whole series of waves, known as a Fourier transform. These component waves have a range of different wavelengths so they are exactly in phase only in one small region, as shown in *Figure 1.4*. Elsewhere, they are out of phase and cancel

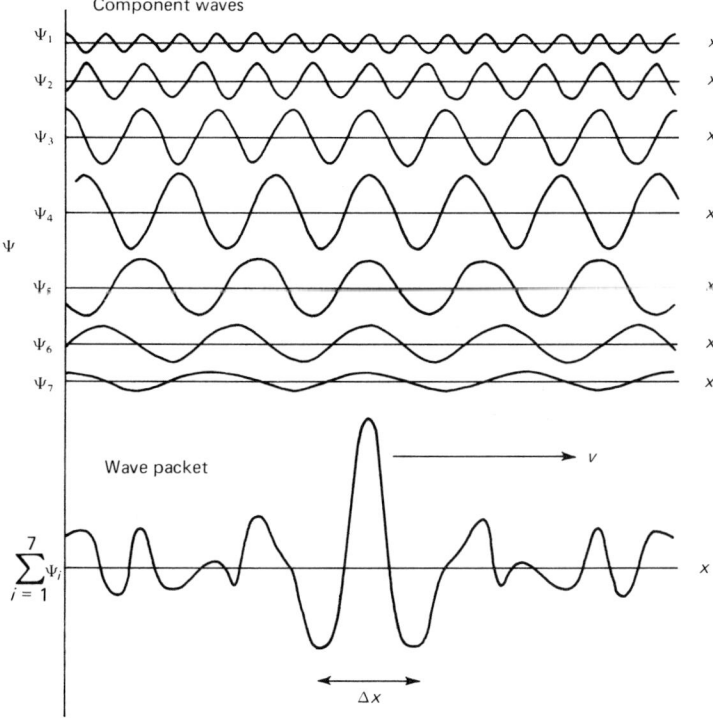

Figure 1.4 Superposition of waves to form a wave packet

each other to produce zero effect. Hence the wave function is more localised in space but loses its well-defined single wavelength. The end product is known as a *wave packet*. It is localised to within an interval of space, Δx, and has a range of wavelengths, $\Delta \lambda$. If $\Delta \lambda$ is increased to become very ill-defined, Δx decreases to be more sharply defined, as in the classical model of a particle. If $\Delta \lambda$ is decreased to zero, Δx increases to infinity, as in the classical model of a continuous wave. In general, there is a fundamental relationship between the uncertainty in x, that is, Δx, and the uncertainty in a simultaneous measurement of λ, that is, $\Delta\lambda$. This can be derived by means of standard Fourier analysis. It shows that the ratio of Δx to $\Delta\lambda$ must always exceed $1/2\pi$, i.e.

$$\frac{\Delta x}{\Delta\lambda} \geqslant \frac{1}{2\pi}$$

Converting from wavelength to momentum, by using equation 1.19, gives a fundamental postulate of quantum theory, known as the *Heisenberg Uncertainty Principle*, namely

$$\Delta p . \Delta x \geqslant \hbar \qquad (1.23)$$

which states that it is impossible to specify momentum and position simultaneously with an accuracy better than Δp and Δx, respectively, where the product of Δp and Δx must exceed $\hbar$. The same is true of other pairs of coordinates, including energy and time.

A rather crude visual model of a particle or photon can be constructed by imagining a point object vibrating, in three dimensions, about a moving point in space. It is vibrating so quickly that it is 'seen' as a blurred image, like a long-exposure photograph of a vibrating string, so its exact position is ill-defined. As it moves through space, it traces a rather poorly defined wave profile.

On this basis, two particles can *overlap* to the extent that their wave functions can occupy, at least partly, the same regions of space, provided their behaviour patterns are not totally identical. Free particles, or photons, are unlikely to coincide so exactly, but total overlap is a real possibility for particles contained in the same bound system and must be formally prohibited by the *Pauli Exclusion Principle*. This states that no two similar particles, in the same system, can have identical wave functions.

Bound systems

The wave function of a particle in free space is a special solution of a wave equation known as the *Schrödinger equation*. The same equation can be solved for a particle in a bound system except that, in this case, the movement of the particle is confined to a small region of space by an attractive, binding force between it and at least one other particle. An atom is a bound system in which each electron is bound by the Coulomb force between it and the nucleus. A nucleus is a bound system in which each nucleon is bound by the nuclear force between it and the other nucleons.

In some respects, the bound particle system is similar to the classical vibrating string. The wave functions form a discrete set of allowed vibrational modes, or *eigenfunctions*, ψ_n, and to each eigenfunction there corresponds a well-defined wavelength *eigenvalue*, λ_n. From equation 1.21, it follows that the particle can have only one of a discrete set of allowed energy eigenvalues, E_n, each associated with a particular eigenfunction ψ_n. Each eigenfunction represents an allowed state, or *eigenstate*, which may or may not be occupied by a particle, so it is known as a *state function* rather than a wave function. If a particle is in the nth eigenstate, its energy is E_n and its spatial distribution is described by $|\psi_n|^2$.

For a given system, the eigenfunctions all have the same general form and differ from each other only in terms of their parameters, such as λ_n or E_n. The number n which is used as a label for the eigenvalues is an integral part of the eigenfunction and it is the fact that n can have only certain allowed values that produces a discrete rather than a continuous range of eigenvalues. This effectively quantises the property of the system and n is known as a *quantum number*. For a vibrating string, the eigenfunctions are sine functions of the form (equation 1.8) $a \sin(n\pi x/L)$ and the allowed wavelength eigenvalues are given by $2L/n$ *(Figure 1.3)*, where L is the length of the string. Since n can only have the values 1, 2, 3, 4, ..., these eigenfunctions and eigenvalues form a discrete allowed set.

In a particle system, other properties besides energy are quantised and each tends to have its own quantum number. The effect can be illustrated by reference to the electron eigenstates of the *hydrogen atom* which, as for other bound systems, require four quantum numbers, as follows:

(1) The energy eigenvalues are determined by a *principal quantum number n* according to the formula

$$E = -13.6/n^2 \qquad n = 1, 2, 3, \ldots \tag{1.24}$$

where the negative sign implies a bound state, i.e. energy must be *given to* the electron to release it from the atom and leave it with zero energy.

(2) The orbital angular momentum L of the electron is also quantised, according to the formula

$$L = [l(l+1)]^{1/2} \hbar \qquad l = 0, 1, 2, \ldots (n-1) \tag{1.25}$$

where l is the *orbital angular momentum quantum number* and $\hbar$ is equal to $h/2\pi$.

(3) The z component of the electron's orbital angular momentum, L_z, is quantised according to the formula

$$L_z = m\hbar \qquad m = 0, \pm 1, \pm 2, \ldots \pm l \tag{1.26}$$

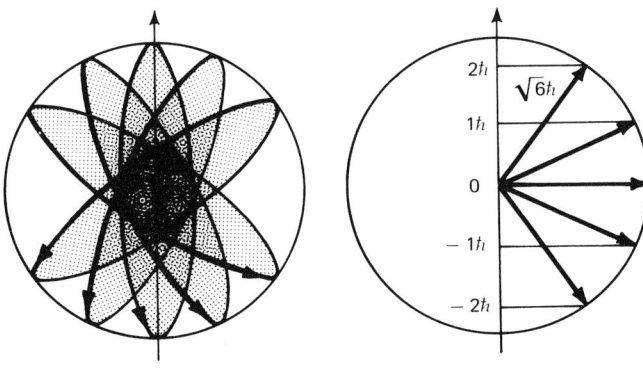

Orbital diagram   Vector diagram

Figure 1.5 Space quantisation of the orbital angular momentum vector

16   Basic concepts

This means that the vector denoting orbital angular momentum is *space quantised* so that it can take up only discrete orientations with respect to some direction, defined to be that of the z axis. The effect is illustrated in *Figure 1.5*. For $l = 2$, the magnitude of the angular momentum vector is $[2(2 + 1)]^{1/2}\hbar$, i.e. $(\sqrt{6})\hbar$, and its possible orientations are those whose projections onto the z axis are $2\hbar, \hbar, 0, -\hbar$ and $-2\hbar$.

In fact, it is never possible to measure orbital angular momentum, only its projection in the direction of a magnetic field, which can, of course, be the z axis. For this reason, m is known as the *magnetic quantum number*.

(4) The electron also has intrinsic spin, S, about its own internal axis and this, too, is quantised into two possible values, given by

$$S = [s(s + 1)]^{1/2} \hbar \qquad s = \pm\tfrac{1}{2} \tag{1.27}$$

Again, this is space quantised so that its measured projection onto the z axis is $\pm\tfrac{1}{2}\hbar$. An electron with a measured spin of $+\tfrac{1}{2}\hbar$ is said to have spin 'up' while a value of $-\tfrac{1}{2}\hbar$ is described as spin 'down'. s is known as the *spin quantum number*.

The hydrogen atom eigenfunctions are associated Laguerre polynomials which incorporate the quantum numbers and form a discrete set by virtue of the allowed values of these quantum numbers. The same is true of other bound systems but, in general, different types of function are required and, sometimes, different quantum numbers. In all cases, however, a particular set of quantum number values uniquely defines one allowed particle state and, by the Pauli Exclusion Principle, each state may contain only one particle. Thus each particle in a bound system has a unique set of four quantum numbers.

If a system contains several particles, the spatial distribution of these components is represented by the sum of the square moduli of their eigenfunctions. This provides a complicated set of diffuse, overlapping patterns that is difficult to visualise and not very useful as a model. The diffuseness is a direct result of the Heisenberg Uncertainty Principle, which defines a related uncertainty in position and momentum. The same principle shows that energy and time uncertainties are similarly related by the expression

$$\Delta E . \Delta t \geqslant h \tag{1.28}$$

For a stationary state, however, time is completely undefined, so the energy eigenvalues can be exactly specified. As a result, a bound system is better represented by a graph of its allowed energy eigenvalues, each one corresponding to a well-defined energy eigenstate. Such graphs are usually described as *energy level diagrams* and are, in effect, models of bound systems.

There are two standard ways of drawing energy level diagrams and both are illustrated in *Figure 1.6*. In one of these, energy eigenvalues are depicted as horizontal lines against a vertical energy scale. These represent total energies; that is, kinetic and potential energies. The other approach includes a separate graph of potential energy against the radial distance of the particle from the centre of the system. The shape of this potential function, $V(r)$, depends on the size of the system and on the type of attractive force involved but it is always *negative* and referred to as a *potential well*. Particles in a bound system have negative potential energies because energy must be *given* to them to remove

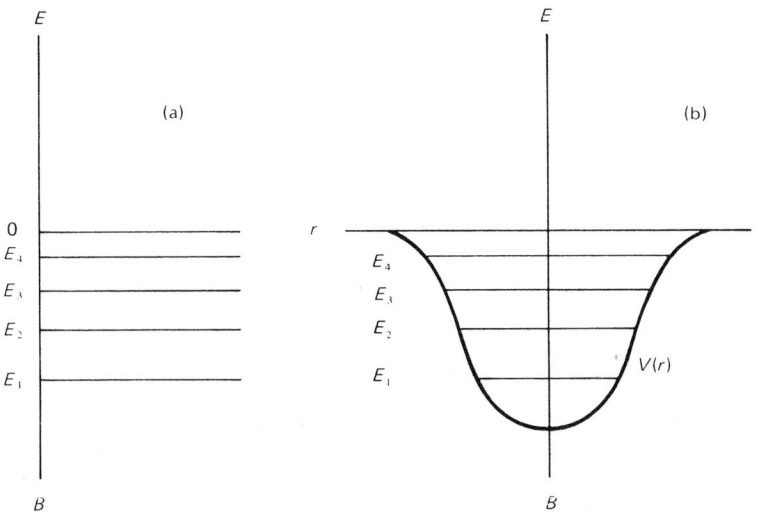

Figure 1.6 Energy level diagrams for a bound system. (a) One-dimensional graph of E; (b) graph of E against coordinate r

them from the system and leave them with zero kinetic and potential energy. The total energies, E_n, are also negative, otherwise the particles in these states would have sufficient kinetic energy to escape.

The energy required to remove a particle from a bound system is known as its *binding energy*, B_n, and is given by

$$B_n = -E_n \qquad (1.29)$$

B_n is therefore a positive number and increases as E_n decreases. If, for example, E_n is -1 eV, then B_n is 1 eV. If E_n is -10 eV, B_n is 10 eV. The lowest energy state is the most tightly bound state and the highest energy state is the least tightly bound state.

As mentioned above, it is possible to remove a particle from a bound system by giving it energy B_n, so that it finally ends up as a free particle with zero energy. The question is, what has happened to the energy used up in the process? The answer is that it has been converted to mass, according to the Einstein equation, $E = mc^2$. Consequently, the mass of a free particle is greater, by B_n/c^2, than that of a bound one. The mass of the whole system is smaller than that of the sum of its component masses, in the free state, by an amount which is the total binding energy of the system. Binding energy is derived from mass.

BIBLIOGRAPHY

ACOSTA, V., COWAN, C. L. and GRAHAM, B. J., *Essentials of Modern Physics*, Harper & Row, New York (1973)
DICKE, R. H. and WITTKE, J. P., *Introduction to Quantum Mechanics*, Addison-Wesley, Reading, Massachusetts (1960)
MATTHEWS, P. T., *Introduction to Quantum Mechanics*, McGraw-Hill, New York (1968)
OREAR, J., *Fundamental Physics*, John Wiley, New York (1967)
STRALEY, J. W., *Basic Physics: A Linear Approach*, Prentice-Hall, Englewood Cliffs, New Jersey (1974)

Chapter 2
The structure of matter

2.1 ATOMS

The most useful classical model of the atom is the *Rutherford model* in which the system is represented as a central, positively charged *nucleus* surrounded by a number of orbitting, negatively charged *electrons*. The charge on the nucleus is $+Ze$ and there are Z electrons, each with a charge $-e$, so the atom has net electrical neutrality. The electronic charge, e, is equal to 1.602×10^{-19} C and the number Z is known as the *atomic number*. The value of Z uniquely determines the nature of the atom, and atoms of the same type, that is, with the same Z number, are said to be of the same *element*. Elements are denoted by standard chemical symbols. Hydrogen, for example, is represented by H and its atoms have $Z = 1$. Carbon, C, has atoms with $Z = 6$. A complete list of elemental symbols and Z numbers is given in Appendix 1.

The diameter of a medium-sized atom is about 10^{-10} m, while that of its nucleus is only about 10^{-14} m. However, most of the atomic mass is concentrated in the nucleus, which has a typical mass of the order of 10^{-26} kg, compared to the electron rest mass of about 10^{-30} kg, so, in the Rutherford model, most of the atom is empty space. The electrons are imagined to move in well-defined circular or elliptical orbits and to be held in the bound system by the attractive electrostatic, or Coloumb, force between them and the nucleus.

There are two problems with this model. First of all, orbiting electrons are being accelerated and should lose energy, by emitting electromagnetic radiation, and gradually spiral into the nucleus. Secondly, the evidence of atomic spectroscopy is that the electrons must have discrete allowed energy values and the Rutherford model does not predict this effect. For these reasons, the model has been superseded by the *Bohr model*, which introduces some concepts of quantum theory, in the form of two postulates:

(1) Atomic electrons exist in stable bound states whose energies have a limited number of allowed values, E_n. They form a discrete set, E_1, E_2, E_3, etc. rather than a continuous range. Energy can be radiated only when an electron makes a transition from one bound state to a lower energy state, in which case one *photon* of electromagnetic radiation, of energy $h\nu$, is emitted from the atom. In the reverse process, an electron can be raised to a higher energy state only if it absorbs a photon whose energy is equal to the difference in the energies of the two electron states. In general

$$h\nu = E_2 - E_1 \tag{2.1}$$

where $E_2 > E_1$.

(2) These allowed orbits are determined by the fact that the orbital angular momentum, L, of each electron is quantised into units of $\hbar$ ($= h/2\pi$); that is,

$$L = n\hbar \qquad n = 1, 2, 3, \ldots \qquad (2.2)$$

where n became known as the *principal quantum number*.

These postulates suggest that atomic electrons exist in non-radiating, stationary *eigenstates* characterised by discrete energy *eigenvalues*, E_n, and discrete orbital angular momentum eigenvalues L_n. It is now recognised that the second postulate refers to *measured* angular momentum and should apply to L_z, the z component of orbital angular momentum. It may be rewritten as $L_z = m\hbar$, where m is the *magnetic quantum number* and has values different from those of the principal quantum number. Nevertheless, Bohr was able to use his postulates to construct an atomic *energy level diagram*. He assumed a simple Coulomb force between electron and nucleus, ignoring interelectron forces, so his results really apply only to the hydrogen atom, for which they agree with the predictions of modern quantum theory. They are less accurate for heavier atoms with more than one electron. His formula for the energy eigenvalue of the nth energy state, E_n (eV), is

$$E_n = -13.6 Z^2/n^2 \qquad (2.3)$$

so that energy level spacing decreases as n increases.

In addition, Bohr constructed a *spatial* or *orbital* model of the atom in which electron states are represented as discrete, circular orbits whose radii, r_n, become more widely spaced as n increases and are given by

$$r_n = 5.3 \times 10^{-11} \times n^2/Z \quad \text{metres} \qquad (2.4)$$

This is a two-dimensional diagram of a three-dimensional system and it is a semi-classical model in that it takes no account of position uncertainty, but it is nonetheless a useful model. Both the energy level diagram and the orbital model are shown in *Figure 2.1*, where they are used to illustrate the processes described in Bohr's first postulate.

If an electron in a particular energy state absorbs a photon whose energy, $h\nu$, is *exactly* equal to the energy difference $E_4 - E_3$, say, between that state and a higher one, the electron is excited to the higher state. The process is known as *excitation*. The reverse process, in which the electron sheds its superfluous energy, in the form of a photon, and de-excites to the lower state, is known as *de-excitation*. If an electron in the nth state is given more energy than its binding energy, $B_n = -E_n$, it is ejected from the atom. The process is known as *ionisation* and it leads to the formation of an *ion pair*, comprising the residual atom, which now has a net positive charge and is a *positive ion,* and the *free electron*. Since energy is not quantised in the free state, ionisation does not require an exact amount of energy, only that the energy given to the electron should exceed its binding energy. As pointed out in Chapter 1, this energy is converted to mass, according to the Einstein equation $E = mc^2$, so that the mass of a free electron is greater than that of a bound one by an amount B_n/c^2. Any excess energy remains as the kinetic energy of the free electron. It is also possible for a neutral atom to capture a free electron, in which case it becomes a *negative ion*.

20   The structure of matter

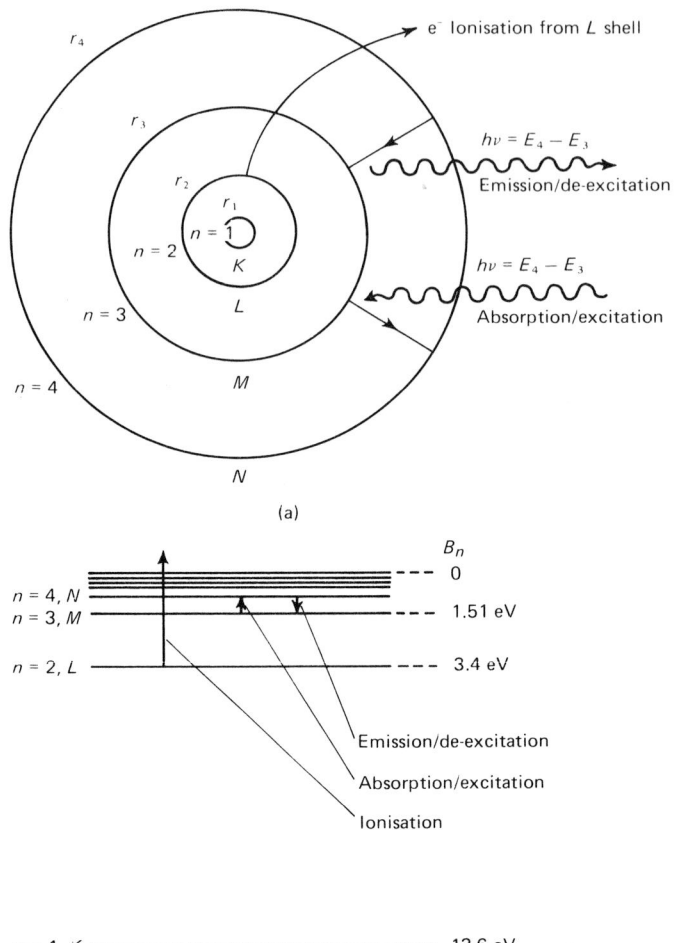

Figure 2.1 (a) Bohr orbital representation and (b) energy level diagram for the hydrogen atom

The full quantum treatment of the hydrogen atom involves solving the *Schrödinger equation* for a Coulomb potential function of the form $V(r) = -e^2/r$ to produce a set of eigenfunctions, $\psi_{n,l,m,s}$, which require four quantum numbers. These are n, the *principal quantum number*, which can have the values 1, 2, 3, 4, etc.; l, the *orbital angular momentum quantum number*, which can have the values 0, 1, 2, 3, etc. to $n-1$; m, the *magnetic quantum number*, which can have the values $-l$ to $+l$, in integral increments; and s, the electron *spin quantum number*, which can have the values $+\frac{1}{2}$ or $-\frac{1}{2}$. A specific set of quantum number values uniquely defines one allowed electron state and, by the Pauli Exclusion Principle, this can be occupied by only one electron. It turns out, however, that the energy values of these electrons depend only on the principal quantum number, as predicted by the Bohr postulates, so two or more electrons may have the same energy values, that is, they can exist in the same energy state. These states are said to be *degenerate* since they fail to identify separate electron states.

This model may be described as the *Shell Model* of the atom since it predicts well-defined energy shells. Each n value represents one energy shell and is given a special name. These shells, and their contained electron eigenstates are as follows:

K shell: $n = 1$: $l = 0$; $m = 0$ $s = \pm\frac{1}{2}$: 2 states

L shell: $n = 2$: $l = 0$; $m = 0$ $s = \pm\frac{1}{2}$: 2 states   8 states
$\phantom{L\ \text{shell}: n = 2:}\ l = 1$; $m = 0, \pm 1$ $s = \pm\frac{1}{2}$: 6 states

M shell: $n = 3$: $l = 0$; $m = 0$; $s = \pm\frac{1}{2}$: 2 states
$\phantom{M\ \text{shell}: n = 3:}\ l = 1$; $m = 0, \pm 1$; $s = \pm\frac{1}{2}$: 6 states   18 states
$\phantom{M\ \text{shell}: n = 3:}\ l = 2$; $m = 0, \pm 1, \pm 2$; $s = \pm\frac{1}{2}$: 10 states

N shell: $n = 4$: $l = 0$; $m = 0$; $s = \pm\frac{1}{2}$: 2 states
$\phantom{N\ \text{shell}: n = 4:}\ l = 1$; $m = 0, \pm 1$; $s = \pm\frac{1}{2}$: 6 states
$\phantom{N\ \text{shell}: n = 4:}\ l = 2$; $m = 0, \pm 1, \pm 2$; $s = \pm\frac{1}{2}$: 10 states   32 states
$\phantom{N\ \text{shell}: n = 4:}\ l = 3$; $m = 0, \pm 1, \pm 2, \pm 3$; $s = \pm\frac{1}{2}$: 14 states

and so on, to the O shell, P shell, and Q shell.

In all atoms, the electrons tend to occupy the lowest possible energy states. An atom in this equilibrium condition is said to be in its *ground state*. If the electrons have excess energy the atom is in an *excited state*. In the ground state of the hydrogen atom, the single electron is in one of the two degenerate states of the K shell, with a binding energy of 13.6 eV. This is the energy normally required to ionise a hydrogen atom and is known as its *ionisation potential.*

The Shell Model can be applied to atoms with more than one electron by taking account of electron–electron interactions. The theoretical approach is known as the Hartree–Fock treatment but for most practical purposes it is sufficient to modify the hydrogen atom model on the basis of experimental observations. These are obtained by means of *atomic spectrometry*. In emission spectrometry, the energies of emitted photons are measured to deduce the energy level structure of the atom, while in absorption spectrometry, the specific energies absorbed from a continuous range of photon energies are noted. Both methods produce a discrete set of photon energies that is uniquely characteristic of the element whose atoms are being investigated, and yield almost identical results. There are slight differences owing to the fact that the atom recoils in both mechanisms so that in emission spectrometry, photon energies are rather less than the available energy, $E_2 - E_1$, while in absorption spectrometry they have to be rather larger than $E_2 - E_1$ to give the atom some recoil energy.

In both cases, the end product is an *energy level diagram* for the atom but this differs, in some important respects, from that of the hydrogen atom. The main difference is that the degeneracy with respect to the l quantum number is removed, so that electron states with the same n number but different l numbers have different energies and, effectively, form energy sub-shells. The effect increases with Z number and n number, with the result that the outer energy shells of heavy elements are determined more by l numbers than by n numbers.

For this reason, these l states are given specific names, like the n shells, as follows:

$l = 0$: s state

$l = 1$: p state

$l = 2$: d state

$l = 3$: f state

and so on, through s, h, j, etc.

It also becomes necessary to develop a full *spectroscopic notation* in which the n, l and s quantum numbers of electron states are specified. This is an extension of the n shell notation used for the hydrogen atom and usually has the general form

$$nX_{l+s}$$

where X denotes one of the l-state quantum numbers s, p, d, etc., and the sum, $l + s$, is equal to the total angular momentum j of the electron. Because of space quantisation, the j vector can take up any of $2j + 1$ orientations with respect to the z axis of the system, each corresponding to a different magnetic number, m_j. Consequently, the above term represents $2j + 1$ degenerate states. Thus, for example, the $n = 3, l = 2, s = +½$ state is denoted by $3d_{5/2}$ and contains six electronic states, all with the same energy.

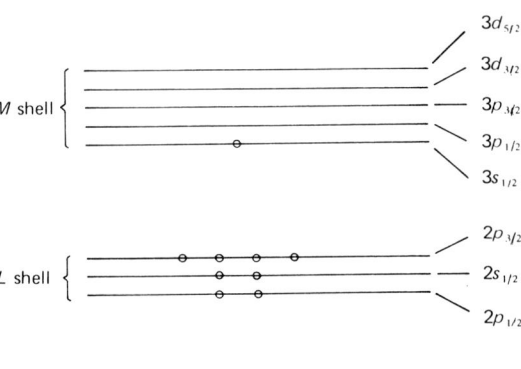

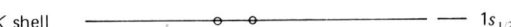

Figure 2.2 Energy level diagram of the ground state of an isolated sodium atom

The terminology is illustrated, for a sodium atom, in *Figure 2.2*. In the ground state, sodium's 11 electrons occupy the lowest 11 energy states. There are two electrons in the $1s_{1/2}$ state, filling the K shell. There are two electrons in the $2s_{1/2}$ state, two in the $2p_{1/2}$ state and four in the $2p_{3/2}$ state, filling the L shell. The 11th electron is in the $3s_{1/2}$ state of the M shell. The sodium atom is said to have the *electron configuration* $(1s)^2 (2s)^2 (2p)^6 (3s)^1$. The electron configurations of all the elements are listed in Appendix 1.

In any interaction between an atom and some other system, such as another atom or ionising radiation, it is the outermost, least tightly bound electrons that are involved. These electrons determine the properties of the atom and, therefore, of the element to which it belongs. The inner electrons form a relatively inert core. For this reason, the energy level diagrams of atoms and atomic aggregate systems generally exclude the lower, *core* energy shells. Excitation and de-excitation processes in the sodium atom, for example, are usually represented in a diagram such as *Figure 2.3*. This diagram starts at the 3s state of the outer electron. It is the energy level diagram of a single electron but it describes most of the atomic processes, including spectrometric effects. Orbital diagrams also

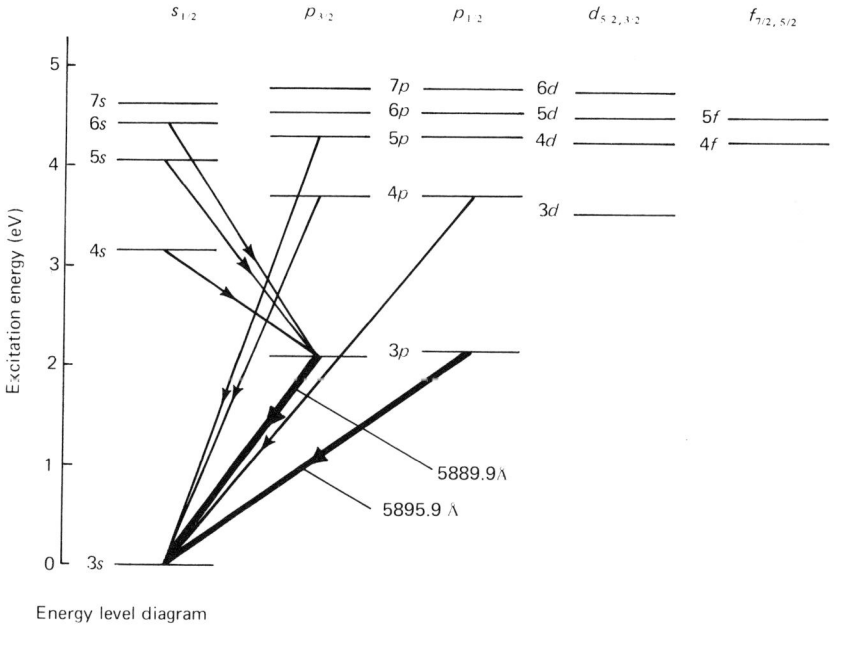

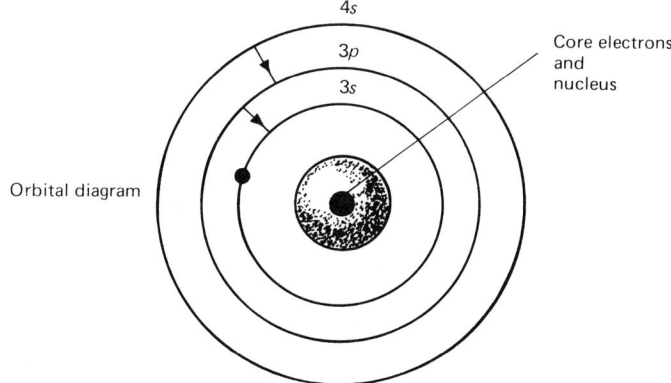

Figure 2.3 Energy level diagram and orbital diagram of the excited states of an isolated sodium atom

depict only the outer electron states, as shown in *Figure 2.3*. In fact, it is the essence of the Shell Model that the behaviour of the whole atom can be described in terms of that of one electron (or a few electrons) in the potential well produced by the nucleus and the inner, core electrons.

An immediate consequence of this observation is that the properties of elements and their atoms are mainly determined by the *number* of electrons in the outer shell and by their *configuration*. The number determines the *valency* of the element and the configuration determines the ease with which electrons can be excited, removed and made to react with external bodies; in other words, the *reactivity* of the element. For example, carbon, silicon, germanium, tin and lead all have four outer electrons with $(s)^2 (p)^2$ configurations. They are all tetravalent and form similar chemical compounds. Density increases with Z number, of course, and the energy gap above the $(s)^2 (p)^2$ shell decreases with increasing Z number, as the n number of the shell increases, and shell spacing decreases as $1/n^2$.

As another example, those elements with *closed-shell* configurations and relatively large energy gaps above these closed shells, form inert, unreactive materials in the form of monatomic gases. These are the *noble gases* with outer electron configurations as follows:

Helium,	He,	$Z = 2$:	$(1s)^2$
Neon,	Ne,	$Z = 10$:	$(2s)^2 (2p)^6$
Argon,	Ar,	$Z = 18$:	$(3s)^2 (3p)^6$
Krypton,	Kr,	$Z = 36$:	$(4s)^2 (4p)^6$
Xenon,	Xe,	$Z = 54$:	$(5s)^2 (5p)^6$
Radon,	Rn,	$Z = 86$:	$(6s)^2 (6p)^6$

It can be seen, therefore, that there is a certain periodicity in the properties of the elements such that, as Z increases, the elemental properties change in a recurrent, periodic fashion. This effect can be systematically presented by arranging the elements in the form of a table. Horizontal rows contain elements whose outer electron configurations increase from one electron to a closed shell, filling each shell in successive rows, while the vertical columns contain elements of the same *group* with the same outer electron configurations and similar properties. This table is known as the *Periodic Table* of the elements and is shown, along with some relevant properties of the elements, in Appendix 1.

A typical periodic property is *ionisation potential*, which is the energy required to remove the outermost electron. This increases from left to right across the Periodic Table — from a single-electron configuration to a closed shell — and it also decreases from top to bottom, as more shells are filled and the outer shell binding energy decreases as the shell number, n, increases. A similar pattern is observed for *excitation potentials*, which are the energies required to excite the outer electron to its first excited state. For closed-shell atoms, these are of the order of 10–15 eV so that the noble gases tend to absorb and emit radiation at characteristic ultra-violet wavelengths of the order of 1000 Å. Other atoms have smaller excitation potentials, typically in the visible-light range.

2.2 ATOMIC SYSTEMS

The only elements which exist naturally as single atoms are the noble gases. Their atoms have closed-shell structures and tend neither to lose nor to gain additional electrons, so they do not readily form bonds with other atoms. They are stable, non-reactive elements. As mentioned above, they have high ionisation and excitation potentials.

All other elements show a tendency to gain, lose or share outer electrons in order to make a closed-shell configuration. In doing so, their atoms form bonds with other atoms. Elements with a small number of electrons outside a closed-shell configuration, such as elements in Groups I and II of the Periodic Table, easily lose these outer electrons to become positive ions. They are said to be *electropositive*. Elements at the other end of the table, in Groups VI and VII, for example, are a few electrons short of a closed shell and tend to capture additional electrons to become negative ions, and are described as *electronegative*. Elements in the middle groups make up closed shells by sharing their outer electrons with other, similar elements.

If these processes can be completed, with only a few atoms, to produce a stable closed-shell configuration for each atom, a *molecule* is formed. Such molecules do not interact with each other very much and normally exist as gases. If, however, the closed shells do not represent major closures, with large energy gaps above them, there is some intermolecular cohesive force and a liquid or an amorphous solid is formed. If the atoms cannot form a good closed-shell configuration with a limited number of atoms, the result is a matrix of indefinite size. This is the structure of a crystalline solid.

Molecules

A diatomic hydrogen molecule, H_2, is formed when two hydrogen atoms come close enough for each atomic electron to be electrostatically attracted to the other nucleus. It is this attractive force which binds the molecule. It is opposed, of course, by the repulsive electrostatic force between the two nuclei, so an equilibrium separation is established such that these two forces are in balance. This is an example of a *covalent bond* in which the two electrons are shared by both nuclei. One electron is in the $1s_{+1/2}$ state, with spin 'up', while the other is in the $1s_{-1/2}$ state with spin 'down', so, effectively, they produce a closed K shell.

Other elements and compounds form covalent-bonded molecules by sharing their outer, *valence* electrons. These include nitrogen, N_2, oxygen, O_2, and the halogens, F_2, Cl_2, Br_2 and I_2, as well as the organic, carbon-based materials such as ethanol, C_2H_5OH, and methane, CH_4. These molecules are mainly electronegative since they are able to attach free electrons to form negative molecular ions. From a spectroscopic point of view, however, they behave like closed-shell atoms. Electron transitions take place between the closed shell and the next energy shell above that but they are smaller than the transitions in a noble-gas structure. Consequently, the atomic spectra of these materials are in the visible-light region rather than the ultra-violet range.

In addition, the atoms themselves have discrete allowed energy states within the molecules and these add to the electronic binding energies to establish a fine

26 *The structure of matter*

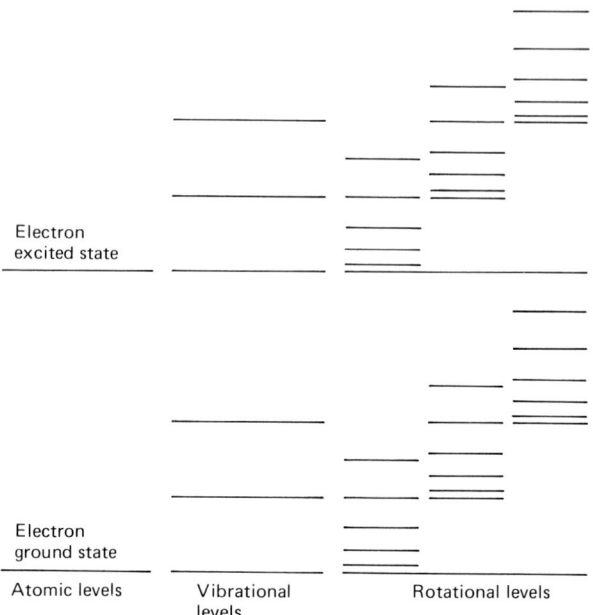

Figure 2.4 Energy level structure of a diatomic molecule

structure in the valence electron energy levels. These atomic motions are due to *vibrations* along the interatomic axes and to *rotations* about the centre of the molecule. Since the molecule is a larger system than a single atom, its stationary-state wavelengths are larger, and its energy eigenvalues correspondingly smaller than those of the atom. Hence the fine-structure spacing, as illustrated in *Figure 2.4*, is of the order of 1 eV. Molecular spectra, involving transitions between vibrational and rotational states, are typically in the infra-red wavelength range.

Solids

Basically, there are two types of solid: *amorphous* materials and *crystals*. The former include *polymers*, which are large molecules, weakly linked to each other, and *glasses* or *ceramics*, which are molecular but have some local organised structure similar to that of crystals. The crystalline solids are taken to include crystal *insulators* and *semiconductors* as well as *metals*. Insulators are formed by ionic or by covalent bonds. An *ionic bond* is produced when an electropositive and an electronegative atom approach each other. The former loses its valence electrons to the latter so that both establish closed-shell configurations. The electropositive atom becomes a positive ion while the electronegative atom becomes a negative ion and they are held together by electrostatic attraction balanced by the repulsion between the positive nuclei. Each ion attracts other ions of the opposite sign and these, in turn, attract further ions, so a crystal matrix is produced, with positive ions interspaced by negative ones. The result is illustrated for sodium chloride, NaCl, in *Figure 2.5*.

Atomic systems 27

The atoms of elements in or near Group IV in the Periodic Table have too many valence electrons to release to other atoms. They form *covalent bonds* by sharing their electrons with adjacent atoms and, again because of the large numbers of electrons involved, they have to share with more than one atom. These atoms, in turn, share electrons with further atoms, and so on, to establish a crystal matrix. A typical matrix, that of carbon (diamond), is illustrated in *Figure 2.5*, in which electrons are represented by connecting lines. It is important to note that both diagrams shown in *Figure 2.5* are two-dimensional models. They do not indicate the real, three-dimensional crystal structures.

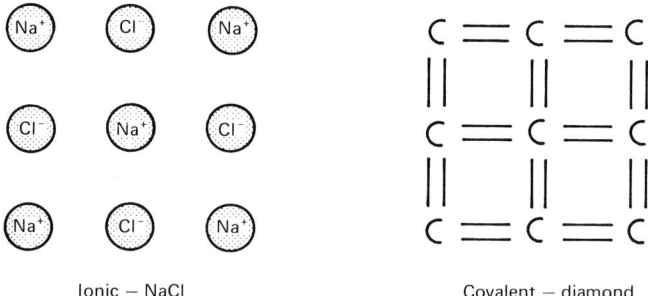

Figure 2.5 Spatial diagrams of ionic and covalent bonds in solids

The effect of both types of solid structure bonds is that atomic electrons, at least in the outer shells, interact with other electrons throughout the crystal. This is similar to the effect of interelectron forces in an atom, which produce a fine structure in the basic energy levels. In a solid, there are many more electrons, typically of the order of 10^{23}, so what would be a set of fine structure lines becomes a broad energy band of touching energy levels. The solid is described by a *band structure diagram* instead of a discrete energy level diagram, especially for the outer, valence electrons. The inner, core electron shells are less affected in this way owing to the screening effect of the outer electrons. The atomic spectrum of a gas has a characteristic line structure because all the atoms or molecules have the same energy levels. The atomic spectrum of a solid has a more continuous nature because all the atoms have slightly different energy levels.

The band structures of sodium chloride (ionic) and diamond (covalent) are illustrated in *Figure 2.6*. In the former the upper band is the empty, $3s$ state of sodium, while the lower band is the full, $3p$ state of chlorine. In diamond, the situation is more complex because the atomic $2p$ and $2s$ levels split and cross over as the interatomic separation decreases. At the equilibrium separation, the upper band contains one electron state from the $2p_{1/2}$ level and one from the $2s_{1/2}$ level, while the lower band contains the other $2p_{1/2}$ and $2s_{1/2}$ states, and is filled by the two shared electrons.

In both cases, the lower band contains the valence electrons and is known as the *valence band*. Since there are no vacant states in this band, electrons cannot move through the crystal from one valence level to another. Electrons can move through the crystal, as an electric current, only if they are raised to the empty states in the upper, *conduction band*. Since this transition requires a considerable energy absorption, of the order of 7—10 eV, these materials do not readily conduct electricity and are insulators. They are also transparent to visible light, because photon energies of 2—3 eV are insufficient to induce absorption transitions.

The energy gap between the valence and conduction bands is known as the *forbidden energy gap*, E_g. It is the solid-state equivalent of the ionisation potential. If an electron is raised to the conduction band, an ion pair is formed with a virtually free electron in the conduction band and a vacancy (that is, a fixed positive ion) in the valence band, so the process is one of *ionisation*.

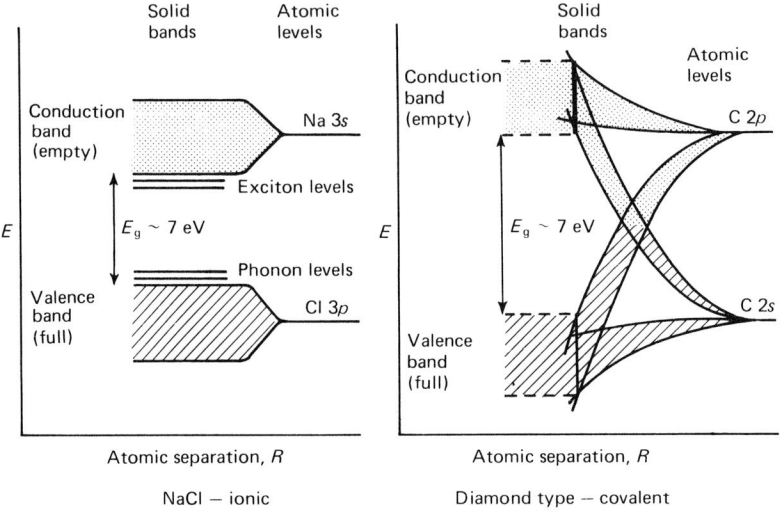

Figure 2.6 Energy band structures of NaCl (ionic) and diamond (covalent) insulators

Excitation processes, in which atomic vibrations provide a set of discrete levels just above the valence band, can also take place. These levels are known as *phonon* or *restrahl* levels. They do not produce ion pairs but they do allow for the absorption of some infra-red wavelengths. The other process involves the excitation of an electron to a set of discrete *exciton* levels just below the conduction band. In these states, the electron is not quite freed from the positive ion. The electron–positive ion combination is itself referred to as an exciton and, under certain circumstances, it can move slowly through the solid.

Semiconductors are similar to covalently bonded insulators except for the fact that their forbidden energy gaps are very small. Silicon and germanium are the best-known semiconductors. They are Group IV elements, with spatial and energy band structures similar to that of diamond, but their energy gaps are only 1.12 and 0.67 eV, respectively. As a result, thermal collisions at room temperature are sufficient to ionise electrons into the conduction band, so these materials do conduct electricity, although not as well as a metallic conductor. In addition, they absorb at visible-light wavelengths, so they are opaque. They are also photoconductive, since absorption of visible light provides additional electrons in the conduction band.

The process of electrical conduction in a semiconductor is rather different from that which takes place in a conductor. The electrons, raised to the conduction band, become *negative-charge carriers*, as in a metal, but they leave positive ions fixed in the crystal lattice. These ions provide electron state vacancies in the valence band which allow some electron movement through the valence levels. When an electric potential is applied to a semiconductor, the conduction band electrons move in one direction and the valence band electrons move, in the same direction, into any available vacancies. In doing so, they leave further vacancies which are filled by more electrons, and so on. The net effect is that these positive vacancies, or *positive holes* appear to move in the opposite direction and behave like *positive-charge carriers*.

Conductors, or *metals*, are formed by *metallic bonds*. They comprise those elements whose outer electrons are very weakly bound and become free, in a solid, owing to the weak attractive forces of nearby atoms. The result is the formation of a matrix of positive ions bound by electrostatic forces to a gas-like environment of free electrons.

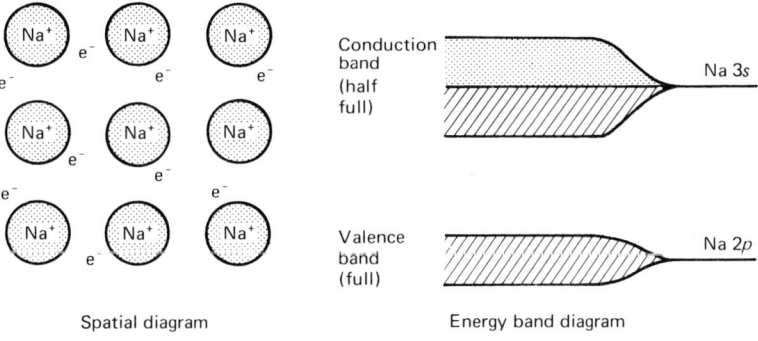

Figure 2.7 Spatial and energy band diagrams for sodium metal

As for other types of solid, an energy band structure is established by the mutual interactions between outer electrons. In sodium metal, as illustrated in *Figure 2.7*, the 3s atomic state, which contains only one electron per atom, is broadened into a conduction band with two states per atom. This band is therefore only half full and electrons in it are easily raised to vacant states and moved through the solid. When an electric potential is applied, these electrons are swept to one electrode and replaced, of course, by the same number of electrons from the external circuit. Because of the small electron transitions involved, metals readily absorb visible light and infra-red radiation.

2.3 NUCLEI

The atomic nucleus contains two types of particle, namely the *proton*, p, and the *neutron*, n. They are approximately equal in size and much larger than the electron. A proton, in the free state, has a rest mass of 0.67×10^{-27} kg, that is, an energy equivalent of 938.26 MeV, which is 1836 times the 0.511 MeV of the electron. It also has a positive charge, $+e$. A neutron has a slightly larger rest mass of 1.675×10^{-27} kg, or, in terms of energy equivalence, 939.55 MeV. It has no electric charge but, in certain circumstances, it behaves like a combination of a proton and an electron. Free neutrons, for example, are radioactive and decay, with a half-life of 12 min, into protons and electrons. In many respects, however, neutrons and protons are quite similar and they are both referred to as *nucleons*.

The number of protons in a nucleus is the *atomic number*, Z. This determines the number of electrons in a neutral atom and, therefore, the properties of the element. The number of neutrons in a nucleus is the *neutron number*, N, and the total number of nucleons is the atomic *mass number*, A, equal to $Z + N$. As mentioned above, an element is a species of atom. Its chemical symbol is uniquely determined by its atomic number, Z, and its properties, apart from those which depend on atomic mass, are also determined by its atomic number. In the same way, a *nuclide* is a species of nucleus but its properties are determined by its atomic number and its mass number (or neutron number). Since nuclear properties depend on Z and A, both of these numbers must be presented when specifying a particular nuclide. The standard terminology is obtained by prefixing the chemical symbol by the Z and A numbers, in the general form $^A_Z X$, where X represents the appropriate chemical symbol. Since Z and X contain the same information, however, the former may be omitted. Thus, for example, the nuclide of potassium, with 40 nucleons — 19 protons and 21 neutrons — is represented by $^{40}_{19}K$, or ^{40}K. Alternatively, it may be described as potassium-40.

A nuclide is said to be *stable* if it tends to remain intact for a geologically long period of time; that is, about 10^{10} years. There are about 284 stable nuclides belonging to 83 elements. The others are unstable and tend to break up, or *decay*, by emitting a particle of ionising radiation and changing into a different type of nuclide. This transmutation must occur because the Z and/or A numbers must change. Such nuclides are known as *radioactive nuclides*, or *radionuclides*, and the decay phenomenon is known as *radioactivity*. It is a statistical process. Not all the nuclei, in a given specimen material, decay at the same time, and certainly not immediately, but they have a definite constant probability of decaying. This probability is measured by the time taken for half of the nuclei in the sample to decay, this time being called the *half-life*, $T_{1/2}$, of the nuclide concerned. Short half-lives denote very unstable nuclides while longer ones indicate less unstable nuclides.

Nuclides with the same Z number, and, therefore, of the same element, but different N and A numbers are described as *isotopes* of the element. Isotopes are chemically similar but they have quite different nuclear properties. In particular, some of them may be *radioactive isotopes*, or *radioisotopes*. These can replace stable isotopes in chemical compounds without affecting the properties of the compound, so that the progress of the radioisotope can be monitored, with radiation detectors, through chemical or biological processes. The compound is known as a radioactively *labelled compound*, or a *radiopharmaceutical*, and it acts as a radioactive *tracer*.

Some examples of isotopes of hydrogen and carbon with their nucleon configurations are as follows:

Hydrogen: ^{1}H: $Z = 1$, $A = 1$, $N = 0$: 1 p
^{2}H: $Z = 1$, $A = 2$, $N = 1$: 1 p + 1 n (deuterium, ^{2}D)
^{3}H: $Z = 1$, $A = 3$, $N = 2$: 1 p + 2 n (tritium, ^{3}T, radioactive)

Carbon: ^{10}C: $Z = 6$, $A = 10$, $N = 4$: 6p + 4n (radioactive)
^{11}C: $Z = 6$, $A = 11$, $N = 5$: 6p + 5n (radioactive)
^{12}C: $Z = 6$, $A = 12$, $N = 6$: 6p + 6n
^{13}C: $Z = 6$, $A = 13$, $N = 7$: 6p + 7n
^{14}C: $Z = 6$, $A = 14$, $N = 8$: 6p + 8n (radioactive)
^{15}C: $Z = 6$, $A = 15$, $N = 9$: 6p + 9n (radioactive)
^{16}C: $Z = 6$, $A = 16$, $N = 10$: 6p + 10n (radioactive)

There is a distinction to be made between a nuclide and an isotope in that the latter term is used when a comparison is being made with other isotopes of the same element. A radionuclide is described as a radioisotope when it is being compared, usually, to a stable isotope. There are two other terms in general use. *Isobars* are nuclides with the same A number but different Z and N numbers. For example, $A = 40$ isobars include $^{40}_{17}$Cl (radioactive), $^{40}_{18}$Ar, $^{40}_{19}$K (radioactive), $^{40}_{21}$Sc. The nuclei of these isobars have approximately the same mass. *Isotones* are nuclides with the same N numbers but different Z and A numbers. For example, $N = 20$ isotones include $^{36}_{16}$S, $^{37}_{17}$Cl, $^{38}_{18}$Ar, $^{39}_{19}$K, $^{40}_{20}$Ca and $^{41}_{21}$Sc (radioactive).

Nuclear models

In Chapter 1, it was pointed out that the most advanced model of a particle or photon is the wave packet model. It was also shown that the most useful models are the continuous-wave and the discrete-particle model, despite the disparity between them. When a particle or photon behaves like a particle, it is treated as a particle. When it behaves like a wave, it is treated as a wave.

A similar situation arises with respect to the nucleus. There is a single Unified Model but it is largely of academic interest. For most practical purposes, it is more convenient to select one of a number of simpler models and, although there is some overlap, these tend to be specific to particular nuclear properties. Static properties and the main features of radioactive decay mechanisms are covered by the Liquid Drop Model, the Shell Model and its derivatives, and the Collective Model. Properties involving interactions with other particles or photons are described by the Optical Model or the Compound Nucleus Model.

The *Liquid Drop Model* is an almost classical model in which the nucleus is represented as a homogeneous, incompressible liquid and the nucleons as component droplets (*Figure 2.8*). It predicts the *total nuclear binding energy, B*, and is used, therefore, to calculate nuclear masses, by means of the Einstein mass—energy relationships, and the energy limitations of the radioactive decay processes.

The evidence for this model is considerable. It includes the fact, illustrated in *Figure 2.9*, that the average binding energy per nucleon, B/A, is fairly constant, at about 8 MeV per nucleon, over a wide range of nuclei. This is exactly as would be expected of a liquid drop, in which the component droplets interact only with adjacent droplets, so that, however many there are, each one is bound by a limited number of neighbours and its binding energy has a saturated, constant value. At the surface, of course, droplets not completely surrounded are less tightly bound, so, in small nuclei with large surface-to-volume ratios, the value of B/A decreases, as shown in *Figure 2.9*. This hypothesis is also compatible with the observation that the range of the nuclear binding force is very small, of the order of one fermi (1F = 10^{-15} m).

Further evidence is presented by the fact that nuclei are found to be basically spherical in shape with radii given by the general formula

$$R = R_0 A^{1/3} \tag{2.5}$$

where the constant R_0 has a value of about 1.2×10^{-15} m, or 1.2 F. This formula shows that the nuclear volume is proportional to the number of nucleons contained, as is the case for an incompressible liquid. It agrees with the 'hard-packed' image presented by *Figure 2.8* and is quite different from the Bohr and Rutherford models of the atom.

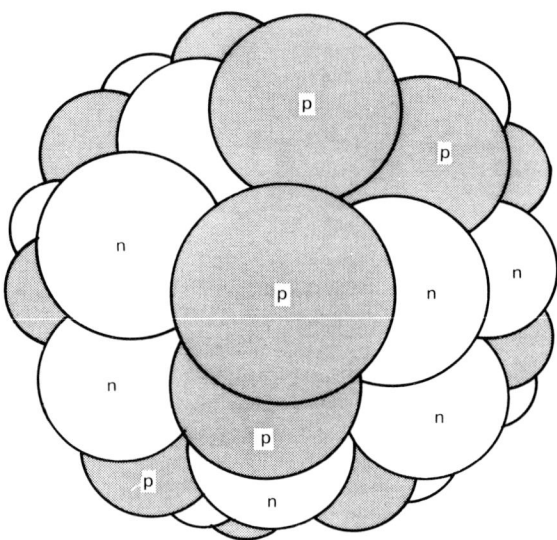

Figure 2.8 The Liquid Drop Model of the nucleus

Weizsacker used this model to formulate an expression for the total binding energy B of a nucleus of atomic number Z and mass number A, as follows:

$$B = a_1 A - a_2 A^{2/3} - a_3 \left[\frac{Z(Z-1)}{A^{1/3}}\right] - a_4 \left[\frac{(N-Z)^2}{A}\right] \pm \frac{a_5}{A^{3/4}} \tag{2.6}$$

The first term indicates that, to a first approximation, binding energy is proportional to the number of nucleons, A. The remaining terms reduce this estimate to account for surface effects, Coulomb repulsion between protons, the fact that N and Z tend to be equal at low A numbers, and the observation that even numbers of nucleons are more stable (that is, more tightly bound) than odd numbers. The parameters a_1 to a_5 are obtained by fitting the formula to experimental measurements, so it is a semiempirical formula.

The mass of a nucleon in a nucleus is less than that of a nucleon in the free state by an amount that is equivalent to its binding energy, according to the Einstein formula, $E = mc^2$. Consequently, the mass, $^A_Z m$, of a nucleus is equal to the sum of the free-state rest masses of its components, less the total binding

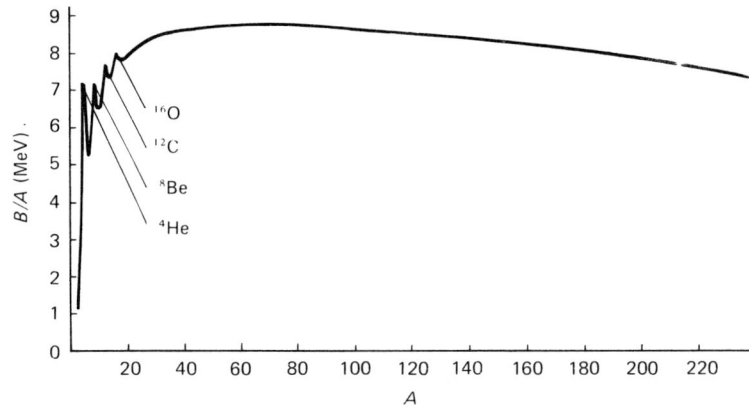

Figure 2.9 Binding energy per nucleon as a function of mass number

energy of the system. If m_p is the mass of a proton and m_n is the mass of a neutron, the Weizsacker semiempirical mass formula (SEMF) predicts the nuclear mass to be

$$^A_Z m = Zm_p + Nm_n - B/c^2 \tag{2.7}$$

In the radioactive decay process, a nucleus can emit a particle if, and only if, the nuclear mass exceeds the sum of the masses of the daughter nucleus and the particle by an amount equal to the binding energy of the particle. It is this requirement which limits radioactive emissions to alpha particles, beta particles and, in some cases, positrons, as will be described in a later chapter. It also shows that alpha decay is energetically impossible except from nuclei with Z numbers greater than about 82. In general, there is an intrinsic relationship between nuclear stability and binding energy. Nuclei with large binding energies tend to be stable. Those with small binding energies tend to be radioactive and the smaller the binding energy, the shorter the decay half-life.

In practice, it is atomic rather than nuclear mass which is measured but this is essentially the same as that of the nucleus. The unit of mass is the *atomic mass unit* (amu) which is defined to be one-twelfth of the mass of a ^{12}C atom. The energy equivalent of an atomic mass unit is 931.44 MeV.

The nuclear *Shell Model* and the *Collective Model* are used to predict nuclear *energy level diagrams*. They are similar, in some respects, to the atomic shell model that is the basis of atomic spectroscopy. The nuclear models predict and represent internal nuclear transitions involving the emission and absorption of gamma radiation, so they provide a theoretical basis for *nuclear spectroscopy*. Thus, the Liquid Drop Model predicts nuclear ground state energies while the Shell and Collective Models predict excited state energies.

Although it is difficult to conceive of Bohr-type orbits in such a densely packed system as the nucleus, there is ample evidence that the nucleons exist in well-defined, discrete energy states. This hypothesis is supported by a vast amount of nuclear reaction data as well as the observed energies of radioactive emissions, especially gamma ray photons, whose energies are given by a formula similar to the first Bohr postulate, namely, $h\nu = E_2 - E_1$. There is further evidence in the fact that nuclides with proton or neutron numbers equal to one of a set of *magic numbers* exhibit unusual stability. These numbers are 2, 8, 20,

50, 82 and 126. They correspond to the closed-shell atomic structures of the noble gases at electron numbers 2, 10, 18, 36, 54 and 86. Shell closure occurs at numbers different from those of atoms because the nucleus is a different type of system. Also, its effects are less pronounced because there are two types of particle present and closure in one, say the protons, leaves the other type, the neutrons, with an active configuration. There are some nuclides, such as $^{208}_{82}$Pb, which are doubly magic and very stable.

Neutrons and protons exist in separate sets of allowed nucleon states, each characterised by four quantum numbers. The Pauli Exclusion Principle prohibits any two protons, or neutrons, from having the same quantum numbers but a proton and neutron can have the same numbers, since they are different particles. There are two energy level diagrams for the nucleus. One describes neutron states and the other refers to proton states but, apart from the additional Coulomb effects in the latter, the two diagrams are very similar. *Figure 2.10* shows these nucleon levels, with major energy gaps above the magic number configurations. Spectral notation is the same as for atoms, except that the K shell, N shell, etc., notation is dropped. The principal quantum number n is less important in the nucleus, and can be exceeded by the values of l, the orbital angular momentum

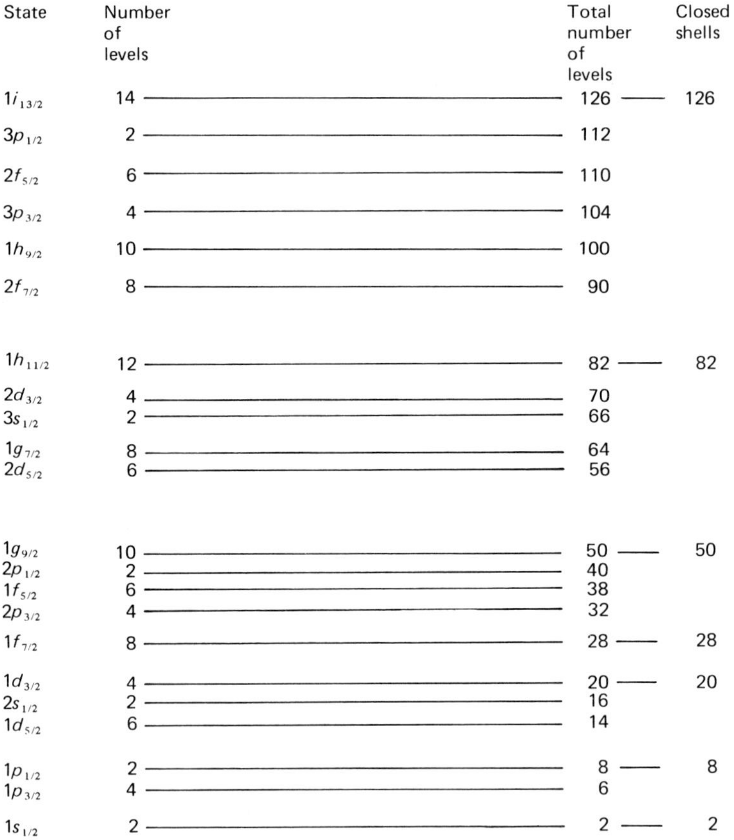

State	Number of levels	Total number of levels	Closed shells
$1i_{13/2}$	14	126	126
$3p_{1/2}$	2	112	
$2f_{5/2}$	6	110	
$3p_{3/2}$	4	104	
$1h_{9/2}$	10	100	
$2f_{7/2}$	8	90	
$1h_{11/2}$	12	82	82
$2d_{3/2}$	4	70	
$3s_{1/2}$	2	66	
$1g_{7/2}$	8	64	
$2d_{5/2}$	6	56	
$1g_{9/2}$	10	50	50
$2p_{1/2}$	2	40	
$1f_{5/2}$	6	38	
$2p_{3/2}$	4	32	
$1f_{7/2}$	8	28	28
$1d_{3/2}$	4	20	20
$2s_{1/2}$	2	16	
$1d_{5/2}$	6	14	
$1p_{1/2}$	2	8	8
$1p_{3/2}$	4	6	
$1s_{1/2}$	2	2	2

Figure 2.10 Allowed energy levels for each type of nucleon in a nucleus

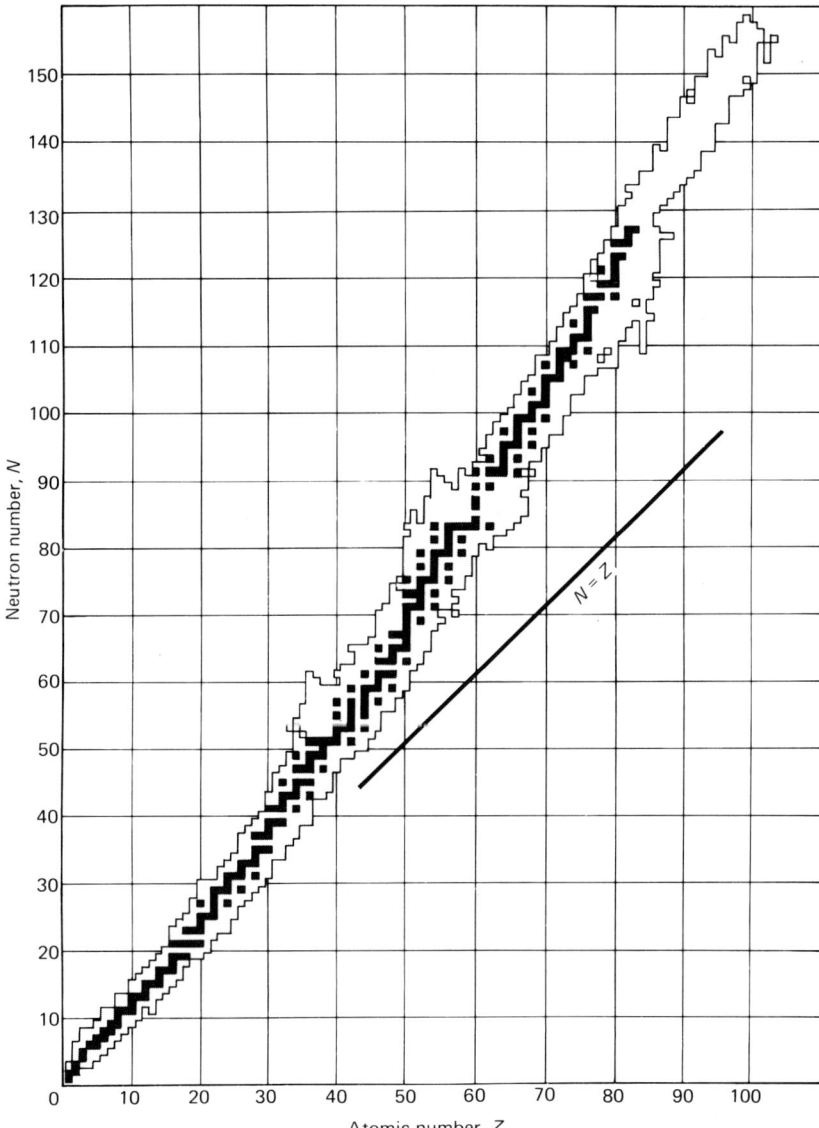

Figure 2.11 Chart of the Nuclides. (■) Stable nuclides; (□) unstable nuclides

quantum number. Thus, the lowest energy level is the $1s_{1/2}$ level, containing two nucleon states. The next level is the $1p_{3/2}$, then the $1p_{1/2}$ level, and so on. A $1p$ state is forbidden in the atom by the requirement that l cannot exceed $n - 1$.

As for atoms, and for molecules and solids, an important feature of this nuclear model is the fact that closed nucleon shells form an inert core and the nuclear properties are determined by the configuration of the 'outer', least tightly bound nucleons. One consequence of this is a periodic variation in nuclear properties analogous to that described in the Periodic Table of the elements. In this case, however, the periodicity is in two dimensions; namely, in Z number and in N number. It is represented in a *Chart of the Nuclides*, which is a graph of Z against N. Each entry in this chart describes one nuclide and, typically, the nuclear properties of that nuclide are entered, in a matrix element, at the appropriate Z and N number location. A small-scale version of this chart is shown in *Figure 2.11*.

A second result of being able to represent a nuclide by its least tightly bound nucleons is that the excited states of the whole nucleus are the vacant states to which these outer nucleons may make transitions. These are the states used in nuclear spectrometry to describe alpha particle, beta particle and gamma ray transitions. Typically, the core levels are omitted from the energy level diagram and the ground state of the nucleus is denoted by the state occupied by the outer nucleons. This convention is illustrated in the energy level diagram of $^{64}_{30}$Zn, presented in *Figure 2.12*, which is analogous to the atomic level diagram for sodium given in *Figure 2.3*, and to the band structure diagram in *Figures 2.6* and *2.7*.

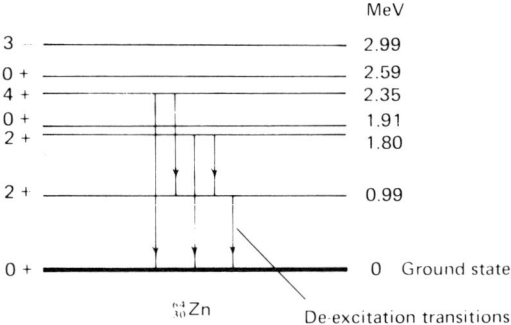

Figure 2.12 Nuclear energy level diagram for $^{64}_{30}Zn$

The only difficulty is generated by the strong nucleon—nucleon force compared to the interelectron force in an atom. If the nuclide has a closed-shell structure it is well described by the Shell Model. If it has a few nucleons more than or less than a closed shell, internucleon forces can be ignored and an extension of the Shell Model, known as the Single-Particle Model, is employed. If, however, there are many extra-core nucleons, they must be treated as a combined surface effect, in the Collective Model. These transitions from one model to another make it relatively difficult to predict nuclear configurations.

Although some theoretical problems remain, nuclear energy levels can be measured and energy level diagrams are available for most nuclides. These provide a basis for the description and prediction of a number of nuclear processes, including radioactive decay.

BIBLIOGRAPHY

ANDERSON, J. C. and LEAVER, K. D., *Materials Science,* Nelson, London (1969)
ENGE, H., *Introduction to Nuclear Physics*, Addison-Wesley, Reading, Massachusetts (1966)
KITTEL, C., *Introduction to Solid State Physics,* John Wiley, New York (1968)
ROSENBERG, H. M., *The Solid State,* Clarendon Press, Oxford (1975)
SPROULL, R. L., *Modern Physics,* John Wiley, New York (1956, 1963)
WERT, C. A. and THOMSON, R. M., *Physics of Solids*, McGraw-Hill, New York (1964)

Chapter 3
Ionising radiations

3.1 INTRODUCTION

Ionising radiations are particles or photons which tend to produce large numbers of ion pairs when they interact with matter. One way of classifying them is to recognise three basic *sources* of ionising radiations. They derive from, or are, *atoms, nuclei* or *fundamental particles*, although there is some overlap between the last category and the first two. This is, basically, the system adopted in this chapter. An alternative approach, which is more appropriate to the practical aspects of radiation detection, is to classify these radiations in terms of their *interactions with matter*. On this basis, there are three types of ionising radiation: *charged particles, photons* and *uncharged particles*. The most important example of the last category is the *neutron*.

The distinction between ionising and non-ionising radiations is not always a clear one. It depends on the nature of the interactions with matter and it may depend on the energy and intensity of the radiations. Most of the charged particles are *directly ionising*. Each particle collides with and ionises a large number of atoms, provided it has sufficient energy. The other radiations are *indirectly ionising*. Neutrons, for example, interact with atomic nuclei to produce charged-particle reaction products which then directly ionise atoms. All neutrons are capable of generating high-energy reaction products, so all neutrons, whatever their energies, are ionising radiations. The same is true of positrons, which react with atomic electrons to produce high-energy photons capable of further ionisation. Photons are also indirectly ionising, but only if each of them has sufficient energy to liberate one atomic electron and give it enough residual energy to create further ion pairs.

Charged particles and photons interact mainly with atomic electrons. They are able to produce ion pairs if the energy of each particle or photon exceeds the *ionisation potential* of the material being irradiated. This criterion defines a minimum energy requirement but it is not sufficient to distinguish between ionising and non-ionising radiation. In order to do so, a number of statistical effects must be taken into consideration.

First of all, ionisation potential varies from one element to another, so that radiation capable of ionising one element may be unable to ionise another. As shown in Appendix 1, ionisation potentials range from the 24.58 eV of helium, whose two electrons form a stable, closed K shell, to the 3.98 eV required to remove the single P-shell electron of caesium. Photons of visible light, with wavelengths in the range 4000–7000 Å, have a maximum energy, given by equation 1.22, of 3.1 eV and are unable to ionise any atomic materials in single events. More generally, photons with more than about 15 eV are, in principle, capable of generating some ion pairs in about half of the elements. Solids such as films, semiconductors and other photosensitive materials are ionised by visible light, but these are special cases.

A second factor derives from the random variation in collision angle between the incident radiation and the target atomic electron. Most collisions are bound to be glancing ones, in which only partial energy transfer takes place, leading to excitation rather than ionisation. As a rule, about half of the energy of incident radiation is spent on excitation, so the average energy, per particle or photon, required to produce an ion pair is about twice the ionisation potential of the target material. This is known as the *mean energy per ion pair*, ϵ, and varies, of course, from one material to another. In a noble gas, apart from helium, it is about 30 eV. In an insulator, it is about 10 eV and in a semiconductor it is only 2–3 eV.

Charged particles such as electrons, and also photons, do produce ion pairs if their energy is greater than ϵ, but not in sufficient quantity to qualify them, generally, as ionising radiations. In addition, they are very weakly penetrating radiations and ionise atoms only in the material in which they are actually formed. A more realistic threshold, above which several ion pairs per particle or photon can be created, is about 100–1000 eV. As shown in *Figure 1.2*, this means that, in general, X-rays are regarded as ionising radiations but ultra-violet rays are not. The only exceptions to this rule are, first, ultra-violet photons produced inside a radiation detector which do create further ion pairs and, secondly, very high intensities of ultra-violet light which lead to considerable, if inefficient, ionisation.

Figure 1.2 also shows that the highest-energy, shortest-wavelength electromagnetic radiations are the *gamma rays*. These extend down to about 12 keV, that is, 1 Å. Overlapping these, and extending down to about 100 eV per photon, are the *X-rays*. High-energy X-rays are often referred to as 'hard' X-rays, while the low-energy end of the range is known as 'soft' X-radiation. Hard X-rays are identical, in all their properties, to gamma rays but the distinctive terminology is retained. Gamma rays are defined as electromagnetic radiation originating from a nucleus or a fundamental particle. Photons deriving from atoms are always described as X-rays. A similarly dual terminology is applied to electrons. If emitted from atoms, they are known as *electrons* but, if they come from nuclei, they are called *beta particles*. The terminology for fundamental particles is less obvious but, as a useful rule, beta particles are always accompanied by uncharged particles, known as *neutrinos*, while electrons are not.

3.2 X-RADIATION

Visible light is produced in two ways: by electronic transitions between discrete, bound-state atomic energy levels, and by transitions within the continuous range of energies of a free electron or electrons in the conduction band of a metal. The first process generates a *line spectrum* which is characteristic of the emitting atoms and can be used, as in atomic spectroscopy, to identify elements. The second mechanism produces a *continuous spectrum*, like that of white light. In both cases, photon energies are of the order of a few electronvolts and are easily absorbed in matter. Most materials are optically opaque because their outer atomic electrons have large numbers of vacant energy states to which they may be excited on absorbing a few electronvolts. Transparent crystals are exceptions in this respect, because they have energy band structures, with filled valence bands and empty conduction bands, typical of insulators, with energy gaps larger than a few electronvolts.

In 1895, Röntgen discovered that some of the electromagnetic radiation produced in a gas discharge tube was much more penetrating than visible light and ultra-violet radiation. He observed that this radiation could be transmitted through optically opaque objects, such as his hand, to produce a shadow-image on film of the internal structure of these objects. The radiation became known as *X-radiation*, or X-rays, and the images are now called X-radiographs. As for visible light, there are two types of X-ray spectra, namely characteristic and continuous. They are more penetrating than visible light, because their photon energies are much higher, usually well in excess of 100 eV, and can be lost in single collisions with atomic electrons only if these are the relatively inaccessible inner electrons. The outer electrons tend to absorb small fractions of the photon energies.

Characteristic X-rays

Photons with energies greater than 1 keV can, like those of visible light, be produced by atomic de-excitation, in which case they have a line energy spectrum which is uniquely characteristic of the element involved. They are known as *characteristic X-rays* and form the basis of a number of analytical techniques, such as X-ray fluorescence. They are generated whenever certain conditions are satisfied, either intentionally, in an X-ray machine, for example, or as a side effect in devices such as gas discharge tubes and television sets.

One requirement is that the energy level structure of the emitting element must be capable of containing such a large quantum jump. Hydrogen, for example, has a K-shell binding energy of only 13.6 eV and does not produce X-rays. However, binding energy increases with Z number, being roughly proportional to Z^2 (equation 2.3) and exceeds 1 keV for Z numbers greater than about 11, so this requirement is for a relatively high Z number. In addition, a similar amount of energy, in excess of about 1 keV, must be given to the atom in order to eject an inner electron or, at least, excite it to an outer, vacant energy state, before the requisite de-excitation can take place. This energy is usually supplied in collisions with incident particles or photons which must, therefore, have this much energy. A third condition is that the intensity of these exciting radiations must be fairly high, because collision with an inner electron is an event with relatively low probability.

The de-excitation process is similar to that described for hydrogen, in a previous chapter, except that many electrons, instead of only one, are involved. Thus, a cascade of electron transitions takes place rather than a single one. A K-shell vacancy, for example, may be filled by an L-shell electron, producing a K_α spectral line, the L-shell vacancy, thereby created, may be filled by an M-shell electron producing an L_α line, and so on. In general, the lines which are obtained most frequently are the K_α and K_β lines, followed by the L_α, L_β and L_γ

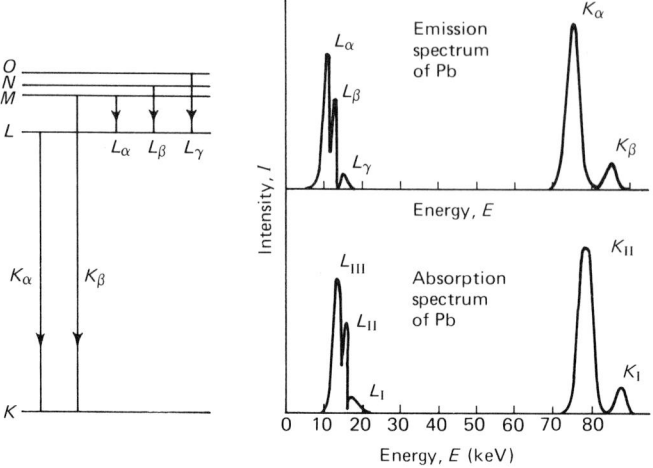

Figure 3.1 *Atomic energy levels and spectra*

lines. The relative intensities of the K_α to K_β lines are 6:1 and those of the L_α to L_β to L_γ lines are 9:6:1, as illustrated, for molybdenum, in *Figure 3.1*. A complete set of characteristic emission line energies is given in Appendix 2.

Bremsstrahlung

Free electrons can also experience changes in their energy. These occur when the electron is accelerated or decelerated and are analogous to the excitation and de-excitation processes, respectively, which affect electrons bound in atomic systems. When an electron moves through matter it is slowed down and deflected in a series of collisions, mostly with atomic electrons or nuclei. In each of these decelerations a photon of electromagnetic radiation is generated (*Figure 3.2*). This is called *bremsstrahlung* or 'braking radiation'. Since the free electron does not have discrete, allowed energy states, these photons can have any energy from near zero to the maximum electron energy, although low energies are favoured

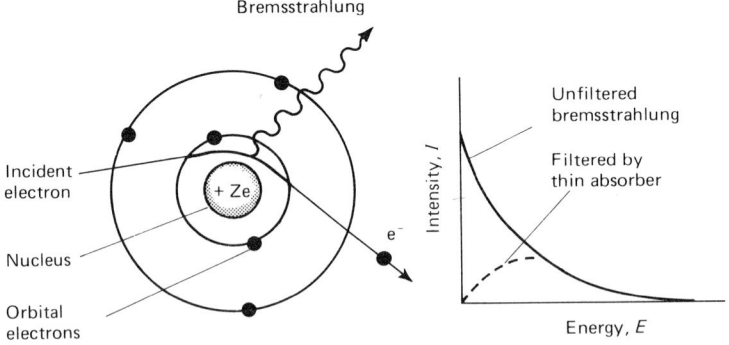

Figure 3.2 *Bremsstrahlung production and its energy spectrum*

by the collision statistics. The process is similar to that which generates white light and the result is a *continuous energy spectrum* or 'white' X-radiation, provided, of course, the electron energies are high enough. Photon energies may extend into the mega-electronvolt range for electrons from an accelerator.

When bremsstrahlung passes through a particular material, a characteristic absorption spectrum can be produced by preferential absorption of photon energies corresponding to quantum jumps within the atoms present. This process produces *absorption edges* at characteristic energies, as illustrated, for molybdenum, in *Figure 3.1*. Absorption energies always exceed emission energies corresponding to the same electron transitions because the former must include some energy to allow the atom to recoil, conserving momentum, while the latter lose some energy to atomic recoil. Emission and absorption line energies are listed in Appendix 2.

The production of X-rays: X-ray machines

A basic version of the modern X-ray tube is illustrated diagrammatically in *Figure 3.3*. Electrons are produced by thermionic emission from a tungsten cathode. These are focused and accelerated by a large potential difference to strike a target anode with very high energy and intensity. In diagnostic radiographic machines, applied potentials may exceed 100 kV so that electron energies

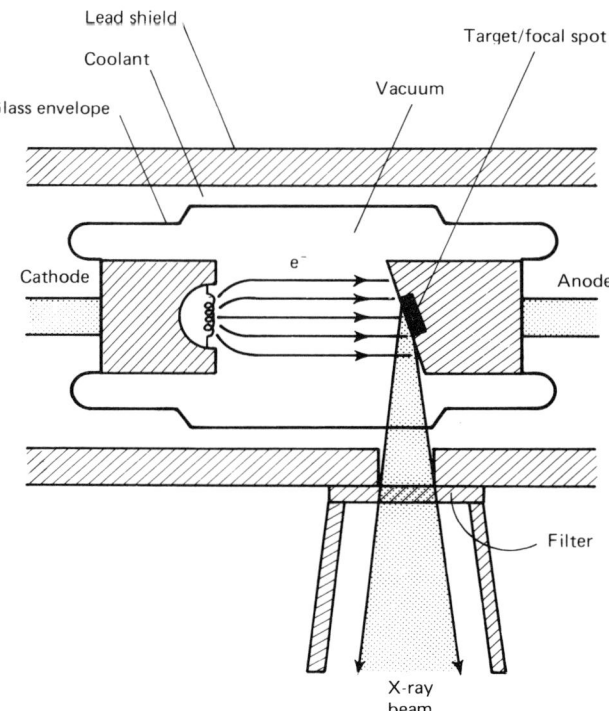

Figure 3.3 Basic structure of an X-ray tube

exceed 100 keV. In crystallographic machines, smaller voltages are typical. In general, the voltage is not uniform but has a considerable a.c. component so that the peak kilovoltage, or kVp, is quoted.

The target may be tungsten, molybdenum, copper or some other metal with a reasonably high Z number and, preferably, good thermal conductivity and a high melting point. Most of the electron energy is dissipated as heat and normally less than 1% appears as X-ray energy. This fraction increases with electron energy and target Z number, as bremsstrahlung output increases. The whole assembly is enclosed in a glass envelope, evacuated to avoid electron scattering at air

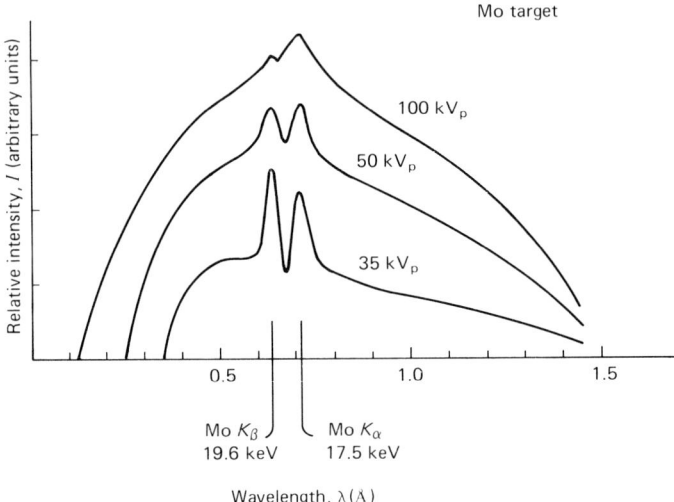

Figure 3.4 *Photon spectra for an X-ray machine*

molecules. Since the X-rays are emitted through this glass envelope, the lowest photon energies are filtered off by absorption. Filtration can be increased by inserting foils of metal, such as aluminium, in the window diaphragm of the tube. Typical emission spectra for a molybdenum anode are shown in *Figure 3.4*. Despite its inefficiency, the machine is the best source of intense X-ray beams suitable for radiography, radiotherapy and crystallography, although accelerators may also be used for therapy.

The production of secondary X-rays

X-radiation can be produced by irradiating a suitable target material with photons or charged particles. The former create excited atoms by photoelectric processes (which will be described in some detail in Chapter 4) and the latter do the same by direct collisions. The excited target atoms then de-excite to provide the characteristic X-radiation of the element involved. If the primary source is a photon emitter, some backscattered photons are obtained from the target material. If it is a particle emitter, some bremsstrahlung may be generated besides the secondary X-rays. Similar processes form the basis of the *X-ray fluorescence* technique used to analyse the elemental composition of the target material.

The production of X-rays by radioactive nuclides

Radionuclides which decay by the emission of gamma radiation may, as an alternative, de-excite via the process of *internal conversion* (IC) in which an atomic electron receives the gamma energy and is ejected from the atom. The probability that an internal-conversion electron will be emitted instead of a gamma ray photon is measured by the internal-conversion coefficient, values of which are listed in Appendix 3. Another nuclear process in which atomic electrons are involved is the form of beta decay known as *electron capture* (EC) in which an atomic electron is absorbed directly into the nucleus. Both processes tend to involve the inner electrons 'closest' to the nucleus, especially K-shell electrons, so that subsequent atomic de-excitation provides characteristic X-rays. Because of these decay modes, many radioactive sources generate characteristic X-radiation which may be almost monoenergetic (or monochromatic). These are, of course, characteristic of the *daughter nuclide*. For example, electron capture in iron-55 produces X-rays characteristic of manganese-55.

In addition, sources of energetic beta particles usually produce some *bremsstrahlung* via interactions between the beta particles and other materials present in the source. Some of these interactions may result in characteristic line emission.

3.3 ELECTRONS

Ionising electrons can be obtained, from atoms, in a number of ways. In the present context, the most important of these are as follows:

Thermoelectrons

The conduction electrons of a metal exist in an almost free state in the conduction band. Very little energy, in the form of thermal energy or an applied potential difference, is required to raise these electrons to vacant states in the conduction band where they move easily through the metal and, with sufficient additional energy, of the order of a few electronvolts, they may overcome the small surface barrier, called a *work function*, and be released from the metal. The same is true of electrons in gases. Both may be regarded as *thermoelectrons*, since the energy is usually supplied as direct heat or as a heating current.

To become ionising radiations, such electrons have to be accelerated, by applied potentials, to higher energies, but this is the most convenient method of producing intense beams of electrons and is used, for example, in X-ray machines, television receivers and electron accelerators.

Photoelectrons

In the *photoelectric effect*, an incident gamma or X-ray photon ejects an inner electron from an atom, creating an ion pair and giving all of its energy to the electron. This *photoelectron* has a net energy equal to that of the photon less its atomic binding energy so that, with incident photon energies of the order of 1 MeV, similarly high electron energies are obtained. This is a very simple and inexpensive method of producing high-energy electrons although by comparison with the thermionic process it generates only very small quantities of electrons. It is more important as a photon detection mechanism and, as described above, as a source of X-rays.

Auger electrons

Excited atoms usually decay by the emission of electromagnetic radiation but sometimes the de-excitation energy is transferred to one of the atomic electrons which is then emitted instead of a photon. Actually, it is a direct energy transfer process, but it can be imagined that a photon is produced, in the first instance, which collides with the electron on its way out of the atom, ejecting the latter as an *Auger electron*. It is an alternative to X-ray emission and the pure atomic analogue of internal conversion.

If, for example, an atom de-excites by means of an electron transition from the L to the K shell then, instead of emitting a K_α photon, it may emit an Auger electron from the L shell. Consequently, these electrons have a line energy spectrum characteristic of the element concerned but their energies are very small and they are not usually regarded as ionising radiations. Their most important effect on the present discussion is the fact that they reduce the number of X-ray photons to be expected of a particular source. This is measured by the *fluorescent yield*, which is the fraction of all de-excitation processes which yield photons.

Internal-conversion electrons

The process of internal conversion has been described above as a means of causing atomic excitation and subsequent X-ray emission. It is also a method of producing high-energy, but low-yield, electrons with 'nuclear' energies of the order of mega-electronvolts. It is more important as an alternative to gamma ray emission since it replaces a photon with an electron of equal energy. Electron energies can be measured much more accurately, using a magnetic spectrometer, than can those of gamma rays, so this provides a useful means of studying photon energies indirectly. It is also important in that it reduces the number of gamma ray photons to be expected of a particular source, sometimes by a considerable amount. *Internal-conversion coefficients,* α, define the fractional probability of observing an electron instead of a photon.

3.4 POSITIVE IONS

In principle, at least, positive ions of any element can be created by stripping off one or more atomic electrons, in thermal or other processes, then accelerating the residual ion up to an ionising energy. Ions of the lighter elements are increasingly being used in high-energy and nuclear physics because their relatively low masses enable them to be more easily accelerated. However, the most important ions are the special category obtained from the hydrogen isotopes, namely, the proton, the deuteron, and the triton, as well as the alpha particle, which is a doubly ionised helium atom.

The *proton*, p or $_1^1\text{H}^+$, is obtained by removing the single electron from a hydrogen atom, leaving the nucleus. It has a positive charge, $+e$, a mass of 1.6725×10^{-27} kg and a rest mass–energy of 938.26 MeV. It is, of course, a fundamental particle. The *deuteron*, d, is a deuterium nucleus, $_1^2\text{H}^+$ or $_1^2\text{D}^+$. It is a composite particle containing one proton and one neutron rather loosely bound together. The *triton*, t, is the nucleus of the third hydrogen isotope, tritium, $_1^3\text{H}^+$ or $_1^3\text{T}^+$, and is a composite of one proton and two neutrons. It is radioactive, decaying with a half-life of 12.35 years and emitting a 0.0186-keV (maximum) beta particle to become $_2^3\text{He}^+$, which is sometimes referred to as a *helion*. All three particles are produced by ionising the appropriate isotope but as will be discussed later they are also formed, in much smaller quantities, in nuclear reactions.

3.5 RADIOACTIVE DECAY THEORY

Unstable nuclides are known as *radioactive nuclides* or *radionuclides*. They decay, or disintegrate, losing excessive mass–energy by the emission of particles and/or photons. There are two types of radionuclide. First, there are the *natural radionuclides*, which exist in nature and can be obtained by chemical or isotope separation methods. Mainly, they emit *alpha particles, beta particles* and *gamma radiation*, although some decay by spontaneous fission. Secondly, there are the *artificial radionuclides*, which are manufactured in nuclear processes, are generally less stable, and emit a wider range of types of radiation. A table of the properties of radioactive nuclides is provided in Appendix 3.

General formulae

Radioactive decay is a *statistical process*. Given a single unstable nucleus, it cannot be predicted when it will disintegrate. It is only possible to define a *probability*, λ, that it will decay in the next second. λ is called the *decay constant*, or the *disintegration constant*. It is a constant property of a particular nuclide and is totally independent of the physical environment of the nuclei or of their age. Thus, if a nucleus does not happen to decay in, say, a period of one year, the probability that it will do so in the next second is the same as it was in the first second of the year. The units of λ are s^{-1}.

On the other hand, with a *radioactive source* comprising a statistically large number of nuclei, it can be predicted, with a certainty defined by statistical error, that in one second a fraction λ of these nuclei *will* decay. Thus, with N_0 nuclei, the number decaying in one second is $N_0\lambda$, leaving $N_0 - N_0\lambda$ nuclei so that, in the next second, $(N_0 - N_0\lambda)\lambda$ nuclei decay, and so on. Clearly, the number of nuclei present at any time is a variable, N, which decreases with time. In fact, if there are N_0 nuclei present when the timing measurement commences, i.e. at time $t = 0$, then the number N remaining at any time t later is given by an exponential formula of the form

$$N = N_0 e^{-\lambda t} \tag{3.1}$$

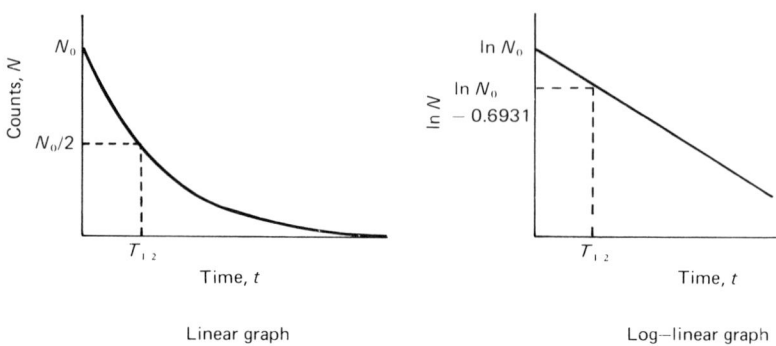

Figure 3.5 Exponential nature of radioactive decay

This *exponential decay law* can be expressed graphically, as in *Figure 3.5*. If N is plotted against t, a curved graph is obtained. Alternatively, the natural logarithm of N can be plotted against time, t, to obtain a straight line, since

$$\ln N = \ln (N_0 e^{-\lambda t})$$
$$= \ln N_0 + \ln (e^{-\lambda t})$$
$$= \ln N_0 - \lambda t$$

The same effect is obtained by plotting N and t on log–linear graph paper. In both cases a straight line is produced with a gradient, $-\lambda$, that is more easily measured than in the exponential graph.

A number of parameters can be defined which are, in general, more useful and more easily measured than the decay constant. They include activity, half-life and mean life.

The *activity*, A, of a radioactive source material is the number of nuclei in the specimen that decay in unit time; that is, the average number of disintegrations per second. It is given by

$$A = N\lambda \qquad (3.2)$$

and, like N, decreases exponentially with time. The *specific activity* is the activity per gram or kilogram of material.

The most widely used unit of activity is the curie (C or Ci), although the SI unit is the becquerel (Bq). These are defined as

$$\begin{aligned} 1\ \text{Ci} &= 3.7 \times 10^{10}\ \text{disintegrations per second} \\ 1\ \text{Bq} &= 1\ \text{disintegration per second} \end{aligned} \qquad (3.3)$$

A curie is an extremely large activity. For most purposes, activities, or *source strengths*, ranging from a millicurie (1 mCi) to a microcurie (1 µCi) are used, where 1 mCi is 10^{-3} Ci and 1 µCi is 10^{-6} Ci.

It is important to note that one disintegration does not necessarily yield one particle or photon. The average number of radiations emitted by a source material, in each disintegration, depends on the decay mechanism. The activity must be multiplied by this number in order to estimate the number of emissions per second from a radioactive source. Data relating to the nature and abundance of decay products, for a range of nuclides, are given in Appendix 3.

Half-life, $T_{1/2}$, is the mean time required for the activity, A, of a radioactive source to decrease by a factor of 2. It is also the time taken for the number of radioactive nuclei, N, to reduce by the same factor. Hence,

$$\frac{N}{N_0} = \frac{1}{2} = \exp(-\lambda T_{1/2})$$

giving

$$T_{1/2} = \frac{\ln 2}{\lambda} = \frac{0.6931}{\lambda} \qquad (3.4)$$

This parameter is shown in the graphs of *Figure 3.5*.

Mean life, τ, is the average lifetime of a radioactive nucleus. Using the usual averaging procedure, this can be shown to be equal to the reciprocal of the decay constant, λ, i.e.

$$\tau = \frac{1}{\lambda} \qquad (3.5)$$

3.6 ALPHA PARTICLES

In the *alpha decay process* two neutrons and two protons combine, within the nucleus, and are emitted as a single composite particle called an *alpha particle* or *α particle*. This is identical to the nucleus of a helium atom, ^4_2He, and is often represented by ^4_2He instead of α. With two protons, it has a double positive charge, $+2e$, and, with four nucleons, it has a large mass of 4.003 amu. Thus,

$$\alpha = 2n + 2p = {}^4_2\text{He} \qquad (3.6)$$

Alpha decay can be represented by a nuclear equation of the form

$$^A_Z X \rightarrow {}^{A-4}_{Z-2} Y + {}^4_2\alpha \qquad (3.7)$$

where $^A_Z X$ represents the *parent* nuclide, $^{A-4}_{Z-2}Y$ represents the *daughter* nuclide and the alpha particle can be represented by $^4_2\alpha$, α or ^4_2He. Note that A and Z numbers are conserved in such equations; that is, the total numbers of nucleons and protons (hence, also, neutrons) are the same before and after the event, but a new nuclide is formed. A typical example of this transmutation effect is

$$^{238}_{92}\text{U} \rightarrow {}^{234}_{90}\text{Th} + \alpha$$

in which thorium nuclei are generated from uranium.

Because of its large mass, an alpha particle requires a great deal of energy to move at relativistic velocities. In fact, it needs about 1600 MeV to develop a velocity of $0.66c$. For most purposes, it can be treated with *non-relativistic* formulae.

Alpha decay is energetically possible only from nuclides in which the binding energy, B_α, of the alpha particle in the nucleus is positive. Thus a necessary condition is given by the equation

$$B_\alpha = ({}^{A-4}_{Z-2}m + {}^{4}_{2}m - {}^{A}_{Z}m)c^2 \geqslant 0 \tag{3.8}$$

In general, this condition is satisfied only if the N and Z numbers of the parent nuclide are greater than about 82, so the process occurs only in heavy nuclides. However, the decay probability, λ, depends on a number of other factors. The particle has to be formed in the nucleus, it has to reach the nuclear surface, then it has to escape from the attractive nuclear force. This force can be represented by a potential energy barrier, that is, a graph of the potential energy of the alpha particle against its position. It can be shown that the *potential barrier* is invariably greater than the energy of the particle so that, classically, alpha emission is impossible. The fact that it does take place is sometimes ascribed to a *tunnel effect*, in which the alpha particle is assumed to tunnel through the insuperable barrier. A more realistic explanation is the quantum one, that the position of the particle is diffuse enough to overlap the barrier, giving a small probability that it will be found on the outside.

Having overcome the barrier, the particle exists in an unbound, positive energy state and emerges with a net positive energy, E_0, called the *disintegration energy*. In order to conserve linear momentum, the nucleus must recoil, when it ejects the alpha particle, and take some of the disintegration energy. The net kinetic energy, E_α, of the alpha particle is given by

$$E_\alpha = \frac{E_0}{1 + (m_\alpha/m_d)} \tag{3.9}$$

where m_α is the mass of the alpha and m_d that of the daughter nucleus. For heavy nuclei, however, m_d is much larger than m_α and this correction can be ignored, so that E_α is assumed to be equal to E_0.

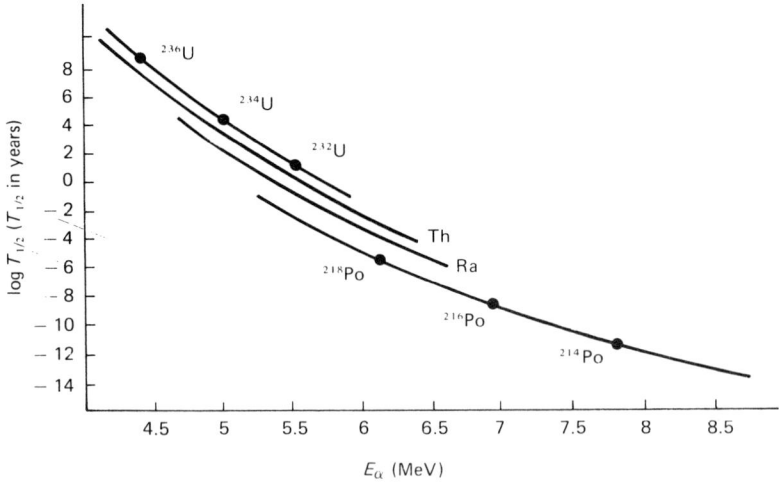

Figure 3.6 Half-life as a function of alpha particle energy

Not surprisingly, the greater the value of E_0 (and E_α), the more likely is the alpha to overcome the surface barrier and escape from the nucleus. Hence the decay constant, λ, increases rapidly with E_α, or the half-life, $T_{1/2}$, decreases rapidly with E_α. This fact is expressed by the experimental law of *Geiger and Nuttall*:

$$\log \lambda = A \log E + B \tag{3.10}$$

where A and B are constants.

The same relationship can be presented more accurately as a graph, usually of $\log T_{1/2}$ against E_α, as shown in *Figure 3.6*. This clearly illustrates the extreme sensitivity of $T_{1/2}$ to E_α. While alpha particle energy increases from 4 to about 9 MeV, the half-life ranges from 10^{10} to 10^{-14} years. The graph is drawn for even–even nuclei in which it is relatively easy for a pair of neutrons and a pair of protons in the highest, least tightly bound energy states to come together as an alpha particle. In other nuclei, this event is much less likely and decay constants are reduced by orders of magnitude.

A quantum theory of alpha decay has been developed by Gamow. His formula for the decay constant, for even–even nuclei, is of the form

$$\lambda = 10^{21} \exp\left(-4\frac{(25 - E_0)^{3/2}}{E_0}\right) \text{ s}^{-1} \tag{3.11}$$

Energy spectra

The alpha particles emitted from a source containing only one nuclide should, and do, have the same energy, E_α, provided the decay is from the ground state of that nuclide to the ground state of the daughter nuclide. This energy can be measured in certain types of radiation detector and a graph of count rate against energy can be plotted. This *corresponds* to an *emission spectrum* for the alpha particles. It is not exactly equivalent to the emission spectrum because statistical effects in the detector (which will be discussed later) broaden what would be a narrow spectral line into a more diffuse line which appears as a peak in the graph of count rate against alpha energy. The detected spectrum has the form shown in *Figure 3.7*.

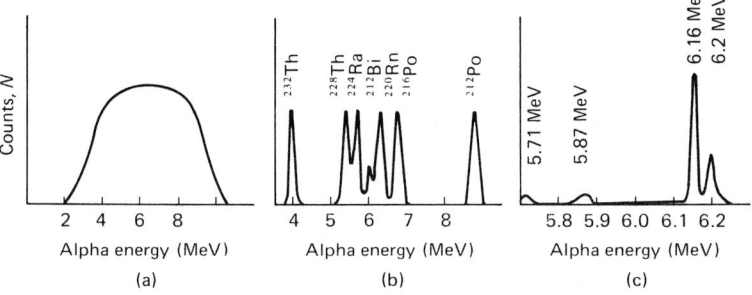

Figure 3.7 Alpha particle energy spectra. (a) ^{232}Th and chain products – unresolved; (b) ^{232}Th and chain products – resolved; (c) ^{212}Bi spectrum – fine structure

50 *Ionising radiations*

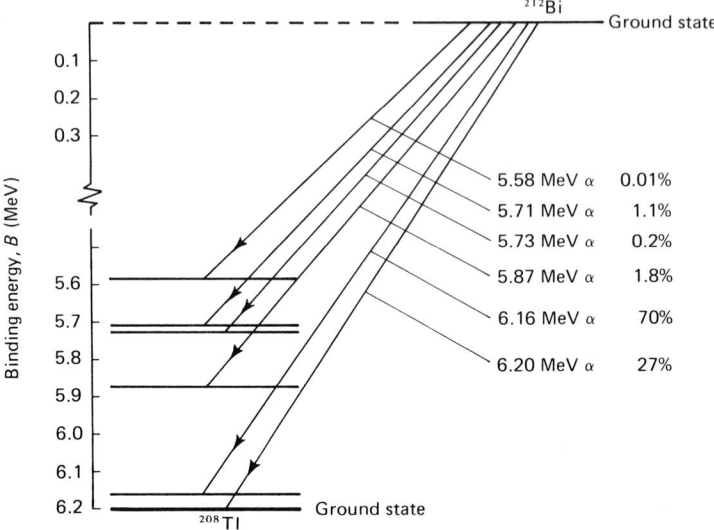

Figure 3.8 Decay scheme of ^{212}*Bi*

In general, however, the daughter nuclide will also decay to another nuclide and so, too, will its daughter decay, and so on, forming a *chain*. Hence any alpha-radioactive source will usually contain several alpha emitters and the energy spectrum will have several peaks, as illustrated in *Figure 3.7*.

If any one of these peaks is examined more closely, that is, with better energy resolution, it will be seen to have a *fine structure* and comprise several peaks instead of only one. This effect is also shown, for ^{212}Bi, in *Figure 3.7*. It is evidence that the nucleus, like the atom, has discrete energy states, as predicted by the Shell Model and the Collective Model. As an example, ^{212}Bi (or Th C, as it may be called) decays to ^{208}Tl (Th C″) such that decay is from the ground state of ^{212}Bi to the ground state or to any of five excited states of ^{208}Tl. Decay can proceed to any of these six eigenstates giving six fine-structure peaks whose relative intensities match the relative probabilities (or decay constants) of the six decay modes illustrated in the *decay scheme* of *Figure 3.8*.

3.7 BETA PARTICLES

There are three different processes which are all described as beta decay. These are β^--decay, β^+-decay and electron capture, EC. They have in common the feature that a neutron, in the nucleus, changes into a proton, or vice versa. The force which produces this effect is not the nuclear, strong interaction, nor is it the electrostatic or the gravitational forces, which must also be present in the nucleus, but a fourth type of force called the *weak interaction*.

In β^--decay, a neutron changes into a proton, which remains in the nucleus, and a β^- *particle* and a *neutrino, ν*. The β^- particle is identical to an electron and often referred to simply as a beta or a β particle. The neutrino, ν, is a particle with zero mass and zero charge. Neither the β nor the ν can be contained within a nucleus and are immediately ejected, sharing the total disintegration energy less that required for nuclear recoil, which is, in this case, very small. The process can be represented as

$$n \rightarrow p + \beta^- + \nu \qquad (3.12)$$

and also applies to a free neutron which decays with a half-life of 12 min.

In nuclear terminology, β^--decay increases the Z number by one and the process can be denoted by

$$^A_Z X \rightarrow ^A_{Z+1} Y + \beta^- + \nu \qquad (3.13)$$

An example is

$$^{234}_{90} Th \rightarrow ^{234}_{91} Pa + \beta^- + \nu$$

Both the β^- and the neutrino are *relativistic particles*. The former travels at two-thirds the velocity of light with only about 170 keV. The latter, having no mass, travels at the velocity of light and is, of course, very hard to detect. It interacts with matter more weakly even than the photon or the neutron.

In β^+-*decay*, a proton changes into a neutron, which remains in the nucleus, and a β^+ *particle* and a *neutrino*. The β^+ particle is known as a *positron* and the process may be described as positron decay. The positron is a positively charged electron, and is identical to the electron in all respects except that its charge is $+e$. It is the *antiparticle* of the electron, or β^-, which is sometimes called a *negatron*, and, when two particles come together, they *annihilate* to form pure electromagnetic energy equal to their total masses.

Positron decay can be represented as

$$p \rightarrow n + \beta^+ + \nu \qquad (3.14)$$

but it must be noted that this is energetically impossible for a free proton, i.e. a hydrogen nucleus, whose mass is smaller than that of a neutron. It takes place in the nucleus by 'borrowing' additional mass—energy from the nuclear binding energy. In nuclear terms, the daughter nuclide has a Z number one less than that of the parent, i.e.

$$^A_Z X \rightarrow ^A_{Z-1} Y + \beta^+ + \nu \qquad (3.15)$$

An example is

$$^{22}_{11} Na \rightarrow ^{22}_{10} Ne + \beta^+ + \nu$$

The positron and the neutrino are relativistic, of course, and the positron is very short-lived (less than 1 μs). It combines with an electron to annihilate, usually within the source material, and form two gamma ray photons. To conserve linear momentum, these photons are emitted in virtually opposite directions and each has an energy of about 0.511 MeV, the rest mass—energy of each of the beta particles.

The β^+ is the antiparticle of the β^- because they annihilate on contact. All other particles have antiparticles. A negative proton, p^-, is the antiparticle of the proton. An antineutron has no charge, of course, but it can be imagined to comprise an antiproton and a positron. The neutrino also has an antineutrino, $\bar{\nu}$, which is extremely difficult to distinguish from the neutrino. Although not specified in the above equations, the neutrino is emitted in β^+-decay and the antineutrino in β^--decay.

In *electron capture*, EC, a nucleus captures an atomic electron. Since only neutrons and protons can exist within the nucleus, the electron is immediately absorbed by a proton, which becomes a neutron. It may be imagined that the nuclear eigenfunction momentarily overlaps that of the electron and, since the K-shell electrons are closest to the nucleus, these are the ones which are mostly absorbed. In this case, the process is known as *K capture*. Generally, it can be represented as

$$p + e^- \rightarrow n + \nu \tag{3.16}$$

and it should be noted that e^- is used rather than β^- because an atomic electron is involved, and once again a neutrino (not an antineutrino) is emitted. This is not the only radiation produced, because the atom is left in an excited state and promptly de-excites, emitting characteristic X-radiation, as described above.

In nuclear terms, the Z number decreases by one, giving the general representation (ignoring X-rays)

$$^A_Z X + e^- \rightarrow ^A_{Z-1} Y + \nu \tag{3.17}$$

An example is

$$^{55}_{26}Fe + e^- \rightarrow ^{55}_{25}Mn + \nu$$

The beta decay processes are also known as *isobaric processes* since they do not alter the A number of the nuclide. In the Chart of the Nuclides, isobars lie in diagonal lines across the line of stable nuclides. As indicated in *Figure 3.9*, β^--decay changes a nuclide downwards, towards the right, while β^+-decay and electron capture change it upwards, towards the left of the curve. This figure includes a graph of mass–energy against Z number drawn along the isobar $A = 60$ to illustrate a general effect. The graph shows a minimum value of mass–energy at the stable nuclide (or nuclides, if there are two). Those nuclides on either side of the minimum have excess energy which is sufficient to release a beta particle from the nucleus. β^--decay proceeds from left to right and β^+-decay from right to left. Hence, beta-unstable nuclides above the line of stability decay by β^--emission and those below it decay by β^+-emission.

Energy spectra

In electron capture, all of the disintegration energy, E_0, is given to a neutrino, which is almost undetectable, so the only useful radiation generated is the atomic de-excitation X-radiation produced by the filling of the K or L shell vacancy. Thus, characteristic X-ray spectra are produced by electron capture. In positron emission the positron is slowed down and annihilated almost immediately and within a few millimetres, so the radiations obtained outside the source are two

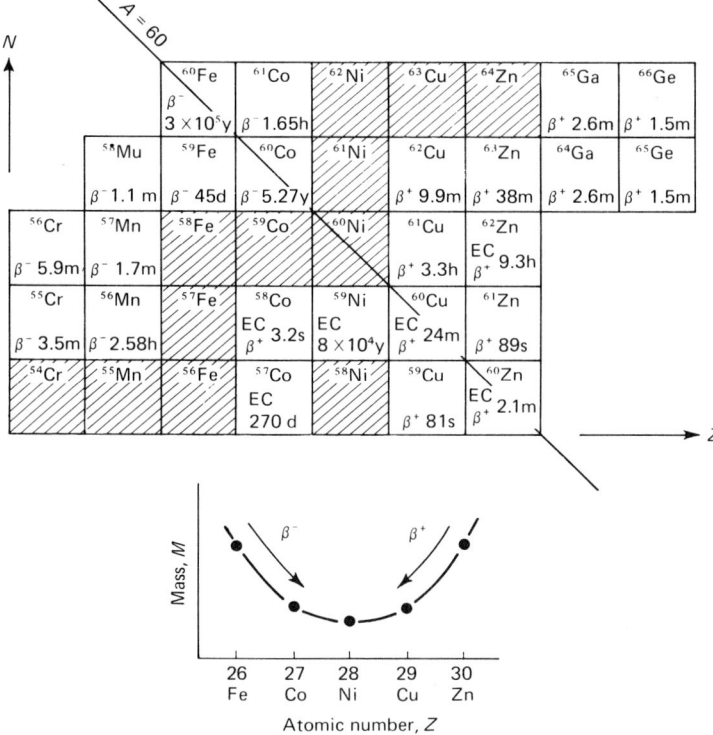

Figure 3.9 Beta decay criteria, illustrated for A = 60 isobars

0.511-MeV gamma ray photons per disintegration. Only in β^--decay is a beta energy spectrum observed.

In this process, the nucleus, in its ground state, may decay to the ground state of its appropriate isobar so that the disintegration energy has a discrete value equal to the mass–energy difference between these two nuclides. As in alpha decay, excited states of the daughter nuclide may be involved to produce a fine structure in this disintegration energy, E_0. There is a major complication,

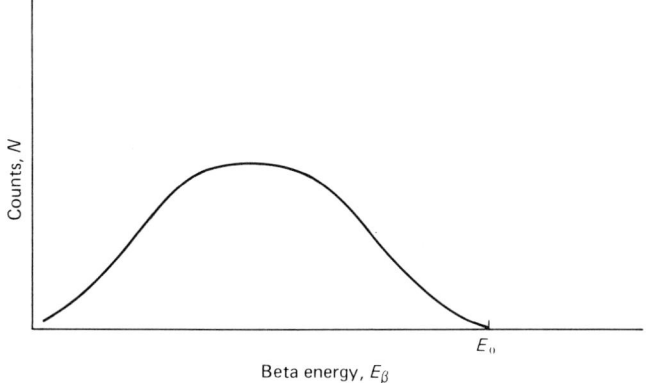

Figure 3.10 Beta particle energy spectrum

however, since this energy must be shared, quite randomly, between the beta particle and the neutrino. As a result, the beta energy, E_β, varies continuously from zero, when all of the disintegration energy is given to the neutrino, to E_0, when none of the disintegration energy is given to the neutrino. The beta energy spectrum is a *continuum* extending from zero to E_0, which is sometimes referred to as E_{max}, with a mean value at about half of E_0, as illustrated in *Figure 3.10*.

This spectrum was observed long before a neutrino was detected (in 1956) and was a source of considerable debate. It appeared that some energy was lost, since the beta seldom had as much as it ought to. The *Pauli neutrino hypothesis* first suggested the presence of another, undetected, particle which shared the available energy.

β^--decay schemes for ^{131}I and ^{64}Cu are shown in *Figure 3.11* and the latter includes an electron capture process going to an excited state of ^{64}Ni and a

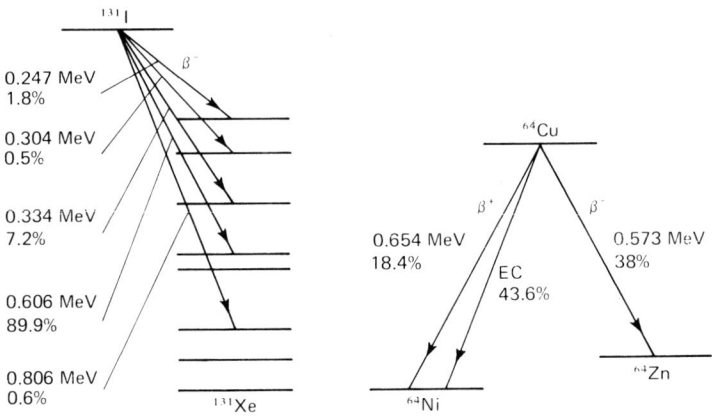

Figure 3.11 Decay schemes of ^{131}I and ^{64}Cu

β^+-decay process to the ground state of the same nuclide. In general, not only in beta decay but in other processes such as alpha and gamma decay, some transitions are favoured, or *allowed*, compared to others. For ^{131}I, for example, 90% of the disintegrations produce a 0.6-MeV beta particle but only 7.2% proceed via the 0.334 MeV route and none makes it to the ground state of ^{131}Xe. This last mode is known as a *forbidden* transition. It is forbidden, as the alternatives are relatively allowed, by *selection rules* which are, simply, the laws governing conservation of energy, momentum, angular momentum, and so on. These laws are more easily satisfied by some transitions than by others. Hence the relative *abundances* of the different radiations.

Because of the continuous nature of the beta energy spectrum, it is extremely difficult to recognise the presence of more than one disintegration energy. If, for example, two energies are present, the lower one is the end point of a continuum which is buried in that of the larger energy. The standard solution to this type of problem (as for the half-life graphs) is to plot variables which result in straight-line graphs. In this case a complicated function, $(P(p)/p^2)^{1/2}$, is plotted against beta energy, E_β. $P(p)$ is the probability that the beta will have

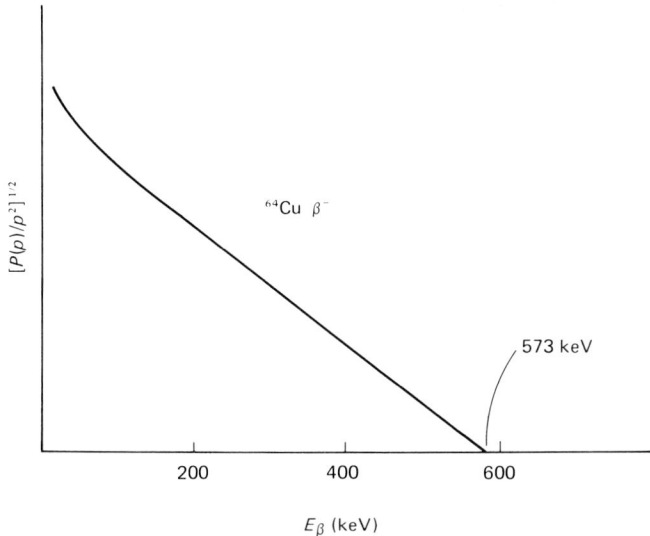

Figure 3.12 Kurie plot for ^{64}Cu

a momentum p. Graphs of this type are called *Kurie plots* and are illustrated in *Figure 3.12*. If two or more disintegration energies are present, this fact shows up clearly as different gradients.

3.8 GAMMA RADIATION

Alpha decay, beta decay and all other nuclear processes can leave a nucleus in an excited energy state. The nucleus then de-excites, in the same way as does an excited atom, by an *isomeric* transition (or a cascade of several transitions) to its ground state, emitting one (or more) gamma ray photons whose energy, $h\nu$, is given by a Bohr-type formula

$$E_\gamma = h\nu = E_2 - E_1 \tag{3.18}$$

where E_2 and E_1 are the nucleon energy eigenvalues either singly, if a Shell Model is appropriate, or collectively, if a Collective Model applies to the nucleus involved.

A nucleus in an excited state is called a *nuclear isomer* of the nuclide and can be represented by a term of the form $^A_Z X^*$. If the excited state is long-lived, then it may be referred to as a *metastable state*. Such nuclides are denoted by the superscript 'm'. For example, technetium-99m (or ^{99m}Tc) is a metastable state of ^{99}Tc, with a half-life of 6.02 h. In general, the gamma decay process can be represented as

$$^A_Z X^* \to {^A_Z X} + \gamma \tag{3.19}$$

Apart from the decay of metastable states, gamma decay is virtually immediate following some other process and its half-life is that of the other process which produces the isomeric condition. Two of the most widely used gamma ray sources are ^{137}Cs (E_γ = 0.662 MeV) and ^{60}Co (E_γ = 1.17, 1.33 MeV). Their decay schemes are illustrated in *Figure 3.13*.

Caesium-137 decays, by β^--emission, in 95% of its disintegrations, to an excited state of ^{137}Ba which then γ-decays to the ground state. The other 5% of all disintegrations go direct to the ground state. Hence, on average, there is 1 beta particle and 0.95 of a photon per disintegration. For example, 1 μCi of ^{137}Cs generates 3.7×10^4 β^- particles and about 3.5×10^4 photons per second.

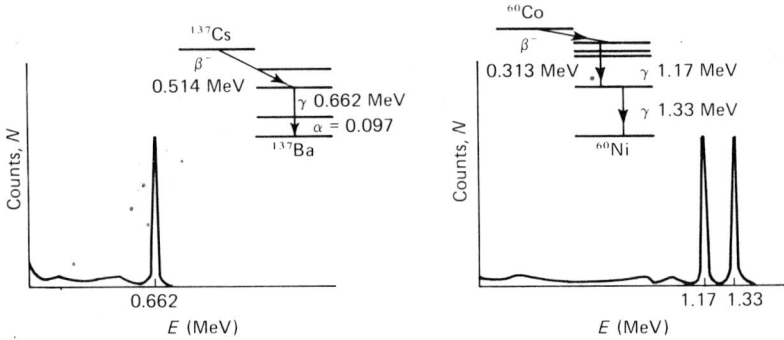

Figure 3.13 Gamma ray spectra for ^{137}Cs and ^{60}Co. α is the internal-conversion coefficient

Cobalt-60 β^--decays in virtually 100% of its disintegrations to an excited state of ^{60}Ni which then decays, via two successive isomeric transitions, to its ground state. Hence, one disintegration of ^{60}Co gives one beta particle and two photons.

Selection rules apply to gamma decay as to other nuclear processes. The decay of ^{60}Co to the ground state of ^{60}Ni is forbidden and decay to the first excited state is almost completely forbidden, while γ-decay from the second excited state to the ground state of ^{60}Ni is also forbidden. Metastable states are those with no easily allowed de-excitation transitions.

Energy spectra

Gamma ray spectrometry is a much more widely used technique than alpha or beta spectrometry. The main reason for this is the fact that a much wider range of nuclides can be made to emit gamma radiation but it is also true that photons are more easily extracted from the source material. The gamma energy spectrum is as characteristic of the nucleus as the X-ray spectrum is of the atom, and is used to identify or trace the location of radionuclides. Unlike the X-ray equivalent, gamma ray spectrometry easily distinguishes between different isotopes of the same element. Gamma ray spectra should comprise discrete lines (or peaks), as in X-ray and visible-light spectra, but the situation is greatly complicated by the detection process. This will be discussed in a later chapter, but it is possible to obtain clear line spectra with at least one type of detector (the semiconductor) and these are shown, for ^{137}Cs and ^{60}Co, in *Figure 3.13*.

Internal conversion

This process is an alternative to gamma decay and is the nuclear analogue of the atomic Auger process. A nucleus about to emit a gamma photon interacts instead with an atomic electron, usually in the nearby K shell, and ejects this from the atom. Thus, the nucleus de-excites isomerically and an atomic electron is emitted with an energy equal to E_γ, the energy which would have been removed with the photon. In addition, the atom is left in an excited state, so the electron is accompanied by characteristic X-radiation.

The probability that this alternative process will take place is denoted by the *internal-conversion coefficient*, α, and may be very large so that the majority of decays proceed by electron instead of photon emission. α is defined as the ratio λ_c/λ_γ, where λ_c is the decay constant for internal conversion and λ_γ that for γ-decay. These are partial decay constants and their sum determines the half-life of the decay process. Alternatively, as in Appendix 3, λ_c and λ_γ are quoted separately as percentages of all decays, so it is easy to determine how many photons and electrons are generated, but some care must be taken with energy level diagrams. In the diagram for ^{137}Cs (*Figure 3.13*), for example, the internal-conversion coefficient is indicated as e^-; $\alpha = 0.097$, so almost 10% of the photons are replaced by electrons and, correcting the above calculation, only about 85% of disintegrations produce gammas.

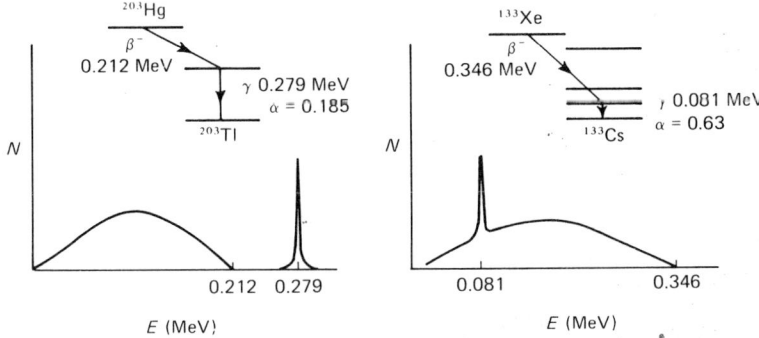

Figure 3.14 Beta spectra for ^{203}Hg and ^{133}Xe with internal-conversion electron peaks. α is the internal-conversion coefficient

Because different detection systems are involved, internal-conversion electrons are seldom detected along with gamma radiation but they are observed along with beta particles, provided, of course, that the nuclide under examination decays both by gamma and by beta emission. They are easily identified because, unlike betas, they do not share their energy with neutrinos so they form discrete spectral lines superimposed on the continuous beta spectrum. Two examples of internal-conversion lines are shown in *Figure 3.14*, for ^{203}Hg and ^{133}Xe, along with the decay schemes for these nuclides.

3.9 NEUTRONS

The neutron, n or $^1_0$n, is a fundamental particle with a mass number, A, of 1, and no charge. Nevertheless, it can be regarded as a composite particle, namely, as a combination of a proton and an electron. Its mass, equivalent to 939.55 MeV, is large enough to support this hypothesis and in the free state it does decay, with a half-life of 12 min, to give a proton, beta particle and neutrino as described by equation 3.12 above.

Because they are uncharged, neutrons are not easily distinguished from photons and, for this reason, they were not identified until 1932, by Chadwick. They share another property with photons in that it is necessary to distinguish between low-energy, or *thermal* neutrons, and high-energy, or *fast* neutrons, corresponding to X-rays and γ-rays. A thermal neutron, n_{th}, is in thermal equilibrium with its environment (like a gas molecule). At 'room temperature' its kinetic energy is about one-fortieth of an electronvolt, i.e. 0.025 eV. Fast neutrons, n_f, are more energetic and may have energies of tens of mega-electronvolts. For accurate calculations, it is usual to define several energy groups, often seven, ranging from thermal upwards. Epithermal neutrons have slightly higher energies than thermal ones.

Neutrons, especially thermal neutrons, have an important property in that they *activate* materials through which they pass. In other words, they make these materials radioactive. This is true of other radiations, such as protons and gamma rays, but the effect is much stronger for neutrons. (For this reason, they represent a unique health hazard, in that they activate human tissue.) Neutrons provide the basis for manufacturing radionuclides and for techniques such as activation analysis.

Neutron sources

Basically, there are three types of neutron source: chemical sources, accelerator sources and reactor sources.

Chemical sources

Chemical, or 'radioactive', sources are mixtures of a radionuclide and a target material, apart from the important exception of ^{252}Cf (*see below*). Radiations emitted by the radionuclide interact with the nuclei of the target, in *nuclear reactions*, to produce neutrons. Like the other neutron sources they produce gamma rays as well as neutrons but they differ from these alternatives in a number of respects. They are relatively inexpensive, physically small and require almost no maintenance, apart from shielding and handling aspects. Their output, however, is smaller than that of machines or reactors.

A widely used mixture is that of ^{241}Am and beryllium. The americium generates alpha particles which react with the beryllium targets to produce neutrons according to the equation

$$^{9}_{4}Be + ^{4}_{2}\alpha \rightarrow ^{12}_{6}C + ^{1}_{0}n \qquad (3.20)$$

The half-life of the americium and, therefore, of the neutron source, is 433 years and the neutron yield per curie of ^{241}Am is about 2.2×10^6 n s^{-1}. The emission spectrum for a ^{241}Am–Be source is shown in *Figure 3.15*.

There are a number of other mixtures which are commercially available, with various yields and half-lives, but the highest yield is obtained from *californium-252*. This decays by *spontaneous fission* in which the nucleus breaks up into two

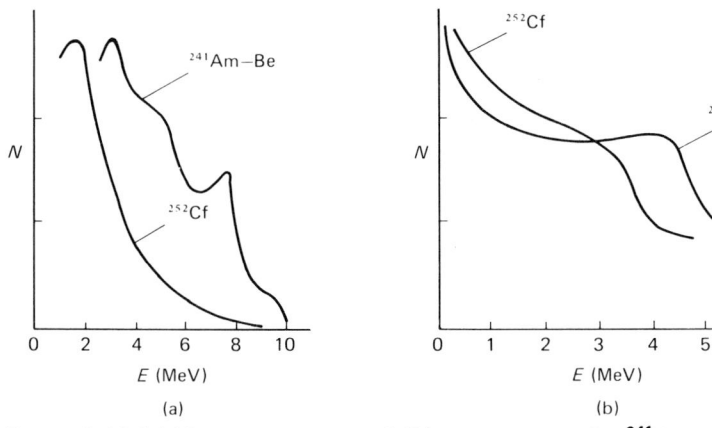

Figure 3.15 (a) Neutron spectra and (b) γ-ray spectra for ^{241}Am–Be and ^{252}Cf sources

roughly equal halves and a few surplus neutrons — since the daughter nuclides cannot contain the neutron excess of the parent. The half-life for fission is about 85.5 years but ^{252}Cf also decays by alpha emission, with a half-life of 2.73 years, and this shorter time determines the useful life of the source. The neutron yield is about 2.3×10^6 n s^{-1} μg^{-1}.

Accelerator sources

Accelerators are machines which use electric and/or magnetic fields to accelerate charged particles up to high energies. They include linear accelerators, Van de Graaff machines, cyclotrons and synchrotrons. The charged particles are directed onto some target material in which neutron-producing reactions take place. The machines are, of course, used for other purposes. Mostly, the charged particles themselves are the desired end product but high-energy gamma rays, or bremsstrahlung, can also be generated.

In high-energy and nuclear physics, large accelerators are used to produce high-energy and, usually, monoenergetic neutrons, with, perhaps, energies of tens or hundreds of mega-electronvolts. This is the only way to obtain such high-energy neutrons. In a typical process, deuterons are accelerated into a target, where a stripping reaction takes place in which the proton is removed, leaving the neutron.

Smaller machines are used to produce a somewhat lower and non-adjustable neutron energy. These provide charged particles with just enough energy to penetrate a target nucleus where a nuclear reaction takes place and generates neutrons. These machines are usually purpose-designed *neutron generators* and a typical process provides deuterons with energy of about 50 keV, sufficient to penetrate tritium target nuclei where the following reaction takes place:

$$^2_1H + ^3_1H \rightarrow ^4_2He + ^1_0n \tag{3.21}$$

or

$$^2_1D + ^3_1T \rightarrow ^4_2He + ^1_0n$$

This is a *fusion reaction*, known as the D–T reaction. The neutrons are emitted with energies of about 17.5 MeV.

Reactor sources

Nuclear reactors are, at present, based on the process of *nuclear fission*; in this case, neutron-induced fission. A neutron is captured by a fissile material, such as $^{235}_{92}$U, which then undergoes fission, breaking up into two roughly equal nuclides, such as $^{94}_{38}$Sr and $^{138}_{54}$Xe, plus a few neutrons and a large amount (about 200 MeV) of energy. The neutrons induce further fissions and a self-sustaining *chain reaction* can be established. The fission process results in the emission of fast neutrons but these are easily slowed down, or *moderated*, to thermal energies.

In general, reactors can be used as sources of fast neutrons, with energies of about 18 MeV, but these are not so monoenergetic as accelerator-produced neutrons. Consequently, reactors are more often used as sources of *thermal neutrons*. They produce these in very large numbers compared with other neutron sources and fluxes of the order of 10^{14} n cm^{-2} s^{-1} are common. Compared with chemical and accelerator sources, reactors are *high-intensity* sources of thermal neutrons.

3.10 FUNDAMENTAL PARTICLES

Fundamental, or *elementary*, particles are essentially single particles rather than composites of two or more particles. It may be that they are not fundamental, in the sense of being unable to be further subdivided, and there is a hypothesis that at least some fundamental particles can be represented in terms of even more 'fundamental' particles called *quarks*.

Despite this uncertainty, there is a clear distinction between fundamental and composite particles. The latter include alphas, deuterons and tritons which are made up of well-defined and easily observed particles called nucleons. The nucleons themselves are fundamental as are electrons, neutrinos and photons, discussed in earlier sections of this chapter. The present intention is to introduce other types of fundamental particle.

It is interesting to note that stability is neither a prerequisite nor a definition of a fundamental particle. Neutrons, for example, decay to a proton, a beta particle and a neutrino but should not be regarded as a composite of these three particles — they are formed in the decay process. Alpha particles, on the other hand, are quite stable but definitely composite.

There are three types of fundamental, or elementary, particle: *photons, leptons* and *hadrons*. *Photons* have been discussed above. They are the lightest fundamental particles, with zero rest mass, and travel at velocity c in a vacuum. They are not charged, but transfer electric and magnetic energy from one charge to another. Although massless, they have an intrinsic angular momentum, or spin, $\hbar$, and are known as *bosons*, these being particles with zero or integral spin quantum numbers. They have no antiparticles.

Particles with half-integral spin quantum numbers are called *fermions*, and *leptons* are the lightest of these. They include the *electron/beta*, e/β, the *muon*, μ, the *electron-neutrino*, ν_e, and the *muon-neutrino*, ν_μ. The electron/beta and the electron-neutrino have been discussed above. The muon-neutrino is similar to the electron-neutrino in all respects except one — it is produced in muon decay instead of beta decay. The muon can have a charge of $+e$ or $-e$ and is represented as μ^+ or μ^-, respectively. It has a mass equivalent of 105.66 MeV and is very unstable, decaying with a half-life of 1.6×10^{-6} s, as follows:

$$\mu^\pm \to e^\pm + \nu_e + \nu_\mu \tag{3.22}$$

The *hadrons* include a large number of particles, classed as *mesons* and *baryons*. The mesons are lighter and they are bosons with zero spin. The baryons are heavier and are fermions with spins of $\frac{1}{2}\hbar$ in all cases except that of the omega particle, Ω, with a spin of $\frac{3}{2}\hbar$. From a general point of view, the most important baryons are the neutrons and protons, while the most interesting mesons are the *pions*, π. The pion can have a charge of $+e$, $-e$, or zero, represented as π^+, π^- or π^0 respectively. The mass equivalent of the charged pion is 139.58 MeV and that of the uncharged pion is 134.99 MeV. They are unstable, as are all hadrons except the proton, decaying as follows:

$$\begin{aligned} \pi^\pm &\to \mu^\pm + \nu_\mu & T_{1/2} &= 1.78 \times 10^{-2} \text{ s} \\ \pi^0 &\to \gamma + \gamma & T_{1/2} &= 0.7 \times 10^{-16} \text{ s} \end{aligned} \tag{3.23}$$

3.11 RADIATION SOURCES

The major sources of ionising radiations are machines (including nuclear reactors), radioactive sources and cosmic radiation. In general, machines are large and relatively expensive but the energy, nature and intensity of their radiations can be controlled and they can, of course, be switched off. Their characteristic advantage is the very high intensity of radiations which they produce and they are suitable for techniques such as activation, therapy and radiography. Radioactive sources are small and inexpensive by comparison and they require very little maintenance. They cannot be switched off but they do not break down. They produce relatively low-intensity radiations that are characteristic of the material used and are suitable for techniques such as isotope imaging and radiotracer work. Cosmic radiations are, in general, regarded as background rather than useful radiations.

The 'active ingredient' of a radioactive source is a radionuclide, which may be natural (found in nature) or artificial (manufactured).

Natural radionuclides

The age of the earth is about 10^{10} years, so any radioactive nuclides created during the formation of this planet will still exist only if their half-lives are within a few orders of magnitude of this period of time. Apart from the parents of radioactive chains, discussed below, there are about 20 such nuclides but those with very long half-lives have such low activities that they are virtually stable, while those with shorter half-lives are present only in minute quantities. None of them is particularly useful as a radioactive source, but some, including ^{40}K, ^{87}Rb, ^{138}La, ^{150}Nd, ^{176}Lu and ^{187}Re, do provide measurable activities, usually as background radiation.

Other radionuclides present in the environment have been formed more recently than the earth and have quite high specific activities. Some of these, such as ^{14}C used in dating techniques, have been formed by cosmic ray interactions with stable nuclides and others include persistent man-made radionuclides, such as ^{90}Sr, ^{131}I and ^{40}K, produced as fall-out in nuclear explosions. In general, however, the most useful natural radionuclides are those produced by the decay of long-lived parent nuclides.

In such a process, a long-lived parent nuclide A decays to B which decays to C, and so on to generate a *radioactive chain* or *series* according to the formula

$$A \xrightarrow{\lambda_A} B \xrightarrow{\lambda_B} C \xrightarrow{\lambda_C} D \text{ etc.} \tag{3.24}$$

It does not matter how large are the values of λ_A, λ_B, λ_C, etc. provided that the initial decay constant, λ_A, is small and, of course, the half-life for this stage is long. Nuclides B, C, D, etc. will be produced continually. The condition is known as (secular) *equilibrium* in which the activities $N_A\lambda_A$, $N_B\lambda_B$, $N_C\lambda_C$, etc. are all equal. In other words, the activity of any daughter nuclide is determined by the rate at which it is produced by the parent of the chain.

In nature, there can be only four separate chains, with no transitions between any two. The reason for this is the fact that alpha decay changes the A number by four while beta and gamma decay leave it unaltered. Thus all the A numbers of one chain must satisfy one and only one of the formulae

$$\begin{aligned} A &= 4n \\ A &= 4n + 1 \\ A &= 4n + 2 \\ A &= 4n + 3 \end{aligned} \tag{3.25}$$

where n is an integer.

Details of these radioactive chains are included in Appendix 3. Their major features are as follows:

Name of series	Type	Longest-lived nuclide	Final stable nuclide
Thorium	$4n$	^{232}Th: $T_{1/2} = 1.4 \times 10^{10}$ years	^{208}Pb
Neptunium	$4n + 1$	^{237}Np: $T_{1/2} = 2.1 \times 10^{6}$ years	^{209}Bi
Uranium	$4n + 2$	^{238}U: $T_{1/2} = 4.5 \times 10^{9}$ years	^{206}Pb
Actinium	$4n + 3$	^{235}U: $T_{1/2} = 7.1 \times 10^{8}$ years	^{207}Pb

In fact, only three of these chains are still in existence since the neptunium series does not have a sufficiently long parent half-life. Any of these three can be used as radioactive sources but they suffer from three disadvantages. First, they emit a wide range of types and energies of radiation because of the large number of radionuclides present. Secondly, their specific activity is limited by that of the parent nuclide. Thirdly, they have long half-lives and this excludes them from some applications, especially in nuclear medicine. These can be eliminated, to a large extent, by chemically separating out shorter-half-lived nuclides from the chain. The process is called *milking* and the parent nuclide is therefore referred to as a 'cow'. Thus, for example, ^{234}Th, which has a half-life of 24 days and emits only beta and low-energy gamma radiation, can be obtained from the uranium series.

Artificial radionuclides

Artificial radionuclides are more widely used than natural radionuclides, mainly because they can be tailor-made to meet the specific requirements of the user. They can be prepared from virtually any element and in an isotopically pure form, that is, with only one radionuclide present, hence they offer a wide choice of half-lives, specific activities and specific emissions as well as chemical properties. They are easily incorporated in a range of materials, including those used as chemical and biological tracers. If necessary, they can be milked *in situ* to provide extremely short half-lives of the order of a few hours.

Artificial radionuclides are produced in *nuclear reactions* in which stable nuclides are bombarded by charged particles, from an accelerator, or by neutrons. Most of them are obtained by neutron irradiation, usually in a nuclear reactor. For example, cobalt-60 is produced by neutron irradiation of the stable cobalt-59 in the reaction

$$^{59}_{27}\text{Co} + ^1_0\text{n} \rightarrow ^{60}_{27}\text{Co} + \gamma$$

This is an *activation process*. If N stable nuclei of a particular type are placed in a flux of Φ particles cm^{-2} s^{-1} and each has a fixed probability σ of undergoing a nuclear reaction in unit time, then the number of reactions induced in unit time and the number of radioactive nuclei produced per second is $N\Phi\sigma$. The parameter σ is called the reaction *cross-section* and will be defined in a later section. Radionuclides are produced at a constant rate but they also decay radioactively during the process so the net activity A, as a function of time, is given by

$$A = N\Phi\sigma(1 - e^{-\lambda t}) \tag{3.26}$$

The number of radionuclides present increases exponentially, as illustrated in *Figure 3.16*. After a few half-lives, the activity reaches a saturation value, $N\Phi\sigma\lambda$, which cannot be exceeded. There is no point, therefore, in activating a specimen for longer than two or three half-lives.

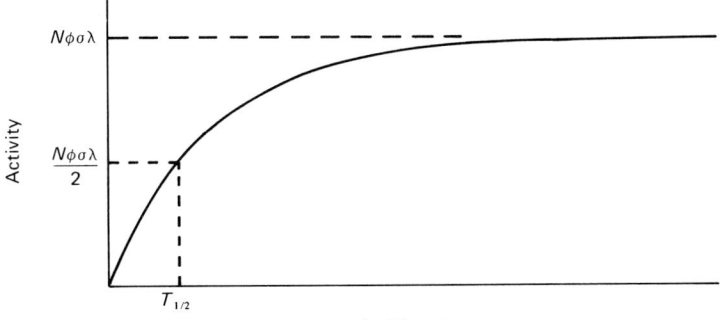

Figure 3.16 Induced radioactivity as a function of irradiation time

Radioactive sources

There are three types of radioactive source, namely, sealed, abrasion-resistant and open sources. In *sealed sources*, the radioactive material is completely contained in a metal or plastic *capsule* so that the radiations, but not the source material itself, can get out. They are regarded as 'clean' sources which do not contaminate their environment with radioactive materials. They can be handled without protective gloves but forceps may be used to maintain some distance between the source and the user's hand and minimise the radiation dose. They should be tested regularly to ensure that no leakage takes place.

It is important to note that sealed sources may not emit all the characteristic radiations of the radionuclide contained since the capsule, or planchet, may be thick enough to absorb some of these, especially alpha particles and low-energy gamma and X-radiation. Beta emitters are usually provided with thin exit windows of gold foil.

Abrasion-resistant sources take the form of bonded deposits on metal bases or gold-covered deposits on foils. They stand up to wipe tests and are virtually equivalent to sealed sources, apart from small edge effects, if they are cut. They are, typically, alpha and X-ray sources which need very thin exit windows.

Open sources come in a variety of forms, but usually as a solution. They must be handled carefully, since they can contaminate environment and personnel, especially if they are volatile. They are used mainly for chemical and biological applications, radiotracer experiments, autoradiography, isotope scanning and other techniques where the source has to be incorporated in something else.

In contrast to sealed sources, abrasion-resistant and open sources emit all the characteristic radiations of the radionuclide. Thus, for example, a source presented as an alpha emitter, such as ^{210}Pb, also emits beta, gamma and X-radiation. As further examples, ^{137}Cs and ^{60}Co are sold as standard sources of gamma radiation but both also emit beta particles. One of the most important principles of radiation detection is that sources tend to emit more than one type of radiation and the detector may not distinguish between these. In order to be sure that the detector is responding to a particular type of radiation it is necessary to consult data such as those presented in Appendix 3, to see which radiations are emitted by the source, and to appreciate the absorptive effects in the source, the detector and any intervening matter. These latter aspects will be the subject of the following chapter.

Background radiations

Background radiations may be defined as any radiations which are present and, usually, detected, in a particular experimental situation but are not produced by the source under investigation. They are undesirable because they are counted along with the radiations being studied and tend to obscure experimental results. Normally, the count rate due to background radiations is relatively small and represents a minor correction factor but there are two situations in which the background becomes significant, namely, in areas where background is exceptionally high or in low-activity experiments where source strength is small.

The first of these occurs in high-radiation environments near accelerators, reactors and in laboratories where other sources are present. In this case, an improved signal-to-background ratio can be achieved by *shielding* the work space with appropriate materials. In a teaching laboratory, for example, detector and sources are usually located in lead castles to eliminate radiations from extraneous sources.

In the study of low-activity sources the main problems are from cosmic radiation, natural long-lived radionuclides in structural materials, especially metals, and ^{40}K in the human body. These may produce between 30 and 60 counts per minute in, for example, a Geiger–Müller counter.

BIBLIOGRAPHY

LEDERER, C. M., HOLLANDER, J. M. and PERLMAN, I., *Table of Isotopes,* John Wiley, New York (1967)
SEGRÉ, E., *Nuclei and Particles,* W. A. Benjamin, Menlo Park, California (1965, 1977)
STERN, B. E. and LEWIS, D., *X-Rays*, Pitman, London (1970)

Chapter 4
Interactions with matter

4.1 GENERAL THEORY

When ionising radiations pass through matter they may interact with whole atoms, electrons, nuclei or nucleons. In this context, the radiations can be referred to as *incident* particles or photons, and the particles with which they interact are usually described as *target* particles. The interaction process is often described as a *collision*, but this term has a well-defined classical meaning which does not transfer completely to quantum systems. In particular, it does not easily apply to waves, or to particles with indeterminate positions, and it implies physical contact which need not take place. Many interactions involve long-range forces, such as the electrostatic force, that do not require close contact. With these reservations, however, the interactions may be regarded as collisions between incident and target particles.

Interactions with individual nucleons are observed only for high-energy incident radiations whose associated wavelengths are of the same order of magnitude as the diameter of a nucleon. In general, they are not used in radiation detection except in special techniques of high-energy and nuclear physics. Interactions with whole atoms are confined to very weak interactions, such as in the Čerenkov effect or to low-energy radiations whose associated wavelengths compare in magnitude with that of the atom. In this case, the interaction may be *coherent* or *incoherent*. The first of these involves simultaneous interaction with several atoms in such a way as to produce wave effects, namely, superposition and diffraction patterns. The second is purely a particle effect.

At the present state of the art, radiation detection employs, almost exclusively, interactions with electrons and with nuclei. Interactions which involve nuclei as target particles are collectively known as *nuclear reactions*. There is no similar term for electron interactions although basically they comprise the same types of reaction. In radiation detection the most important interactions are absorption processes and scattering processes, as illustrated in *Figure 4.1*.

In *absorption*, or *capture* interactions, the incident particle or photon is completely absorbed by the target particle and effectively ceases to exist as an independent object. If it is a photon that is absorbed, the target receives pure energy. An atomic electron is raised to a higher bound state, forming an excited atom, or to a free state, forming an ion pair. A nucleus is raised to an internally excited state. Both systems may subsequently decay by isomeric transitions, emitting new photons, and the nucleus may disintegrate into various reaction products. Particle absorption is mainly confined to nuclei (an exception being positron annihilation with an electron) in which case a new nuclide is formed since mass, as well as energy, is absorbed.

(a)
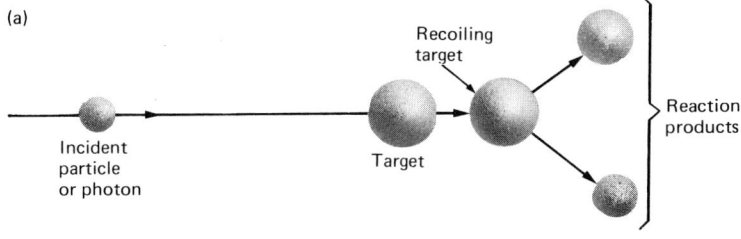

(b)
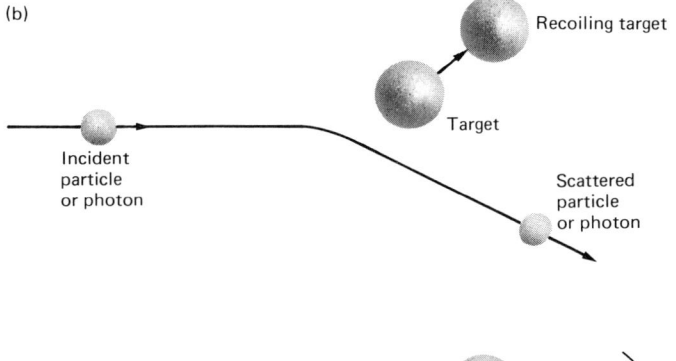

(c)
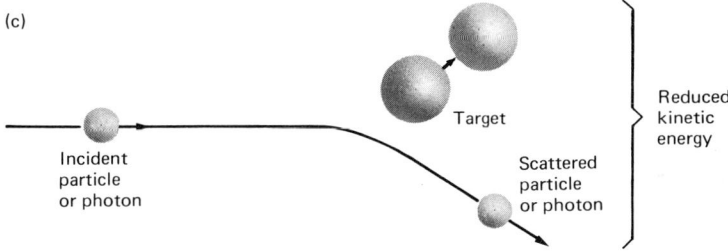

*Figure 4.1 Interactions of radiations with target particles.
(a) Absorption; (b) elastic scattering; (c) inelastic scattering*

In *scattering* interactions, the incident particle or photon collides with a target particle; the latter recoils, taking some energy, and the radiation is deflected in a new direction with appropriately reduced energy. In *elastic scattering*, kinetic energy is conserved and the particles involved behave like classical colliding particles. In *inelastic scattering*, this is not so, and some kinetic energy is converted to excitation energy, either in a nucleus or an atom. As far as the target particle is concerned, the results of inelastic scattering are similar to those of photon absorption.

These interactions are not equally important from the point of view of radiation detection. Some of them are more useful for studying the structure of materials, target particles and incident particles as well as the forces which exist between these components. They include a variety of processes, such as Rutherford scattering of alpha particles to investigate atomic structure, the scattering of light photons from a surface to study the surface, electron microscopy using electrons instead of light photons, the coherent scattering of photons to form

68 Interactions with matter

diffraction patterns in X-ray crystallography, differential photon absorption in X-radiography and many nuclear and nucleon reactions used in high-energy physics. Radiation detection requires interactions which have a relatively high probability of taking place and, usually, stop the radiations or absorb much of their energy.

Thus, there are two major reasons for studying the interactions of radiations with matter. One is to study the matter and the other is to study the radiations — often, it is true, with a view to using them to study matter. The present text is concerned with the latter objective and will concentrate on the most relevant interactions. Briefly, these are electron absorption and scattering processes, for the detection of charged particles and photons, nuclear absorption reactions for thermal neutron detection and nuclear scattering reactions for fast neutrons.

Forces

All interactions involve forces, either attractive or repulsive. Absorption processes are strongly affected by this property. Positively charged particles, for example, are electrostatically repelled by nuclei and require a great deal of energy to initiate a nuclear reaction. Thus, nuclear reactions are less important for the detection of charged particles than for neutrons. For the same reason, particle absorption by an electron is, at present, restricted to the absorption of its antiparticle in positron annihilation. Scattering interactions are less affected by repulsive or attractive characteristics, which merely determine the direction of scatter, as shown in *Figure 4.2*.

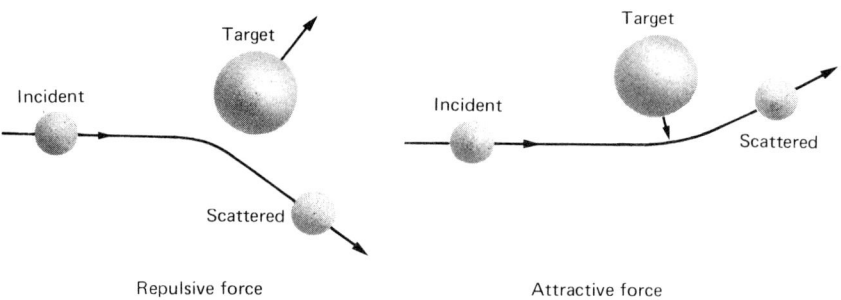

Figure 4.2 Scattering interactions

In general, there are four types of force and two or more may be experienced simultaneously. These are:

(1) *Gravitational force*. This exists between any two masses. It is the oldest-known force but the least understood theoretically. It obeys the inverse-square law, that is, it is proportional to the reciprocal of the square of the distance between the two masses, so it is a long-range force. It is attractive but it is by far the weakest force and, although present in all particle–particle interactions (excluding neutrinos), it has a negligible effect.

(2) *Weak interaction*. This is about 10^{12} times as strong as the gravitational force but is still a weak effect. It is the force which produces beta decay and affects some pion and muon interactions. It has virtually zero range so the particles can be imagined to be in physical contact. It is not generally useful in radiation detection.

(3) *Electromagnetic interaction*. This is, very approximately, ten orders of magnitude stronger than the weak interaction. It includes electrostatic, or Coulomb, forces, magnetic forces and all interactions involving electromagnetic radiations. It is a long-range, inverse-square-law force which can be attractive or repulsive and is the main form of interaction of photons and charged particles with matter.

(4) *Strong interaction*. This is the force between nucleons which binds the nucleus together. It is, therefore, attractive and is about ten times stronger than the electromagnetic force. It has a very short range, of the order of 1 F, and is the major contribution to nuclear reactions. It is responsible for neutron absorption and scattering processes.

Nuclear reactions

Any interaction that involves a nucleus as a target particle is a nuclear reaction and can be represented by an equation of the general form

$$x + X \rightarrow y + Y \tag{4.1}$$

where x denotes the incident particle or photon, X the target particle, y the product or scattered particle and Y the recoiling or product nucleus.

In *absorption reactions*, an intermediate system, called a *compound nucleus*, is formed. If this is represented by X*, then the reaction actually proceeds as follows:

$$x + X \rightarrow X^* \rightarrow y + Y \tag{4.2}$$

There are three possibilities; X* may be stable, in which case the reaction stops there, X* may be a radioactive nuclide, in which case the decay to y and Y takes place in accordance with the laws of radioactive decay, or X* may be completely unstable and decay to y and Y virtually immediately. The first of these processes may be described as *transmutation*, the second as *activation* and the third is often referred to, simply, as a *nuclear reaction*. The compound nucleus is formed, usually, in an excited energy state and, even if it is a stable nuclide, it will tend to shed excess energy by isomeric de-excitation. Where it decays by particle emission, it does so with total independence from its mode of formation and, generally, decay products are different from the incident and target particles.

Examples of these reactions include the following:

(a) Transmutation: $^{1}_{1}p + ^{55}_{25}Mn \rightarrow ^{56}_{26}Fe$ (stable)

(b) Activation: $^{1}_{0}n + ^{59}_{27}Co \rightarrow ^{60}_{27}Co$ (radioactive)

(c) Absorption–emission: $^{4}_{2}\alpha + ^{14}_{7}N \rightarrow ^{18}_{9}F \rightarrow ^{1}_{1}p + ^{17}_{8}O$
(nuclear reaction)

In scattering reactions, x and y are identical, as are X and Y, and there is no compound nucleus formed. If the interaction is an inelastic one, the scattered particle is represented by x'.

For all nuclear reactions there is a convenient shorthand notation, X(x,y)Y. Such a reaction may be referred to as an '(x,y) reaction'. For example, (c) above is an (α,p) reaction which can be denoted by the full expression $^{14}_{7}N(\alpha,p)^{17}_{8}O$.

In absorption reactions, the sum of the masses of the initial particles is not equal to that of the final particles. Some mass is converted to energy, or energy to mass, according to the Einstein formula, $E = mc^2$. This energy difference is called the Q *value* of the reaction and, ignoring excitation energies, if present, is given by

$$Q = (m_x + m_X) - (m_y + m_Y) \qquad (4.3)$$
$$= T_y + T_Y - T_x$$

where T represents kinetic energy. Although the target, X, is almost certainly moving when it is struck by the incident radiation, x, the *average* collision involves a stationary target, so its kinetic energy is ignored in the above equation.

Q is therefore the net gain in useful, kinetic energy. If it is positive, as in the fission and fusion processes, then energy is created and the reaction is said to be *exoergic*. If it is negative, energy is converted to mass and the reaction is *endoergic*. In the latter case, the energy can only be supplied by the incident particle which must, therefore, have a minimum *threshold energy* to generate the required mass and enable the reaction to take place. If the incident particle is a positively charged one, it also needs some energy to overcome the repulsive Coulomb force between it and the nucleus so that, even for exoergic reactions, some initial kinetic energy, usually of the order of tens of kilo-electronvolts, is required. Neutrons, of course, do not experience this potential barrier and easily initiate exoergic reactions.

Cross-sections

When incident radiation enters a piece of matter it may or may not interact with one or more target particles. Whether it does so or not depends upon the energy of the radiation. It also depends on a number of other factors, including the strength and range of the interaction force, the sizes and structures of incident and target particles and the types of interaction which may take place. There is a well-defined probability that, within a given thickness of material, some interaction will take place. It is extremely difficult to calculate this interaction probability from the quantum theory of interacting systems but it is relatively easy to measure it and measured values are, in fact, widely used to test theoretical methods and hypotheses. The probability is measured by, and proportional to, the *interaction cross-section*, σ.

Cross-section is defined as an *area* whose magnitude is proportional to interaction probability. It can be imagined to be an effective area surrounding the target particle, normal to the direction of incidence of radiation or, perhaps, as the cross-sectional area of an interactive, spherical volume centred on the target particle. An interaction takes place if, and only if, the incident radiation arrives within this area.

General theory

This is an extremely classical model and is not entirely accurate, but it can be extended to a more abstract, mathematical representation that is quite compatible with quantum theory. This can be done by considering, in the first instance, a small, elemental thickness of matter, dx, containing n target particles per unit volume, each with a cross-section σ. dx is small enough to ensure that none of these cross-sectional areas overlap, so that, if radiations are incident

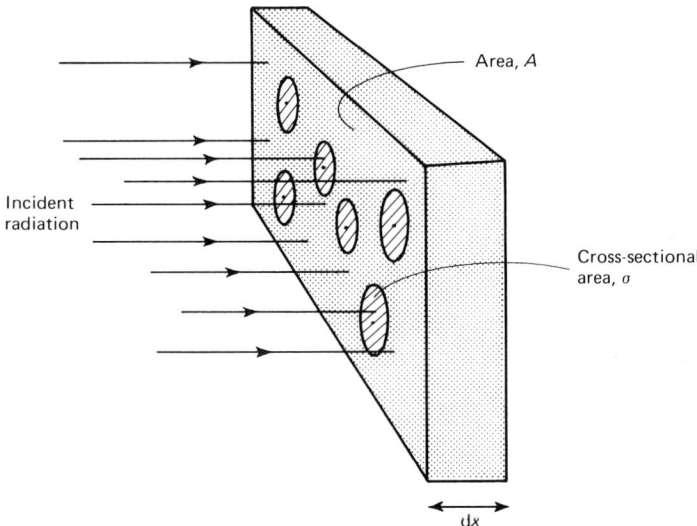

Figure 4.3 *Physical model of interaction cross-section*

normally on this matter, then over an exposed area A the total interactive area is $n\sigma A\,dx$ (*Figure 4.3*). A may be the area of a beam or flux of radiations or it may be chosen to be unit area within a beam. In either case, the fraction of radiations which undergo an interaction, that is, the interaction probability, P, is given by

$$P = \frac{\text{interactive area}}{\text{exposed area}} = \frac{n\sigma A\,dx}{A} = n\sigma\,dx \tag{4.4}$$

Although σ still has the dimensions of an area, this is a model-independent definition in which it is related to a pure probability P. It can be visualised as a probability rather than a physical area. It includes energy effects and has zero value for incident energies less than the threshold energy of a nuclear reaction. In general, it is a property of the radiation as much as of the target matter. For example, the same target material will generally have different cross-sections for different types and energies of incident radiation. It will also have different cross-sections for different types of interaction, but these are additive, so that the total interaction probability is measured by a *total cross-section*, σ_T, which is the sum of *partial cross-sections* for all possible interactions. If, for example, these refer to absorption, σ_A, elastic scattering, σ_E, and inelastic scattering, σ_I, then

$$\sigma_T = \sigma_A + \sigma_E + \sigma_I \tag{4.5}$$

Units

Cross-sections can be measured in square nanometres (10^{-18} m^2) or in square fermis (10^{-30} m^2) but the most common unit is the *barn* (b), where

$$1 \text{ b} = 10^{-28} \text{ m}^2 = 10^{-24} \text{ cm}^2 \tag{4.6}$$

Values of σ are given for a specific target particle; for example, per atom, σ_a, or per electron, σ_e. If the target particle is a nucleus, then n, in equation 4.4, is the number of nuclei, and therefore the number of atoms, per unit volume, given by

$$n_a = N_A \rho / M \tag{4.7}$$

where N_A is Avogadro's number, 6.02×10^{23} atoms per g-mole, ρ is the density of the material (g cm^{-3}) and M is the atomic mass (g per g-mole), often replaced by A, the atomic mass number.

The total number of electrons per unit volume is obtained by multiplying the above expression by Z, the atomic number, i.e.

$$n_e = N_A \rho Z / M \tag{4.8}$$

Differential cross-sections

If θ is the angle between the direction of incident radiation and that of the scattered radiation or of the radiation emitted following an absorption interaction, then the interaction cross-section is a function of θ and can be written as $\sigma(\theta)$. Similarly, it is a function of the energy, E, of the scattered or product radiation and can be written as $\sigma(E)$.

$d\sigma(\theta)/d\theta$ is the differential cross-section per unit angle. It describes the probability, $d\sigma$, that an interaction will take place and result in a product radiation being emitted within the range θ to $\theta + d\theta$. The integral of $d\sigma(\theta)/d\theta$ with respect to θ gives the total cross-section, σ.

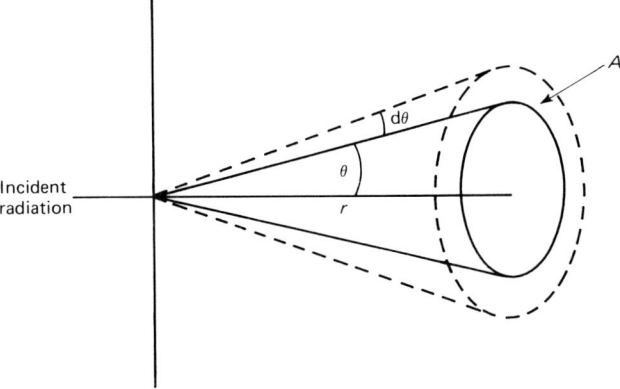

Figure 4.4 Differential scattering at an angle θ

$d\sigma(E)/dE$ is the differential cross-section per unit energy interval. It describes the probability, $d\sigma$, that an interaction will take place and result in a product radiation being emitted within the energy range E to $E + dE$. Its integral with respect to E is σ.

$d\sigma(\theta)/d\theta$ is inconvenient for some applications and may be replaced by $d\sigma(\theta)/d\Omega$, the differential cross-section per unit *solid angle* (usually per steradian). These two concepts are related (*Figure 4.4*). Using the terminology defined in this figure, the elemental solid angle, $d\Omega$, is defined to be the area A divided by r^2 (ignoring the slight obliquity of A to r). Hence

$$d\Omega = \frac{A}{r^2} = r\,d\theta \cdot 2\pi r \sin\theta / r^2 = 2\pi \sin\pi \cdot d\theta$$

Hence

$$\frac{d\sigma(\theta)}{d\theta} = \frac{d\sigma(\theta)}{d\Omega} \cdot 2\pi \sin\theta \tag{4.9}$$

As an example, isotropic scattering distributes particles or photons equally in all directions. This is defined by $d\sigma(\theta)/d\Omega$ being a constant. $d\sigma(\theta)/d\theta$, however, is not a constant but a sine function.

4.2 ATTENUATION

As radiation penetrates matter, particles or photons may be removed from the incident beam either by absorption interactions or by those scattering interactions which deflect the particle or photon out of the beam or reduce its kinetic energy below some critical value. In general, radiation intensity and/or energy decreases with distance, x, into an interactive material. The decrease is referred to as *attenuation* and may be more specifically defined as intensity attenuation or energy attenuation. *Intensity* is usually defined to be the number of particles or photons arriving at a given point per unit area, across the beam, and per unit time. *Energy* is the energy of a single particle or photon in the beam.

The number of interactions which take place within a given distance in a material per unit time may be described as the *interaction rate*. As a fraction of incident beam intensity, this is the interaction probability used above. It depends upon interaction cross-section, as shown by equation 4.4, but it also depends upon the beam intensity. If this is reduced, by attenuation, the interaction rate will decrease with x. In fact, the number of interactions, dI, which occur in a small element of radiation path length, dx, is given by equation 4.4, which can be rewritten as

$$\frac{dI}{I} = n\sigma\,dx \tag{4.10}$$

where I is the radiation intensity entering the element dx. This is a differential equation and it can be solved (in principle, at least) for any specific situation, to show how intensity I varies with penetration distance x into a medium. I, as a function of x, can be described as an *attenuation function* or, if expressed as a graph, as an *attenuation curve*. Interaction rate at any point x is then given by equation 4.10. The attenuation function is determined by the type of radiation and matter involved, but it is possible to identify three functions which cover most situations. These may be referred to as *differential, Gaussian* and *exponential* attenuation.

Differential attenuation

Differential attenuation takes place in a very *thin* material, sometimes known as a 'dx' material. A thin material can be defined as one for which the product $n\sigma\,dx$ is much smaller than unity and it satisfies the condition used in equation 4.4 to define cross-section, namely, that there should be no overlapping cross-section areas. Thus, a thin material, in terms of radiation interactions, may be physically thick, that is, with a relatively large value of dx, provided that n and σ are so small that the interaction probability $n\sigma\,dx$ is small.

The number of interactions taking place in a thin material is given by equation 4.10. If the radiation intensity arriving at the surface of the material is I_0 then the number leaving it, I, is given by

$$dI = -I_0 n\sigma\,dx$$

$$\therefore I - I_0 = -I_0 n\sigma\,dx$$

$$\therefore I = I_0(1 - n\sigma\,dx) \tag{4.11}$$

where the negative sign shows that dI is a reduction in intensity, or in other words that I is less than I_0. In addition, the cross-section, σ, refers to those processes which *remove* radiations from the beam, so equation 4.11 is a simple form of attenuation function.

Equation 4.11 can be used only for thin absorbers but it applies to any type of radiation. It may be used as a crude approximation to calculate the attenuation of quite thick absorbers.

Gaussian attenuation

There is no generally recognised term for this type of attenuation so one has to be adopted for present purposes. It may be described as 'Gaussian' attenuation because the attenuation function can be represented, approximately, in terms of the Gaussian distribution function of statistics.

Gaussian attenuation occurs most frequently as a description of the attenuation of *charged particles* in matter. Most of these undergo *scattering interactions* which do not remove them from the beam, so a characteristic feature is that the *beam intensity remains constant* for some distance into the medium. Each particle experiences a succession of collisions which remove small fractions of its energy, so a second characteristic feature is that the *radiation energy decreases* with distance, x, into the medium.

Eventually, the particles run out of energy and are stopped, that is, removed from the beam. Because of the statistical nature of the scattering process, the distance penetrated varies from one particle to another but there is a well-defined average penetration, called the *mean range, R*, of the radiation. This range is the third characteristic feature of Gaussian attenuation. Another useful parameter is the *mean free path*, λ, which is the average distance travelled by one particle between successive collisions. The number of collisions is therefore, on average, the range R divided by λ.

The statistical probability that a particle will penetrate a distance x into the medium can be described, approximately, by the well-known Gaussian distribution function. Consequently, this function describes the 'removal' interaction rate as a function of x and the beam intensity, I, as a function of distance, x, is given by the original intensity, I_0, less the Gaussian function. The mean range, R, is, of course, the mean x value of the Gaussian function and at this point the beam intensity is reduced to half of its original value, so the attenuation function can be written as

$$I = I_0 - \frac{I_0}{2} \exp[-(x-R)^2/2\sigma^2]$$
$$= I_0 \{1 - \tfrac{1}{2} \exp[-(x-R)^2/2\sigma^2]\} \qquad (4.12)$$

where σ is the *standard deviation* of the distribution.

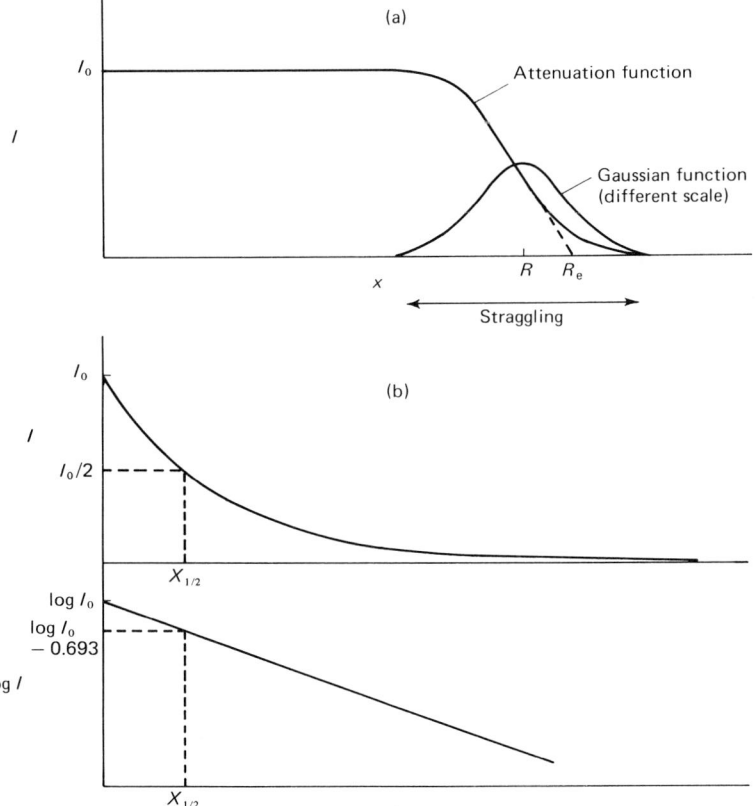

Figure 4.5 *(a) Gaussian and (b) exponential attenuation*

For most purposes, this is an unduly complicated function and the graphical representation shown in *Figure 4.5* is preferred. In any case, the Gaussian function is only approximate, so graphed data may be more accurate. The attenuation curve can be used to illustrate two more parameters. One of these is the *extrapolated range*, R_e, which is the x value determined by extrapolation of the decreasing attenuation curve and is often more easily measured than the mean range. The second is *straggling*, or the variation in the range of individual particles. Straggling is measured by the width of the Gaussian function and is greater for light, easily deflected particles than for heavy ones. Hence the range of heavy particles, such as alpha particles, is defined more precisely than it is for light particles, such as beta particles.

Exponential attenuation

Exponential attenuation is typical of *photons*, or other radiations which are completely removed from the incident beam, usually in *absorption interactions* but also in scattering interactions that happen to remove photons. Since each single interaction effectively absorbs a photon, the *beam intensity decreases* into the medium, there is *no definable range*, and the *radiation energy remains constant*. Each photon (or particle) can interact only once so its energy remains unchanged until it does so.

The attenuation function can be obtained by solving the differential equation 4.10. Intensity, I, is a variable with the initial value I_0 at $x = 0$, dI is the differential reduction in I, and the cross-section, σ, is a constant because the beam energy and all other affective parameters are constant. The function produced in this way is

$$I = I_0 \, e^{-n\sigma x} \tag{4.13}$$

showing that beam intensity decreases exponentially with x. This function is plotted, as an attenuation curve, in *Figure 4.5*. As for the radioactive decay curve, it may be plotted as a graph of $\log I$ against x. The gradient of this line is $-n\sigma$ and the intercept with the I axis is at $I = I_0$.

In passing, it may be noted that the series expansion of the exponential function is

$$e^{-n\sigma x} = 1 - n\sigma x + \frac{(n\sigma x)^2}{2} - \frac{(n\sigma x)^3}{3.2} + \ldots \tag{4.14}$$

If $n\sigma x$ is very small, this reduces to $1 - n\sigma x$ and inserting this expression in equation 4.13 gives equation 4.11, for thin absorbers. Thus, as indicated above, equation 4.11 can be used for an approximate treatment of exponential attenuation, provided $n\sigma x$ is very small.

Attenuation coefficients

In exponential attenuation, the characteristic parameter, obtained from measurement and used in calculations, is the product $n\sigma$. It is convenient, therefore, to replace this product by a single parameter, called the *attenuation coefficient*. If the distance, x, is measured in units of length, the *linear attenuation coefficient*, μ, is defined (for cross-section per atom) by

$$\mu = n\sigma = \frac{N_A \rho \sigma}{A} \qquad (4.15)$$

and equation 4.13 becomes

$$I = I_0 e^{-\mu x} \qquad (4.16)$$

μ has the dimension of inverse length. If n is in cm^{-3} and σ is in cm^2 (converted from barns), then μ is in cm^{-1}.

Frequently, it is more convenient to measure absorber thickness in units of mass per unit area, e.g. in g cm^{-2}, instead of in units of length. This is known as the *equivalent thickness* of the material. If x is thickness, in cm, m is equivalent thickness, in g cm^{-2}, and ρ is the density, in g cm^{-3}, then

Volume of 1 cm area $= x$ (cm^3)
Mass of 1 cm area $= \rho x$ (g)

i.e.

Equivalent thickness $m = \rho x$ $\qquad (4.17)$

For example, the equivalent thickness of 1 μm of Al ($\rho = 2.7$ g cm^{-3}) is given by

$$m = 2.7 \text{ (g cm}^{-3}) \times 0.0001 \text{ (cm)}$$
$$= 0.27 \text{ mg cm}^{-2}$$

If equivalent thickness is used in the attenuation formula of equation 4.16 instead of thickness, μ must have the units of area per unit mass, e.g. cm^2 g^{-1}. In this case, it is known as the *mass attenuation coefficient* and is written as μ_m. Using the above formula, it can be seen that

$$\mu_m = \frac{\mu}{\rho} \qquad (4.18)$$

Hence, for cross-section per atom,

$$\mu_m = \frac{n\sigma}{\rho} = \frac{N_A \sigma}{A} \qquad (4.19)$$

For composite materials, containing more than one element, the net attenuation coefficient is the appropriate sum of the separate contributions. This can be determined in terms of linear or mass attenuation coefficients but the latter produce the simpler formula

$$\mu_m = \alpha_1 \mu_{m_1} + \alpha_2 \mu_{m_2} + \alpha_3 \mu_{m_3} + \ldots = \sum_i \alpha_i \mu_{m_i} \qquad (4.20)$$

where α_i are the *fractional mass abundances* of the component elements.

As an example, water contains two hydrogen atoms per oxygen atom. The fractional mass abundance of hydrogen, α_H, is 1/9 and that of oxygen, α_O, is 8/9, so that the mass attenuation coefficient of water is

$$\mu_{H_2O} = \frac{1}{9}\mu_H + \frac{8}{9}\mu_O$$

where μ_{H_2O}, μ_H and μ_O represent mass attenuation coefficients.

To complete this discussion of exponential attenuation, two further definitions are required, as follows:

Half-value thickness, $X_{1/2}$, is the distance in which the radiation intensity reduces by a factor of two. For a Gaussian attenuation this would be the range parameter. For exponential attenuation it is analogous to the half-life of radioactive decay and given by

$$\frac{I}{I_0} = \frac{1}{2} = \exp(-\mu X_{1/2})$$

$$\therefore X_{1/2} = \frac{\ln 2}{\mu} = \frac{0.6931}{\mu} \tag{4.21}$$

Mean free path, λ, is the mean distance travelled by a photon before being removed from the incident beam. It is analogous to the mean life parameter of radioactive decay and is given by

$$\lambda = \frac{1}{\mu} \tag{4.22}$$

For particles, the mean free path usually describes the average distance between collisions.

Geometry

The attenuation formulae developed above are reasonably good approximations to most practical situations but they do not include the effects of beam–absorber geometry. They describe the attenuation of incident radiation at a depth x in a homogeneous infinite slab absorber. They do not include edge effects or lateral transfer of radiations. In other words, the treatment so far has been one-dimensional, apart from the acknowledgement that scattering can simulate absorption in removing radiations from the beam.

The treatment can be improved by considering some of the effects of beam–absorber geometry. Of the many possible arrangements, four represent the major features and are described in *Figure 4.6*. These are:

(1) *Narrow beam geometry,* or good geometry, refers to a narrow, collimated beam of radiation. Scattering effectively removes a particle or photon from the beam so the equations derived above apply exactly with total (that is, absorption plus scattering) cross-sections.

Attenuation 79

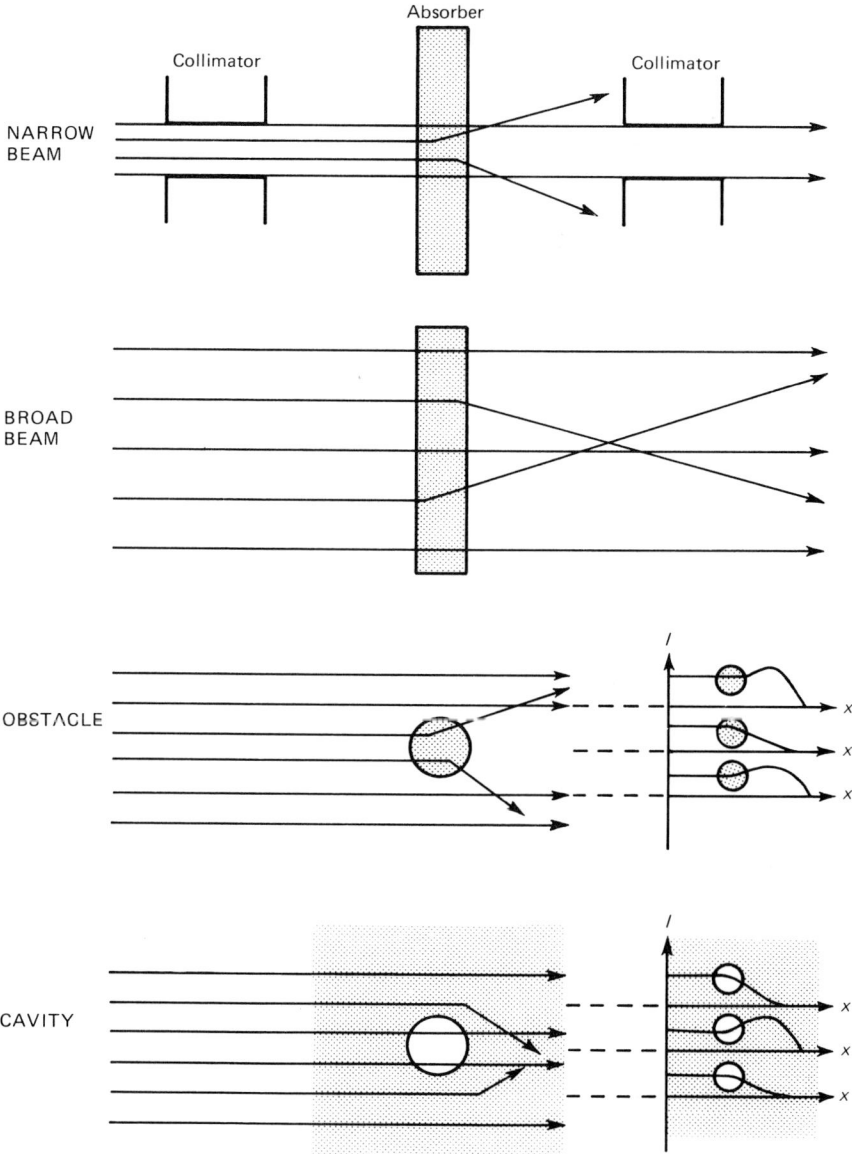

Figure 4.6 Effects of geometry on radiation scattering

(2) *Broad beam geometry*, or bad geometry, relates to a wide beam of incident radiation, as used, for example, in X-radiography. In this case, absorption removes radiation from the beam but scattering does not — it only moves it from one point to another within the beam. The attenuation formulae apply, provided absorption cross-sections are used, but, unless these are large fractions of total cross-sections, little effect is observed.

(3) *Obstacle geometry* describes the attenuation of a uniform broad beam by an inhomogeneous medium containing a relatively small obstacle, whose absorptive power is greater than that of the surrounding medium. Attenuation in the obstacle is greater than in the adjacent material so the radiation intensity in the shadow region behind it is less than in other regions in the same lateral plane and the beam is no longer homogeneous. This effect is enhanced by scattering interactions. Since the obstacle scatters more radiation than its environment, there is a net transfer of radiation out of the shadow and into adjacent areas. It is possible to estimate the amount of lateral transfer from a knowledge of the angular properties of the processes involved, but the calculation can be a difficult one. The effect can be represented by multiplying the attenuation function by a build-up factor, B, which is a function of distance, x. For exponential attenuation, this gives the general result

$$I = BI_0 e^{-\mu x} \tag{4.23}$$

where B is greater than unity outside the shadow of the obstacle and less than unity inside it.

(4) *Cavity geometry* describes the attenuation of a uniform broad beam by an inhomogeneous medium containing a relatively small cavity, or region of reduced interaction cross-section. It has the opposite effect to that obtained by an obstacle and may be regarded as focusing the radiation, to some extent, into the region immediately behind the cavity. Obstacle geometry and cavity geometry are particularly relevant to radiation attenuation in the human body and in radiation detectors, since both present inhomogeneous structures. Both geometries refer to relative interaction cross-section and not necessarily to relative density.

4.3 CHARGED PARTICLES

From the point of view of radiation detection, the most important interactions of charged particles involve the *electrostatic force*. At low energies, electrons are the target particles and *ionisation* and *excitation* processes are dominant. At higher energies, especially for beta particles, *bremsstrahlung* predominates and nuclei are the target particles. In both cases, because of the long range of the electrostatic force, incident particles can interact with a target without passing very close to it. Unless the medium is very thin, of the order of the mean free path of the radiations, the interaction cross-section is virtually unity and not a particularly useful parameter. In addition, the processes involved are *scattering interactions*, with small energy transfer and small deflection angles, so charged particles tend to exhibit all the characteristic features of *Gaussian attenuation*, including a well-defined range parameter.

Besides ionisation, excitation and bremsstrahlung, the *Čerenkov effect* is present for extremely energetic, i.e. relativistic particles, and, for all antiparticles, including positrons, *annihilation* is the only interaction which need be considered. A number of other interactions, including Rutherford scattering, Coulomb scattering and nuclear reactions, can be observed. These are valuable for investigating nuclear structure but play little part in the detection of charged particles.

Ionisation and excitation

For all heavy charged particles and for betas with less than about 1 MeV of energy, ionisation and excitation constitute the major interaction mechanism. As the particles pass through matter, they undergo a succession of long-range electrostatic collisions with atomic electrons. Depending upon the distance between particle and target and on the relative motion of these, the electron may be knocked out of the atom, creating an ion pair, or it may only be excited to a higher bound state, creating an excited atom. Unless they are removed, by an electric field, the ion pairs rapidly dissipate their energy as thermal energy and recombine, or are otherwise neutralised, to form atoms. The excited atoms de-excite, emitting low-energy photons that either escape or are quickly captured, with the same end result. The net effect is that a trail of ion pairs is formed along the track of the incident radiation, then these are neutralised and energy is dissipated as heat and light. If a large amount of light escapes the medium, the latter is said to *fluoresce*. If the light escapes more slowly, owing to slow atomic de-excitation, the material is said to *phosphoresce*. The radiations themselves eventually come to rest; alpha particles collect two electrons to form helium gas and beta particles are captured by atoms and ions of the medium.

The process is illustrated in *Figure 4.7* for alphas and betas. The former, being much heavier than the atomic electrons, are not deflected very much and thus

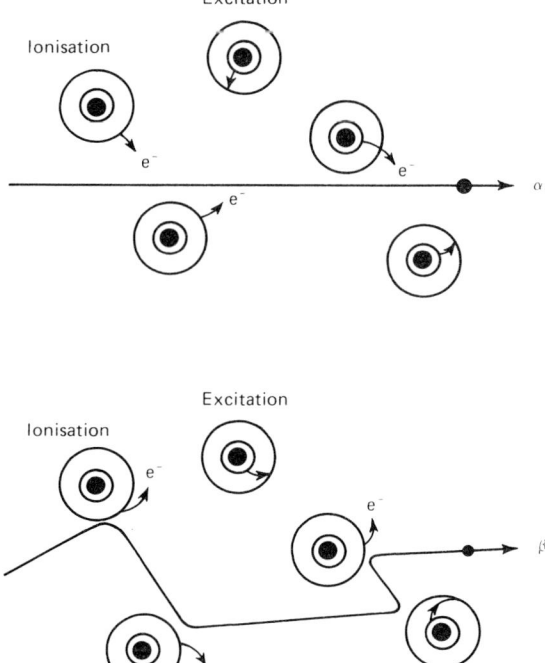

Figure 4.7 Ionisation and excitation processes due to passage of α and β particles

form straight tracks. Since they are doubly charged, they also form very dense tracks and lose energy in a short distance. Betas, on the other hand, are easily deflected. They produce meandering tracks, with relatively light ionisation, and travel further into the medium. The electrons liberated by heavy charged particles may have sufficient energy to form short ion trails of their own. These are referred to as *delta rays* (δ-rays) and are illustrated in *Figure 4.8*.

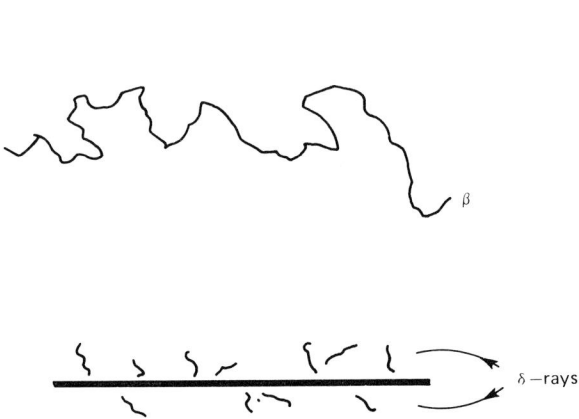

Figure 4.8 Ion tracks produced by charged particles

As a general rule, about half of the energy deposited produces ionisation and half leads to excitation. For this reason, the mean energy deposited per ion pair formed, that is, the mean energy required to form an ion pair, ϵ, is roughly twice the ionisation potential of the atoms in the medium. For example, the ionisation potential of argon is 15.7 eV but the mean energy per ion pair is about 26.4 eV. Thus, if a 1-MeV alpha particle deposits all of its energy in argon, about $10^6/26.4 = 37\,880$ ion pairs are produced. The same alpha particle would generate about 378 800 ion pairs in a semiconductor ($\epsilon \simeq 2.6$ eV). To a first approximation, these figures are independent of the type of radiation involved.

Quantitative studies of charged-particle interactions are based on two parameters. One of these is the *range*, which determines how far a given type of radiation, with a particular energy, will travel and, for example, whether or not it will be stopped in a radiation detector or an absorber. *Mean range*, R, and the *extrapolated range*, R_e, have been defined above. Another version of this parameter is the *residual range*, which is the mean distance a particle can still travel after having passed through some material. It is useful in dealing with the common situation in which radiation has to penetrate a source encapsulation, intervening air and structural materials before entering the sensitive volume of a radiation detector.

The other parameter measures the amount of energy deposited, or the number of ion pairs produced, per unit path length. Various definitions are used, as follows:

Linear stopping power, dE/dx, or *specific energy loss* is the total energy deposited per unit path length, measured in a straight line. It is given, for example, in keV cm^{-1}.

Mass stopping power, dE/dm, is the same as linear stopping power except that equivalent thickness, dm, is used instead of length, dx (although the term 'dx' may be used for both). If ρ is the density of the absorber, then, from equation 4.17,

$$dE/dx = \rho \cdot dE/dm \qquad (4.24)$$

Linear energy transfer, LET, is the energy deposited locally, that is, by ionisation and excitation, per unit path length. It does not include bremsstrahlung, as do the above definitions but, in most cases and especially at fairly low energies, it is equal to dE/dx.

Specific ionisation, dN/dx, is the number of ion pairs produced per unit path length and equal to LET/ϵ, where ϵ is the mean energy deposited per ion pair formed.

There is a general formula for LET, or dE/dx without bremsstrahlung, as a function of the properties of the incident particle and the absorbing medium. It is known as the *Bethe–Bloch formula* and an approximate, non-relativistic version is

$$-\frac{dE}{dx} = \frac{2\pi e^4}{m} \cdot \frac{Mz^2}{E} \cdot n \cdot \ln\left(\frac{kE}{MZ}\right) \qquad (4.25)$$

where the negative sign shows an energy decrease as the variable, x, increases, e is the electronic charge, m is the electron rest mass and k is a constant. The term Mz^2/E refers to the particle, M being its mass, z its atomic number and E its energy. n refers to the absorber and is the *effective* number of electrons per unit volume. In general, n is less than the total number of electrons per unit volume, given by equation 4.8, since the inner electrons are less likely to be involved in the collision process. The last term, being logarithmic, varies more slowly than the other variable terms and may be disregarded in approximate calculations.

In practice, the Bethe–Bloch formula is inaccurate for very low energies and generally it is easier and more accurate to use tabulated or graphed values of dE/dx. Some of these are provided in Appendix 4. For very rough estimates, however, proportional relationships can be used to derive values of dE/dx from available data. Thus, for *different particles*, same absorber,

$$-\frac{dE}{dx} \propto \frac{Mz^2}{E}$$

and, for *different absorbers*, same particles,

$$-\frac{dE}{dx} \propto n$$

so that

$$-\frac{dE}{dx} \propto \frac{Mz^2}{E} \cdot n \qquad (4.26)$$

where n is rather less than the value given by equation 4.8.

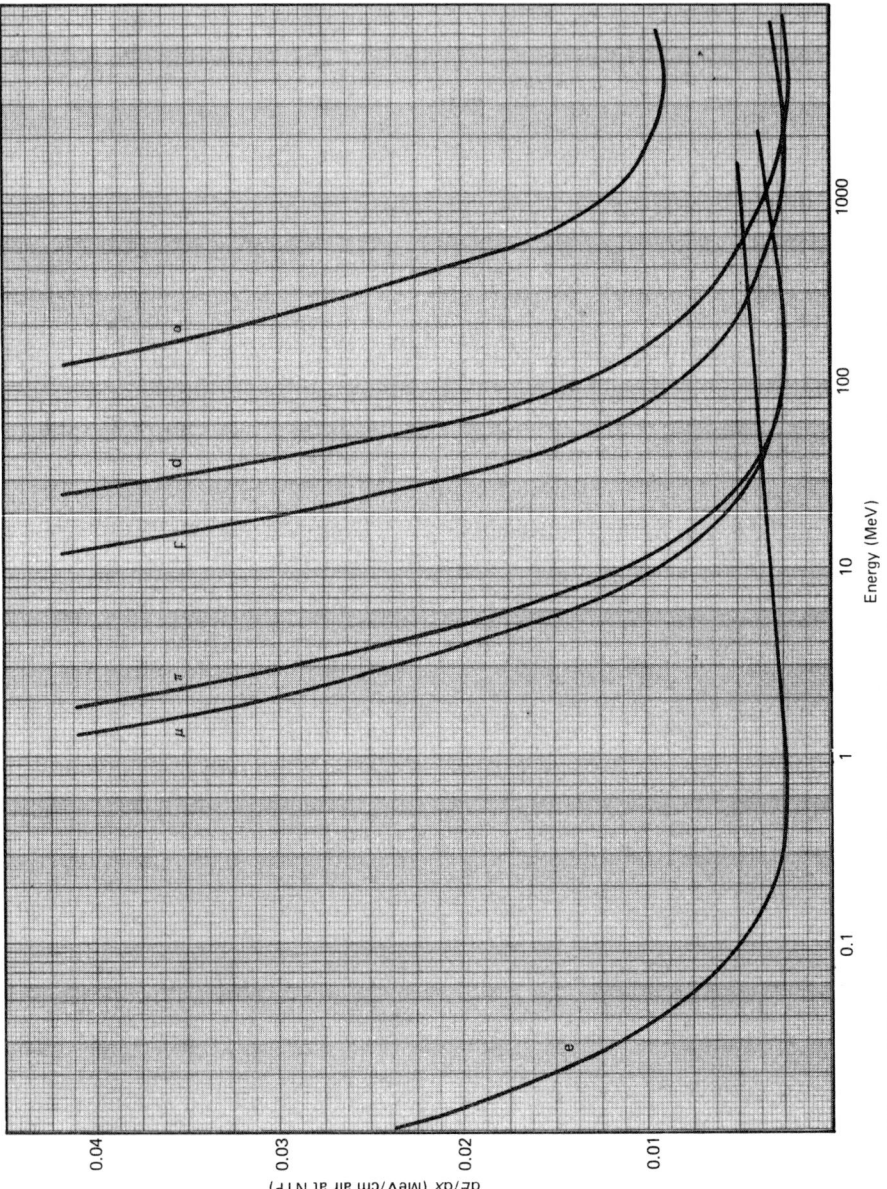

Figure 4.9 Specific energy loss of charged particles as functions of energy, in air

The Mz^2/E dependence is illustrated in *Figure 4.9*, showing $-dE/dx$ as a function of energy, E, for various particles. The same effect is observed in *Figure 4.10*, which shows that as a charged particle penetrates into matter, its energy gradually decreases and $-dE/dx$ correspondingly increases, reaching a maximum towards the end of the track before falling to zero as the particle is stopped. This graph is known as a *Bragg curve*.

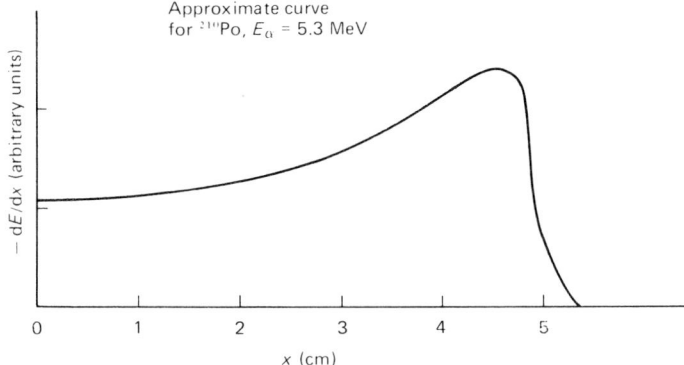

Figure 4.10 Bragg curve: specific energy loss as a function of distance penetrated by a radiation beam

For general purposes, the range parameter, R, is more useful than dE/dx. It is obtained by integrating the reciprocal quantity, dx/dE, with respect to energy, E, which decreases from an initial value, E_0, to zero, when the particle is stopped at an x value equal to the range, R. Again, tabulated or graphed data, some of which are given in Appendix 4, are the most accurate sources of information, but some estimates can be made by means of proportionalities and empirical relationships as follows:

For *different particles*, same E_0, same absorber,

$$R \propto \frac{1}{Mz^2} \qquad (4.27)$$

For *alpha particles*, range, R_α, is given by an empirical formula,

$$R_\alpha = 3.2 \times 10^{-4} \cdot (A^{1/2}/\rho) R_{air} \qquad (4.28)$$

where A is the atomic mass number of the absorber and ρ is its density, while R_{air} is the range in air of an alpha with the same energy, as given in *Figure 4.11*. R_α and R_{air} are both expressed in units of length, while ρ is in g cm^{-3}. For example, the range in air of a 5.3-MeV, ^{210}Po alpha particle is 3.8 cm. In aluminium ($A = 27$, $\rho = 2.7$ g cm^{-3}) it is 0.23 mm.

For *electrons*, a number of empirical range formulae exist. They also apply to *beta particles*, in which case they give the maximum range corresponding to the maximum beta energy. The formulae due to Glendenin are

$$R = 0.542E - 0.133 \qquad E > 0.8 \text{ MeV}$$
$$R = 0.407E^{1.38} \qquad 0.15 \text{ MeV} < E < 0.8 \text{ MeV} \qquad (4.29)$$

where E is measured in mega-electronvolts and R in g cm^{-2}.

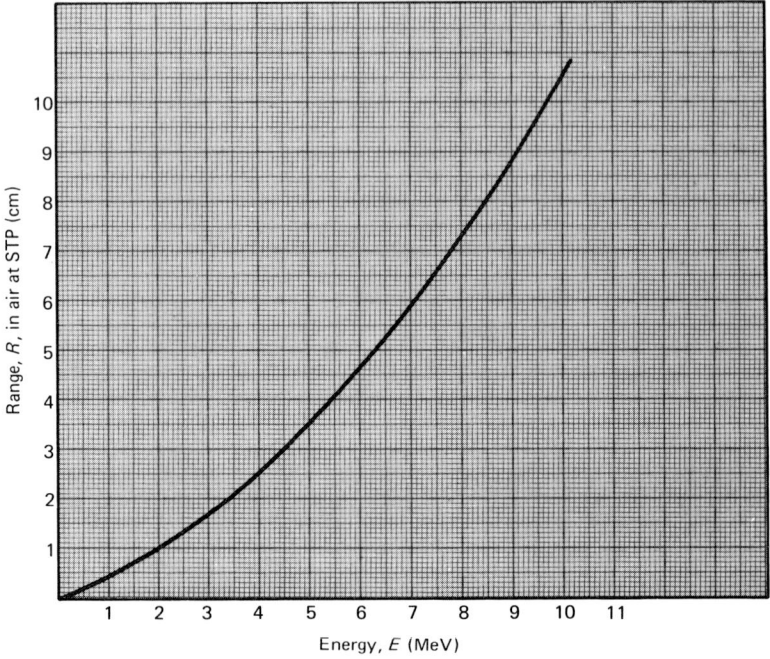

*Figure 4.11 Range–energy curve for alpha particles in air. From Jesse, W. P. and Sadauskis, J., Phys. Rev., **78**, 1 (1950)*

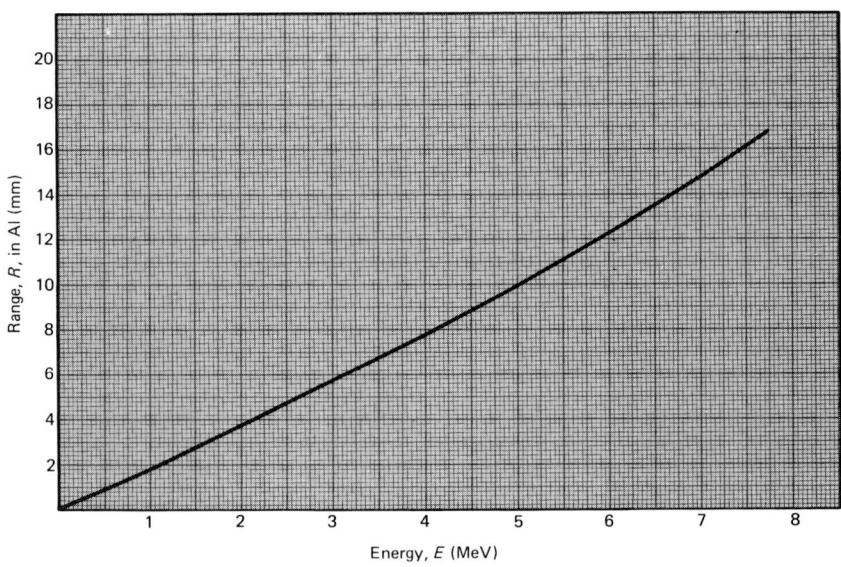

Figure 4.12 Range–energy curve for beta particles or electrons in aluminium

It is useful to note that the range, in equivalent thickness, of electrons and beta particles is largely independent of the nature of the absorber. Thus, a graph of R against E for one medium, as given for aluminium in *Figure 4.12*, can be used for any material.

Bremsstrahlung

Bremsstrahlung, or 'braking radiation', is electromagnetic radiation emitted whenever charged particles undergo sharp accelerations or decelerations. These can be caused by electrostatic collisions with nuclei or, less frequently, electrons, in an absorbing medium. The process is analogous to atomic de-excitation except that the particle energy states are not bound, but free ones, so the photons produced may have any of a continuous range of energies instead of discrete, characteristic values. Thus, the bremsstrahlung energy spectrum is a continuum in the X-ray and gamma ray region.

The effect is much more pronounced for beta particles and electrons than for the heavier particles because the former are more easily deflected. For most purposes, bremsstrahlung can be ignored for heavy particles. The process is illustrated in *Figure 4.13* for a particle of charge $-ze$ and a target nucleus of charge Ze. The Coulomb attractive force between these two particles is zZe^2/r^2 and the deceleration of the incident particle is zZe^2/Mr^2, where M is its mass.

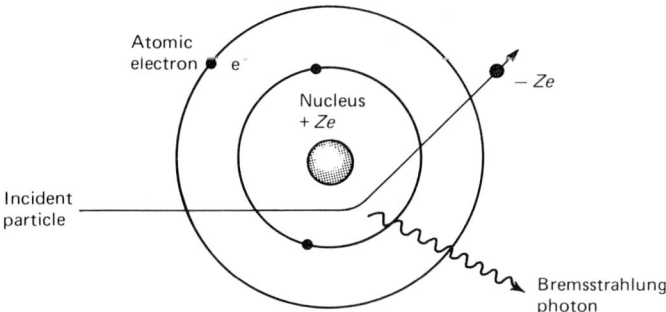

Figure 4.13 Bremsstrahlung process

It can be shown from classical theory that, in all cases of decelerating charges, energy is emitted at a rate which is proportional to the square of the deceleration (or acceleration) of the charge. For bremsstrahlung, this gives

$$-\frac{dE}{dx} \propto (zZ/M)^2 \qquad (4.30)$$

Since dE/dx is proportional to $1/M^2$, it is much greater for electrons than for heavier particles. It is also proportional to the energy, E, of the incident particle, so a more complete relationship, separating the properties of the particle and the absorber, is given by

$$-\frac{dE}{dx} \propto \frac{z^2 E}{M^2} \cdot Z^2 \qquad (4.31)$$

The total energy loss is due to ionisation and excitation as well as bremsstrahlung and is given by the general formula

$$-\frac{dE}{dx} = -(dE/dx)_{ion} - (dE/dx)_{brem} \qquad (4.32)$$

At low energies, the ionisation and excitation term, $(dE/dx)_{ion}$, predominates and decreases with increasing energy. At higher energies, the bremsstrahlung term is the larger and increases with energy. As illustrated in *Figure 4.9*, there is a *minimum ionising* value. For betas, this occurs at about 1 MeV and at lower energies bremsstrahlung can be neglected. For other particles the minimum ionising value is much larger. In general, a useful formula for comparing the two processes is

$$\frac{(dE/dx)_{brem}}{(dE/dx)_{ion}} = \frac{EZ}{800} \qquad (4.33)$$

where E is in mega-electronvolts.

Čerenkov radiation

When a charged particle moves through matter it produces excited atoms as well as ion pairs. In some materials, known as scintillators, these de-excite immediately, mainly emitting visible light. In others, the de-excitation processes are non-radiative. In all materials, however, the charged particle can produce a relatively weak emission of visible light, known as Čerenkov radiation. This process depends upon a form of excitation in which atoms are temporarily polarised. A positive charged particle, for example, repels atomic nuclei and attracts electrons creating vibrating electric dipoles which emit electromagnetic radiation. Normally, these atomic dipoles are randomly out of phase, so their radiations cancel each other to result in no net emission, but, if the particle exceeds the velocity of light in the medium, coherent emission takes place. This is Čerenkov radiation.

The Čerenkov effect is not an important energy loss mechanism, since it does no more than is already accounted for by excitation processes, but it does serve as the basis for a type of radiation detector, known as a Čerenkov counter, and will be discussed, later, in that context.

Annihilation

Antiparticles interact with matter in much the same way as particles do, until they happen to collide with their corresponding particles, whereupon they annihilate. The positron is a special case because matter abounds with its corresponding particle, the electron. For this reason, positrons do not penetrate very far in matter, usually less than a few millimetres, before coming close enough to an electron to annihilate.

In this interaction, the positron is captured by an electron to form a bound, positron–electron system, called *positronium*, for a very short period of time, of the order of 10^{-7} s or less. The two particles then combine, or annihilate.

In effect, they cancel each other, converting their mass into pure electromagnetic energy, according to the Einstein formula, $E = mc^2$.

Since the system is virtually at rest prior to annihilation, there is no kinetic energy to add to the mass–energy, nor is there any linear momentum. Consequently, in order to conserve energy and momentum, two photons are generated,

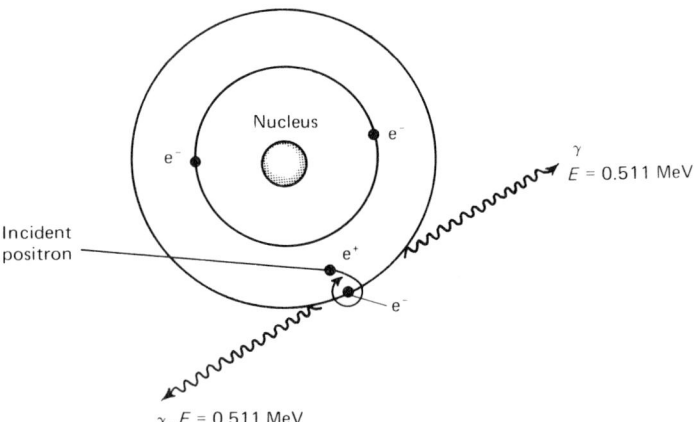

Figure 4.14 *Annihilation of a positron*

each with an energy of 0.511 MeV and travelling in opposite directions, as illustrated in *Figure 4.14*. Generally, annihilation takes place within the source material itself, so positron emitters, such as ^{22}Na, can be regarded as photon emitters.

4.4 PHOTONS

X- and gamma rays interact with atomic electrons and with nuclei in a variety of different processes. These are extremely valuable for studying atomic and nuclear structure but, for radiation detection, only three need be considered and all of these, conveniently, interact with and liberate *atomic electrons*. They are the *photoelectric effect, Compton scattering* and *pair production*.

In each event, a photon ejects an electron in a primary ionisation process and the electron continues, usually in a completely different direction from that of the photon, to produce further ion pairs, in secondary ionisation processes such as those already described for charged particles.

The photoelectric effect and pair production are absorption interactions in which the photon is completely captured. They tend to involve the inner electrons, mainly in the *K* shell, so the electron is ejected with the photon energy minus *K*- or, less frequently, *L*-shell binding energy, and the atom subsequently de-excites, emitting X-rays characteristic of the *K, L, M* shell, etc. in cascade, along with a few Auger electrons. Compton scattering does not completely absorb the photon but it does tend to scatter it through a large angle and reduce its energy considerably, so it effectively removes the photon from the incident beam, except in very broad beam geometry.

Thus, photons exhibit all the characteristics of exponential attenuation. There is no well-defined range and intensity falls off exponentially with penetration distance. Since a single interaction removes a photon from the incident beam, the energy of those remaining is a constant and so, too, is the interaction cross-section. This is the sum of partial cross-sections due to photoelectric capture, σ_{PE}, Compton scattering, σ_C, and pair production σ_{PP}, per electron.

$$\sigma = \sigma_{PE} + \sigma_C + \sigma_{PP} \tag{4.34}$$

Similarly, the net attenuation coefficients, linear or mass, are given as the sum of partial coefficients,

$$\mu = \mu_{PE} + \mu_C + \mu_{PP} \tag{4.35}$$

where

$$\mu_i = n\sigma_i = \frac{N_A \rho Z}{A} \cdot \sigma_i$$

and

$$I = I_0 e^{-\mu x}$$

The photoelectric effect

In this process, an incident photon collides with, and transfers all of its energy to an atomic electron. The photon is totally absorbed and the electron is emitted from the atom with an energy E_e, equal to the photon energy E_γ less the atomic binding energy E_B:

$$E_e = E_\gamma - E_B \tag{4.36}$$

These electrons are known as *photoelectrons*. Since the atom is ionised, the process is one form of *photoionisation*. Compton scattering and pair production are others. *Photoexcitation* may also take place but, because of the exact energy match required by the Bohr formula, $h\nu = E_2 - E_1$, it is a low-probability event compared with ionisation and is not generally regarded as a photoelectric effect.

The photoelectric effect is not confined to X-ray and gamma ray photons. It is also experienced by low-energy photons in the ultra-violet and visible-light range. In fact, the photoelectric capture cross-section increases rapidly as photon energy decreases, so it is the dominant absorption process for visible-light photons and the basis, for example, of the conversion of light into electric current in a photocell or a photomultiplier tube.

For energy and momentum to be conserved, the nucleus must recoil and, for this reason, the process takes place more readily at inner than at outer electron shells. As a general rule, photoelectrons tend to be emitted from the most tightly bound shell that can be ionised by the incident photon. If the photon has more energy than the K-shell binding energy, E_K, then about 80% of the interactions involve K-shell electrons and most of the remaining 20% eject L-shell electrons. A *K*-shell interaction is illustrated in *Figure 4.15*. It is followed, virtually immediately, by cascade de-excitation emitting characteristic X-rays and Auger electrons, while the photoelectron is creating further, secondary, ion pairs, excited atoms and de-excitation photons.

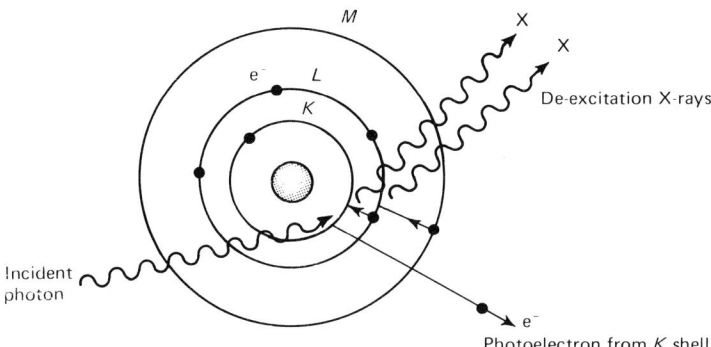

Figure 4.15 Photoelectric effect

There are formulae for photoelectric effect cross-sections but these are complicated and less accurate, for practical work, than experimental data such as those provided in Appendix 4. As for charged particles, it is sufficient, for most purposes, to use proportionalities, as follows:

For the same material, different photon energies,

$$\mu_{PE} \propto E_\gamma^{-7/2}$$

For the same photon energy, different materials,

$$\mu_{PE} \propto \rho Z^5$$

Hence

$$\mu_{PE} \propto E_\gamma^{-7/2} \rho Z^5 \qquad (4.37)$$

The Z^5 dependence is quite approximate; the power varies from 4 to 5 with photon energy, but it is, nevertheless, a very strong dependence. These proportionalities mean that (a) photoelectric capture predominates, for most materials, for low photon energies, including X-ray and low-energy gamma ray energies, and (b) photons are absorbed much more strongly in high-Z materials, such as lead, than in low-Z materials. Thus, lead shielding is used to screen personnel from electromagnetic radiations and high-Z materials are used to detect them. The variation of μ_{PE} with energy is illustrated, for lead, in *Figure 4.16*. The sharp discontinuities occur at *absorption edges* where the photon energy becomes large enough to excite the next most tightly bound electron shell.

The angular distribution of photoelectrons can be described by means of the differential cross-section, $d\sigma(\theta)/d\Omega$, which, in this instance, is defined to be the

92 *Interactions with matter*

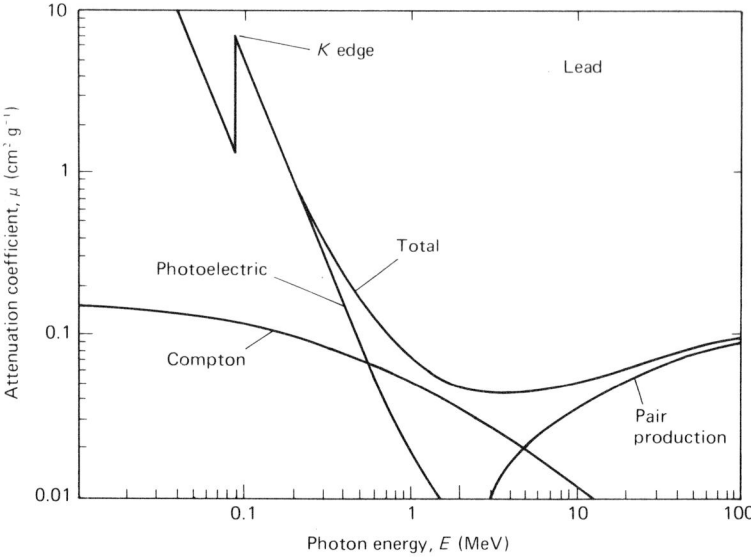

Figure 4.16 Attenuation coefficients as functions of photon energy for lead

probability that a photoelectric capture event will take place and an electron emitted into a unit solid angle about the angle θ. This function is shown in *Figure 4.17*. The graph indicates that, at low photon energies, photoelectrons tend to be emitted at right angles to the direction of the incident photon, i.e. at about $\theta = 90°$, and hardly at all in the forward direction of $\theta = 0$. As energy increases the maximum-intensity angle moves forward to smaller values of θ.

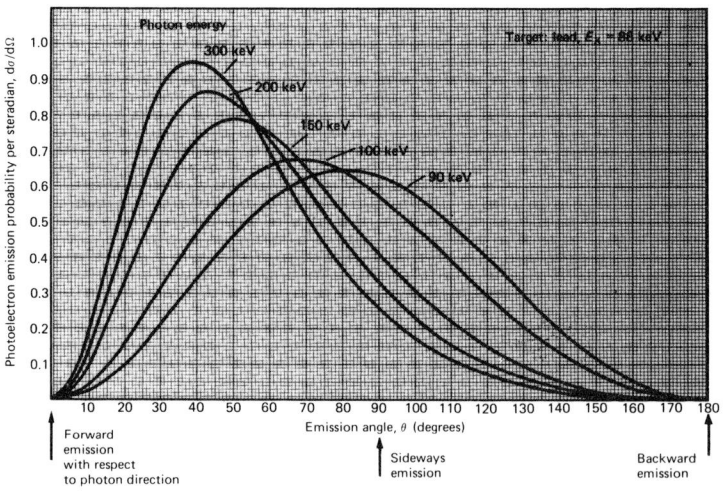

Figure 4.17 Angular distribution of photoelectrons

Mathematically, this distribution can be represented, approximately, by an expression of the form

$$\frac{d\sigma(\theta)}{d\Omega} = \frac{A \sin^2 \theta}{(1 + E_\gamma - 2\sqrt{E_e} \cos \theta)^4} \quad (4.38)$$

where $E_e = E_\gamma - E_B$ is the energy of the photoelectron and E_γ is the photon energy, both in mega-electronvolts. A is a normalisation factor.

Compton scattering

In this process, the incident photon collides with an electron in the outermost atomic shell. Because the binding energy of this electron is very small, the interaction can be regarded as an elastic scattering one and the kinematics of the reaction are similar to those observed for a collision between two particles, confirming the validity of the particle representation of a photon. The photon is scattered at an angle θ with respect to its original direction (*Figure 4.18*) and its energy is reduced from $h\nu$ to $h\nu'$. Thus its frequency is changed and its wavelength increased from λ to λ'. The electron is scattered at an angle ϕ and its energy is virtually $h\nu - h\nu'$.

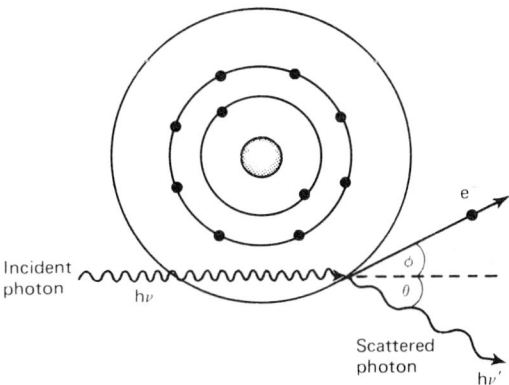

Figure 4.18 Compton scattering process

Following the interaction, the scattered photon may experience another interaction, which may be a photoelectric capture or a second Compton collision, or it may escape from the medium. The electron may produce further ion pairs via the mechanisms already described for charged particles, but can be assumed to be stopped in the medium, unless this is very thin. Thus, the usual situation is one in which the energy deposited in the medium is that given to the electron, unless the photon undergoes a second capture event.

There is a full theoretical treatment of Compton scattering, called the Klein–Nishina theory, but once again it is sufficient for present purposes to extract some proportional relationships. The cross-section, σ_c, per electron, is independent of the Z number of the absorber but is proportional to its density, ρ. It is inversely proportional to photon energy, $E_\gamma = h\nu$, i.e. for the same material, different photon energies,

$$\mu_c \propto E_\gamma^{-1}$$

and for the same photon energy, different materials,

$$\mu_c \propto \rho$$

Hence,

$$\mu_c \propto E_\gamma^{-1} \cdot \rho \tag{4.39}$$

The energy dependence is illustrated, for lead, in *Figure 4.16*. Because μ_c does not decrease, with increasing energy, as rapidly as μ_{PE} does, Compton scattering tends to be the major interaction mechanism at intermediate energies, while photoelectric capture dominates at low energies.

The energy, $h\nu - h\nu'$, or $E_\gamma - E_\gamma'$, that is deposited in the material in the interaction can be calculated by applying the conservation laws for energy and linear momentum. The simplest expression for the result of this calculation is for the increase in wavelength, $\Delta\lambda$, given by

$$\Delta\lambda = \lambda' - \lambda = \frac{h}{mc}(1 - \cos\theta) \tag{4.40}$$

m being the rest mass of an electron.

A more useful formula is obtained by converting from wavelengths to photon energies to get

$$E_\gamma' = \frac{E_\gamma}{1 + (E_\gamma/mc^2)(1 - \cos\theta)} \tag{4.41}$$

and, for the kinetic energy of the electron, E_e,

$$E_e = E_\gamma - \gamma_\gamma$$

$$= E_\gamma \frac{(E_\gamma/mc^2)(1 - \cos\theta)}{1 + (E_\gamma/mc^2)(1 - \cos\theta)} \tag{4.42}$$

These functions show that the minimum value of E_e is a zero when $\theta = 0°$ and the electron recoil angle, ϕ, is 90°. The maximum value of E_e is always less than the photon energy. It corresponds to a head-on collision in which the photon is scattered backwards, that is, through an angle, θ, of 180°. In this case, $\cos\theta = -1$ and equation 4.40 reduces to

$$E_\gamma' = \frac{E_\gamma}{1 + 2E_\gamma/mc^2} \tag{4.43}$$

This value of E_γ' is the smallest energy that a photon can have after a Compton collision and the smallest possible difference between the initial photon energy and the electron energy. If energies are expressed in mega-electronvolts, mc^2 becomes 0.511 MeV and equation 4.43 further simplifies to

$$E_\gamma - E_e = E_\gamma' = \frac{E_\gamma}{1 + 4E_\gamma} \tag{4.44}$$

for a head-on collision.

This equation is an important one in gamma ray detection and it is represented graphically in *Figure 4.19*. An interesting feature is that, as photon energy increases beyond about 1 MeV, the value of E_γ' tends to a constant 0.25 MeV, or 250 keV.

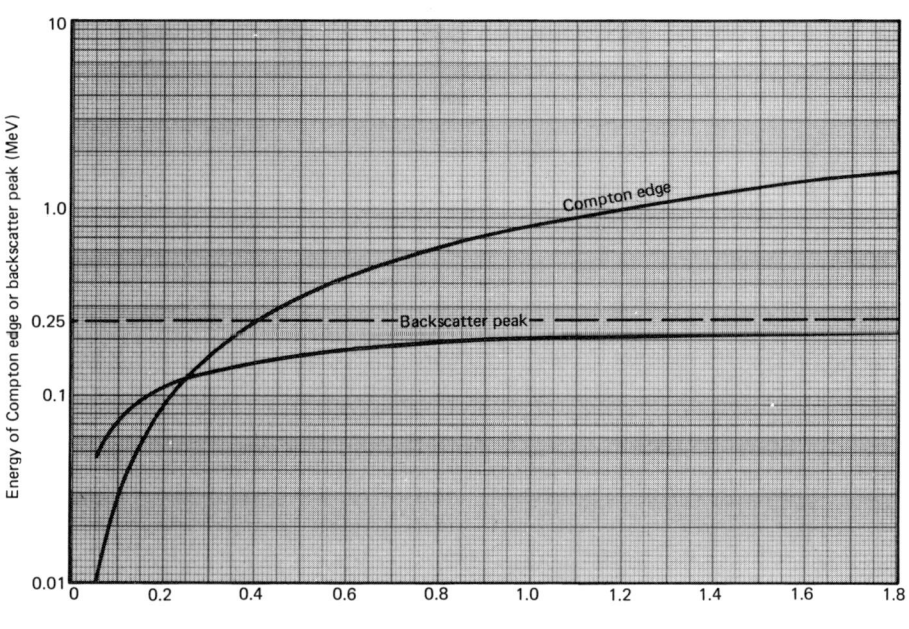

Figure 4.19 Compton-scattered photon (backscatter peak) and electron (Compton edge) energies as functions of incident photon energies

There are four *differential cross-sections* for Compton scattering: two for angular distributions and two for energy distributions. The former are described in *Figure 4.20*. The differential photon cross-section, $d\sigma(\theta)/d\Omega$, shows an equal amount of forward- and backscattering at low energies but the emphasis tends towards forward scattering as energy increases. The electron cross-section, $d\sigma(\phi)/d\Omega$, shows an equal preference for sideways ($\phi = 90°$) and forward scattering at low energies with forward scattering dominating at higher energies.

If the functions presented in *Figure 4.20*, giving the relative numbers of photons and electrons scattered at different angles, are multiplied by those of equations 4.41 and 4.42, giving the energy—angle relationships, the results are the differential cross-sections with respect to energy; that is, the relative probabilities that photons and electrons will end up with particular energy values. The cross-section for photons, $d\sigma(E_\gamma)/dE_\gamma$, is described in *Figure 4.21*. As expected, this shows a sharp cut-off at the low-energy end of each curve where the photon has its minimum energy, given by equation 4.44. Since this energy corresponds to a backscattered photon, the cut-off feature is known as the *backscatter peak*. Cross-sections for electrons, $d\sigma(E_e)/dE_e$, are illustrated in the same figure and these show sharp terminations at the high-energy ends of the curves corresponding to the maximum energies given to recoil electrons in head-on collisions in which the photon is backscattered and the electron forward-scattered. This feature is

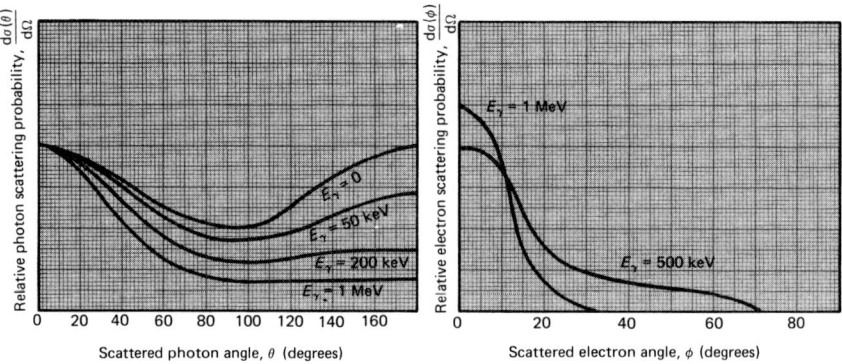

Figure 4.20 Differential Compton scattering cross-sections with respect to scattering angle. (a) Angular distribution of Compton-scattered photons; (b) angular distribution of Compton-emitted electrons

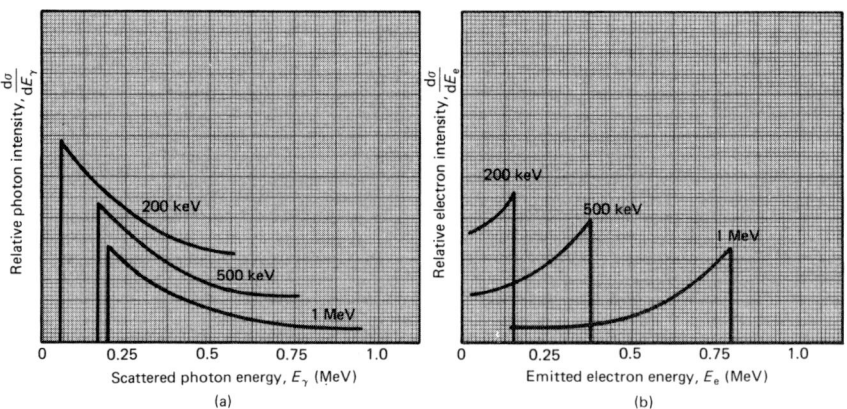

Figure 4.21 Differential Compton scattering cross-sections with respect to energy. (a) Energy distribution of Compton-scattered photons; (b) energy distribution of Compton-emitted electrons

known as the *Compton edge*. For a given absorber and photon energy the sum of the backscatter peak energy and the Compton edge energy must, of course, be equal to the energy of the original photons (equation 4.42) and both can be deduced from *Figure 4.19*.

The functions represented in *Figure 4.21* have been introduced as differential cross-sections but, when a large number of events are recorded, they denote relative intensity as a function of energy. In other words, they are *energy spectra*. If a radiation detector is set up to measure the energy deposited in it by Compton interactions, it will produce a graph of $d\sigma(E_e)/dE_e$, the electron energy distribution. If it is set up to measure the energies of photons scattered into it, then the result will be $d\sigma(E_\gamma)/dE_\gamma$. In practice, if Compton scattering is present, both distributions will be observed in a gamma ray energy spectrum.

Pair production

In this process, illustrated in *Figure 4.22*, an incident photon undergoes an electromagnetic interaction with an atomic nucleus and is converted into an electron–positron pair. It is a creation operation, exactly the opposite of the

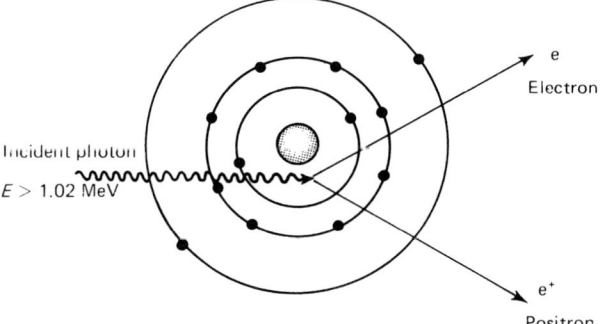

Figure 4.22 Pair production process

annihilation process, and mass is created from energy. Since the rest mass of each particle is 0.511 MeV, pair production can take place only if the photon energy exceeds a threshold value of 1.022 MeV. The positron annihilates with another electron, almost immediately, to form two 0.511-MeV photons. These may escape from the medium or interact in Compton or photoelectric processes.

Unlike the other two photon interactions, pair production has a cross-section which increases, although slowly, with photon energy, so this interaction tends to be the dominant one at high energies. This effect is shown in *Figure 4.16*. The proportional relationships for pair production are, for the same material but different photon energies,

$$\mu_{PP} \propto \ln E_\gamma$$

and for the same photon energy but different materials,

$$\mu_{PP} \propto \rho Z^2$$

Hence,

$$\mu_{PP} \propto \ln E_\gamma \cdot \rho Z^2 \tag{4.45}$$

4.5 NEUTRONS

Neutrons interact with *atomic nuclei* in both absorption and scattering processes. The cross-sections, per nucleus, σ_a and σ_s, for both types of interaction, are strongly dependent upon neutron energy to such an extent that it is standard practice to identify a number of *energy groups* and to treat each one separately. In radiation detection, two groups are usually sufficient and these are effectively regarded as separate types of particle. The low-energy group may be defined as containing *thermal neutrons*, n_{th}, with an energy of about 0.025 eV, or it may be defined as containing *slow neutrons* and taken to include the slightly higher energies of *epithermal neutrons* extending up to about 1 eV. Whichever definition is adopted, the remaining higher-energy groups are described as *fast neutrons*, n_f.

The reason for this dichotomy is that slow, or thermal neutrons are detected by means of absorption events and fast neutrons are detected by scattering processes, leading to two types of neutron detector. This is a simple approach, since both groups undergo both types of interaction, but it is a sufficient one for detection purposes. The absorption processes are often described as nuclear reactions although strictly, this term applies to all interactions with nuclei.

It is interesting to note that a similar attitude can be adopted towards photons. X-rays and low-energy photons tend to interact via the photoelectric capture process while higher-energy photons are affected mainly by the Compton scattering interaction. The major difference is that photon cross-sections vary smoothly with energy and Z number of the absorber whereas neutron cross-sections vary quite irregularly with energy and often dramatically from one element to another, or even from one isotope to another of the same element. This is especially true at low energies, where absorption predominates, and as a result there are in general no simple formulae or proportionalities which completely describe slow-neutron reactions.

Most neutron detectors are thin absorbers so that *differential attenuation*, as defined by equation 4.16, can be used, as a good approximation, to calculate interaction rates. There are two problems, however. First of all, as pointed out in Chapter 3, neutron sources are not generally monoenergetic. Cross-sections are available in data tables and in graphs (Appendix 4) but they are functions of energy, and the cross-section value to be used has to be obtained as the average value over the neutron energy spectrum, weighted by the shape of the spectrum. This can be a difficult calculation and, for this reason, mean values of thermal and fast cross-sections (σ_{th} and σ_f) are used in practice. It has to be accepted that this is not a very accurate approach and it will be observed that the values quoted for these cross-sections tend to vary slightly from one data source to another.

The second problem is that neutrons of all energies are easily scattered and, in general, there are almost as many neutrons going through a detector, or absorber, 'backwards' as there are going forwards. If the neutrons are truly monodirectional the beam intensity is defined as *neutron current*, I, or current density. This is the number of neutrons crossing unit area in one direction in unit time. The number of neutrons crossing unit area in any direction in unit time is known as the *neutron flux*, ϕ, or, more accurately, the neutron flux density. Neutron flux is generally the more useful concept since it includes all neutrons which may interact in the medium and reduces to current for a monodirectional beam. Consequently, equation 4.11 can be stated, in terms of flux, as

$$dI = \phi_0 n\sigma \, dx = \phi_0 \mu \, dx \qquad (4.46)$$

where dI is the interaction rate in the differential absorber. μ is the absorption coefficient but in neutron physics it may be described as the *macroscopic cross-section* and written as Σ.

In thicker materials, in which neutrons make only one interaction before being removed from the incident beam, a monodirectional and monoenergetic neutron current exhibits *exponential attenuation* as given by equation 4.16, $I = I_0 e^{-\mu x}$. Absorption removes thermal neutrons and scattering removes fast neutrons. The attenuation coefficient is given, for each case, in terms of cross-section per nucleus, as

$$\mu_a = n\sigma_a = \frac{N_A \rho}{A} \sigma_a$$

$$\mu_s = n\sigma_s = \frac{N_A \rho}{A} \sigma_s$$

(4.47)

A more general case occurs for thick materials, whose smallest dimension exceeds about 10 cm, when the neutrons are neither monodirectional nor monoenergetic, and when broad beam geometry applies so that scattering does not remove neutrons. This can be treated by representing the neutron energy as a coarse histogram of contiguous energy groups. In radiation detection two groups are used but, for bulk shielding or absorption calculations, as many as seven groups may be defined, each of which behaves independently. Scattering removes neutrons from one group and adds these to the lower group. The top group is attenuated exponentially. All other groups gain neutrons and lose some so their attenuation is described by two exponential functions, of the form (for infinite slab geometry)

$$\phi_2 = \phi_1 \exp(\mu_1 x) + \phi_2 \exp(-\mu_2 x)$$

(4.48)

where ϕ_1 and ϕ_2 are the fluxes in groups 1 and 2 and μ_1 and μ_2 are known as the *removal coefficients* for each group.

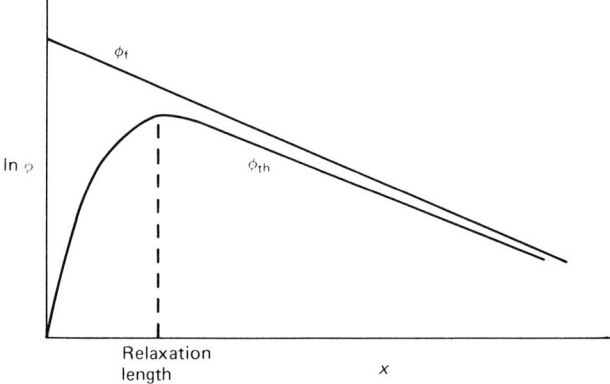

Figure 4.23 Neutron attenuation for two energy groups

The point to note is that the flux of a low-energy group can increase with distance, x, into the attenuating medium if it receives more neutrons than it loses. This is illustrated, for two groups, in *Figure 4.23*, showing the absorption of an initially fast neutron beam. Thermal flux rises to a maximum in a short distance, called the *relaxation length*, then decreases exponentially in parallel with the fast flux. In water, the relaxation length is about 8 cm for a chemical neutron source.

Thermal neutrons

Thermal neutrons diffuse through matter like gas molecules, making a large number of elastic scattering collisions with atomic nuclei. Electrons are too small to make much impression. Since, on average, they gain as much energy as they lose in these collisions, elastic scattering does not absorb energy. However, they do spend a lot of time in close proximity to nuclei so there is a relatively large probability that they will be captured in absorption interactions and produce nuclear reactions. If a reaction results in the immediate emission of charged particles (photons are less frequent and less convenient because they may travel some distance from the capture location before being stopped), it provides a detection mechanism for thermal neutrons in terms of ion pairs produced by these particles. The interaction probability, σ_a, depends on the neutron energy and the nature of the target nuclide.

(1) *Energy dependence.* At or near the thermal energy, reaction cross-section is proportional to the time spent by a neutron in passing a nucleus. Thus, it is inversely proportional to the neutron velocity and also to the square root of its energy, i.e.

$$\sigma_a \propto \frac{1}{v} \quad \text{i.e.} \propto \frac{1}{\sqrt{E}} \tag{4.49}$$

This feature is common to all absorbers and is illustrated in the graphs of *Figure 4.24*. In addition, some absorbers, such as indium and cadmium, show *resonance capture* effects which are superimposed on the σ_a–E graph as sharp peaks, usually in the epithermal region. Neutrons with these energies exactly match the nuclear shell levels and are more easily captured. It is a resonance effect similar to the line absorption spectra obtained for the interactions of photons with atoms.

(2) *Absorber dependence.* Because neutron absorption is a nuclear effect it is extremely sensitive to nuclear structure. The absolute values of the cross-sections shown in *Figure 4.24* depend on the nuclide involved, as do the resonance energies. Cross-sections are particularly large for nuclides with one neutron less than a closed shell, or sub-shell configuration; that is, one less than a magic or semi-magic number. They are particularly low, on the other hand, for nuclides with closed neutron shell configurations. Resonances occur in the capture cross-sections when the neutron energies are such that when captured, they fit exactly into an energy state of the compound nucleus.

An important result of this dependence on nuclear structure is the fact that σ_a varies considerably from one isotope to another of the same element. The value of σ_a for natural boron, for example, is about 700 barns but for boron-10, which is 20% abundant in natural boron, it is about 3800 b. One consequence of this feature is that *neutron radiography* can pick out different isotopes of the same element, or elements with adjacent Z numbers, whereas X-radiography shows a gradual change in absorption with Z number. The effect is described by the absorption coefficients shown in *Figure 4.25*. Some data are also specified, for

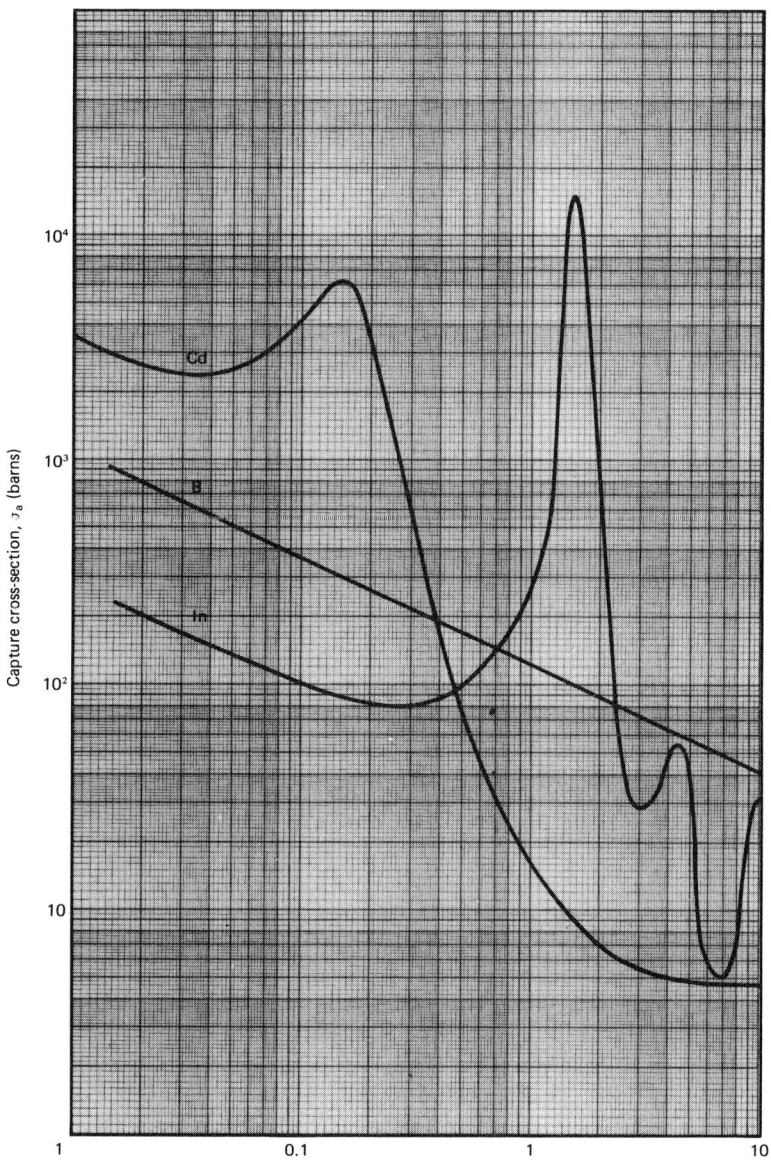

Figure 4.24 Thermal-neutron capture cross-sections as functions of energy

particular isotopes, in *Table 4.1*. These are the isotopes most widely used in radiation detectors. The highest thermal cross-sections are observed for ^{157}Gd, ^{155}Gd and ^{149}Sm. These produce gamma ray photons but have very high internal-conversion coefficients, so the electrons emitted can be used to detect the neutrons. The only problem with these nuclides is their high cost. Boron, lithium and cadmium are the cheapest and most widely used materials.

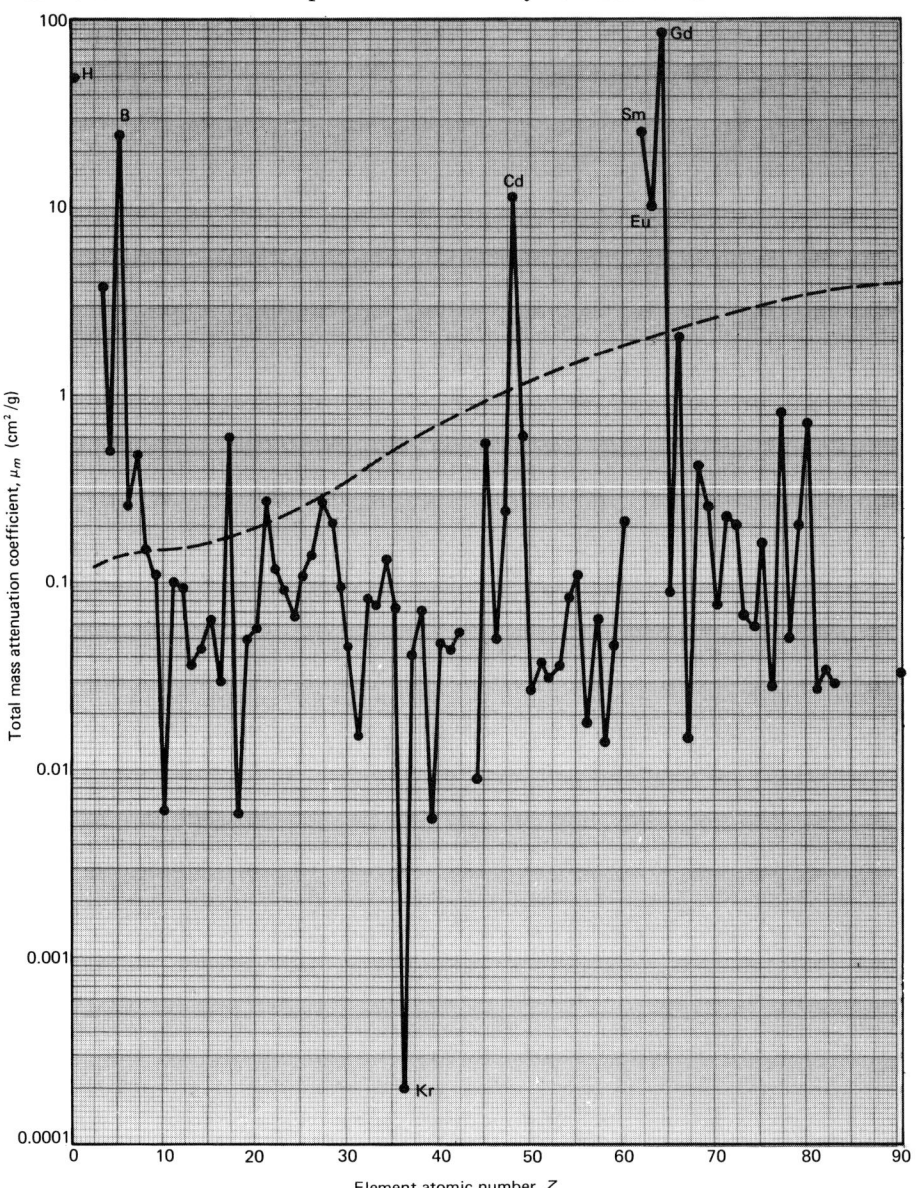

Figure 4.25 Total mass attenuation coefficient for thermal neutrons (continuous graph) and for 1-Å X-rays (dashed graph) as a function of absorber atomic number. From Thewlis, J., Brit. J. Appl. Phys., 7, 345 (1956)

Table 4.1 Thermal-neutron capture cross-sections for isotopes yielding immediate decay products

Target isotope	Natural abundance (%)	Reaction	Cross-section (barns)
^{6}Li	7.52	^{6}Li$(n, \alpha)^3$H	910
^{10}B	18.8	^{10}B$(n, \alpha)^7$Li	3 800
^{113}Cd	12.26	^{113}Cd$(n, \gamma)^{114}$Cd	20 000
^{149}Sm	13.8	^{149}Sm$(n, \gamma)^{150}$Sm	40 800
^{155}Gd	14.73	^{155}Gd$(n, \gamma)^{156}$Gd	61 000
^{157}Gd	15.68	^{157}Gd$(n, \gamma)^{158}$Gd	240 000
^{235}U	0.72	Fission	683

Fast neutrons

Fast neutrons do generate some absorptive nuclear reactions but much less frequently than thermal neutrons. They also undergo inelastic scattering interactions, in which the nucleus is left in an excited state and de-excites isomerically, emitting gamma radiation, but this, too, is a relatively low-probability event. The most important interaction is that of *elastic scattering*, in which neutron and nucleus collide like billiard balls. The nucleus recoils, usually with sufficient energy to produce ion pairs which can be used to detect the neutron. The neutron is deflected with reduced energy.

Apart from the low-probability, fast nuclear reactions, the only way to stop fast neutrons is to slow them down to thermal energies, by means of a succession of scattering interactions, then absorb them. The slowing down process is called *thermalisation* or *moderation* and the material in which it takes place is called a *moderator*. In thermal nuclear reactors, neutrons have to be moderated in order to produce further fission processes and moderators are selected for their ability to moderate without capturing neutrons. Heavy water (D_2O) and graphite (C) are used for this purpose.

The moderating properties of a medium depend on its atomic mass number, A. In any elastic scattering collision, energy transfer is determined by the relative masses of the particles involved. If the mass of a target nucleus is much greater than that of an incident neutron, it will not recoil very much and not accept much energy. If, however, these masses are equal, all of the neutron energy may be transferred, in a head-on collision. These results are due to the need to conserve energy and linear momentum. Analogous observations may be made by throwing tennis balls at (a) bowling balls and (b) other tennis balls. In all cases, particles lose energy and are moderated more effectively in particles of similar size than in a medium of heavier particles. For neutrons, this means that low-A materials, such as hydrogen, are much better at stopping fast neutrons than high-A elements, such as lead. This is exactly the opposite to the situation that applies to photons, for which high-Z elements are more absorptive than low-Z ones. It has the somewhat unexpected result that wood is a better shield against fast neutrons than is lead, which is virtually transparent to fast neutrons and also, by the way, has a very low absorption cross-section for thermal neutrons.

Mathematically, for an incident neutron energy E_n and a target nuclear mass m, the energy, E_n', of a neutron scattered at an angle θ is given (in the normal laboratory system of coordinates) by

$$\frac{E_n'}{E_n} = \frac{m^2 + m_n^2 + 2mm_n \cos\theta}{(m + m_n)^2} \qquad (4.50)$$

To make comparisons between different materials, it is convenient to measure energy loss for head-on collisions in which $\theta = 180°$ and energy loss is a maximum. Replacing nuclear mass, m, by the atomic mass number, A, which is good enough for comparisons, gives

$$\frac{E_n'}{E_n} = \left(\frac{A-1}{A+1}\right)^2 \qquad (4.51)$$

For example:

    In hydrogen $(A = 1)$, $E_n'/E_n = 0$ (total loss)

    In helium $(A = 4)$, $E_n'/E_n = 0.36$

    In carbon $(A = 12)$, $E_n'/E_n = 0.71$

    In uranium $(A = 238)$, $E_n'/E_n = 1.0$ (virtually no loss)

Fast-neutron detectors are made of materials of low A number, as are neutron shields, and usually backed by a thermal-neutron absorber, such as ^{10}B or ^{6}Li, to remove thermalised neutrons.

Activation

Some neutron capture reactions form compound nuclei which do not decay immediately. This process is known as *neutron activation* and the compound nuclei become radioactive nuclides. Decay times vary from almost zero to many years, depending on the relative stability of the nucleon configuration in the compound nucleus. Typical processes are

$$\sigma_{act} = 0.18\text{ b}: {}_0^1n + {}^{27}Al \longrightarrow {}^{28}Al \xrightarrow{\beta^-} {}^{28}Si: T_{1/2} = 2.3 \text{ min}$$

$$\sigma_{act} = 22.5\text{ b}: {}_0^1n + {}^{59}Co \longrightarrow {}^{60}Co \xrightarrow{\beta^-,\gamma} {}^{60}Ni: T_{1/2} = 5.3 \text{ years}$$

Data for other reactions are provided in *Table 4.2*.

The activity produced by irradiation in a flux ϕ for a time t is given by equation 3.26, as

$$A = N\phi\sigma_{act}\left[1 - \exp(-\lambda t)\right] \qquad (4.52)$$

Table 4.2 Thermal-neutron activation data for some nuclides

Target isotopes	Abundance (%)	Product isotope	$T_{1/2}$	Major radiations emitted (energy in MeV)	σ_{act} (barns)
^{27}Al	100	^{28}Al	2.3 min	β^- (2.85), γ (1.78)	0.18
^{31}P	100	^{32}P	14.3 d	β^- (1.71)	0.2
^{40}Ar	99.6	^{41}Ar	1.8 h	β^- (1.2), γ (1.3)	0.61
^{55}Mn	100	^{56}Mn	2.6 h	β^- (2.85), γ (0.85, 1.8, 2.1)	12.8
^{59}Co	100	^{60}Co	5.3 years	β^- (0.31), γ (1.17, 1.33)	22.5
			10.5 min	β^- (1.55), e^- (0.05), X (0.007)	0.73
^{63}Cu	69.1	^{64}Cu	12.9 h	β^- (0.57), β^+ (0.66), e^- (1.33), γ (0.51), X (0.008)	4.3
^{65}Cu	30.9	^{66}Cu	5.1 min	β^- (2.63), γ (1.04)	2.1
^{75}As	100	^{76}As	12.5 h	β^- (2.97), γ (0.56, 0.6, 1.2)	4.3
^{107}Ag	51.8	^{108}Ag	2.4 min	β^- (1.64), β^+ (0.9), X (0.021)	30
^{109}Ag	48.2	^{110}Ag	24 s	β^- (2.87), γ (0.66)	96
		^{110m}Ag	255 d	β^- (0.53, 0.09), e^- (0.09, 0.1), γ (0.66 to 1.4)	2.3
^{113}In	4.28	^{114m}In	50 d	e^- (0.164, 0.19), γ (0.19, 0.6, 0.7), X (0.024)	61
		^{114}In	72 s	β^- (1.99), β^+ (0.42), X (0.023)	2
^{115}In	95.7	^{116m}In	54 min	β^- (1.0), γ (0.42, 0.82, 1.1, 1.3, 1.5, 2.1)	145
		^{116}In	14 s	β^- (3.3), γ (1.3)	52
^{127}I	100	^{128}I	25 m	β^- (2.1), γ (0.44), X (0.028)	6.8
^{197}Au	100	^{198}Au	2.7 d	β^- (0.96), e^- (0.33, 0.4), γ (0.41)	96
^{208}Pb	52.3	^{209}Pb	3.3 h	β^- (0.635)	0.0005

where λ is the decay constant of the nuclide formed and the number of active nuclei produced is A/λ. N is the number of nuclei irradiated.

If a specimen is activated for a time t_1, then allowed to decay for a time t_2 (for example, during transport from the activation area), then the net activity is given by

$$A = N\phi\sigma_{act} [1 - \exp(-\lambda t_1)] \exp(-\lambda t_2) \tag{4.53}$$

There are two reasons for activating stable nuclides. One is to measure the neutron flux, ϕ, and the other is to measure the number N of stable nuclei present in the material. Both procedures can involve thermal- or fast-neutron flux, ϕ_{th} or ϕ_f, and σ_{act} must be used appropriately. Thermal activation is the more common since it generally has the larger cross-section and produces greater activity.

The first of these techniques employs foils of materials such as gold, manganese, indium and cadmium. Measurement of induced activity and the use of equation 4.53, with measured values of N and σ, give the value of neutron flux, ϕ. If foils whose materials have resonance capture peaks are used, the measured flux is for the specific resonance energy. Several foils can be used, with resonances at different energies to measure neutron energy spectra.

The second technique is *neutron activation analysis*. In this case A is measured, while ϕ and σ_{act} are known, so that N is obtained. This is the most accurate method of measuring the elemental composition of most materials. One important exception is lead, whose capture cross-section is very small. Neutron transfer radiography is similar to activation analysis except that the positions of activated nuclei are imaged instead of their energies being measured.

BIBLIOGRAPHY

GRODSTEIN, G. W., *National Bureau of Standards Circular 583* (1957)
KATZ, L. and PENFOLD, A. S., 'Range–Energy Relations for Electrons and the Determination of Beta-Ray End-Point Energies by Absorption', *Rev. Mod. Phys.*, **24**, 28 (1952)
MARION, J. B. and YOUNG, F. C., *Nuclear Reaction Analysis,* North-Holland, Amsterdam (1968)
SEGRE, E., *Nuclei and Particles,* W. A. Benjamin, Menlo Park, California (1965, 1977)
SIEGBAHN, K. (Ed.), *Alpha-, Beta-, and Gamma-Ray Spectrometry,* North-Holland, Amsterdam (1965)
THEISEN, R. and VOLLATH, D., *Tables of X-Ray Mass Attenuation Coefficients,* Verlag Stahleisen, Düsseldorf (1967)

Chapter 5
Detector properties

5.1 GENERAL FEATURES

A complete radiation detection system is made up of a *radiation detector* and a *signal processing system*. In general, the radiation detector consists of the following components (*Figure 5.1*):

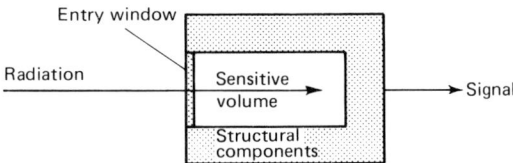

Figure 5.1 Basic structure of a radiation detector

(1) *A sensitive volume* in which the informative interactions with incident radiations take place.
(2) *Structural components* including a physical structure to maintain the sensitive volume and electrical supplies, if required. Normally, the structural components completely enclose the sensitive volume so an essential feature is an *entry window* which is transparent to the radiations being detected.
(3) *A data output mechanism* or readout facility, which is the means of extracting information from the sensitive volume and presenting it to the signal processing system.

The signal processing system converts the detector output into directly useful information. It may be nuclear electronics, a dark room or a human brain but, whichever form it takes, it does affect the properties and behaviour of the detection system.

Common examples of radiation detectors are the eye and a camera, used to detect visible light, and a television screen, used to detect electrons. All of these have sensitive volumes (retina, film and fluorescent screen), structural components and data output mechanisms. All three have signal processing systems to convert output signals into information (optic nerves and brain, dark room, eyes).

Basically, there are two ways of approaching a study of radiation detection. The traditional method of introducing the subject, especially to physics students, is to classify material in terms of *detector type*, that is, in terms of the structure and theory of operation of specific devices. On the other hand, those who use

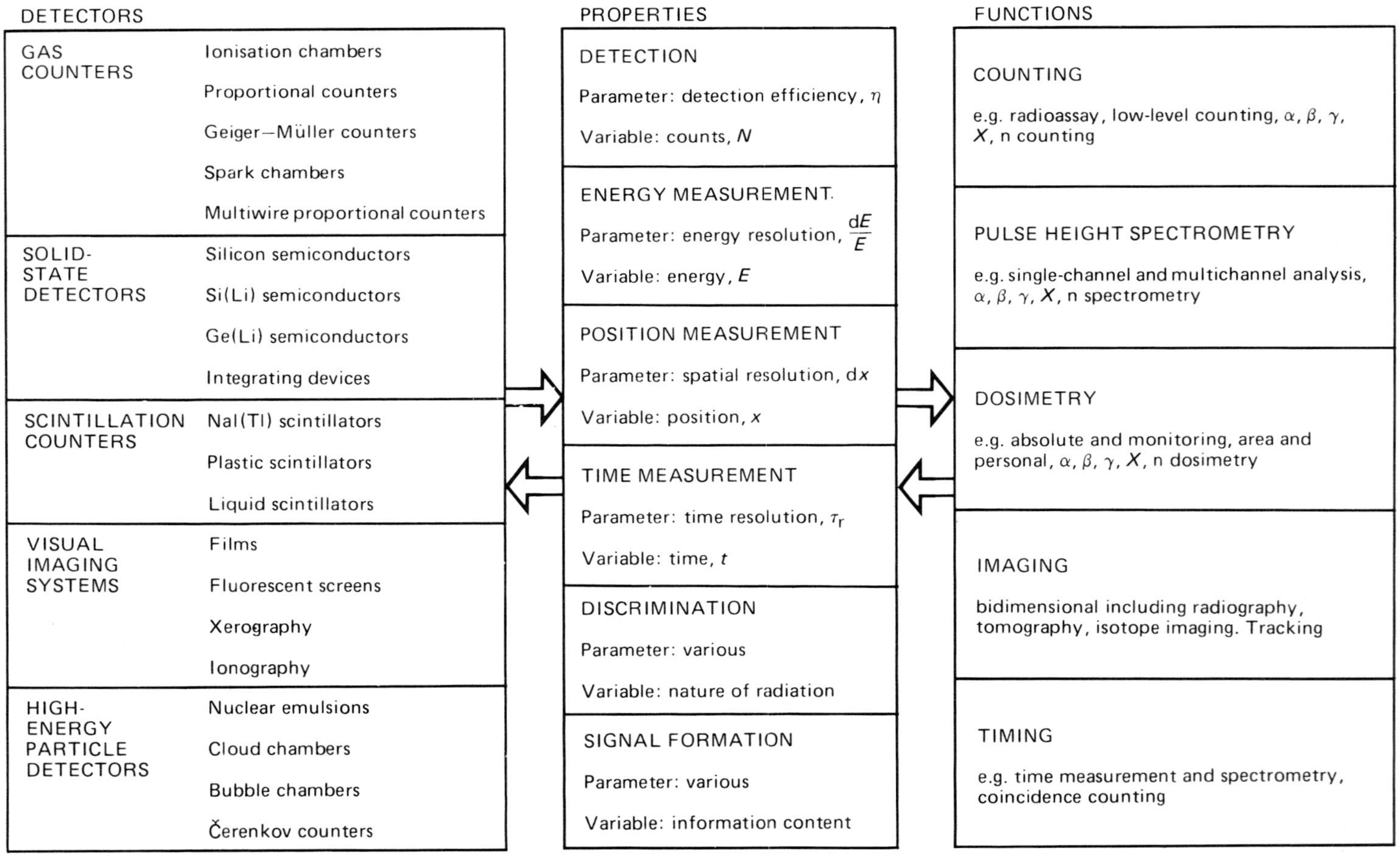

DETECTORS		PROPERTIES	FUNCTIONS	
GAS COUNTERS	Ionisation chambers Proportional counters Geiger–Müller counters Spark chambers Multiwire proportional counters	**DETECTION** Parameter: detection efficiency, η Variable: counts, N	**COUNTING**	e.g. radioassay, low-level counting, α, β, γ, X, n counting
SOLID-STATE DETECTORS	Silicon semiconductors Si(Li) semiconductors Ge(Li) semiconductors Integrating devices	**ENERGY MEASUREMENT** Parameter: energy resolution, $\frac{dE}{E}$ Variable: energy, E	**PULSE HEIGHT SPECTROMETRY**	e.g. single-channel and multichannel analysis, α, β, γ, X, n spectrometry
SCINTILLATION COUNTERS	NaI(Tl) scintillators Plastic scintillators Liquid scintillators	**POSITION MEASUREMENT** Parameter: spatial resolution, dx Variable: position, x	**DOSIMETRY**	e.g. absolute and monitoring, area and personal, α, β, γ, X, n dosimetry
VISUAL IMAGING SYSTEMS	Films Fluorescent screens Xerography Ionography	**TIME MEASUREMENT** Parameter: time resolution, τ_r Variable: time, t	**IMAGING**	bidimensional including radiography, tomography, isotope imaging. Tracking
HIGH-ENERGY PARTICLE DETECTORS	Nuclear emulsions Cloud chambers Bubble chambers Čerenkov counters	**DISCRIMINATION** Parameter: various Variable: nature of radiation **SIGNAL FORMATION** Parameter: various Variable: information content	**TIMING**	e.g. time measurement and spectrometry, coincidence counting

Figure 5.2 Types of detector, fundamental properties and some typical functions

detection systems tend to identify specific *functions*. They set out to perform some operation, such as gamma ray spectrometry or isotope imaging, rather than to use a particular type of detector. Essentially, the first approach is that of the designer and the second is that of the user of detection systems, but both are necessary ingredients of a full treatment of the subject. Designers need design criteria and users need to know what to buy.

These two aspects of radiation detection are related to each other through the *fundamental properties* of detection systems. Detector type determines the characteristic properties and these, in turn, determine the functions that can be performed. To reverse the logic, functions define required properties and these in turn define the detectors that can be used. *Figure 5.2* lists types of detectors, properties and functions. The present approach will be to discuss the properties first, using a 'black box' treatment, before going on to describe functions and detector types. This provides a rational basis for the discussion of these last two aspects.

It is possible to define six basic properties, as listed in *Figure 5.2*. Each has an *associated parameter*, although not always a quantitative one, and an *associated variable*. The former provides a comparative measure of the property and the latter defines its area of application. In this chapter, each property will be dealt with in turn.

5.2 DETECTION

This is the most elementary property of a detection system. It refers to the ability of the system to detect ionising radiations and implies, also, the ability to count them. The associated parameter, which measures the ability of the system to detect and record events, is the *detection efficiency*, η. The associated variable is the number of *recorded counts*, N. Usually the objective is to measure the strength, S, of a particular source, or the intensity, I, of incident radiations. Alternatively, it may be to maximise the number of counts obtained from a given source, in which case the detection efficiency may be referred to as the *sensitivity* of the system.

Detection efficiency, η, is the fraction of all the radiations emitted from a source which produce some *recorded* interaction in the sensitive volume of the detector. Generally, it is a function of radiation type and energy as well as of the detection system. This definition requires that interactions must be recorded and provides a more useful parameter than the alternative, which does not take into account the fact that some interactions in the sensitive volume may not be observed. Using the terminology defined above, η is given by

$$N = \eta S \tag{5.1}$$

but S must be converted from curies to the number of radiations, of the type and energy being detected, which are emitted by the source during the measurement time. This conversion is effected by means of the following equation, derived from the definition of the curie:

$$S = 3.7 \times 10^{10} \cdot r \cdot A \tag{5.2}$$

where A is the source activity in curies and r is the relative abundance of the radiation involved; that is, the average number of particles or photons per disintegration. Values of r have to be obtained from decay schemes as described in Chapter 3. They are given in Appendix 3.

The variable involved in the measurement of source strength is N, the recorded number of *counts*, or, if counting is over unit time, the recorded *count rate*. Each count records the detection of one particle or photon. As in any measurement process, there are two separate sources of experimental error, or uncertainty. These relate to:

(1) *absolute measurement*, involving calibration or calculation and, in this case, determined by the accuracy with which *detection efficiency* is known; and
(2) *relative measurement*, involving resolution or observation and, in this case, determined by uncertainty in the measurement of *count rate, N*.

For example, if the length of an object is measured with a ruler, there is an absolute measurement error, related to the accuracy with which the ruler is calibrated, and a relative error, in the ability of the observer to resolve between adjacent marks on the ruler and decide which one corresponds to the length of the object.

Source strength is given by equation 5.1, which may be rewritten as

$$S = N/\eta \tag{5.3}$$

Hence the total error, dS, in the determination of S is obtained by adding contributions from the measurement of N and η, namely, dN and $d\eta$. In counting measurements, most of the errors are *statistical*, so these uncertainties must be added in quadrature, i.e.

$$(dS)^2 = (dN)^2 + (d\eta)^2 \tag{5.4}$$

Calculation of detection efficiency

It is extremely difficult to calculate accurately the detection efficiency of a system. On the other hand, an approximate calculation is quite simple and it is informative, since it illustrates the effects of various factors and develops the ability to understand and control an experimental arrangement. As defined above, detection efficiency is made up of four factors:

(1) *Geometrical attenuation factor, G*: the fraction of all the emitted radiations which are emitted in the direction of the sensitive volume of the detector.
(2) *Material attenuation factor, M*: the fraction of those radiations emitted in the direction of the sensitive volume which actually reach it.
(3) *Interaction efficiency, R*: the reaction or interaction probability, which is the fraction of radiations arriving in the sensitive volume that react with it.
(4) *Data recording efficiency, D*: the fraction of radiations interacting with the sensitive volume which produce recorded events.

In the most general case, the source and detector are of undefined shape. The factors G, M, R and D are not independent of each other and the efficiency, η, is

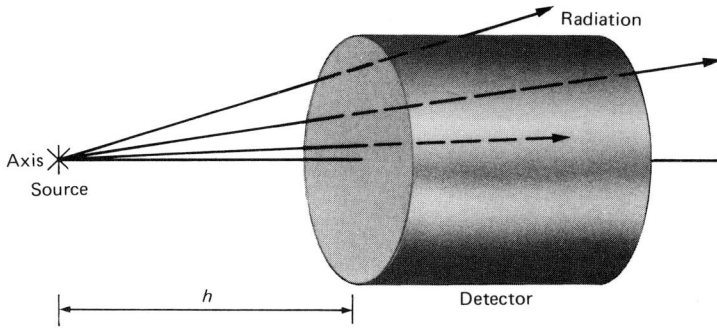

Figure 5.3 Detection geometry for a point source on the axis of a cylindrical detector

not simply the product of these four factors. The situation can be illustrated by what is, in fact, a relatively simple arrangement, in which a point source is located on the axis of, and at a distance h from, a cylindrical detector. As shown in *Figure 5.3*, the axial radiations pass through a greater detector thickness and have a higher interaction probability, R, than those which intercept the corners of the detector. R is a function of geometry.

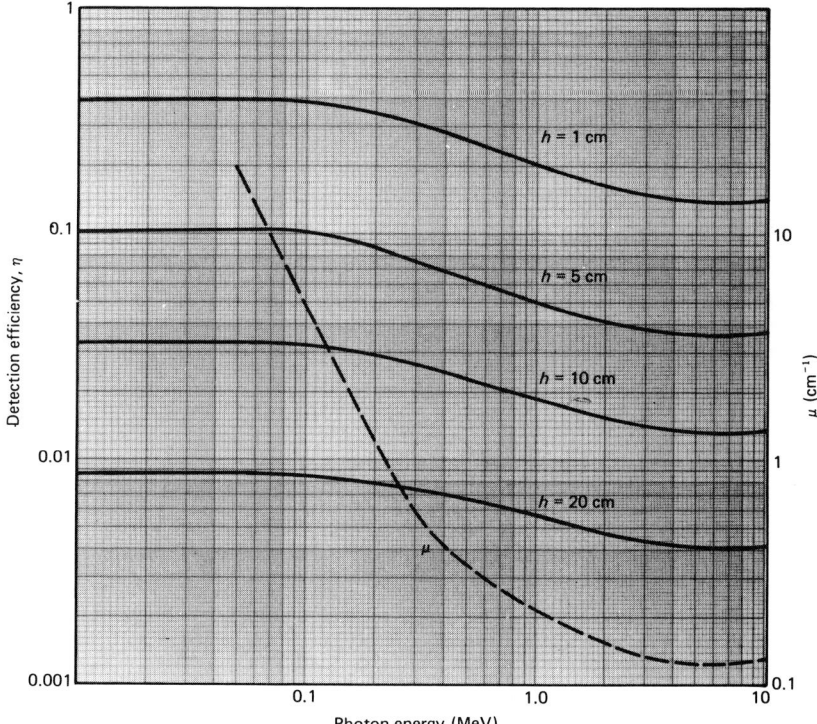

Figure 5.4 Detection efficiencies (geometrical factor, G, multiplied by interaction efficiency, R) for a 3 in × 3 in NaI(Tl) detector for a point axial source

This is a typical source–detector arrangement and graphs and tables are available showing detection efficiency as a function of h. These are the results of measurements or numerical integrations. Data for sodium iodide scintillation counters are presented in *Figure 5.4*. These depict the values of the product GR as a function of incident photon energy. It will be observed that above a certain energy these curves tend to follow the variation of attenuation coefficient, μ, with energy.

Approximate calculations can be completed for a number of situations on the assumption that

$$\eta = GMRD \qquad (5.5)$$

The simplest arrangement is that of a point source on the axis of a uniformly thick detector (including a cylindrical one) at a considerable distance from it. Thus h is assumed to be much larger than the thickness, x, of the detector's

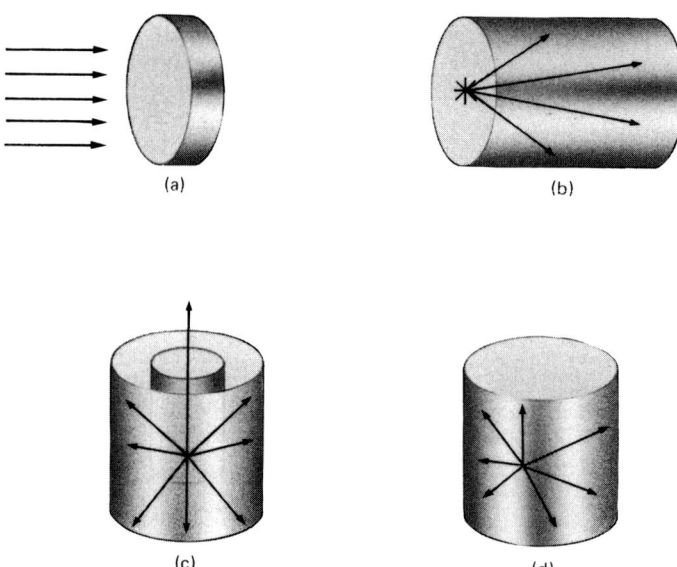

Figure 5.5 Special detector geometries. (a) Parallel-beam geometry; (b) 2π geometry; (c) almost 4π geometry; (d) 4π geometry

sensitive volume, and the radiations form a parallel beam so that all of them traverse the same path length, x, in the detector. The arrangement is illustrated in *Figure 5.5* and the four factors are determined as follows:

The *geometrical attenuation factor*, G, is given by the usual inverse-square law. The fraction of radiations intercepted is the area A of the detector face divided by the area $4\pi h^2$ of a sphere of radius h about the source point. Thus,

$$G = \frac{A}{4\pi h^2} \qquad (5.6)$$

The *material attenuation factor*, M, depends on the type of radiation involved and its interaction with any intervening matter, as discussed in Chapter 4. Normally, attenuation takes place in the source material and its encapsulation, if any, in the air, and in the detector entry window. There may be some build-up factor due to scattering into the detector of radiations that would otherwise have missed it.

The *interaction probability*, R, is also determined from data and formulae relating to the interaction of radiations with matter. For photons, for example, the number entering the detector is I_0 and the number leaving the other side is $I \, (= I_0 e^{-\mu x})$, so the value of R is given by

$$R = \frac{I_0 - I_0 e^{-\mu x}}{I_0}$$

$$= 1 - e^{-\mu x} \qquad (5.7)$$

For very thin detectors, with a small attenuation coefficient, the approximation $R = \mu x$ is usually good enough. The same is true of neutrons. For charged particles, R is close to unity unless the detector thickness is of the order of the mean free path of the particles, in which case some will escape without interacting.

The *data recording efficiency*, D, is made up of a number of factors of which the most important are due to *dead-time losses, coincidence counting losses* and incomplete capture, defined by *peak-to-total ratio*.

Dead-time losses derive from the fact that, following the detection of a particle or photon, and while the event is being registered, the detector and/or the associated electronics may be insensitive for a short time. This is called the *dead time* or the *paralysis time*. Other interactions occurring within this time are not recorded. Thus, during a *real time* of 1 second, during which N counts are recorded, the detector has a *live time* of only $1 - N\tau$ seconds, where τ is the dead time of the detection system. If the *interaction rate* is n, given by

$$n = GMRS \qquad (5.8)$$

then the recorded count rate is given by

$$n = \frac{N}{\text{live time}} = \frac{N}{1 - N\tau} \qquad (5.9)$$

and the factor D for the effect of dead-time losses only is given by

$$D = \frac{N}{n} = 1 - N\tau \qquad (5.10)$$

Thus, for a recorded count rate, N, the source strength, S, is given by

$$S = N/GMRD$$

$$= N/GMR(1 - N\tau) \qquad (5.11)$$

If, however, the source strength is known, then the recorded count rate, N, is given by

$$N = GMRDS$$

$$= GMRS(1 - N\tau)$$

i.e.

$$N(1 + GMRS\tau) = GMRS$$

$$\therefore N = \frac{GMRS}{1 + GMRS\tau} \qquad (5.12)$$

In most cases, dead-time losses constitute the major contribution to the D factor but the dead time, τ, is very small. For a Geiger–Müller counter dead time of about 10^{-4} s, dead-time losses become significant only for count rates approaching 10^3 or 10^4 counts per second. There is, of course, a saturation effect. If, in this example, the count rate exceeds 10^4 per second, the formula does not apply. The detector merely counts once after every dead time and records a count rate of 10^4 per second, whatever the interaction rate.

Coincidence counting losses occur when two or more particles or photons interact almost simultaneously with the detector; that is, within a finite *resolving time* τ_r that is a characteristic parameter of the system. This is a *random coincidence* effect that becomes significant, like dead-time losses, at high count rates. Normally, τ_r is much smaller than τ so these losses are included in the dead-time losses, but they lead to a double-sized event and, if the system does not record these, then both counts are lost.

Random coincidence is observed when one interaction takes place within a time interval $\pm\tau_r$ of another interaction. Hence, there is a total 'sensitive' time of $N \cdot 2\tau_r$ per second. This is the fraction of live time during which coincidences can occur and it is the fraction of the recorded counts, N, that is lost. Thus,

$$\text{Coincidence count rate} = 2N^2\tau_r$$

and

$$\text{Net recorded count rate} = N - 2N^2\tau_r \qquad (5.13)$$

Peak-to-total ratio measures the fraction of recorded events which involve complete, as opposed to partial, absorption of radiations. In gamma ray spectrometry, it is often convenient to record only complete capture events (due to photoelectric capture) and ignore others. In this case the effective detection efficiency, η', is less than η and is given by

$$\eta' = P\eta \qquad (5.14)$$

where P is the peak-to-total ratio.

There are three other source–detector arrangements which occur quite frequently and are amenable to approximate calculation. These are described in *Figure 5.5*.

In the first of these, the point source is touching the detector surface, i.e. $h = 0$. The detector subtends a solid angle of 2π steradians at the source and intercepts half of the emitted radiations. This is known as *2π geometry*. The geometrical factor, G, is equal to 0.5 and is constant, while $R = 1.0$ for all radiations for which the radius, r, of the detector is a stopping thickness, since this is the minimum path length.

The second example shows a *well-shaped detector* containing the source. In this case, the detector intercepts almost all of the emitted radiations so it is almost a 4π geometry condition, with $G \simeq 1.0$. This arrangement is used to maximise detection efficiency, especially for unsealed sources, which must be contained in a test tube.

The third example is of a true *4π geometry* in which the source is actually inside the sensitive volume. In this case, $G = 1$ and, since there is no intervening matter, $M = 1$. It is used for weakly penetrating radiations, such as the 18-keV beta particles from tritium, which would not pass through an entry window. In such instances, the interaction probability is also unity, i.e. $R = 1$.

Measurement of detection efficiency

In practice, the only accurate way of determining detection efficiency is to measure it. Basically, there are two methods, as follows:

Calibration against a standard source

If a source of known strength S_0 produces a count rate N_0, then the detection efficiency, η, is equal to N_0/S_0, and this value of η applies to all other situations in which the factors G, M, R and D remain unchanged. For this to be the case means that the experimental arrangement, including source dimensions, source detector geometry and any backscattering materials, must be unaltered. It also requires source emissions of the same type and energy, and, ideally, from the same source material, to maintain constant values of G, M and R; and similar source strengths, to avoid altering the value of D.

These are extremely restrictive conditions. They imply that it is impossible to measure an unknown source strength, S, by calibration against a known source, S_0, unless the value of S is the same as that of S_0, which would be, to say the least, quite fortunate. In practice, of course, an exact correspondence is impossible, and unnecessary, since corrections can be made by using the formula for η and the above theory. Although this is approximate, for absolute calculations, it is usually accurate for minor corrections.

For example, if S and S_0 differ only in strength then the only variable factor is the data recording efficiency, D, which is easily adjusted. If the source radiations are of the same type but different energies, corrections may be made to the M and R factors by means of the appropriate attenuation formulae. If the source radiations are of different types, the calculation is generally impossible. To be useful, a calibrating source of strength S_0 must emit the same type of radiations as the unknown source, S, but it may emit additional types of radiation if these can be eliminated by absorbers. If S is a ^{22}Na source emitting 0.511-MeV gammas and S_0 is a ^{137}Cs source emitting 0.662-MeV gammas and some betas, then the betas must be absorbed and the factors M and R adjusted for the different photon attenuation coefficients.

Coincidence techniques

If a suitable calibrated source is not available, then a source of unknown strength can be used if it emits two particles or photons simultaneously and if these can be detected as *true coincidences*. Such sources can be calibrated by means of a number of coincidence techniques and, provided the above conditions are met, they can be used to calibrate detector efficiency for other sources.

One method of measuring source strength is to use two detectors, of unknown efficiencies η_1 and η_2, to count coincident emissions, such as the two 0.511-MeV photons from ^{22}Na. If the respective count rates are N_1 and N_2 and the source strength is S, then

$$N_1 = \eta_1 S \quad \text{and} \quad N_2 = \eta_2 S$$

If the true coincident count rate is N_{12} then this is given by

$$N = \eta_1 \eta_2 S$$

since the count rate in one detector is $\eta_2 S$ and there is a probability η_1 that the other will record the second event simultaneously. Thus

$$S = \frac{N_1 N_2}{N_{12}} \tag{5.15}$$

and if N_1, N_2 and N_{12} are measured, S is obtained without any knowledge of the values of η_1 and η_2.

Interaction efficiency

It is difficult to make general comparisons between the detection efficiency of one system and that of another because this parameter depends on experimental conditions to such a large extent. In particular, it is sensitive to the prevailing geometry, material attenuation and dead-time losses. For this reason, it is common practice to compare systems in terms of their interaction efficiencies, R; that is, the fraction of incident radiations that produce interactions. Alternatively, the product RD may be used; that is, the recorded interaction efficiency, or the fraction of incident radiations that produce recorded interactions. The distinction is not always made clear, because it is not often significant, but, unfortunately, both parameters are sometimes referred to as 'detection efficiency'. Normally, it is clear from the text which parameter is being referred to.

When photons are being detected, the recorded interaction efficiency is more often described as the *quantum efficiency* (QE) or the *detective quantum efficiency* (DQE). Another alternative is to define the *intrinsic efficiency* of a detector; that is, the efficiency of the detector itself, excluding the extrinsic effects of geometry and material attenuation. These terms, and the recorded interaction efficiency, mean the same thing. They differ from the detection efficiency, η, in that they compare the detector response with the quantity of incident radiation rather than the source strength. Quite often, this is all that is required and in some situations, for example in dosimetry and in measuring radiation levels in reactors and particle accelerators, it is all that can be measured. In common with detection efficiency, all of the terms yield pure numbers, expressed as fractions or percentages. Each is the ratio of the number of recorded counts to the radiation quantity, stated as the number of incident particles or photons.

In the discussion so far, radiation quantity has been described by means of its intensity, I, which is rather loosely defined as the number of particles or photons incident upon unit area. This approach has simplified the treatment of radiation detection, but it should be pointed out that more rigorous definitions exist. For example, the *particle flux density*, ϕ, is the number of particles passing through a unit area, in any direction, in unit time, and the *particle fluence*, Φ, is the number passing through unit area, in any direction, over a prescribed time. Fluence is the time integral of flux density and either can be implied by the intensity, I. They are used when the radiation is not monodirectional, as for example in dosimetry and reactor flux monitoring. When there is a well-defined direction, as in the beam produced by a particle accelerator, *particle current* is a more useful term and refers to the number of particles passing through unit area, in one direction, in unit time.

In some situations, radiation quantity is not measured in terms of numbers of particles or photons. Instead, it is measured by the effect that it would have if absorbed in matter. This is particularly the case in dosimetry and in some imaging processes, and the units will be described in detail in the section on dosimetry in the following chapter. Meanwhile, as a brief preview, it is sufficient to point out that these units are related to *absorbed dose, D, exposure, X*, and *dose equivalent*, DE. Absorbed dose is the energy absorbed from radiation by unit mass of material and the radiation is quantified in terms of the absorbed dose which it would produce. Exposure is the charge deposited by photons in unit mass of dry air at STP and electromagnetic radiation can be quantified as the exposure which it would produce. Dose equivalent is the absorbed dose multiplied by a *quality factor* to account for the biological effect of the specific type of radiation involved. Again, the radiation can be quantified by stating the dose equivalent which it would produce if allowed to fall on biological tissue.

When radiation quantity is not measured as a number, the recorded interaction efficiency cannot be expressed as a pure number. In such cases, the detector performance is described by its *sensitivity*, which is its response per unit quantity of incident radiation, however specified. The detector response may also be a non-numeric one. For example, the response of a film is measured by its degree of blackening, or optical density. Film performance is usually related to a sensitivity determined by optical blackening and radiation quantity measured in units of exposure.

5.3 ENERGY MEASUREMENT

Some detectors do not measure the energy of radiations. Those which do are referred to as energy *spectrometers*. The ability of a detector to measure energy is quantified by the associated parameter of *energy resolution* and the associated variable is, of course, the *energy* of a particle or photon. As for the detection of radiations, the measurement of their energies involves two processes: *absolute measurement*, with associated calibration errors, and *relative measurement*, which depends upon energy resolution.

118 *Detector properties*

The result of an energy measurement is $E \pm dE$, where dE is the sum of calibration and resolution errors. Resolution is a measure of the ability of the detection system to separate (that is, to distinguish between) two nearly equal energies. It represents the ability of the detector to identify the number of different energies present. This is a prerequisite to the measurement of these energies and the major source of uncertainty, and so, in general, dE can be identified with energy resolution. Calibration errors are much smaller and depend upon the assumption that all energies are resolved.

There are a number of methods of measuring radiation energies. Briefly, these are as follows:

Magnetic deflection:	the most accurate method for charged particles
Crystal diffraction:	the most accurate for photons with energy less than about 300 keV
Range measurement:	based on the range–energy relationship; not very accurate
Čerenkov effect:	accurate for relativistic charged particles
Absorption spectrometry:	most generally applicable

The present discussion is concerned with the last of these. It requires that at least some of the particles or photons must be stopped, and deposit all of their energy in the detector. It also requires that the detector should produce a voltage pulse output signal, per particle or photon, whose size, or height, is proportional to the energy deposited. These are called *analog* or *linear* pulses. The alternative is a voltage pulse whose height is independent of the energy deposited; this is referred to as a *digital* or *logic* pulse. The former type of pulse is a characteristic property of an energy spectrometer.

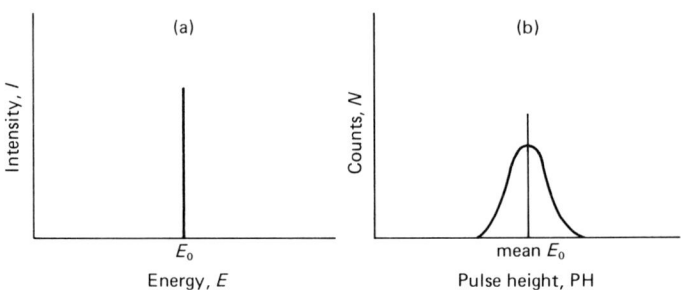

Figure 5.6 (a) Energy and (b) pulse height spectra for monoenergetic radiations

The main reason for a finite, non-zero energy resolution is the fact that most of the processes involved in radiation detection, both in the detector itself and in the output electronics, are statistical. Particles or photons of exactly the same energy produce pulse heights that are not equal but vary randomly about some mean value, usually in a near-Gaussian distribution. The effect is sometimes called *line broadening*, since a monoenergetic line spectrum emitted by a source is recorded as a relatively broad peak in the detector output spectrum. The *emission spectrum*, which is the property being investigated, should always be drawn as a graph of radiation intensity, I, against energy, E, as shown in *Figure 5.6a*. The recorded spectrum is a graph of counts, N, against pulse height, PH, and is known as a *pulse height spectrum*.

Pulse height is proportional to energy, and count rate is proportional to intensity, but both relationships are diffused by statistical uncertainties. Thus, a pulse height spectrum is not exactly the same as an emission spectrum. Clearly, the narrower the peak in a pulse height spectrum, the smaller; that is, the better is the resolution and the simpler is this relationship. However, it can never be absolutely certain that a single pulse height spectrum peak contains only one spectral line so, although the mean pulse height can be measured very accurately, it is the width of the peak which determines the experimental uncertainty. This width is measured by the energy resolution.

Some criterion is required to define resolution quantitatively — like the Rayleigh criterion of optical spectrometry. Several are used, but the most common, standard practice is to quote the *full-width half maximum* (FWHM) because adjacent peaks must be separated by at least this distance if their sum effect is to show two distinct peaks. Resolution can be given as the energy interval, dE, but this varies with energy, and a more constant and convenient

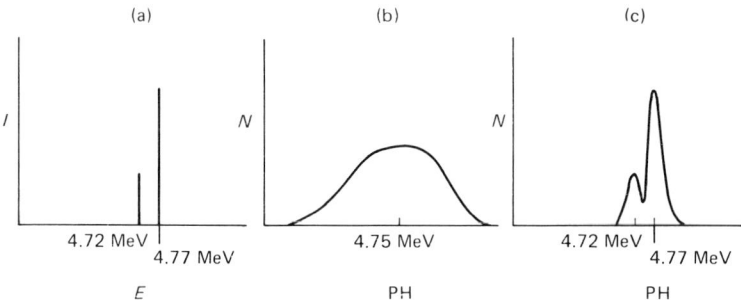

Figure 5.7 Alpha particle energy spectra for ^{234}U. (a) Emission spectrum; (b) scintillator spectrum; (c) semiconductor spectrum

parameter is the ratio dE/E, where E is the energy corresponding to the mean pulse height. This is often given as a percentage.

For example, as illustrated in *Figure 5.7*, a scintillation counter with an energy resolution, dE/E, of about 10%, does not 'see' the fine structure of an alpha decay spectrum whereas a semiconductor detector, with dE/E of about 1%, does.

Intrinsic resolution

The most important contributor to line broadening is intrinsic resolution in the sensitive volume of the detector itself. This places an absolute limit on the attainable resolution since it is a fundamental property of the detector. It derives from statistical variation in the number of ion pairs (and/or excited atoms) produced in the sensitive volume by each incident particle or photon. Assuming the peak to be a Gaussian function, its width is given by its standard deviation from the mean, σ, as

$$\sigma = (\bar{n})^{1/2} \epsilon \qquad (5.16)$$

where $\bar{n}$ is the mean number of ion pairs produced, and ϵ is the mean energy deposited in one recorded count.

Standard deviation is not a very convenient practical measure. It represents half of the peak width at $1/e^{1/2}$ (i.e. 61% of its maximum value). It is related to FWHM resolution by

$$\text{FWHM} = 2.35\sigma \qquad (5.17)$$

so FWHM intrinsic resolution is given by

$$\frac{dE}{E} = \frac{2.35\sigma}{\bar{n}\epsilon} \qquad (5.18)$$

Since $\sigma = (\bar{n})^{1/2}\epsilon$ for a Gaussian distribution, this formula reduces to

$$\frac{dE}{E} = \frac{2.35}{(\bar{n})^{1/2}} \qquad (5.19)$$

In energy terms, using the fact that $\bar{n} = E/\epsilon$, where ϵ is the mean energy deposited per ion pair formed,

$$\frac{dE}{E} = \frac{2.35}{(E/\epsilon)^{1/2}} = \frac{2.35\epsilon^{1/2}}{E^{1/2}} \qquad (5.20)$$

This yields two interesting conclusions. First, at any given energy E, resolution is proportional to $\epsilon^{1/2}$, which is an intrinsic property of the material in the sensitive volume of the detector. The *effective* values of ϵ for a semiconductor, a gas counter and a scintillation counter are very roughly in the ratio of 3 eV : 30 eV : 300 eV, and the characteristic resolving powers of these detectors are in the ratio of 1% : 3% : 10%, at reasonably high radiation energies. In other words, the resolution of a particular detector can be predicted, to some extent, via equation 5.19 (although it must be noted that the scintillator data relate to other effects besides ionisation and excitation).

The second conclusion is that, for any given detector, energy resolution is inversely proportional to $E^{1/2}$; that is, it deteriorates with decreasing radiation energy. This effect is illustrated in *Figure 5.8*, which shows how resolution typically increases with decreasing photon energy, for a number of detectors, and is quite poor for X-rays.

The theory so far is slightly deficient in one respect – it assumes a Gaussian distribution. In fact most pulse height spectrum peaks are asymmetrical and not quite Gaussian. This can be accounted for, retrospectively, by the addition of empirical factors to equation 5.17. There are a number of such factors but the most significant and, for general purposes, the only one which need be used, is

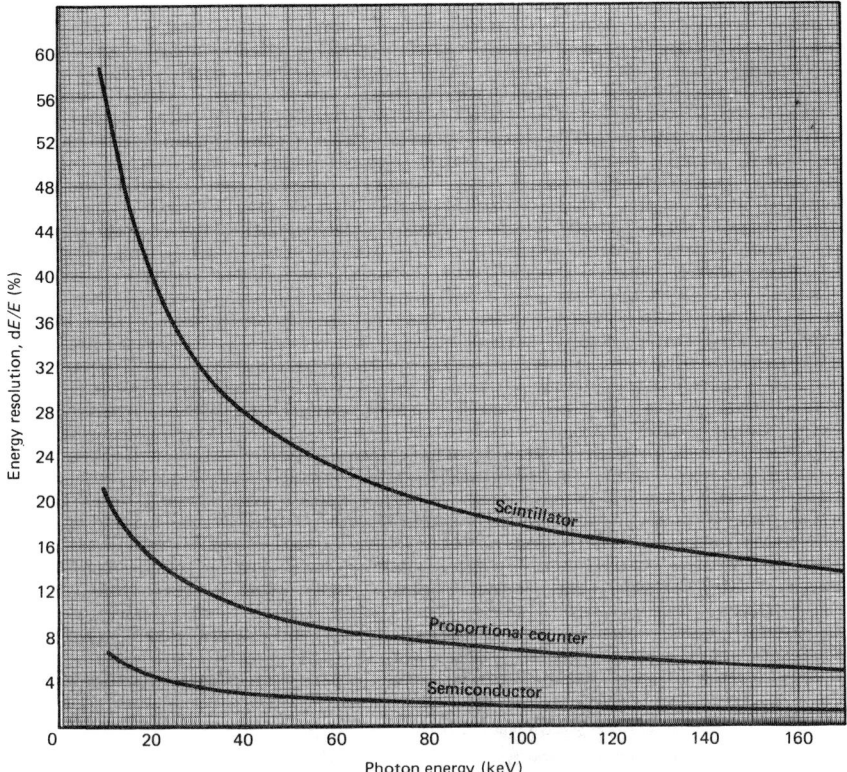

Figure 5.8 Approximate energy resolution as a function of photon energy for different types of detector

the *Fano factor*, F. This is less than unity and converts the variance, σ^2, of a Gaussian distribution to a smaller value $F\sigma^2$, so it improves the resolution by $F^{1/2}$, i.e.

$$\frac{dE}{E} = \frac{2.35 F^{1/2}}{(\bar{n})^{1/2}} \tag{5.21}$$

Extrinsic resolution

The term 'extrinsic resolution' is used to describe all line-broadening effects other than those intrinsic to the signal generation in the sensitive volume of a detector. There are several contributions and they vary in relative size from one type of detector to another but, basically, there are two sources of extrinsic resolution. One is the detector itself, mainly via *leakage currents* through or around the sensitive volume. The other is the output electronics and is mainly due to *electronic noise* currents in the early stages of signal processing where the signal pulses have not been amplified and *signal-to-noise ratio* is smallest.

It is the random fluctuation in these currents that causes line broadening. Current variations are seen by the readout system as a succession of small voltage pulses. Signal pulses may coincide with a peak, or a trough between peaks, or any intermediate position, so random coincidence between signal pulses and leakage or noise pulses increases the spread of the measured pulse height and widens the spectral peak, increasing the resolution.

For any given detection system, the contributions to extrinsic resolution are more or less independent of radiation energy so they add constant increments dE to the total FWHM resolution which are, of course, relatively larger fractions of the total at lower radiation energies. Since they are statistical effects, like experimental uncertainties in counting measurements, they add in quadrature and total resolution is given by:

$$dE^2 = dE^2_{intrinsic} + dE^2_{leakage} + dE^2_{noise} \qquad (5.22)$$

or, more generally, by

$$dE^2 = dE^2_{intrinsic} + dE^2_{extrinsic} \qquad (5.23)$$

Extrinsic resolution is usually quoted in equivalent energy units such as kiloelectronvolts or as equivalent root mean square numbers of ion pairs, FWHM. The net relative resolution, dE/E, is obtained by adding in quadrature, as above, then dividing the square root of the total by E.

5.4 POSITION MEASUREMENT

The variable associated with this property is the spatial position of a particle or photon or, at least, its point of intersection with the detector. This can be a one-dimensional measurement of, say, the x coordinate, along the focal plane of a momentum-analysing magnet. It can be a two-dimensional measurement of x and y coordinates, as in X-radiography, in which case the process can be described as *imaging*. It can be three-dimensional measurement of x, y and z coordinates along the path of a particle as, for example, in a bubble chamber, in which case the process can be described as *tracking*. These terms are not well-established, however, and it is common practice to refer to all three as imaging processes. In this case, it is necessary to distinguish between bidimensional and tridimensional imaging since these involve different techniques and criteria.

As for the other detector properties, there is an absolute part and a relative part to any position measurement. The former involves *calibration* either with respect to an external datum point, as on the focal plane of a magnet, or with respect to an internal reference point, as in determining the amount of distortion present in a radiographic image or in the localisation of a particle track. The relative part of the measurement is defined by the finite size of the recorded event and is determined by the parameter associated with position measurement, which is *spatial resolution*.

Spatial resolution is a measure of the ability of a detection system to separate the positions of two adjacent events. As for energy resolution, it is less important, in general, as an experimental error in position than as a measure of ability to identify how many events have been detected. Both qualities are used in high-energy and nuclear physics, in tracking devices, but usually only the latter is important in bidimensional imaging devices, where the objective is not to measure position but to improve picture quality by resolving details.

There is no single quantitative criterion for measuring spatial resolution. Different criteria are used in different circumstances and may not be explicitly defined. In order to avoid any misunderstanding, it is important to appreciate the reasons for this diversity and to organise these in some logical fashion. This can be done by considering the following dualities:

(1) *Differential/integral resolution*. Some detectors have differential spatial resolution in that they locate events to positions inside their sensitive volumes. These include *tracking devices*, such as cloud chambers and bubble chambers, and *cameras*, which produce bidimensional 'pictures' covering a large area and include films and fluorescent screens. Other devices have integral resolution in that they merely indicate whether or not an event has taken place somewhere in their sensitive volumes. They include Geiger–Müller counters and most semiconductor counters. Such detectors can generate bidimensional images by measuring the radiation intensity at each point, in turn, of the image. In this application, they are known as *scanning devices*.

(2) *Single-particle/beam resolution*. Resolution may be measured for a single particle as the width of the impression recorded for that event. Typically, this applies to tracking devices. Alternatively, the resolution may be measured for a beam of particles. Even if they all arrive at the same point in the detector there is an additional statistical uncertainty as to where the system will record the events. This is, in effect, a line-broadening process.

(3) *Object/image resolution*. Resolution may refer to the separation of points in the object, that is, the source of the radiation, or to the separation of corresponding features in the image produced by the detection system. In bidimensional imaging, the aim is to resolve object details, but image resolution is the more easily measured and the more frequently specified parameter, since it is a relatively constant property of the detection system.

The relationship between object and image resolution is determined by the *magnification* introduced by the detection process; that is, by the ratio of image to object size (and not by any subsequent enlargement). For example, the vertical resolution of a television screen is determined by the number of picture lines used. A 'close-up' picture of an object reveals more detail than a more distant view but the resolution of the screen does not change.

(4) *Intrinsic/extrinsic resolution*. As for energy resolution, it is possible to define intrinsic spatial resolution as that of the sensitive volume by itself, and extrinsic spatial resolution as the effects of all other parts of the detection system. Unlike energy resolution, however, these two components of spatial resolution are not additive. Usually, one or the other is the dominant parameter and the only one to be considered. For example, if a detector with integral resolution, such as a scintillation counter, is used to image radiations, its intrinsic resolution is larger than any other effect but, if radiations are collimated on to some smaller part of its sensitive volume, then it is the extrinsic resolution of the collimator which determines resolution and, in fact, improves it considerably. On the other hand, the resolution of a film is about 1 μm, the grain size, but if it needs a collimator to produce an image, then the much larger resolution of the collimator is the operational parameter.

Measurement criteria for spatial resolution

The above comments describe four different ways of classifying definitions of spatial resolution, implying 16 possible definitions. In practice, only a few are used. The major definitions are as follows:

(1) *Track resolution* is the width of the detector response to a point event. If the positions of the tracks of two separate particles are x and $x + dx$, then the track resolution is the smallest value of dx which allows two events, rather than one, to be observed. A single interaction with a film, for example, causes one grain to become developable so the differential, single-particle, image, intrinsic resolution of the film is the size of one grain. This is the track resolution.

(2) *FWHM resolution* is the full-width half-maximum value of the intensity distribution produced by an incident beam of radiation with a point-sized cross-section. To use the example of a film once again, the FWHM resolution for a beam of particles is the same as the track resolution but for a beam of photons it is much larger than track resolution. Photons arriving at the same point liberate photoelectrons in all directions about the arrival point, so that FWHM resolution is approximately equal to the range of the photoelectrons. FWHM resolution is more relevant to bidimensional imaging than track resolution since it measures the smearing of each image point and the degradation of image clarity. For the record, it is a differential, beam, image resolution but it may relate to intrinsic or extrinsic resolution.

(3) *Spatial frequency*, ν, is the maximum number of transverse lines per unit length which can be separated by the detection system. Normally it refers to lines per centimetre (for example) in the *object* but it may instead refer to the image formed. In the latter case it is related approximately to FWHM resolution, if this is assumed to be the width of a resolvable line. Thus spatial frequency, ν, is given by

$$\nu = \frac{1}{\text{FWHM}} \tag{5.24}$$

where, if ν is in lines cm^{-1}, FWHM is in cm.

5.5 TIME MEASUREMENT

Time measurement relates to the measurement of the time at which a particle or photon interacts with the sensitive volume of a detector. Again, there are two parts to the measurement, an absolute and a relative part. The *absolute measurement* involves the use of a clock mechanism and uncertainties in this component depend upon the accuracy of *calibration* of the clock. This applies, for example, to the measurement of radioactive half-lives, especially short ones, and to time of flight methods in which the velocity of a particle is determined by the time it takes to cover a measured distance. The *relative measurement* depends upon the *time resolution* of the detection system, which is a measure of the ability of the system to distinguish between two events occurring close together in time. This is the associated parameter and, once again, it is important not only with respect to relative time measurement but also in so far as it determines whether the system will observe one or two counts.

Some detectors have *intrinsic time resolution* in that they record events at the same time as they happen. Generally, these are electrical devices which generate voltage output signals. Although there is a small delay between the event and the recording of it, the time relationship between recorded pulses is the same as that between events in the detector. Other systems have *extrinsic time resolution* since they do not, by themselves, record the time of events. Resolution is much poorer for such systems and they are not often used to measure time. A film, for example, records time only if it is used in a camera with a shutter mechanism. It is interesting to note the similarity in this respect between time, position and energy measurement as well as detection efficiency. All have intrinsic and extrinsic capabilities. For energy resolution, this corresponds to the difference between detectors with analog output signals and those with digital ones which require some additional device to measure energy.

Systems with intrinsic resolution produce output signals that are usually in the form of voltage pulses, one per event. Typical analog pulses, as functions of time, are shown in *Figure 5.9*. The pulse shape varies from one detector to another but it usually has an almost linear increase and a roughly exponential decrease. The increase is measured by the *rise time*, which is the time taken for the voltage to increase from 10% to 90% of its maximum height. The decrease tends to be longer and is measured by a *decay constant, RC*, which is the transient time constant for a discharge through a resistance R and a capacitance C given by

$$v = e^{-t/RC} \tag{5.25}$$

Rise time is a truly intrinsic property associated with the detection mechanism in the sensitive volume, but decay time depends upon the total resistance, R, and capacitance, C, of the detector and its output electronics. Briefly, the rise time is the time required for an event to charge up the detector and the decay time relates to the time taken for it to discharge, both through the detector and across any external connections.

There is no established quantitative criterion for time resolution, but it is approximately equal to the rise time plus the decay time. For accurate work it has to be measured as the smallest observable time interval between events. As mentioned above, there are two types of operation. One is to distinguish between successive events in order to count them. The other is to measure the exact time of the event, assuming it to be a single event.

The first of these affects the detection efficiency of the system and has been introduced in Section 5.2. If two pulses arrive within the resolving time, τ_r, they are indistinguishable and only one pulse is recorded. This is known as pulse *pile-up* or *summing*. Since one pulse sits on top of the other, the single recorded pulse is larger than either of its components. If pulse height is being measured, or selected, the result is a pulse of the wrong size and, effectively, both pulses are lost from the useful data.

Pile-up is due to *random coincidence counting*, defined to take place when two or more radiations chance to arrive within the resolving time of the detection system. As described earlier, the random coincidence count rate is equal to $2N^2\tau_r$, where N is the recorded count rate. This formula enables coincidence counting losses to be estimated, if τ_r is known, or assessed as rise time plus decay

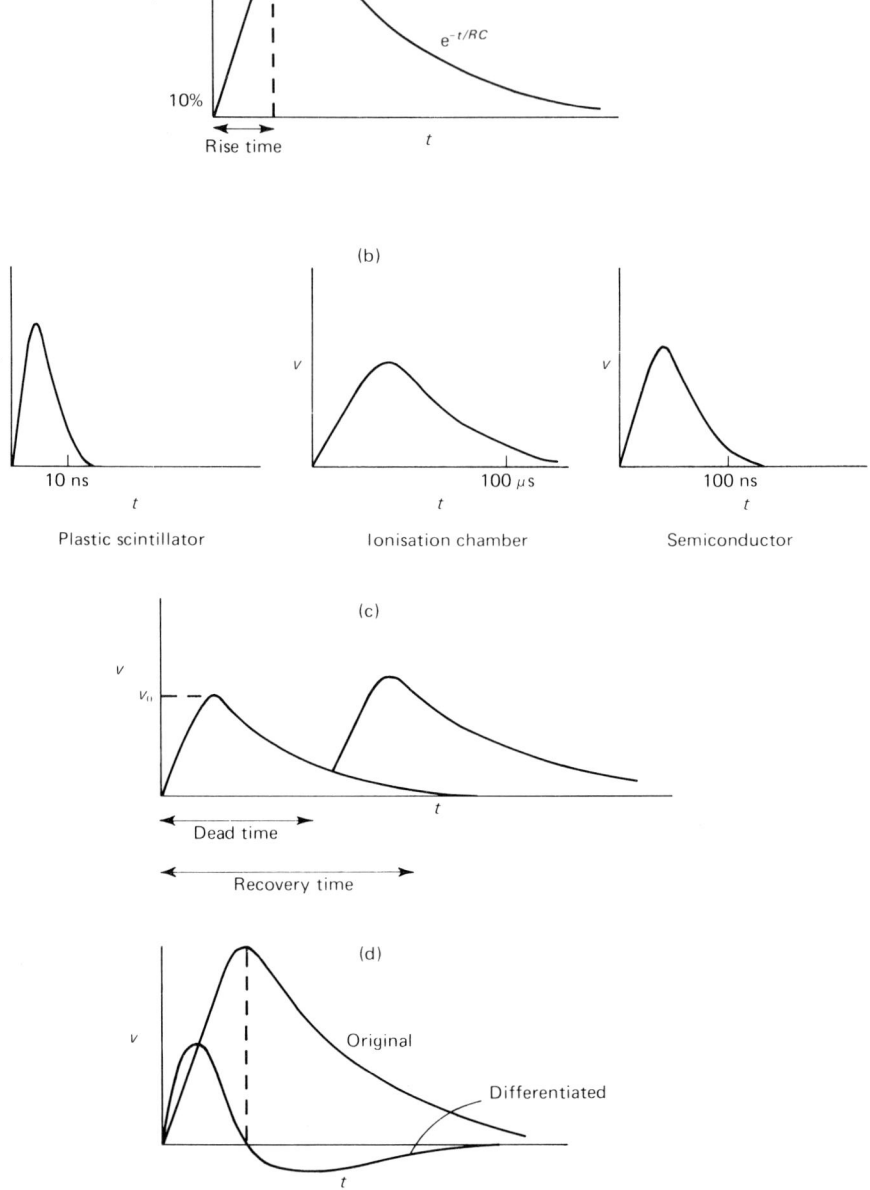

Figure 5.9 Typical signal pulse shapes. (a) Detector pulse; (b) pulses for a plastic scintillator, an ionisation chamber and a semiconductor; (c) overlapping pulses; (d) differentiated detector pulse

time. Alternatively, it provides a method of determining τ_r experimentally, by measuring N and the coincidence count rate.

It is not possible, in a general treatment, to define an exact relationship between the pulse shape parameters, rise time and decay time constant, and the timing parameters τ and τ_r, but it is useful to examine the conceptual relationships. The voltage pulse is usually a transient change in the value of a larger, constant potential which maintains the sensitivity of the detector. It is approximately true that the detector becomes insensitive, or dead, when the pulse exceeds a certain height, say v_0. If a second pulse arrives before the first one reaches this value, it is detected in coincidence, so the resolving time is roughly equal to the rise time (*Figure 5.9*).

If, however, the second pulse arrives during the period of time when the height of the first pulse exceeds v_0 then it is not detected and this period is the dead time. It usually includes much of the long tail of the pulse and is of the same order of magnitude as the decay time constant.

There is a third possibility, that the second pulse arrives immediately after the dead time has elapsed. This is detected but it sits on the tail of the preceding pulse and has the 'wrong' height. For counting and timing applications this effect is not important but for energy spectrometry it gives very inaccurate results. In this case, a longer dead time has to be imposed on the system electronically and the dead time becomes equal to the total length of the pulse. Sometimes this total length is referred to as the detector *recovery time* – it is the time required for the original zero pulse height to be reached.

If it can be assumed – for example, by examining shape or height – that a recorded pulse represents only one event, then the relative accuracy in timing the event is much smaller than resolving time. The pulse peak can be located with an uncertainty less than τ_r and so too can the leading edge. Neither of these is a very convenient reference, so for accurate time measurement the pulse is usually differentiated. This process will be described later. Meanwhile, it is sufficient to point out that a differentiated detector pulse, as illustrated in *Figure 5.9*, is almost the derivative, with respect to time, of that pulse. It is *bipolar*, being positive during pulse rise time, zero at pulse maximum and negative at pulse decay. The point at which it changes polarity and crosses the time axis is a very precise reference point and is known as the *crossover point*. Accurate timing circuits measure the time of this point in a technique referred to as 'crossover pickoff'. There may be some statistical variation in the position of the crossover point, owing to variable ionisation times in the detector and electronic 'jitter', but this is generally quite small.

5.6 DISCRIMINATION

Discrimination describes the ability of a detection system to distinguish between different types of radiation. There is no single associated variable. It may be mass, charge, energy, specific ionisation or any other useful property of the radiation. Discrimination also includes the ability of the system to distinguish between radiation and electronic effects, again without any general, quantifiable variable. Consequently, there is no single associated parameter, but a range of parameters and criteria corresponding to the radiation properties used as a basis for discrimination.

There are two approaches, one based on the preferential *selection* of one type of radiation and the other based on the retrospective *identification* of the radiation detected. In the first case, a system is specifically designed to detect one type of radiation and it has intrinsic discrimination. In the second, a general-purpose system is used, sometimes with a purpose-planned experimental arrangement, and the results are analysed to identify radiation type. Both approaches depend upon the enlightened application of the theory of interactions with matter, of general detector properties such as detection efficiency, energy measurement, position and time measurement, and of the characteristic properties of individual detectors. The second approach also requires the ability to determine which types of radiation are likely to be emitted from a particular source.

There are several possibilities and, in many situations, selection or identification must be achieved in unusual or innovative ways. It is difficult to review all options but some standard methods can be described and these should serve as examples of the general approach.

Electronic effects

These fall into two categories, namely, the effects of healthy circuitry and the effects of malfunctioning, inappropriate or incorrectly set up units. The latter are often difficult to diagnose. Fault recognition requires some knowledge of the properties and behaviour of electronic units, a subject which will be discussed in Chapter 12. Meanwhile, as a general rule, any effects which cannot be explained in terms of radiation properties should be viewed with suspicion and data obtained under such circumstances must be of doubtful quality.

A healthy, properly functioning system can affect experimental results by increasing dead time, resolving time and line broadening. These effects have been described above and should always be considered when selecting equipment or analysing results. Relevant data can be obtained from instrument manuals or from equipment manufacturers but, once again, a full appreciation depends upon the nucleonic theory presented later.

One effect which can and must be examined at this stage is that of *electronic noise*. As mentioned previously, this comprises a large number of very small voltage (or charge) pulses produced at or near the detector end of the system by random variations in leakage current and other noise currents. Individually, these pulses are indistinguishable from weak signal pulses but collectively, they have a number of properties which serve as bases for discrimination.

First of all, they are present when the source of radiations is removed. In situations where the source can be removed, noise effects can be eliminated by counting first with, then without the source and subtracting the latter result from the former.

Secondly, they are very small pulses and, in situations where the signal pulses are relatively large, the noise pulses can be eliminated by passing all pulses through an electronic unit known as a *low-level discriminator*. This stops all pulses smaller than some preselected voltage height, which may be referred to as its *bias* setting. In certain circumstances, this method is preferable to the subtraction technique. Noise pulses increase dead time and random coincidence counting losses and, if these occur in the electronic circuitry rather than in the detector, insertion of a discriminator before the critical unit reduces these losses. With the subtraction technique, corrections may have to be made for dead time and coincidence losses.

Thirdly, noise pulses have a characteristic pulse height distribution, as illustrated in *Figure 5.10*. In a pulse height spectrum, this shows up as a continuum which

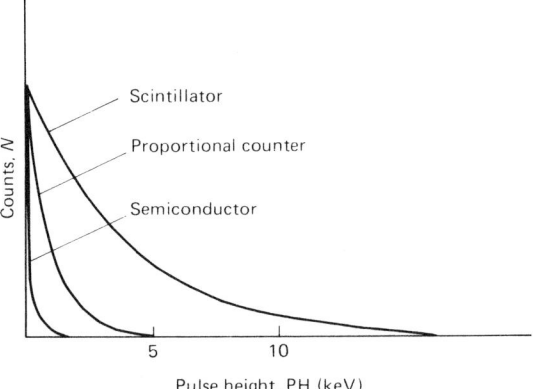

Figure 5.10 Electronic noise spectra

decreases, almost exponentially, with increasing pulse height. In addition, the extent of this continuum is typical of the detector employed. It is largest for a scintillation counter and smallest for a semiconductor detector. For this reason, scintillators cannot generally be used to measure low radiation energies, in X-ray spectrometry for example. Noise spectra are easily recognised in all situations except one, namely, when bremsstrahlung radiation is present. This produces a very similar pulse height spectrum which can be distinguished from a noise spectrum only in that it adds to noise effects, increasing the observed range of the continuum beyond that expected for the detector in use, and, of course, it disappears when the source is removed.

Finally, noise pulses are random in their time relationships. They can be identified and eliminated in situations where two or more detectors are used to observe true coincidence events. The signal pulses in the two detectors are always coincident in time but their noise pulses are not. The output pulses from the detectors may be passed into an electronic unit known as a *coincidence counting unit*. This produces an output signal only when it receives two input signals in coincidence, thereby eliminating virtually all of the noise pulses.

Background radiation

It is almost always the case that radiations under investigation are accompanied by other radiations from the environment, and sometimes from the detector itself. These can be regarded as background radiations. If the effect is significant, the standard procedure is to obtain a count first with, then without the source and calculate the count from the source alone by subtraction. There are, however, a number of situations in which this approach is inappropriate and alternative methods, such as those described in later sections, must be used. These situations include the following:

(1) *Source and environment inseparable.* In this situation, the source cannot be removed from the background radiations. For example the detection of neutrons is usually accompanied by the detection of gamma rays, either from the neutron source or from materials, in the environment and the detector, activated by the neutrons. As another example, the emissions of radionuclides introduced into a human body may be detected along with those of ^{40}K occurring naturally in the body.

(2) *Very high background count.* If the background count rate is very high, it may introduce large dead-time and coincidence counting losses which make a straight subtraction inaccurate and increase statistical errors. As an example, any experiment on the interactions of charged particles from an accelerator is plagued by neutron and gamma fluxes generated by the particle beam in the experimental area.

(3) *Very low source count.* If the count rate for source radiations is of the same order of magnitude as, for example, that due to cosmic rays, natural activity in structural and nearby materials or materials which have been contaminated by radionuclides, then subtraction is inaccurate. The net statistical counting error is the sum of those due to the two counts obtained and each of these is, in any case, relatively large because of the small number of counts. This is a typical problem with carbon dating techniques.

(4) *Varying background count.* If the background count rate cannot be assumed to remain constant while the two counts are taken, subtraction is unreliable. As a check against this possibility, background counts should be obtained before and after the count with the source present. Any count taken over a long period of time is subject to this uncertainty as is any count near an accelerator or neutron source where induced activity may be a function of time.

(5) *Scattered radiation.* In an exhaustive treatment, radiation scattered into the detector from its environment may be regarded as background radiation. If, for example, the object is to obtain an X-ray or neutron radiograph of a specimen then all scattered radiations incident on the detector tend to obliterate the radiographic image.

Selective absorption

One of the most direct methods of eliminating unwanted radiations is to absorb them in a suitable attenuating medium before they reach the sensitive volume of the detector. This approach involves the manipulation of the material attenuation factor, M, on the basis of the theory of interactions with matter and the data provided in Appendix 4. If the radiations are known, the required absorber materials and thicknesses can be determined by reference to these data. If they are not known, they can be identified by measuring cross-sections, ranges and attenuation coefficients and comparing these values with the data. In both cases, it must be remembered that an attenuator or filter designed to stop one type of radiation will have at least some effect on the others. Thus, a few centimetres of air or a few mg cm^{-2} of any solid absorber will stop most alpha particles.

About 1 g cm^{-2} of material will stop most beta particles. Photon intensities will be reduced exponentially, according to their attenuation coefficients and effectively removed by high-Z materials such as lead. X-rays, of course, are more easily absorbed than gammas.

Detector entry windows provide a built-in selection mechanism. Alpha detectors must have no windows or exceptionally thin ones. Most entry windows stop alphas, low-energy betas and soft X-rays. Some, such as the thick windows of scintillation counters and all large semiconductors, stop all radiations except hard X-rays, gammas and, if present, neutrons. General-purpose dosimeters, for example, usually have relatively thick end caps covering thin entry windows. With the cap on, these measure gamma ray dose, but they measure combined photon and beta dose when the cap is removed.

It may be that the radiations to be counted are more easily stopped than the accompanying unwanted radiations. In this case, a count may be taken first with an absorbing filter, then without one, and the difference between these two results yields the count due to the absorbed radiations — after correcting for any attenuation of the more penetrating radiations.

Selective detection

The interaction efficiency, R, of a detector depends upon the thickness and the material of its sensitive volume but it also depends on the type of radiation being detected. Thus, the detection efficiency of a given detector varies from one type of radiation to another and there is, to a certain extent, an intrinsic selective mechanism. Normally, the 'resolution' of this property is not sufficient to isolate only one type of radiation, but it can be improved by the incorporation of suitable materials, in accordance with the theory of interactions with matter. This is the most common method of producing a purpose-designed detection system, and most detectors can be modified to improve selective ability.

Thin detectors, for example, respond to charged particles but not very well to gamma rays or neutrons. High-Z materials detect photons much more efficiently than neutrons. Low-A materials are more effective for fast neutrons, and materials containing ^{10}B, ^{6}Li, ^{3}He, etc. are exceptionally responsive to thermal neutrons. The application of these general principles will be discussed in more detail in the chapters on specific detectors.

Magnetic and electric deflection

Magnetic or electric fields can be used to deflect charged particles out of a collimated beam of radiation. This either removes them from the detected beam or deflects them into the detector. The deflecting force due to an electric field E on a particle of charge q is given by

$$F = Eq \tag{5.26}$$

and the direction of the force is in the same direction as E, for a positive charge, or the opposite direction for a negative charge, as shown in *Figure 5.11*.

The deflecting force due to a magnetic field, or a magnetic induction, B, depends on the charge, q, and on the velocity, v, of the particle. It is given by the formula

$$F = Bqv \tag{5.27}$$

and the direction of the force is at right angles to both B and v, as shown in *Figure 5.11*. One way of determining the direction of a magnetic deflection is to use the so-called 'left-hand rule'. If the thumb, first finger and second finger of

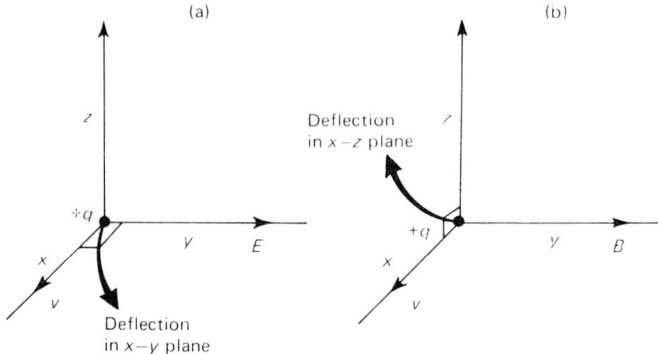

Figure 5.11 (a) Electric and (b) magnetic deflection of a positively charged particle

the left hand are extended at right angles to each other with the first finger in the direction of B, and the second finger in the direction of v, then, for a positive charge, deflection is in the direction of the thumb. For a negative charge, the deflection is in the opposite direction.

For energetic particles, with high velocities, electric deflection is very small. Magnetic deflection, on the other hand, is proportional to velocity and produces much larger effects for energetic particles.

Pulse height spectrometry

A very useful method of retrospectively analysing the nature of radiations being detected is provided by energy spectrometry. The energy measured is one clue, but a more important feature is the shape of the pulse height spectrum. This is

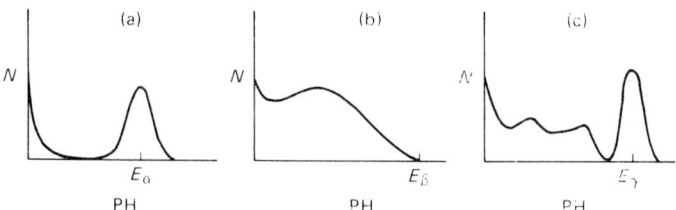

Figure 5.12 Typical pulse height spectra for (a) alpha, (b) beta and (c) gamma radiation obtained with a scintillation counter

quite different for alpha particles (and other heavy charged particles), for beta particles and for photons.

Typical spectra are illustrated in *Figure 5.12*. Alpha particles generate single, well-defined peaks, sometimes showing fine structure if the energy resolution is good. Beta particles have characteristic continuous distributions extending up to the maximum disintegration energy, at which point the beta has all the energy and the neutrino has none. Gamma ray spectra show a well-defined peak corresponding to total absorption, in a photoelectric process, and a low-energy distribution which has a complex but typical shape. This will be described in greater detail in the next chapter. It has a Compton edge and a backscatter peak produced by partial absorption via Compton scattering events in the detector and in its environment, respectively.

dE/dx pulse height spectrometry

If a charged particle is incident on a very thin, or 'dx' detector, the energy deposited, dE, is a very small fraction of the total energy of the particle and is proportional to the specific energy loss, dE/dx, of the particle. Analog output pulses form a pulse height spectrum in which pulse height is proportional to dE/dx. From equation 4.25, therefore, pulse height is proportional to Mz^2 of the incident particle.

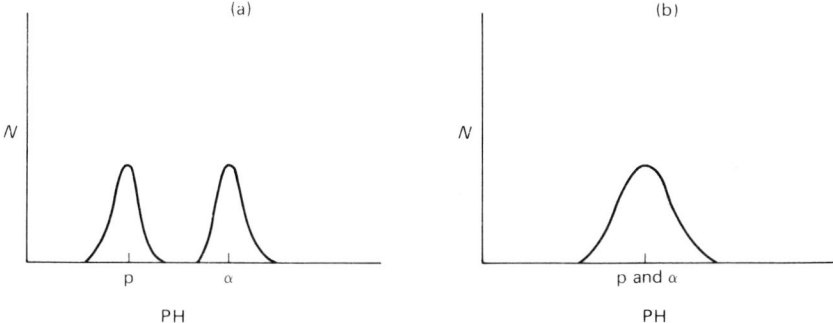

Figure 5.13 (a) dE/dX and (b) E spectra

As an example, if an alpha particle and a proton of the same energy are stopped in a detector their output pulses, if analog, would be the same height. In a dE/dx detector, the alpha leaves much more energy than the proton so its output pulse is much larger. A mixed beam of alphas and protons can be separated on this basis. A dE/dx spectrum has two peaks, one for the protons and the other for the alphas, as shown in *Figure 5.13*, while an E spectrum has only one peak.

There is a technique known as single-channel analysis (SCA) that will be discussed in some detail in a later chapter. Essentially, this involves the selective counting of pulses whose heights fall within a relatively narrow range, or channel. In the example above, the channel can be set to count protons or alpha particles. More generally, single-channel analysis is a method of selecting radiations by means of their dE/dx values, or their energies, if a stopping detector is used.

More positive selection or identification can be obtained by using a dE/dx detector and an E detector together. Since the former is transparent, it is placed in front of the latter, which stops the radiations. The arrangement is usually known as a *particle identification* (PI) system and is illustrated in *Figure 5.14a*. Using the mixture of alphas and protons as an example, the dE/dx detector separates these in terms of pulse height. A single-channel analyser passes on the signals due to one or the other, to a coincidence unit which is also fed by the E detector. Output from the coincidence unit occurs only for alphas, or protons,

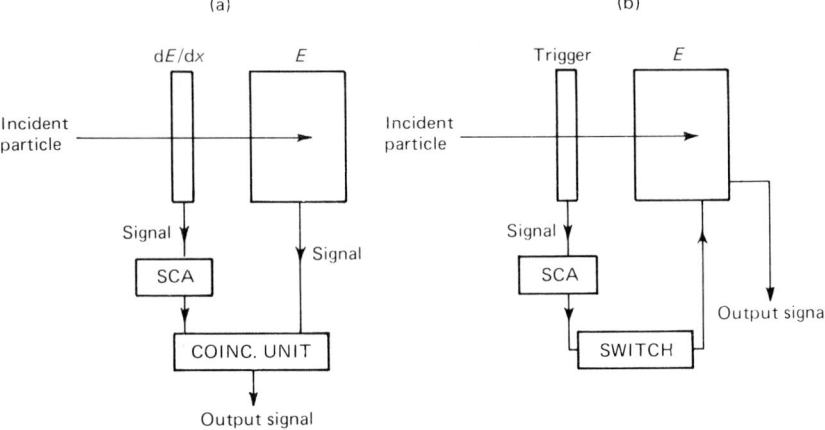

Figure 5.14 Two applications of dE/dX detectors. (a) Particle identification (PI) system; (b) trigger system

depending on which of these is passed by the single-channel analyser, and these can be pulse-height-analysed to form a pure alpha or proton energy spectrum. This is one example of a PI system. There are others, but they all involve double identification by means of a dE/dx and an E detector. Since photons and thermal neutrons interact only once, either in the dE/dx detector or in the E detector, but not in both, they cannot be selected by such systems.

There is another variation of this technique, which is extremely useful in high-energy physics. In this, the leading, dE/dx device is used to switch on the E detector when it registers an interesting event, so the E detector is insensitive to any other type of radiation. This is known as a *trigger system* (*Figure 5.14b*) and the dE/dx detector is called the *trigger detector*. Once again, this approach cannot be used for photons or for thermal neutrons. Fast neutrons can interact twice but they are deflected out of the original line of flight, in the dE/dx detector.

Pulse shape discrimination (PSD)

Analog pulses may contain information regarding the type of radiation detected as well as its energy. This information is provided by the shape of the pulse, which is affected by the specific energy loss, dE/dx, of the radiation. A simple analysis suggests that radiations with large values of dE/dx deposit their energy quickly and in a short path length, leading to faster pulses than those generated

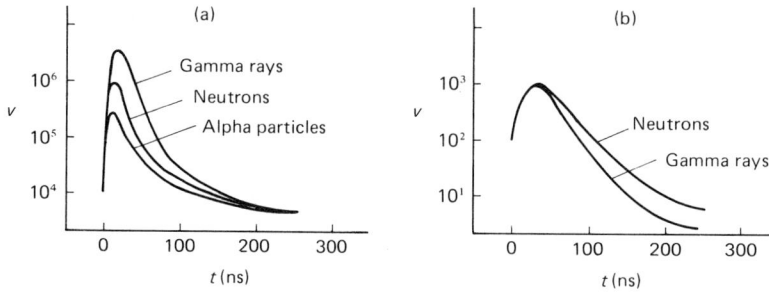

Figure 5.15 Pulse shape discrimination in scintillators. (a) Stilbene; (b) quaterphenyl

by radiations with smaller values of dE/dx. Differences in rise times are usually very small but decay times can be strongly influenced.

In practice, this simple theory is greatly complicated by the detector response mechanism involved, but it remains a fact that some detectors, and especially scintillators, do produce pulses whose shapes depend on dE/dx. Two examples of this effect are provided in *Figure 5.15*. Fast electronic circuitry can distinguish between these different pulse shapes.

Timing methods

There are many methods of identifying or selecting radiations on the basis of their time relationships. They depend, of course, on the ability of a detector to measure the time of an event and on the accuracy with which it does so, but the techniques themselves are basically independent of the type of detector used. In other words, the same techniques may be used with different detector types. The best detector may not always be the one with the fastest response or the smallest resolving time. It may be the one with the largest detection efficiency for the radiations involved, or some other essential property. Consequently, timing methods should be regarded primarily as a general function of detection systems and they will be discussed as such in the following chapter.

Time of flight (TOF) techniques involve the measurement of the time taken for radiations to travel across a flight path from source to detector. They require a *pulsed beam* of radiation in order to define the starting time and they distinguish between different velocities so they are used, for example, to separate photons, which move at the velocity of light, from fast neutrons, which are much slower.

Half-life measurement provides a means of identifying different nuclides in a mixture of radionuclides. This is particularly useful, for example, in activation analysis but, because of experimental limitations, it can be used only for relatively short half-lives.

Coincidence or anticoincidence counting techniques are used, respectively, to select or reject radiations emitted in true coincidence with each other. They apply, therefore, to sources with *true coincidence properties*. These include, for example, positron emitters which generate two simultaneous 0.511-MeV photons. In some cases, however, coincidences apply to the simultaneous response of two or more detectors to the same event.

Time to amplitude conversion (TAC) is an extension of the coincidence method in which the time interval between any two events, not necessarily coincident, is measured and converted to a single voltage pulse whose height is proportional to the time interval. This information frequently serves to identify or select radiations.

Position measurement

There are a number of techniques, based on the measurement of the position or direction of radiations, that can be used to select those radiations of particular interest. As for timing methods, those tend to be system functions rather than detector properties, although they do depend upon the latter, and they will be treated in more detail in the following chapter, as imaging techniques. Meanwhile, a few examples can be given in order to illustrate the relevance to discrimination.

Magnetic analysers involve the use of magnetic fields to deflect charged particles, as described above. The radius of curvature, ρ, of a particle in a field of magnetic induction B is given (for non-relativistic particles) by

$$B\rho = \frac{mv}{q} \tag{5.28}$$

so that measurement of ρ in one dimension along the focal plane of a magnet or by observation of three-dimensional tracks in a tracking device, provides a measure of the momentum, mv, of the particle.

Collimators are used, for example, in radiography and isotope imaging to preserve a point-to-point geometry between the source object and the image. Thus, each point in the detector is exposed to one corresponding point in the object and a representative image is obtained. The collimator eliminates radiations that are scattered en route from source to detector, or emitted from non-corresponding source points so as to obliterate the image.

Telescopes measure the directions of radiations, that is, their positions in three dimensions, and eliminate radiations travelling in directions other than the one selected. A cosmic ray telescope, for example, may comprise several detectors in a straight line with their output signals counted in coincidence. Radiations interacting with all detectors are recorded. Others crossing the telescope and interacting, say, with only one detector are not recorded.

5.7 SIGNAL FORMATION

Signal formation is primarily a property of the detector, rather than of the complete detection system. It describes the ability of the detector to produce an output signal and is determined, first of all, by the intrinsic interaction mechanism within the sensitive volume and, then, by the initial processing mechanisms that convert the interaction into a signal. All detectors produce output signals but the information content and the type of these signals vary from one detector to another. Hence information content and signal type are the variables associated with this property. Since these are largely qualitative, however, there is no single associated parameter.

Information content depends on the five properties discussed above, namely, the ability of the detector to measure radiation intensity, energy, position, time and type. This information is contained within the intrinsic interaction but it has to be converted into a signal to be of any value so the signal formation process also affects the information content of the signal. It does so mainly by selecting or enhancing specific pieces of information, usually at the expense of others, and the criteria for the selection process are determined by the data required.

For example, a scintillator generates a small light flash at the point of interaction with ionising radiation and the intensity of this flash is proportional to the energy deposited. Intrinsically, it is capable of measuring energy and position. The scintillator can be connected to a photomultiplier to form a scintillation counter. This measures energy but not the position of an event within the scintillator. Alternatively, the scintillator may be in the form of a fluorescent screen. This measures position very accurately but loses all information regarding radiation energy. In both cases, signal formation selects a particular type of information.

These are extreme types of scintillation detector. There are other, intermediate examples, such as the gamma camera in which a single scintillator is examined by several photomultipliers. This measures position and energy, but much less accurately than the screen and the single-photomultiplier counter, respectively.

It is an important general rule that optimisation of the measurement of one variable is always at the expense of others. This is often due to signal formation processes but it is also an intrinsic feature of the interaction mechanism. In effect, the six detector properties are not entirely independent of each other.

Signal type is also determined both by the intrinsic interaction mechanism and by the signal processing operation. The former presents a range of options and the latter selects one of them. For example, the scintillation counter generates an electrical output in the form of a voltage pulse, while the fluorescent screen produces a light flash. As a further example, the interaction of radiation with a noble gas provides a charge pulse, a scintillation and, if the energy is large enough, Čerenkov radiation. Any of these effects can be used to form a signal and all of them are, in different types of detector.

Choice of signal type is mainly influenced by the requirements of data retrieval and analysis. Basically, the choice is between electrical or non-electrical signals and between differential or integral output.

Electrical signals include voltage or charge pulses and current or total charge readings. They have the disadvantage that they require some electronic processing, which may add to the experimental uncertainty and the cost of the system. On the other hand, they have the advantage that they are compatible with data analysis techniques and, especially, computer analysis. This is an important and increasing advantage, first because the functions of detection systems are becoming more sophisticated and computer analysis is an essential aspect of many applications in high-energy physics, nuclear physics, medical physics and other areas, and secondly because of the growing need for fully automated detection systems for applications in areas outside physics.

Non-electrical signals are mostly *visible* ones although they include audible signals. They have the advantages of simplicity, reliability and, for imaging purposes, accuracy. Apart from imaging applications, however, they tend to be less quantitative, less amenable to data analysis and less informative.

Differential outputs are those which record individual events. They are obtained from *counters*, which produce voltage or charge pulses, and from tracking devices, which image individual interactions. Characteristically, they measure all variables but their main advantage is their ability to measure the time of events. Their great disadvantage is that their counting efficiency is restricted by dead-time losses and their count rate capacity is limited by this parameter.

Integral outputs describe the time-integrated effects of many events. Electrical signals are converted to total charge or current, to record radiation intensity rather than count rate, and visible signals similarly produce a total or average effect. Typically, this feature is utilised in dosimeters which measure total exposure or dose. Integrated outputs tend, also, to have zero or minimal dead time so they are capable of dealing with very high count rates. This is a characteristic property of films and fluorescent screens and is an advantage for applications such as X-radiography where dead-time losses would reduce the detection efficiency of the system and increase the dose given to the patient, or specimen, to produce a useful image in a short time.

There is a distinction to be made between a *detector* and a *counter*. The former is the more general term, while the latter applies to devices which record individual events. Counters are usually associated with differential, electrical output signals, that is, charge or voltage pulses, but the description may be applied to visible signals, such as those of a bubble chamber, which record individual radiations, or to some integrating systems which produce a current output.

Since the electrical signals generated in a pulse-counting device are usually very small, an integral part of such a counting system is an amplifier. This is divided into two parts. One of these is the *main amplifier*, which provides a high, linear gain and outputs a relatively large voltage pulse, of the order of a few volts, suitable for further processing. The other part is the *preamplifier*, which supplies the main amplifier with a voltage pulse. The preamplifier has two important functions. As the earliest stage in the amplification process, its output experiences the full gain of the main amplifier, so it has to be a *low-noise* device to maintain a good signal-to-noise ratio. In addition, it is designed to match the particular detector used, so that whatever the form of the detector signal, the product of the detector–preamplifier system is compatible with any main amplifier. In this respect, the preamplifier is almost an integral part of the detector.

BIBLIOGRAPHY

MARION, J. B. and YOUNG, F. C., *Nuclear Reaction Analysis,* North-Holland, Amsterdam (1968)
PRICE, W. J., *Nuclear Radiation Detection,* McGraw-Hill, New York (1958, 1964)
SIEGBAHN, K. (Ed.), *Alpha-, Beta-, and Gamma-Ray Spectrometry,* North-Holland, Amsterdam (1965)

Chapter 6
Detector functions

6.1 COUNTING

Counting involves the determination of the number of events occurring in a detector within a prescribed time. If this is unit time, the *count rate* is measured — for example, as counts per second or per minute. The most common objective is to measure a source strength, S, via equation 5.3. The experimental arrangement employed is known as a *counting system* and, in general, it must exhibit at least four of the detector properties described in the previous chapter. It must have a reasonable detection efficiency for the radiations being counted. It must provide a differential and, usually, an electrical signal output to register and record individual particles or photons. It must measure the time of these events in order to define a counting time and it must have some ability to discriminate against unwanted radiations. In some special situations there is also a need to measure energy and/or position, usually in order to select radiation of one type or energy.

Counting systems

A simple counting system is illustrated in *Figure 6.1*. This is suitable for sealed sources of relatively low activity. The source may be handled with *forceps* to

Figure 6.1 A typical counting system. (Courtesy of Nuclear Enterprises Ltd)

preserve some distance between it and the user's hand, thereby utilising some geometrical attenuation to minimise the radiation dose at the hand. The source is placed on a *source holder* and, along with the sensitive volume of the detector, is located within a lead shield known as a *lead castle*. This reduces the dose to the experimenter but its main purpose is to minimise background counts from adjacent experiments and from cosmic radiation. The lead castle reduces the cosmic ray count by a factor of two or three.

The detector is connected by a single electrical lead to a *preamplifier* which, in turn, is fed by two leads, one to supply it with a high voltage (or bias or HT) and the other to take the signal pulse out. If the preamplifier is mounted on the detector, the assembly may be described as a *probe unit*. The remaining electrical components include a high-voltage supply, a power supply for all the units involved, a low-level discriminator to eliminate noise pulses, an amplifier, a *scaler* to count the pulses or a *ratemeter* to record the count rate and perhaps a clock to allow counting to proceed for some preset time. These, and any other units, such as an energy analyser, may be contained in a single piece of apparatus or they may be *modular*, which enables separate units to be removed or exchanged as required.

A similar arrangement can be used for more active sources (of the order of a millicurie) if larger, *remote-handling tools* are employed but for very large source strengths a *remote-handling facility*, with thicker shielding and built-in remote-handling tools, should be used. The basic system can, however, be used for open sources. These may be evaporated, or deposited on metal *planchets* and sealed temporarily with tape. Alternatively, sources in test tubes may be inserted into lead castles designed for that purpose. As a general rule, large and open sources should be handled in an area separate from that in which small, sealed sources are used. The former present special radiation hazards and the latter may be spilled and contaminate equipment. Open sources should, of course, be handled with disposable gloves, and other forms of protective clothing should be worn to avoid contamination of personnel.

This is a very basic counting system. There are many variations. Some of these have been introduced in the previous chapter and others will be discussed, in a more appropriate context, with the treatment of individual types of detector. It is useful, however, to review some of these systems very briefly, in order to establish an overall point of view. In addition, it is worth pointing out that systems and techniques at present based on one type of detector may be transferable to other types of detector.

Special geometries are employed to maximise detection efficiency. In 2π geometry, a point source is located very close to a large detector which subtends a solid angle of 2π steradians at the point. In 4π geometry the source is contained within the sensitive volume, in suspension, solution or in a cavity. This is the arrangement with the maximum possible value of G, the geometrical attenuation factor, namely unity.

Specific radiations require particular types of detection system designed on the basis of the theory of interactions with matter. Alpha particles, for example, must have thin entry windows to penetrate to the sensitive volume. The same is true of soft X-rays. Gamma rays require high Z number detectors, fast neutrons require low A number materials and thermal neutrons require special isotopes such as ^{10}B, ^{6}Li and ^{3}He. The methods available for discrimination have been described in Chapter 5.

Coincidence counting techniques can be applied to any source, whether it is one of direct or scattered radiations, that provides two simultaneous events. The advantages of the method are first that specific decay processes can be identified, and secondly that source strength can be measured without any knowledge of the detection efficiency of the system (Chapter 5, Section 5.2).

Automatic counting techniques are employed, in biological and chemical measurements, when a large number of sources have to be counted under controlled conditions. At present, the most common procedures involve the detection of very low-energy betas from ^{3}H and ^{14}C, so these materials are dissolved in liquid scintillators which are presented, in turn, to a photomultiplier counting unit, by means of an automatic sample changer. These and other systematic counting processes are often described as *radioassay* techniques.

Low-level counting systems are used to measure the activities of very weak sources. The main problem is increased statistical counting errors due to background radiations. For this reason, the source–detector arrangement should be heavily shielded against extraneous radiations and, whenever possible, the experiment should be conducted in a low-activity environment. There are two types of background count which cannot easily be eliminated. These are due to cosmic radiation and to radioactive isotopes occurring naturally in the materials of the detector and the shielding structure.

Cosmic ray counts can be reduced by means of thick lead shielding but a more effective approach is to locate one (or more) other detector above the source–detector assembly so that cosmic rays tend to register in both detectors simultaneously but the low-activity source affects only one at a time. Thus, the two detectors are in line with cosmic radiations but not with the source emissions. If these detectors are connected in anticoincidence, the cosmic ray counts will be rejected.

Residual activity in structural materials can be eliminated only by not using such materials. Lead, as the final product of natural radioactive chains, may be quite active. The best materials are ferrous metals salvaged from artefacts that have neither been exposed to nor processed in an atmosphere contaminated by radioactive fallout from thermonuclear devices. This means pre-1945 steel. Also, if the detection system is sensitive to X-radiation this can present a severe problem since most materials emit some thermally excited X-rays at room temperature.

Dead-time measurement

An important part of any counting procedure is to estimate and allow for dead-time counting losses. A preliminary estimate can be obtained from a knowledge of the detector characteristics but if this appears to suggest significant losses, a more accurate determination must be made.

The standard method employs two sources. One of these is placed in position and produces a recorded count rate of N_1. The second is then placed alongside the first, without disturbing it, so that the count rate, due to both sources, is increased to N. The first source is then removed without disturbing the second one, and the count rate falls to N_2. Apart from dead-time losses, the detection efficiency for each source remains constant, so the interaction rates are additive, i.e. from equation 5.9,

$$n_1 + n_2 = n \tag{6.1}$$

$$\therefore \frac{N_1}{1 - N_1\tau} + \frac{N_2}{1 - N_2\tau} = \frac{N}{1 - N\tau} \tag{6.2}$$

Hence, the values of N_1, N_2 and N can be inserted to obtain τ.

Counting statistics

For every recorded number of counts, N, there is an associated statistical uncertainty. This is partly due to statistical effects in the emission of radiations and partly due to statistical effects in the detection process. The net result is that the variable, N, is not always found to have the same value, even if experimental conditions remain constant, but its value fluctuates about some mean or 'true' value which is the average that would be obtained over an infinite number of measurements. Clearly, this true value cannot actually be measured. All that can be done is to assess how close to the true value is the measured value, N.

If several measurements of N are made, and these give values in excess of about 50, then a graph of the number of times a given value occurs, that is, the

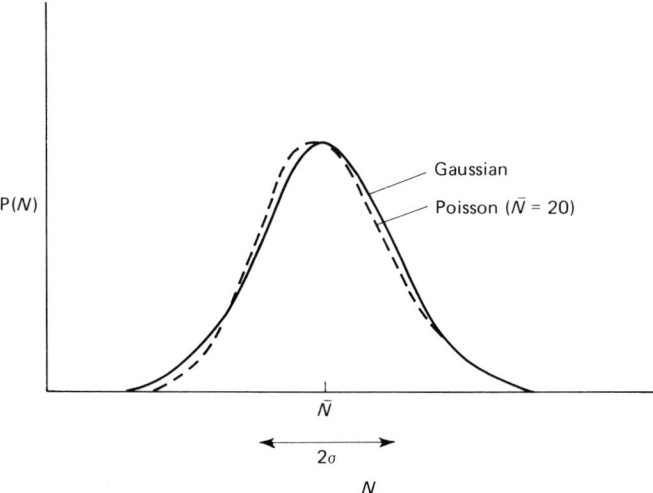

Figure 6.2 *Gaussian and Poisson distributions for counting statistics*

probability $P(N)$ of obtaining a value N, against the values of N, is a Gaussian (or normal) distribution function, as illustrated in *Figure 6.2*. For N values less than about 50, a Poisson distribution is obtained. These functions are of the forms

$$\text{Gaussian:} \quad P(N) = \frac{\exp\left[-(N - \bar{N})^2/2\bar{N}\right]}{(2\pi\bar{N})^{1/2}} \tag{6.3}$$

Poisson: $$P(N) = \frac{(\bar{N})^N e^{-\bar{N}}}{N!} \tag{6.4}$$

In each case, $\bar{N}$ is the mean value of N averaged over all the values obtained. It is not the true value but it is the best estimate available from the results obtained. In most counting procedures, only one value of N is measured. In effect, this is the sum of all the counts used to construct the probability distribution function and corresponds to the mean of this distribution. In any case, it is the only value available and is itself the mean of a distribution of one, i.e. $N = \bar{N}$. It is necessary to estimate how close this value is likely to be to the true value; that is, to allocate a *statistical uncertainty* to the measured value of N.

Statistical uncertainty is described by the width of the distribution function and this can be measured by the *variance*, V, which is the average of the squares of the deviations of all the counts, N, from the mean value, $\bar{N}$. The square root of this parameter is called the *standard deviation*, σ, hence

$$V = \sigma^2 = \frac{\Sigma(\bar{N} - N)^2}{\Sigma P(N)}$$

It can be shown from equations 6.3 and 6.4 that the standard deviation, for a single count, N, is given by

$$\sigma = N^{1/2} \tag{6.5}$$

Thus, the width of the associated distribution and the statistical uncertainty increases with recorded counts, N, but the *relative* uncertainty decreases. This is given by

$$\text{Relative uncertainty} = \frac{\sigma}{N} = \frac{1}{N^{1/2}} \tag{6.6}$$

and is usually expressed as a percentage. If, for example, $N = 100$, there is a relative uncertainty of 10%; if $N = 1000$, it is 3.2%, and if $N = 10\,000$, it is 1%. Accuracy improves with increase in the value of N.

Although statistical uncertainty is measured by the standard deviation, it is not necessarily equal to this parameter. In other words, there is no guarantee that the measured value of N will lie within $+\sigma$ or $-\sigma$ of that value. Only probabilities can be quoted, as follows:

The probability is 0.683 that the true value is in the range $N \pm \sigma$

The probability is 0.954 that the true value is in the range $N \pm 2\sigma$

The probability is 0.997 that the true value is in the range $N \pm 3\sigma$

The end product of a counting measurement is a result of the form $N \pm dN$, where dN is the experimental uncertainty due to statistical and other effects, such as timing errors. If represented in a graph, a line should be drawn through the data point, N, extending a distance dN on each side. This is called an *error bar*.

144 Detector functions

The count rate, N, may not be obtained directly but may be the result of other measurements. In this case, it is given by some function $F(x, y, z)$ of these measured variables, x, y and z. If these are statistical variables the total error is given by adding their uncertainties, σ_x, σ_y and σ_z, in quadrature, i.e.

$$dN = \sigma = \left[\left(\frac{\partial F}{\partial x}\right)^2 \sigma_x^2 + \left(\frac{\partial F}{\partial y}\right)^2 \sigma_y^2 + \left(\frac{\partial F}{\partial z}\right)^2 \sigma_z^2\right]^{1/2} \tag{6.7}$$

where $\partial F/\partial x$ represents the partial derivative of the function, F, with respect to the variable, x, and $\partial F/\partial y$ and $\partial F/\partial z$ have corresponding meanings.

If x, y and z are not statistical the total error, dN, is given by

$$dN = \left|\frac{\partial F}{\partial x}\right| dx + \left|\frac{\partial F}{\partial y}\right| dy + \left|\frac{\partial F}{\partial z}\right| dz \tag{6.8}$$

where dx, dy and dz are the uncertainties in x, y and z, respectively, and the vertical lines represent the modulus, or absolute value, of the partial derivative; that is, negative signs are ignored.

As an example of this theory, the count rate N may be the difference between a count rate N_1 and a background count rate N_2. If $N_1 = 10\,000$ and $N_2 = 200$, the total uncertainty is obtained as follows:

$$N = N_1 - N_2$$

$$\therefore dN = \left[\left(\frac{\partial F}{\partial N_1}\right)^2 \sigma_1^2 + \left(\frac{\partial F}{\partial N_2}\right)^2 \sigma_2^2\right]^{1/2}$$

$$= \left[\sigma_1^2 + \sigma_2^2\right]^{1/2}$$

$$= (10\,200)^{1/2} = 114$$

$$\therefore N = 9800 \pm 114$$

If errors in time measurement are significant, they must be included via equation 6.8.

6.2 PULSE HEIGHT SPECTROMETRY

In pulse height spectrometry, each particle or photon absorbed in the detector generates an output analog voltage pulse whose height is proportional to the energy deposited. A graph of count rate against pulse height is called a *pulse height spectrum* and is equivalent to the radiation energy spectrum in all respects except two. First, line-broadening effects convert spectral lines into peaks. Secondly, radiations may be incompletely absorbed so that pulse heights represent only parts of their energies.

There are several methods of measuring the energies of ionising radiations but for most purposes pulse height spectrometry is the most convenient. In general, it is the simplest approach and the most accurate for photons, which are the most widely used type of radiation. Outside of high-energy physics, where radiations are usually too energetic to be stopped in a spectrometric detector, pulse height spectrometry is the form of energy spectrometry most often used.

Pulse height spectrometers

A typical spectrometer system is illustrated in *Figure 6.3*. It comprises an analog detector, the usual HT and power supplies, preamplifier and amplifier (which must be linear to preserve the proportionality between pulse height and energy), and a *pulse height analyser* (PHA). The pulse height analyser can be a *single-channel analyser* (SCA) or a *multichannel analyser* (MCA).

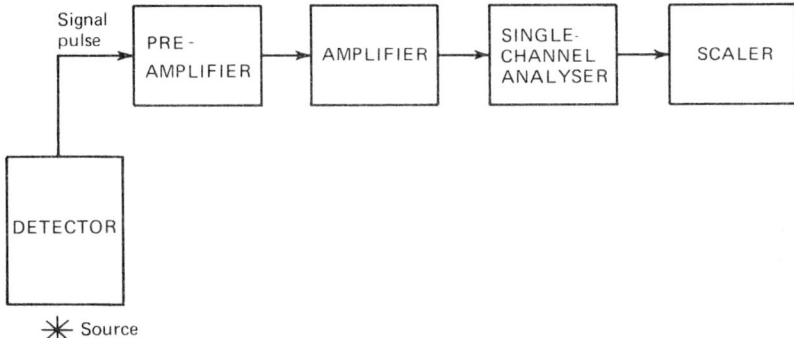

Figure 6.3 Basic structure of a pulse height spectrometer

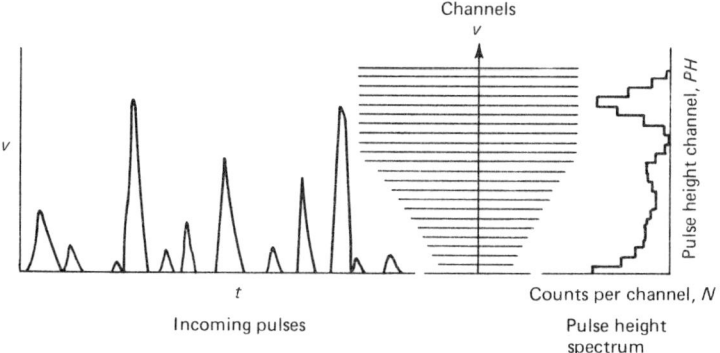

Figure 6.4 Pulse height analysis in terms of discrete energy channels

In single-channel analysis, the single-channel analyser passes and counts only those signal pulses whose heights fall within a narrow range of values. They must exceed a lower limit, or *threshold* voltage, v, but be smaller than an upper limit, $v + dv$, where the voltage interval, dv, is known as the window or *channel width*. If dv is kept constant while v is increased, continuously or incrementally, from zero to some large value, the count rate varies as the number of pulses in each voltage interval. This process is called *scanning* and the result is a graph of count rate against pulse height; that is, a pulse height spectrum. Alternatively, the channel can be set to a fixed position to select a particular range of pulse heights such as those produced by one type of radiation.

146   Detector functions

Multichannel analysis effectively measures the counts in many energy channels simultaneously. Typically, there are 256, 512, 1024, 2048 or 4096 channels, so the procedure is much faster than that of a single-channel analyser. The mechanism involved is illustrated in *Figure 6.4*, which shows analog pulses arriving at an MCA and being sorted into channels according to their heights to form a pulse height spectrum. This is actually a histogram, with one 'step' per channel. If there are only a few channels in a spectral peak the effect becomes quite noticeable and channel width becomes an important contribution to FWHM resolution. For example, the peak of a semiconductor spectrum may cover only three channels. If the peak occurs at channel 100, there is an uncertainty of at least 1% in determining the FWHM peak width, to add to the 1% due to statistical effects. In such cases, the spectrum should be expanded to cover more channels.

Energy measurement

In order to measure the energy associated with a spectral peak, the pulse height spectrum has to be calibrated against known radiation energies. This means that the proportionality constant between pulse height and energy must be determined. Since pulse shape, and height, may depend on the type of radiation detected, the radiations used to generate the calibration peaks must be of the same type as that producing the peak of unknown energy. In order to obtain an accurate calibration, they should also be of similar energy, producing peaks above and below the unknown peak.

There are two methods of calibrating pulse height spectra, namely, a *graphical* and an *algebraic* method. The graphical method is the more accurate but the more time-consuming. It is illustrated in *Figure 6.5*. The peaks due to the known energies, E_1, E_2, E_3 and E_4, are superimposed on the same spectrum as that due to the unknown energy, E. Vertical lines are drawn through the mean positions of the known peaks and the heights of these lines are proportional to the energies

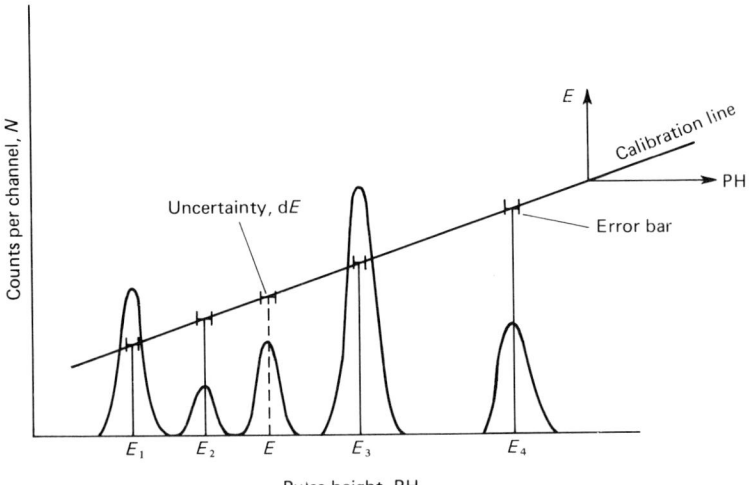

Figure 6.5 Calibration of a pulse height spectrum

E_1, E_2, E_3 and E_4. They bear no relationship whatsoever to the heights of these peaks, which are proportional to count rate, N.

The tops of these vertical lines are then joined by a *calibration line*, as shown. If the detector has a good linear response for the energy range involved, this is a straight line, but in general it does not extrapolate to the origin since detectors tend to be non-linear at very low radiation energies. The calibration line is a graph of energy against pulse height and its point of intersection with a vertical line through the mean position of the unknown peak gives the energy value of that peak.

In the algebraic method, the dispersion, d, of the spectrometer is calculated. This is the energy interval per unit pulse height interval and is given, for example, as keV mV^{-1} or keV per channel. It is the energy interval between two calibration peaks, $E_2 - E_1$, divided by the corresponding pulse height interval, PH$_2$ - PH$_1$, in volts or channels, i.e.

$$d = \frac{E_2 - E_1}{PH_2 - PH_1} \tag{6.9}$$

The energy, E, of the unknown peak at a mean pulse height PH is obtained by reference to one of the calibration peaks, such as E_1, by

$$E = E_1 + d(PH - PH_1) \tag{6.10}$$

As mentioned in Chapter 5, the result of an energy measurement is $E \pm dE$, where dE is the total experimental error. When it is not known how many energies are present, unresolved, in a spectral peak, dE is equal to the FWHM resolution. If, as is often the case with calibration sources, it is known that only one energy is represented in a peak, then dE is much smaller than the FWHM resolution. It is the accuracy with which the mean value of the peak can be located and how close this is likely to be to the true value of energy defined as the average over an infinite number of measurements. This statistical error can be measured by the standard deviation, σ_E, of the mean, $\bar{E}$, of the peak from the true value of energy, E, and is given by

$$dE = \sigma_E = (\bar{E}\epsilon/N)^{1/2} \tag{6.11}$$

where N is the total number of counts in the peak.

Error bars showing the magnitude of this effect can be seen in *Figure 6.5*.

Gamma ray spectrometry

Photon spectrometry, and especially gamma ray spectrometry, is the most widely used pulse height spectrometric technique. It also presents special problems owing to the high probability of partial absorption, and for these reasons is worthy of a separate study.

In order to understand the spectral effects, it is essential to remember that the energies deposited in the detector are primarily those of the *electrons* ejected from atoms in the absorption processes. Thus, in photoelectric capture events the energy deposited is $E_\gamma - E_B$, where E_B is the electron binding energy, and a spectral peak should be formed at a mean pulse height corresponding to an energy $E_\gamma - E_B$. In practice, the atom tends to de-excite, emitting characteristic X-rays, almost immediately. These are usually detected along with the photoelectron to give a mean pulse height equivalent to the full photon energy, E_γ. If this does not happen, the fact will be evident from the formation of a separate, low-energy peak at the characteristic X-ray energy. In any case, the atomic binding energy, E_B, is usually much smaller than E_γ so the photoelectron peak can be assumed to register the photon energy, E_γ. For this reason, it is known as the *photopeak* or the *full absorption peak* and is the part of the spectrum from which the photon energy is determined.

The remaining parts of the spectrum, illustrated in *Figure 6.6*, are mainly due to Compton scattering. The distribution of Compton electron intensity, as a function of energy, for events taking place in the detector's sensitive volume, is described by the differential cross-section, $d\sigma(E_e)/dE_e$. This has been discussed in Chapter 4 and is illustrated in *Figure 4.21*. The intensity increases up to a maximum value at an energy E_c, which is the maximum energy that can be given to a Compton electron, and this spectral feature is known as the *Compton edge*.

A second prominent feature is produced by the photoelectric capture of photons which have been Compton-scattered into the detector from its environment and structural materials. The intensity distribution of these photons is given by the differential cross-section, $d\sigma(E_\gamma)/dE_\gamma$, also discussed in Chapter 4 and illustrated in *Figure 4.21*. This shows a sharp cut-off at the low-energy end corresponding to the minimum energy which a Compton-scattered photon can have when it is scattered backwards. This gives rise to a spectral peak called the *backscatter peak*, at an energy E_s.

Because of the statistical processes contributing to energy resolution, these distributions are less sharp in a pulse height spectrum than shown in *Figure 4.21*, but they remain well defined. The energy, E_s, of the backscatter peak depends on the original photon energy as described in *Figure 4.19*, and is equal to the energy gap, $E_\gamma - E_c$, between the photopeak and the Compton edge. These are useful data which can be used to calibrate a pulse height spectrum. For ^{137}Cs, for example, a photopeak is formed at an energy of 0.662 MeV, a Compton edge is located at an energy of 0.482 MeV and a backscatter peak is found at 0.180 MeV (*Figure 6.6a*). The only problem with this approach is that these *escape peaks* are usually less pronounced than full absorption peaks.

If the photon energy exceeds 1.022 MeV it may undergo pair production interactions, forming an electron and a positron. The latter annihilates virtually immediately to give two 0.511-MeV photons. If one of these escapes from the detector a spectral peak is produced at an energy level of $E_\gamma - 0.511$ MeV. If both escape, a peak is produced at an energy level of $E_\gamma - 1.022$ MeV. These tend to be relatively minor effects, confined to high-energy photons.

A photon pulse height spectrum has a fairly complicated structure and, if several photon energies are present, it may be difficult to distinguish photopeaks from the escape peaks of higher-energy photons. The quality of a spectrum, in this respect, is measured by two parameters. The first of these is the FWHM *energy resolution*, dE/E, which determines the sharpness of the photopeaks. It also determines the height of the photopeaks since a narrow peak is much taller than a broad peak containing the same number of counts. The second parameter can be defined as the peak-to-total ratio; that is, the fraction of all counts contained in the photopeak. This is determined by the relative magnitudes of the photoelectric and Compton interaction cross-sections and is important in counting measurements. It is less useful in spectrometry because it does not completely relate to the prominence of the photopeak. A better parameter is the *peak-to-Compton ratio*, which is the ratio of the height of a photopeak to that of its Compton edge and immediately describes the spectral quality.

The peak-to-Compton ratio, PCR, depends on the energy resolution, the ratio of photoelectric to Compton effect cross-section and the geometry of the detector. Some typical effects are illustrated in *Figure 6.6*. Germanium has a

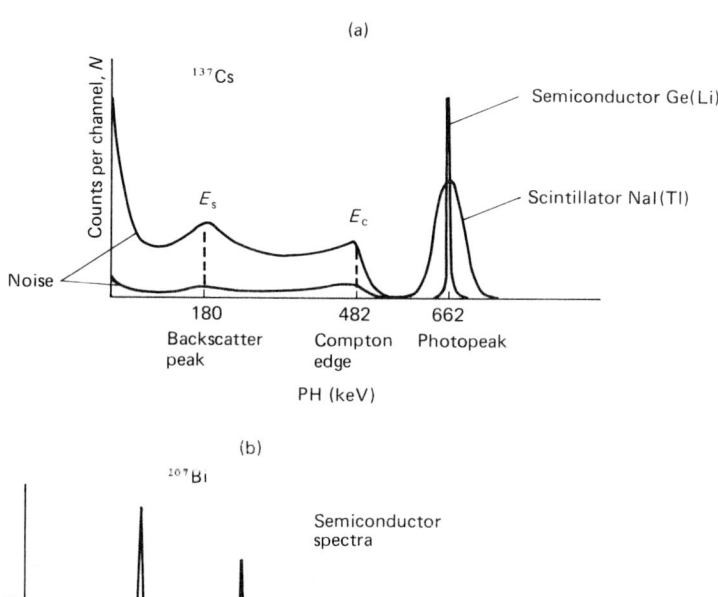

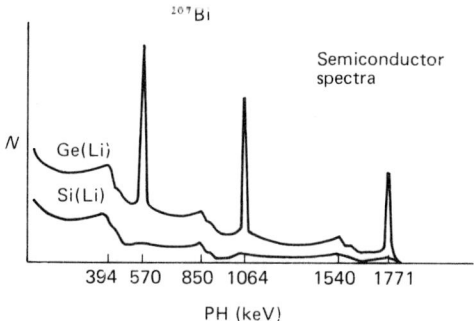

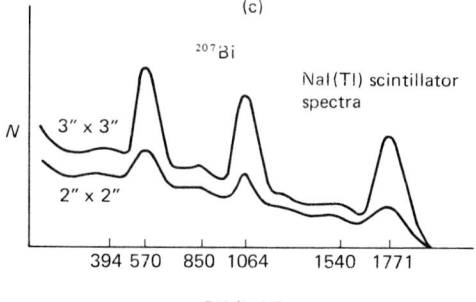

Figure 6.6 Gamma ray pulse height spectra for (a) ^{137}Cs; (b) ^{207}Bi with semiconductor detectors; (c) ^{207}Bi with scintillation counters

much better resolution than sodium iodide so its photopeaks are sharper and taller. Photoelectric capture cross-sections are proportional to Z number to the power of about 5, so the photopeaks in germanium ($Z = 32$) are much larger than the almost non-existent ones in silicon ($Z = 14$). The PCR in a 3 in × 3 in (762 × 762 mm) sodium iodide detector is larger than in a 2 in × 2 in (508 mm × 508 mm) device because Compton-scattered photons are more likely to be completely captured in a second, photoelectric process in the larger detector. Incidentally, the bump on the side of the Compton edge for the germanium spectrometer is due to second Compton interactions producing a second Compton edge effect. This is not observed in the sodium iodide spectrum because the resolution is not good enough.

6.3 DOSIMETRY

The biological effect of ionising radiation depends on

(1) the total energy deposited and, therefore, on the energy and intensity of the radiation;
(2) the spatial distribution of ion pairs formed and, therefore, on the interaction mechanism and the type of radiation; and
(3) the nature of the tissue being irradiated.

These effects can be quantified in terms of units defined by the International Commission on Radiological Units (ICRU). There are three types of measurement, as follows:

Absorbed dose, D, specifies the energy imparted to a region per unit mass of material. It is defined by the formula

$$D = \frac{\Delta E}{\Delta m} \tag{6.12}$$

where ΔE is the energy deposited in an elemental volume and Δm is the mass of that volume. Since absorbed dose can vary rapidly with position, especially in an inhomogeneous medium such as the human body, it is defined for an elemental volume but this is assumed to be large enough to provide a good statistical average. It represents the average value of $\Delta E/\Delta m$ at a point.

Until recently, the standard unit of absorbed dose has been the *rad*, where 1 rad denotes an energy absorption of 10^{-2} J kg^{-1}. This unit is still widely used but it has been replaced by the *gray* (Gy), where 1 Gy represents an energy absorption of 1 J kg^{-1}. Both units are independent of the type of radiation or material involved.

Exposure, X, specifies the charge deposited by photons in unit mass of dry air at STP. It is defined by the formula

$$X = \frac{\Delta Q}{\Delta m} \tag{6.13}$$

where ΔQ is the total charge, in ions of one sign, deposited in an elemental volume and Δm is the mass of that volume. All the electrons liberated by the photons must be stopped in air, although not necessarily in the elemental volume. This means that the volume must be surrounded by air so that it gains as many electrons as it loses and edge effects can be ignored.

The unit of exposure is the *röntgen* (R), which is equal to a charge production of 2.58×10^{-4} C kg^{-1}, of each sign. This rather odd value is derived from Röntgen's original definition as the number of electrostatic units deposited in 1 cm^3 of air at STP (0.001 293 g).

Dose equivalent, DE, specifies the biological effect of a particular type of radiation. It is the absorbed dose multiplied by a *quality factor*, QF, or *relative biological effectiveness*, RBE, which accounts for the ability of the radiation to cause biological damage. Thus, dose equivalent is given by the formula

$$DE = D \times QF \qquad (6.14)$$

The quality factor depends upon and increases with the density of ionisation; that is, with the linear energy transfer (LET) of the radiation. For X-, gamma and beta radiation, it is equal to unity. For thermal neutrons, it is about 3. For alpha particles and for fast neutrons, via their knock-on protons, it is of the order of 10.

The unit of dose equivalent is the *rem*, which is the effect of one rad deposited by X-, gamma or beta radiation. The same amount of energy deposited by thermal neutrons, for example, is three times as damaging to biological tissue, and is measured as 3 rem.

As for activity and absorbed dose, there is a new SI unit of dose equivalent known as the *sievert* (Sv). It is the effect of 1 Gy deposited by radiation with a quality factor of one. These SI units were defined by the Conférence Générale des Poids et Mesures (CGPM) in 1975, and they are summarised in *Table 6.1*.

Dose is a measure of the energy deposited by radiation in matter, irrespective of the nature of either the radiation or the matter. Exposure, on the other hand, defines a quantity of specifically electromagnetic radiation, by its effect on air. Since 1 R produces 2.58×10^{-4} C in 1 kg of air, and the charge on each electron is 1.6×10^{-19} C, then it generates 1.61×10^{15} ion pairs per kilogram of air. The mean energy, $\bar{e}$, required to form an ion pair in air is about 33.7 eV per kilogram of air. This is equal to 0.0087 J kg^{-1}, or 0.87 rad, so 1 R deposits 0.87 rad in air.

Table 6.1 SI units of radioactivity, absorbed dose and dose equivalent

Quantity	SI unit	Non-SI unit	Relationship
Activity	becquerel (Bq) 1 dis s^{-1}	curie (Ci) 3.7×10^{10} dis s^{-1}	1 Bq = 0.27×10^{-10} Ci
Absorbed dose	gray (Gy) 1 J kg^{-1}	rad 10^{-2} J kg^{-1}	1 Gy = 100 rad
Dose equivalent	sievert (Sv) 1 Gy × QF	rem 1 rad × QF	1 Si = 100 rem
Exposure	coulomb kg^{-1} 1 C kg^{-1}	röntgen (R) 2.58×10^{-4} C kg^{-1}	1 C kg^{-1} = 3876 R

Dose calculation

Exposure and dose are both defined for elemental volumes, in which ionising radiations are attenuated differentially, as described in Chapter 4, and given by equation 4.10, viz. $dI = N\sigma I\, dx$. This equation provides the basis for dose calculations but it must be adapted slightly. In the first place, the elemental path length, dx, must be expressed in terms of equivalent mass, dm, in order to calculate the effect on unit mass. dm may be given, for example, in g cm^{-2}. Secondly, there is some ambiguity in the meaning of the intensity, I, which may refer to the number of incident particles or photons in unit time, so that dI defines the interaction rate, or it may refer to the total, integrated intensity, so that dI refers to the total number of counts recorded over the measurement time. In order to distinguish between these two interpretations, it is useful to introduce the concept of *particle fluence*, Φ, which is the total number of incident radiations per unit area from all directions, integrated over the exposure time. This relates specifically to dose rather than dose rate.

For *photons*, equation 4.10 becomes

$$\frac{d\Phi}{dm} = \mu_m \Phi \qquad (6.15)$$

where $d\Phi/dm$ is the number of interactions per unit mass in an elemental volume and μ_m is the *mass attenuation coefficient*, in cm^2 g^{-1}, for example.

A similar equation can be constructed to specify the energy deposited. The total radiation energy incident on 1 cm^2 of matter from all directions is known as the *energy fluence*, F. For monoenergetic radiations, $F = E\Phi$, where E is the energy of an individual particle or photon. For non-monoenergetic radiations, F is the appropriate integral of Φ with respect to E. In either case, energy attenuation is given by

$$\frac{dF}{dm} = \mu_{mE} F \qquad (6.16)$$

where dF/dm is the energy absorbed per unit mass and μ_{mE} is the *mass–energy attenuation coefficient* or the *mass absorption coefficient*.

There is also an absorption coefficient, or linear energy attenuation coefficient, μ_E, corresponding to the linear attenuation coefficient, μ. If E_a is the energy absorbed from one photon, then

$$\mu_{mE} = E_a \mu_m / E$$
$$\mu_E = E_a \mu / E \qquad (6.17)$$

Values of μ_{mE} and μ_m for air are given in *Figure 6.7*. In the energy range where photoelectric capture predominates they are similar in magnitude to μ_E and μ, respectively, but in the higher energy range where Compton scattering is important, E_a is less than E, and the values of μ_{mE} and μ_m are smaller than those of the linear coefficients.

If F and dF are in mega-electronvolts and μ_{mE} is in cm² g⁻¹, then equation 6.16 gives the absorbed dose, for photons, in MeV g⁻¹. To convert to rads, this value must be multiplied by a conversion factor of 1.6×10^{-8}, to give

$$D = 1.6 \times 10^{-8} \mu_{mE} F \quad \text{rads}$$
$$= 1.6 \times 10^{-8} \mu_{mE} E \Phi \quad \text{rads} \tag{6.18}$$

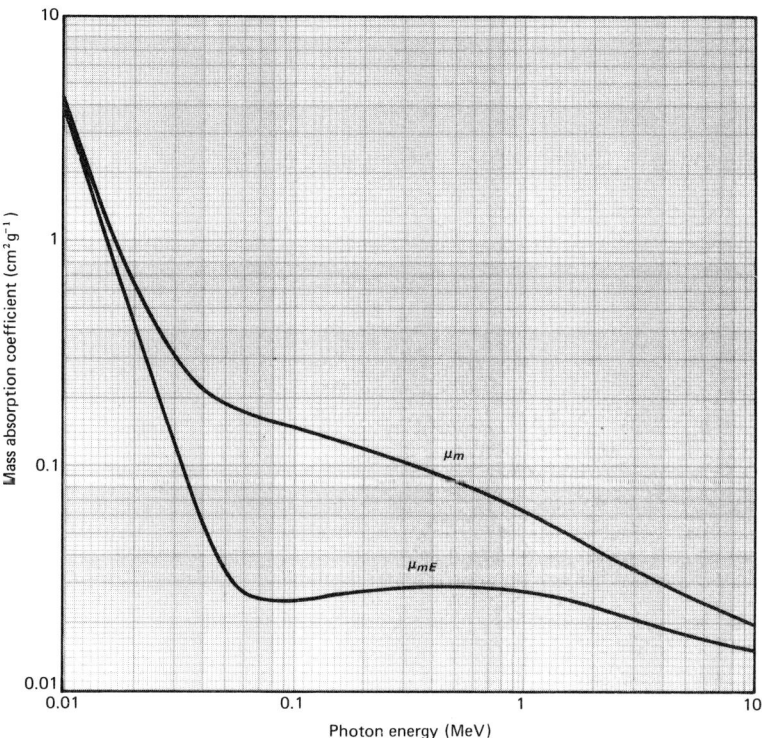

Figure 6.7 *Mass attenuation coefficient, μ_m, and absorption coefficient, μ_{mE}, for air*

For *charged particles*, dF/dm is simply the beam intensity per unit area, Φ, multiplied by the mass stopping power, dE/dm, so, if dE/dm is in MeV g⁻¹ cm⁻²,

$$D = 1.6 \times 10^{-8} \Phi \, dE/dm \quad \text{rads} \tag{6.19}$$

Exposure, X, is equal to the energy, dF/dm, deposited by photons, divided by the energy required to form an ion pair in air, ϵ_{air}, and multiplied by the electronic charge, e. This gives the result in C g⁻¹. To convert to röntgens, it must be divided by 2.58×10^{-7}. Inserting the usual values for ϵ_{air} and e, and replacing dF/dm by $\mu_{mE} E \Phi$, gives the formula

$$X = 1.84 \times 10^{-8} (\mu_{mE})_{air} E \Phi \tag{6.20}$$

where E is in mega-electronvolts, Φ in photons cm^{-2} and μ_{mE} is in cm^2 g^{-1}.

The relationship between dose and exposure can be derived by comparing equations 6.17 and 6.19. The dose produced in air, D_{air}, by an exposure of X röntgens is given by

$$D_{air} = 0.89X \quad \text{rads} \tag{6.21}$$

and the dose, D, that would be produced by the same quantity of radiation in any other material, with an energy absorption coefficient μ_{mE}, is given by

$$D = 0.89\mu_{mE} X/(\mu_{mE})_{air} \tag{6.22}$$

Equations 6.21 and 6.22 can be used to convert from dose to exposure, and vice versa. Equations 6.18, 6,19 and 6.20 can be used to calculate dose and exposure from the energy and intensity of incident radiations. In general, both the energy and the intensity of these radiations depend on geometrical and material attenuation between the source and the elemental volume under examination. These factors can be estimated by means of the attenuation formulae developed in Chapter 4, as used, for example, in the determination of detector efficiency (Chapter 5). Formally, the calculation involves complicated integration with respect to spatial and energy coordinates but, for a point, external source emitting S particles or photons of a single energy, the beam intensity reaching the elemental volume is given by

$$\Phi = GMS \tag{6.23}$$

where G is the geometrical attenuation factor and M the material attenuation factor.

For a point source in air, M can be assumed to be equal to 1.0 and G is $1/4\pi h^2$, where h is the distance from source to elemental volume. From equation 6.21, exposure is given by

$$X = 1.84 \times 10^{-8} SE\mu_E/4\pi h^2 R \tag{6.24}$$

If photons of two or more energies are emitted by the source, the total exposure is the sum of the contributions provided by each set of values of S, E and μ_E. Such calculations have been completed for a range of standard sources for $h = 1$ m and are given in *Table 6.2* as *specific gamma ray constants*, Γ; that is, for these conditions,

$$X = \Gamma \quad \text{R h}^{-1} \text{ Ci}^{-1} \text{ cm}^{-2} \tag{6.25}$$

These data can be used to compare the effects of different sources for other values of h and, if the appropriate modifications are made, for values of M other than 1.0.

As an example of the use of equation 6.25, ^{137}Cs emits 0.85 photons per curie, each with an energy of 0.662 MeV. The mass–energy attenuation coefficient,

Table 6.2 Specific gamma ray constants, Γ, giving the gamma ray dose, in R h^{-1}, absorbed by a 1 cm^2 area 1 m distant from a point 1 Ci source in air

Radionuclide	$T_{1/2}$	Gamma ray energy (MeV) and abundance (%)	Γ
^{22}Na	2.6 years	0.511 (181), 1.275 (100)	1.20
^{24}Na	15.02 h	1.369 (100), 2.754 (100)	1.84
^{51}Cr	27.7 d	0.005 (22), 0.32 (10)	0.018
^{59}Fe	44.6 d	0.143 (1), 0.192 (3), 1.099 (56)	0.60
^{57}Co	270 d	0.006 (55), 0.014 (9), 0.122 (85), 0.136 (11)	0.059
^{58}Co	71 d	0.006 (26), 0.511 (30), 0.811 (99), 0.864 (1)	0.055
^{60}Co	5.27 years	1.173 (100), 1.333 (100)	1.28
^{131}I	8.06 d	0.08 (2), 0.284 (6), 0.364 (82), 0.637 (7), 0.723 (2)	0.22
^{137}Cs	30.1 years	0.032 (8), 0.662 (85)	0.32
^{192}Ir	74 d	0.296 (30), 0.308 (31), 0.316 (83), 0.468 (47)	0.40
^{198}Au	2.7 d	0.412 (95), 0.676 (1)	0.23
^{203}Hg	46.6 d	0.071 (13), 0.279 (82)	0.16
^{226}Ra	27 d	*See* Appendix 3	0.84

μ_{mE}, is obtained from *Figure 6.7* as 0.03 cm^2 g^{-1}. The exposure 1 m from a 1-Ci source is given by

$$X = 1.84 \times 10^{-8} \times 0.85 \times 3.7 \times 10^{10} \times 0.662 \times 0.03/4\pi \times 10^4$$

$$= 0.00009 \text{ R s}^{-1} \text{ cm}^{-2}$$

$$= 0.32 \text{ R h}^{-1} \text{ cm}^{-2}$$

$$= \Gamma$$

given in *Table 6.2*. From this result, the absorbed dose produced by the same quantity of radiation in any material can be calculated from equation 6.22.

Maximum permissible dose

The International Commission on Radiological Protection (ICRP) has recommended maximum levels of dose and dose rate for radiation workers and other personnel. These form the basis of national legislation. They are expressed as *maximum permissible levels* (MPL) or *maximum permissible doses* (MPD), given in units of dose equivalent, that is, in rem, mrem (10^{-3} rem), mrem h^{-1}, and so on. The MPL for classified radiation workers under medical supervision, and for members of the general public, are specified. Basically, the MPL for a radiation worker is 5 rem year^{-1}, after the age of 18, and ten times the dose permitted to the general public. Assuming a 50-week working year, this breaks down to an average MPL of 100 mrem week^{-1} and, for a 40-hour working week, this gives an average dose rate of 2.5 mrem h^{-1}. These levels apply to whole-body exposure when the limits are those of the most sensitive organs, such as the gonads, red bone marrow and eyes. Other organs, especially the extremities, can tolerate much larger exposures of up to 75 rem year^{-1}, or 37.5 mrem h^{-1}.

Dose measurement

A detector used to measure radiation dose is called a *dosemeter* (or dosimeter). A device which is used to ensure that the dose remains within MPL is a *monitor*. In principle, the former makes an absolute measurement and has to be the more accurate, while the latter is a calibrated device of limited accuracy. In practice, there is very little, if any, distinction between the two terms.

Ideally, a dosemeter or monitor should measure dose equivalent and its sensitive volume should be constructed of biological tissue. Except in the sense that experimental animals may be used to observe biological effects, this is not generally possible and dosemeters record absorbed dose or exposure. In the first case, they are based on *tissue-equivalent* materials; that is, materials whose composition and density are similar to that of tissue. These include water, which is not available as a radiation detector, and various forms of plastic and liquid materials. In the second case, exposure can be measured in air sensitive volumes but this is less convenient than using the noble gases, which can be used as *air-equivalent* materials. If the devices are to be employed as monitors, they can be calibrated, in rem, against observed effects in tissue or, more easily, in water. Conversion from exposure to absorbed dose can be made as described above. Conversion from absorbed dose to dose equivalent depends upon the determination of quality factors. Since these vary from one type and energy of radiation to another, it is difficult to calibrate a single device for a wide range of radiations and for accurate work different detectors may be required. A less accurate alternative is a single device with built-in variable absorbers to eliminate one type of radiation. Thus, a thin-windowed detector registers photons and betas but, when a thick cap is placed over the window, it admits only photons. Energy compensation can be engineered in the same way. The best example of this method of selecting radiations is the film badge whose holder contains a variety of entry windows. This will be discussed later.

In general, the basic property required of a dosemeter or monitor is that it should be constructed of tissue-equivalent or air-equivalent material, to measure absorbed dose or exposure, respectively, and calibrated in dose-equivalent units. A second requirement is that it should be sensitive to all types and energies of radiations which might be encountered. A third requirement is that the signal readout should be integral, to provide a record of accumulated dose or of dose rate and the record should be permanent or semi-permanent rather than transient.

More specifically, it is possible to identify particular dosimetric functions which require additional properties. Thus, devices can be classified as absolute dosemeters, area monitors, personal monitors, surface monitors or air monitors.

Absolute dosemeters are detection systems capable of making accurate and absolute measurements of absorbed dose or exposure. They are used to observe the effects of different radiations and provide the basis for the calibration of other dosemeters and monitors. They must be energy-sensitive and able to discriminate between different types of radiation. Their detection efficiencies must be known accurately, and their dead time and coincidence counting losses must be zero or easily measured. In order to provide absolute measurements, their sensitive volumes must be determined precisely.

Area monitors, or dosemeters, are used to measure the level of radiation dose in working environments. These, too, should be accurate and, therefore, specific to particular types and energies of radiations but they can be calibrated, rather than absolute, devices. They should be regularly tested and recalibrated since they are the instruments on which operational radiation safety ultimately depends. They must also be reasonably portable so that they have access to all parts of the working area.

Personal monitors are normally worn on the chest by personnel working in radiation environments, to provide a continuous and cumulative indication of personal exposure. The main requirement is that they should be small. This limits their accuracy, not only because small devices tend to be inaccurate, having small detection efficiencies and relatively large structural deviations, but also because they may not always encounter the highest radiation intensity — which tends to be nearer the hands than the chest. Other requirements, which may not always be satisfied, are that they should be inexpensive and, if possible, reusable; they should be robust, and they should provide immediate, permanent readings.

Surface monitors are used to check exposed working surfaces, clothing, hands, feet, etc. for contamination by unsealed radioactive sources. In particular, two problems have to be overcome. First, it may be difficult to distinguish between surface contamination and general radiation background, especially when the latter is high, such as on the external surface of a supposedly sealed source. Secondly, the radiations may be very weakly penetrating and difficult to count through a detector window. Because of this property, they do not present a serious external hazard but, being unsealed, they can become an internal hazard. Tritium, which is widely used as a radioactive tracer and emits an 18-keV beta particle, is a typical example of this problem. One standard solution to both problems is the *wipe test*, in which the contaminated surface is wiped with a filter paper soaked with an appropriate solvent. This can then be transferred to a radiation detector, outside the high-background area, for measurement.

Air monitors measure atmospheric contamination by radionuclides. One approach is to pass a sample of the air through a filter paper. Particulate contaminants are deposited on the paper, which can be removed to a suitable detector. Gaseous contaminants can be passed through the paper into a detector, for monitoring.

6.4 IMAGING

Imaging can be taken to include all those functions which depend upon measuring the positions of radiations. This is a broad interpretation but useful in the present context, since similar detector properties are involved. The primary requirement

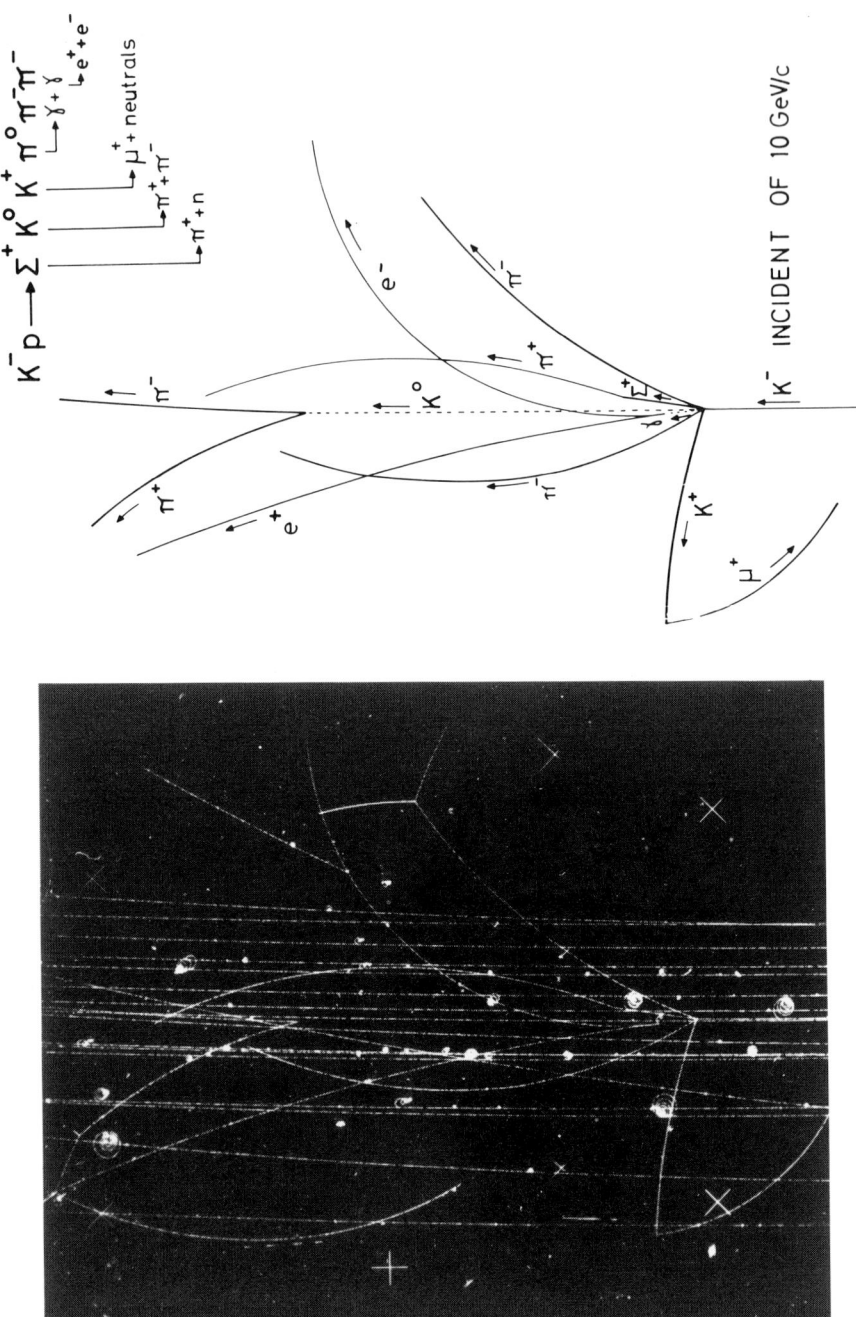

Figure 6.8 Collision of a 10 GeV c^{-1} negative kaon against a proton in the CERN 2-m liquid hydrogen bubble chamber. (PHOTOCERN)

is, of course, the ability to measure position but it is not the only requirement and, in some techniques, spatial resolution may not even be the most important parameter. Each technique has its own characteristic needs.

Tracking

Tracking is a form of three-dimensional imaging in which the tracks of individual radiations have to be observed. It is a highly developed function but, at present, a very specialised one, being restricted to high-energy and nuclear physics applications. Typically, the detection system includes a large and powerful magnet so that charged particles can be identified by the radii of their curved paths in the magnetic field (equation 5.26). The detector may be outside the magnetic field, measuring the positions, along the focal plane of the magnet, at which the particles emerge from the field. In this case it is, by itself, a one-dimensional imaging device. Alternatively, it may be within the field and produce an image of the entire curved path. *Figure 6.8* illustrates the paths of various charged particles generated by kaon absorption in a bubble chamber.

Such tracking systems must have good spatial resolution, both to separate events and to measure their positions. Although a small device can be moved along the focal plane of a magnet, the normal requirement is for intrinsic resolution. In most applications, however, detection efficiency is just as important and the ability to discriminate, using a magnet, is even more important since the major objectives are to stop highly energetic particles or photons and to identify them as well as to measure their positions.

Signal formation must also be considered. Data processing is an integral part of most tracking experiments and visible output signals involve much more laborious processing than electrical signals, which are computer-compatible.

Telescopes may be regarded as tracking systems since they measure or select the three-dimensional directions of incident radiations. In order to define a long path length, they tend to be constructed of several separate devices, which can be located quite far apart, but single devices can be designed to have directional properties.

Radiography

Radiography, and other bidimensional imaging processes, differ from tracking techniques in two major respects. First, they produce an image, within the detector, of an external object rather than of the radiations themselves. In medical radiography the object is a patient and in industrial radiography it is a specimen. Secondly, the plane of the two-dimensional image is generally at right angles to the direction of the beam so the detector records a cross-section of the beam intensity.

A conventional arrangement is illustrated in *Figure 6.9*. It comprises a point source of radiation, an aperture to limit beam width, a filter to reduce the relative intensities of undesired types or energies of radiation and, sometimes, a collimator. Any bidimensional imaging system must have *optical geometry*; that is, some means of focusing radiations in order to preserve a singular correspondence between image and object points. In radiography, optical geometry depends on

160 *Detector functions*

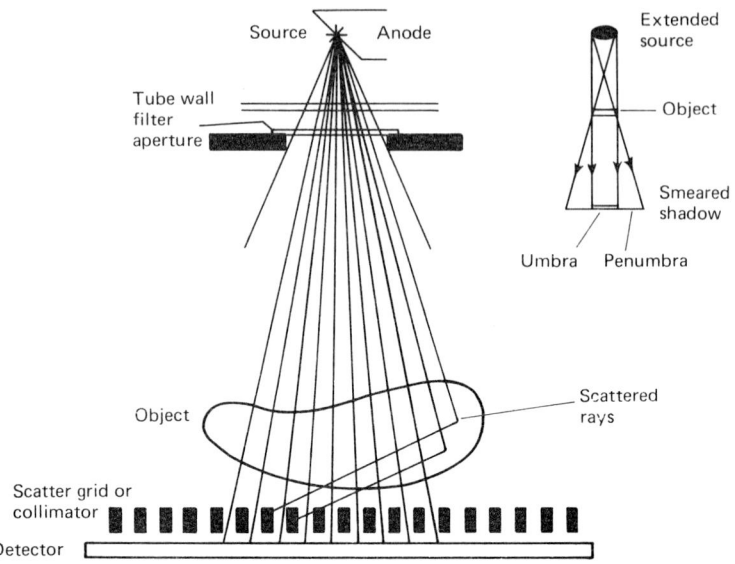

Figure 6.9 Conventional radiographic system

the size of the source point, which should be as small as possible so that all rays from the source that pass through a particular point in the object arrive at the same image point in the detector. In X-radiography, for example, the source is the focal spot of the electron beam on the anode of the generator and is usually about 1 or 2 mm wide. An extended source results in an extended image of each object point, as shown in *Figure 6.9*, increasing the intrinsic spatial resolution and smearing the image. However, if the object is much closer to the detector than it is to the source spot, this effect is minimised and spatial resolution is much smaller than the spot size.

The collimator, or scatter grid, is used to eliminate radiations scattered out of their original directions in the object. These tend to occur in a random distribution over the image, reducing contrast and image quality. The collimator is a series of parallel slits or holes that accept only parallel rays, or it may be focused on the beam spot.

There are two characteristic features of a radiographic image. The first is that each image point is an end-on projection of a straight-line path through the object and the complete radiograph is a two-dimensional projection of the shadows of *all* the internal, object components superimposed on each other. Secondly, because of the divergent beam, a radiographic image is magnified compared to the object and structures nearer the source are magnified more than those closer to the detector, resulting in some image distortion.

X-radiography is the most common radiographic technique because of its medical applications. X-ray absorption in the human body is sufficient to produce a useful range of image intensity and contrast. *Gamma radiography* is more suitable for industrial radiography in which metallic, and other high-density, high Z number objects are to be imaged. Since photon absorption depends on density and Z number, both X- and gamma radiographs identify structural features that differ in these respects and, in particular, emphasise high Z number components.

Fast-neutron radiography does the opposite. Since fast neutrons are preferentially attenuated by low *A* number materials, neutron radiographs tend to image light elements, even when these are screened by heavy elements such as lead. Thermal-neutron absorption cross-sections vary sharply from one element to another, and even from one isotope to another of the same element. *Thermal-neutron radiography* offers the prospect of being able to image specific elements or isotopes in a matrix of other materials of similar atomic and mass number, as illustrated in *Figure 6.10*. The problems with neutrons generally are: (a) it is difficult to obtain a point or collimated source with good optical geometry; (b) neutrons are difficult to detect and to distinguish from photons, and (c) they

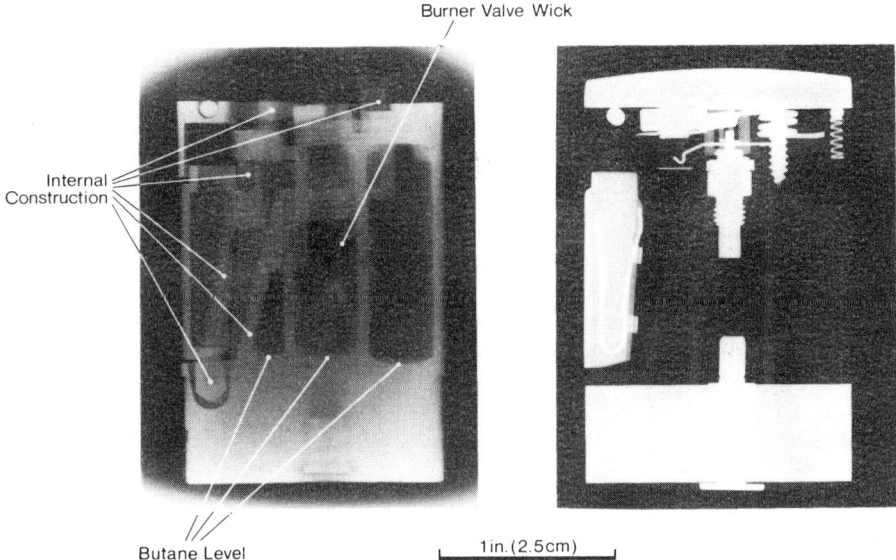

Figure 6.10 *X- and neutron radiographs of a butane cigarette lighter. (Courtesy of AERE Harwell)*

tend to activate materials. So far, neutron radiography has been used for mainly non-medical applications.

Stereoradiography is a special form of radiography in which the depth of a structural feature is measured. An ordinary radiograph is a two-dimensional projection of a three-dimensional object. It measures x and y coordinates but does not differentiate with respect to the depth, or z coordinate, of any part of the object. The technique is illustrated in *Figure 6.11*. Basically, two radiographs of the same object are obtained with the source point in two different positions. The source movement has the effect of displacing the image of a particular object point to one side, and this displacement depends on the depth coordinate. In the analog technique, the geometry is reproduced in a viewing device so that both radiographs, viewed simultaneously, present a stereoscopic image.

162 *Detector functions*

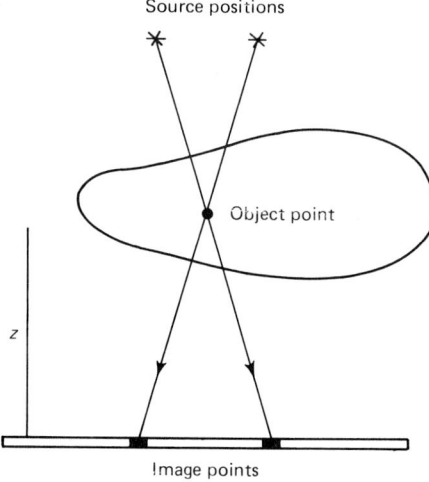

Figure 6.11 Principles of stereo-radiography

Alternatively, in the computed method, image displacement can be measured to determine, by calculation, the depth of any component.

Tomography is another means of obtaining information regarding the depth coordinate but, in this case, by imaging only a selected planar cross-section through the object. If the imaged layer is perpendicular to the radiation beam direction, the technique may be described as *longitudinal tomography* (*Figure 6.12*). If it is parallel to the beam direction, the technique is known as *axial tomography*. Both of these may be analog techniques in which the source and the detector are moved continuously, with respect to the object, in such a way

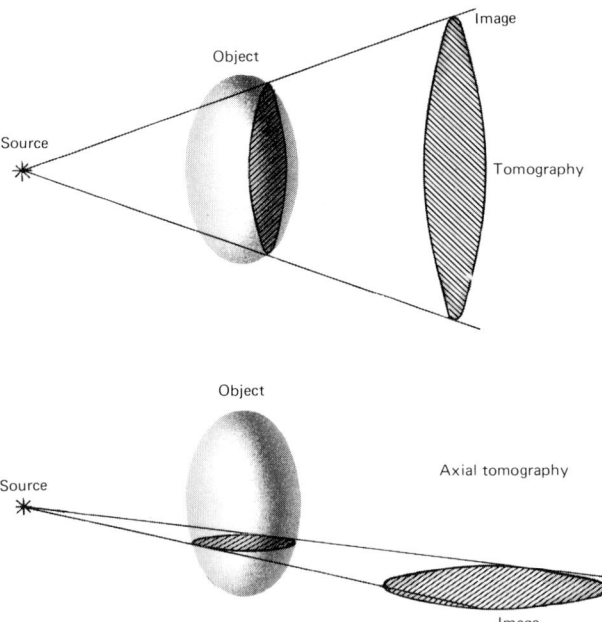

Figure 6.12 Principles of tomography

that a sharp image is generated only for one plane layer. The images of all other layers are defocused by the movement. A successful alternative, at least for axial tomography, is *computed tomography* (CT). In this technique, a flat, fan-shaped beam is transmitted through the object to produce a one-dimensional image of the layer involved. This single projection is not very informative, but, if the source and detector are rotated around the object, a series of projections of the same layer are obtained from which it is possible to reconstruct the internal structure of the layer. The method involves considerable computation and requires some form of on-line or dedicated computer but it does produce high-definition images.

The detector properties required for radiography are good spatial resolution and high detection efficiency. For most clinical applications a resolution of about ten lines per millimetre is sufficient and easily achieved with a number of systems, including films and fluorescent screens. Detection efficiency is determined mainly by dead-time losses since a machine X-ray source demands a very high count rate capability. This requirement is also met by films and fluorescent screens, which have zero dead time, and despite the relatively poor stopping power of these detectors they are preferred to electronic imaging systems whose dead-time losses would make them very inefficient. The exception to this rule is the computed tomography technique, which has an overriding need for computer-compatible output signals. At present, this is best satisfied by a scintillation counter.

Isotope imaging

Whereas tracking systems image radiations and radiographic systems image objects through which radiations are transmitted, isotope imaging systems image radioactive sources. These are radioactively *labelled* compounds, or *radiopharmaceuticals*, introduced into biological or chemical materials. They concentrate in specific structures or locations, via the same natural uptake processes which affect stable isotopes, so that the source distribution becomes identical to that of the structures concerned. This is imaged by means of the radioactive emissions of the source compound. There are several techniques that may be described as isotope imaging. They include the use of cameras and scanning systems as well as autoradiography and radiochromatography.

A *camera*, like its photographic counterpart, images all parts of the radioactive object simultaneously; that is, in a single exposure. In this respect, it is similar to a radiographic system but in other respects it differs considerably from such a system.

Typical camera systems are illustrated in *Figure 6.13*. They consist of an imaging detector and a collimator, which may be a multihole or a pinhole type, to produce optical geometry. They are used, biomedically, to study the uptake of labelled compounds by the organs or tissues in which these compounds concentrate. They have the advantage over radiographic systems of being able to image specific parts of the object although these may be surrounded by tissue of similar density and Z number. As a second advantage, they can perform an analytical function by using successive exposures to study the rate at which the compound reaches or passes through the organ concerned. Iodine isotopes, for example, can be concentrated in the thyroid gland, noble-gas isotopes can be

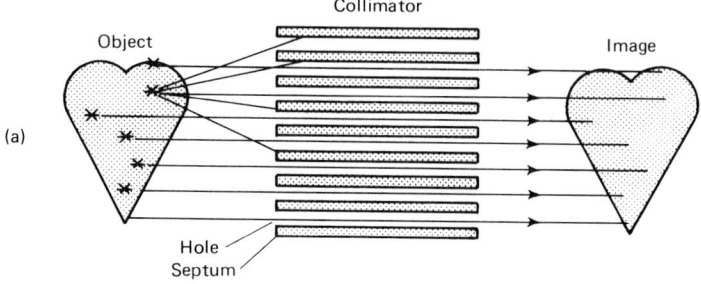
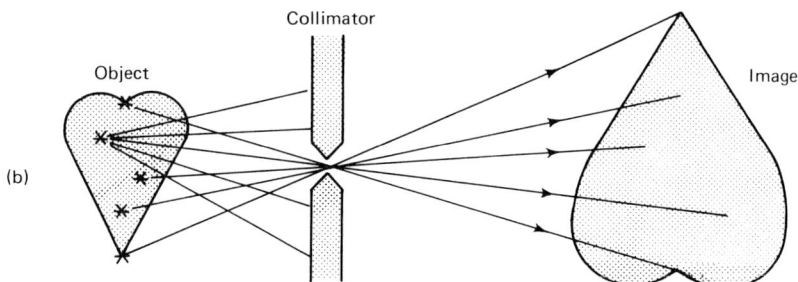

Figure 6.13 Gamma camera systems with (a) multihole and (b) pinhole collimators

inhaled into the lungs, various radionuclides can be introduced into the cardiovascular system, and so on, to study the structure and function of these organs. Dynamic studies of moving systems can also be recorded.

In other respects, camera techniques are inferior to radiography. The images they provide have poorer spatial resolution, contrast and statistical quality. There are a number of reasons for these deficiencies. First of all, the radioactive source may have to be located inside the patient for some time, determined by the radioactive half-life of the source or the biological half-life of the compound in the patient. In either case, the duration of the radiation exposure is much longer than the fraction of a second required for an X-radiograph so, to maintain tolerable dose levels, there is an upper limit to the source strength that can be used, normally in the region of 100 μCi to 10 mCi. This in turn limits the radiation intensity at the camera and, since the image must be obtained in a short time to avoid blurring due to the movement of the patient, the final result is a *statistically poor image*.

A second difficulty arises from the fact that the source material is usually surrounded by a considerable thickness of tissue. This absorbs and scatters radiations, increasing the radiation dose to the patient and reducing the efficiency of the imaging process. In order to produce any image at all, the radiations have to be quite penetrating. X-ray sources can be used for superficial organs but, for most purposes, gamma ray emitters must be employed. The camera is, therefore, known as a *gamma camera* and it must have a high detection efficiency for gamma radiation. This requirement tends to rule out films and other thin detectors which have good intrinsic spatial resolution. Most gamma cameras are based on scintillation counters whose spatial resolutions are of the order of several millimetres and may exceed one centimetre.

A third problem concerns the *optical geometry* of the camera. Since there is no point source to act as a focus, as in radiography, the optical geometry is entirely defined by the *collimator*. *A multihole collimator* consists of a thick absorber perforated by a large number of parallel holes. Basically, each hole views a separate small area of the object, maintaining an approximate point-to-point correspondence between the object distribution and the image. With a *pinhole collimator* the same correspondence is achieved, for all points, through the same hole, which acts as a focal point. For most purposes, the multihole collimator is preferred. It produces a rectilinear projection in which differential magnification is avoided and has a larger field of view than the pinhole, but both collimators represent a compromise between spatial resolution and detection efficiency. Narrow holes provide the best resolution but the poorest efficiency. As a result, the need for a collimator reduces both image quality and detection efficiency.

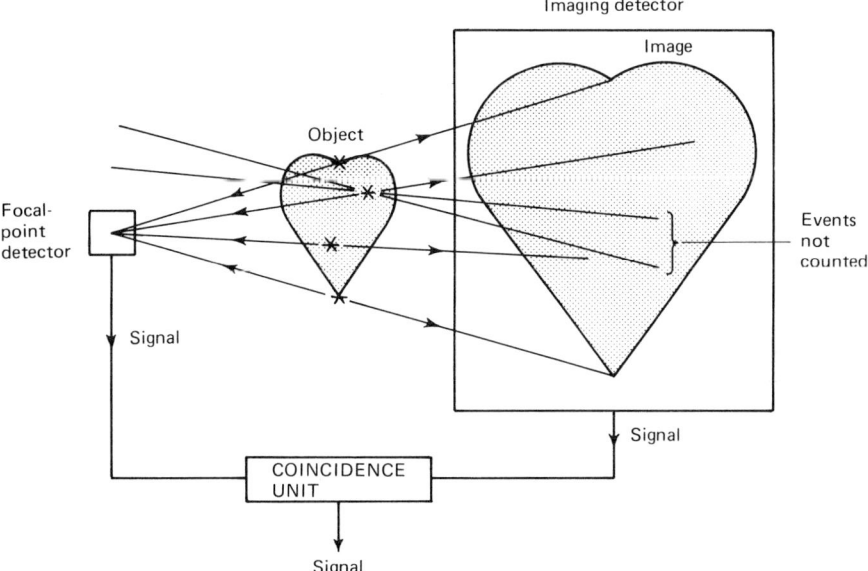

Figure 6.14 Principles of the Anger positron camera

There is one exception to this dependence on a collimator, namely, the *positron camera* devised by Anger. The principle of the positron camera is illustrated in *Figure 6.14*. Optical geometry is provided by the fact that the positrons from a positron-emitting source annihilate very close to their source nuclei and emit two 0.511-MeV photons in opposite directions. A small point-sized detector is located 'behind' the source material and its output signals are counted in coincidence with those of the imaging detector. The only photons recorded by the latter are those emitted along with, and in the opposite directions to, photons detected by the smaller detector. As a result, the only photons

166 Detector functions

contributing to the image are those which appear to diverge from the small detector, which acts, therefore, as a focal point for optical geometry. The point-to-point correspondence is not exact because of the finite range, of the order of 1 mm, of the positrons, but it is good enough to limit image resolution to a few millimetres.

Scanners provide an alternative means of imaging radioisotope distributions. A typical scanning system is illustrated in *Figure 6.15*. Basically, it comprises a

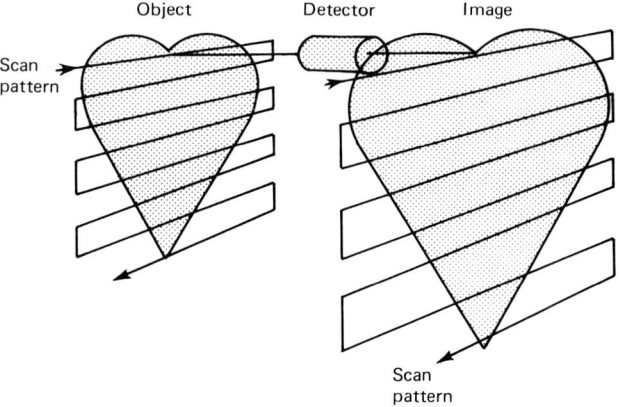

Figure 6.15 Principles of a scanning system

single detector which examines one point in the source through a single or multihole collimator, focused on the point. The whole source is imaged by moving the collimator and detector through a scanning pattern which studies each object point in turn.

The scanning system performs much the same function as does a camera and, like the latter, is normally used to image gamma ray emitters. It has an advantage that the detector does not need to have intrinsic resolution. Spatial resolution is defined entirely by the collimator. In the gamma camera, intrinsic detector resolution is a major contribution to the overall effect, so the resolution obtainable with a gamma scanner can be better. The overall detection efficiency must be poorer than that of a camera but is improved with a focused multihole collimator. The only disadvantage of the scanner is the longer time it requires to complete an image, during which time object movement must be restricted. Dynamic studies of moving objects cannot be undertaken with a scanner.

Autoradiography is a technique in which a very thin labelled specimen is placed in contact with an imaging detector, as illustrated in *Figure 6.16*. In this case no collimator is required and optical geometry is provided by the physical

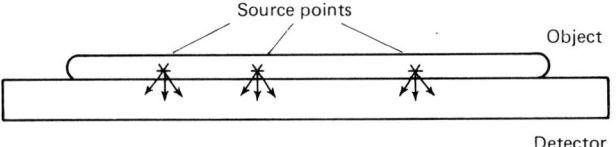

Figure 6.16 Principles of autoradiography

proximity of the source material and the detector. Because of the geometrical attenuation factor (that is, the inverse square law of attenuation), the interaction rate recorded from each source point falls off rapidly with distance from it, so the image of each point is formed in the adjacent region of the detector. If, however, the radiations are penetrating ones, the image is not localised to a point but forms a distributed image, so the spatial resolution is very poor. In order to obtain a high-resolution image, the source emissions must be of very short range. For this reason, the technique is used mainly with low-energy beta emitters such as tritium and carbon-14, in which case spatial resolutions of less than 1 mm are easily obtained.

The detector requirements for autoradiography are large area, very thin entry window and good spatial resolution. At present, films are used, almost exclusively, for this technique.

Radiochromatography is another technique that can be described as isotope imaging. There are several variations of the method, in which chemical compounds are analysed by means of processes which physically separate their components. If some of these components are radioactively labelled, position measurement and identification can be carried out with a radiation detector. In paper chromatography, for example, long strips of filter paper are used. A spot of labelled material is placed on one end and mobilised, with an appropriate solvent, to travel along the paper. This process separates the components into different bands across the paper and these can be identified by detectors with crude imaging capability.

Electron microscopy

There are two types of electron microscope in general use, the *transmission electron microscope* (TEM) and the *scanning electron microscope* (SEM). In both devices, an intense electron beam is generated by thermionic emission from a heated filament. This beam is then focused by deflection in magnetic lenses, in the same way as a light beam is focused by refraction in optical lenses. The whole process is carried out in a vacuum, to avoid the defocusing effects of

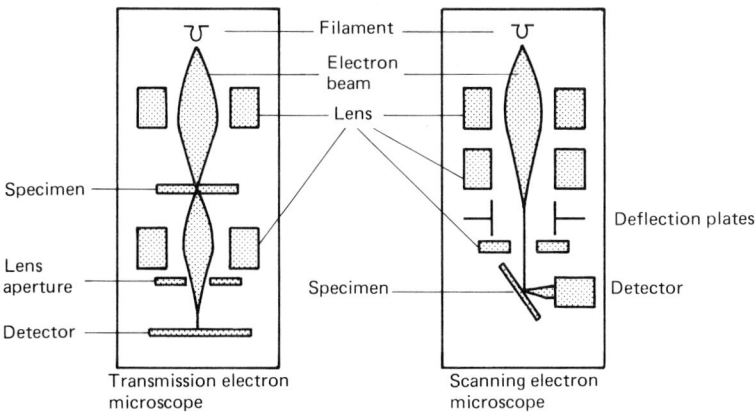

Figure 6.17 Basic principles of electron microscopes

scattering from air molecules, and at some point in the electron optical system, the specimen to be imaged is inserted. The two systems are illustrated in *Figure 6.17*.

In the *transmission electron microscope*, the electron beam is transmitted through a very thin specimen, then expanded to form a greatly enlarged image. The process is analogous to the action of a conventional transmission optical microscope and to the standard radiography technique except that magnifications of up to about 10^6 may be obtained and the attenuation mechanism in the specimen is not absorption but, mostly, inelastic scattering of electrons out of the beam. At present, the best object resolution obtainable, limited mainly by lens aberrations, is of the order of 1 Å, that is, 10^{-4} μm. This requires an image resolution of the order of 100 μm for a gain of 10^6, or 1 μm for a gain of 10^4. Films and fluorescent screens are used as detectors. They have intrinsic spatial resolution, high count rate capability and thin entry windows required and they can be exposed to a vacuum.

In the *scanning electron microscope*, a narrowly focused beam with a diameter of about 10^{-2} μm is incident on the specimen. The beam is made to scan the surface of the specimen by bidimensional deflection produced by electrostatic or magnetic deflecting fields, as in a cathode ray oscilloscope or a television receiver. At each position on the surface of the specimen, secondary electrons are emitted. These are focused, accelerated up to a few kilo-electronvolts and observed by a detector, such as a scintillation counter, whose output is electrical current proportional to the incident current. In this way the structure of the surface is imaged. Object resolution depends on lens aberrations and beam spot size and is of the order of 10^{-2} μm. Magnification may be up to about 10^5. Secondary-electron current may be of the order of 10^{-12} A; that is, about 6×10^6 electrons per second. The detector must be able to measure this current, have a thin entry window and be vacuum-compatible. Image resolution is defined by the secondary-electron focusing system, not the detector.

Image quality

Image quality is defined by the requirements of the user and these are bound to vary from one situation to another. An image that is suitable for one purpose may be inadequate for another and, even when the purpose is prescribed, the personal preference of the user may vary. Consequently, there is no single, general criterion against which image quality can be measured and specified. There are, however, a number of *image quality factors* which are used to describe different aspects of image quality so that users can specify a set of criteria to be met for any particular function.

The two most important image quality factors are *image unsharpness* and *contrast*. The first of these measures the extent to which fine details are blurred owing to the limited *spatial resolution* of the imaging system. The image of a point object is an extended distribution whose density, as a function of spatial coordinates, is described by a *point spread function* (PSF). The FWHM width of this density function, or its standard deviation, can be used to define image unsharpness. If the image is magnified, then the unsharpness should be divided by the magnification factor to obtain the true value.

Image contrast measures the extent to which different intensity levels are represented in the image; that is, the half-tone quality. It is determined by the source strength and the exposure time as well as by the *detection efficiency* of the imaging system. If the counting statistics are very good, as in X-radiography, it is defined as the increase in image density per unit of radiation exposure, or some function of exposure. If the image statistics are relatively poor, as in a typical gamma camera image, then statistical effects predominate, giving the image a grainy quality which reduces the clarity. In this case, a more useful quality factor is the signal-to-noise ratio (SNR). Assuming a Gaussian distribution, the statistical uncertainty in a total image count N is equal to $N^{1/2}$. If N is defined to be the signal, and $N^{1/2}$ to be the noise, then the SNR is given by

$$\text{SNR} = N/N^{1/2} = N^{1/2} \tag{6.26}$$

The object, too, has a signal-to-noise ratio defined in terms of N_0, the number of radiations emitted, and $N_0^{1/2}$, the statistical uncertainty in this number, so the detection efficiency, η, of the system is given by

$$\eta = N/N_0 = (\text{SNR}_{\text{image}}/\text{SNR}_{\text{object}})^2 \tag{6.27}$$

It should be noted that the term 'signal-to-noise ratio' is also used in reference to electronic effects, in which case it is the ratio of the mean signal pulse height to the root mean square height of electrical noise pulses generated in the system.

Unsharpness and contrast, or SNR, both affect the clarity of an image. There is little point in having good unsharpness with poor contrast or statistics, and vice versa. The effects of deterioration in unsharpness and SNR are illustrated in *Figure 6.18*. As a rough guide, some thousands of counts per point spread area are required to resolve features of that order of magnitude.

Since image quality is determined largely by the properties of the imaging system, it is useful to define some *instrumental quality factors*, by which means system performance can be described. The system *transforms* an object into an image and, in the process, loses some of the *information content* of the object. Because of the system limitations, it is not possible to transfer all of the information and, in any case, some of this would be superfluous to requirements. Thus, the non-zero spatial resolution eliminates some fine structure and the non-unity detection efficiency loses some counts.

Energy resolution, time resolution, discriminating ability and signal readout mode may all affect imaging properties but the major detector parameters, in this respect, are spatial resolution and detection efficiency. Neither of these parameters, however, is particularly informative with regard to imaging performance, so more suitable concepts have been developed. Spatial resolution is replaced by the more descriptive *modulation transfer function*, while detection efficiency may be replaced by a *characteristic curve*, for films and similar systems, and by a *figure of merit*, for other systems.

The *modulation transfer function* (MTF) can be defined from a theoretical or an experimental point of view. Theoretically, an object and image can be regarded as functions of intensity against spatial coordinates. Equally, they can be represented by their Fourier transforms, which are functions of *spatial frequency*, ν, measured, for example, in lines per centimetre. The functions indicate the relative proportions of various spatial frequencies in the object and the image, and the ratio of the image transform to the object transform is the modulation transfer function.

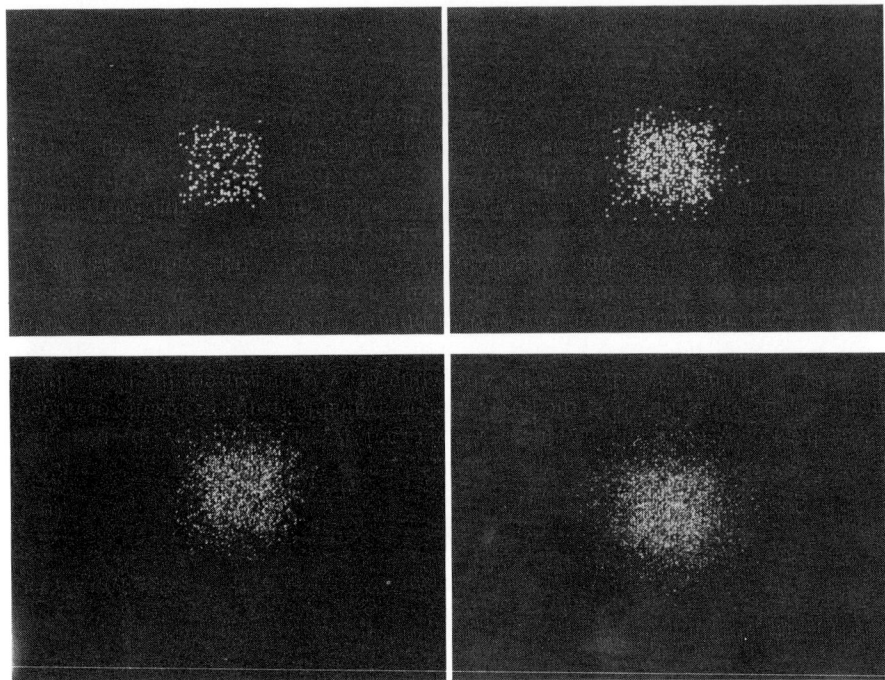

Figure 6.18 The effects of unsharpness and signal-to-noise ratio (SNR) on image quality, illustrated by means of computer-simulated images of a uniform square object. Standard deviation (unsharpness) and total counts (SNR) both increase from top left to bottom right

This is normalised to unity at zero spatial frequency, that is, for a uniform object intensity, and falls off to zero at a higher frequency determined by the spatial resolution of the system.

Experimentally, the MTF can be defined and measured by reference to an object comprising a set of parallel lines with a single spatial frequency. The source density, ρ, is assumed to vary sinusoidally across these lines and the *object modulation*, or relative contrast, m_0, is given by

$$m_0 = \frac{\rho_{max} - \rho_{min}}{\rho_{max} + \rho_{min}} \qquad (6.28)$$

where ρ_{max} is the maximum density, in the middle of a line, and ρ_{min} is the minimum density, between the lines.

The image is also a set of lines, if they are resolved, with an *image modulation*, or relative contrast, m_I, given by

$$m_I = \frac{n_{max} - n_{min}}{n_{max} + n_{min}} \qquad (6.29)$$

where n refers to recorded counts.

The value of the MTF, for the particular spatial frequency used, is the ratio of these two expressions:

$$\text{MTF} = m_I/m_0 \qquad (6.30)$$

The complete MTF is determined by obtaining this ratio for a range of spatial frequencies and plotting these values as a function of ν. As the lines become too close together to be resolved, that is, as the spatial frequency increases, the gaps between them are increasingly filled in by the overlap of the image lines and image modulation deteriorates. The concept of MTF is particularly useful for assessing imaging performance because it makes no sharp distinction between resolution and non-resolution. Instead it defines a more gradual transition between these extremes.

In practice, of course, a real object does not have a conveniently simple line structure, but it can always be regarded as a superposition of many sets of parallel lines representing the various spatial frequencies present. This is a form of two-dimensional Fourier analysis and the representation is the Fourier transform of the object. It is for this reason that the lines should, strictly speaking, have a sinusoidal density profile.

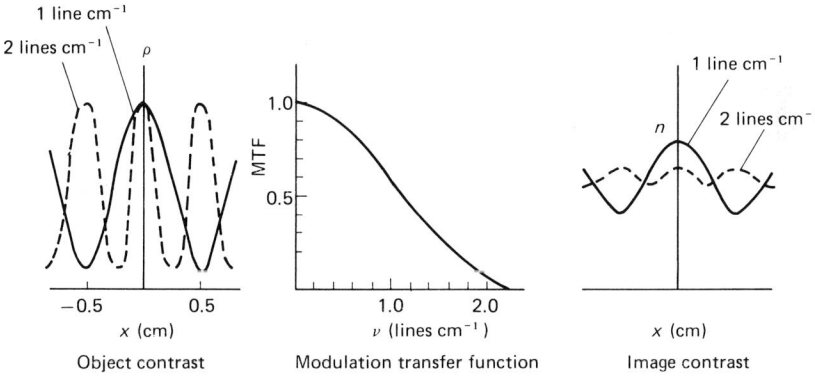

Figure 6.19 *A typical gamma camera modulation transfer function (MTF) with corresponding image and object contrasts*

Figure 6.19 shows a typical MTF for a gamma camera and its effect on object features with spatial frequencies of 1 and 2 lines cm^{-1}. In both cases the object modulation is assumed to be given by

$$m_0 = \frac{0.9 - 0.1}{0.9 + 0.1} = 0.8$$

For the 1 line cm^{-1} features, the MTF is 0.6 so the image modulation is 0.48 and this structure is quite clearly shown. For the 2 lines cm^{-1} features, the MTF is only about 0.05, so image modulation is 0.04 which is insufficient to make these features visible.

A *characteristic curve* is used to describe the response of a film or any other system that saturates after a certain exposure and will be discussed more fully in Chapter 10. Basically, it is a graph of accumulated counts, or optical blackening, against radiation exposure or, more frequently, the logarithm of exposure. The curve incorporates three major parameters. These are the *film speed*, which is the reciprocal of the minimum exposure required to produce an observable effect, the *contrast*, which is its gradient, and the *latitude*, which describes the exposure

172 Detector functions

range that can be encompassed. Electronic systems do not saturate and register counts from one upwards to infinity, so they have unlimited latitude and film speed.

The *figure of merit* concept is not well established and at least two formulations are available. One of these is the product of the detection efficiency and the square of the MTF, measuring the ability of the system to generate image contrast as a function of spatial frequency. Another version uses the formula $S^2/(S + 2B)$, where S is the signal strength and B the background strength, both in counts. In many situations, however, it is sufficient to use detection efficiency or quantum efficiency as an operational parameter.

6.5 TIMING

In terms of basic requirements, there is little difference between the timing function and the timing property. The function may depend on detection efficiency, especially in coincidence counting, and it does, generally, require a differential electrical output signal, but the main requirement is time resolution. There are, however, a number of ways in which the property can be utilised. In other words, there are several variations of the timing function. All of them can be regarded as forms of *time spectrometry* and they perform single-channel analysis, scanning single-channel analysis and multichannel analysis of time intervals.

The analogy with energy spectrometry is complicated by the fact that there is no universal zero time with respect to which the time intervals can be measured. One possibility is to measure count rate as a function of time, so that all events are measured against a common starting point. An alternative is to define a separate starting point for each event. A START pulse is produced when an interaction begins and a STOP pulse shows the completion of the interaction. The time interval between these two pulses supplies information about the interaction within the detection system.

Half-life measurement

A useful example of time spectrometry based on a common zero point is provided by the measurement of short half-lives of radioactive sources, ranging from tens of seconds to a few hours. The source is examined by a detector whose output is fed into a multichannel analyser operating in the *multichannel scaling* (MCS) mode. In this mode, the analyser does not measure pulse height, although a low-level discriminator may be used to eliminate noise pulses. It records all pulses, first in channel 1, then in channel 2, and so on for each channel in turn, spending the same *dwell time* on each channel. The result is a graph of count rate against time showing the exponential decay of the source. The dwell time can be adjusted, for example, from 10 to $10^6\,\mu s$, to match the decay constant of the sample. The same procedure can be carried out with a scanning single-channel analyser in which the channel height and width remain constant. This approach is more suitable for longer half-lives.

Coincidence counting

'Coincidence counting' describes a wide range of techniques based on the identification of true coincidences. These can be accepted, by coincidence counting, or rejected, by anticoincidence counting. Both functions are carried out by a *coincidence unit* which accepts two (or more) input pulses and outputs one pulse if the input pulses are coincident, within a selected resolving time, or, in the alternative mode, anticoincident.

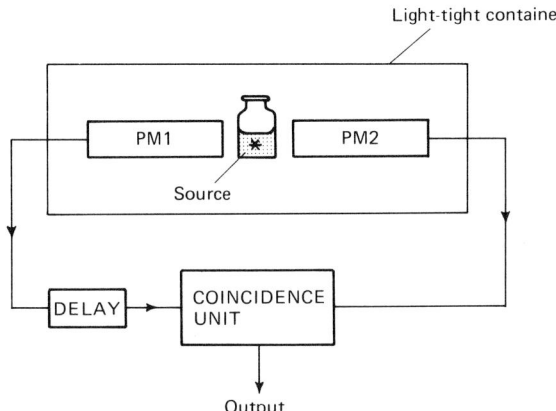

Figure 6.20 Coincidence counting system using a liquid scintillator

A typical application, illustrated in *Figure 6.20*, is the coincidence counting of 18-keV betas from tritium. These do not easily penetrate detector entry windows so the source material is placed in a liquid scintillator, but the light pulses which are produced are so small that they cannot be distinguished from electronic noise. The solution of this problem is to place the scintillator between two photomultipliers which simultaneously record the light flash from each beta particle. The output pulses from the photomultipliers are fed into a coincidence unit whose output pulses correspond to beta pulses rather than noise pulses, which are seldom coincident. A delay unit may be placed in one line to balance the pulse transit times in each line.

Similar systems are used, for example, to identify coincident annihilation photons in the Anger positron camera and to select coincident events in a cosmic ray telescope. Anticoincidence counting is used, for example, in a Compton spectrometer to reject Compton scattering interactions, which produce counts in two detectors, and in low-level counting, where an anticoincidence detector can be used to reject cosmic ray background counts.

Time of flight methods

In time of flight (TOF) methods, the time intervals are the times taken by radiations to traverse a measured flight path. The radiations must be emitted as a pulsed beam so that all the radiations in one pulse start at the same time. This effect can be produced electronically, as in a particle accelerator, by switching the beam on and off, or mechanically, by using a beam-interrupting device known as a beam chopper, which is usually an absorbing disc with a hole in it, rotating in front of the source.

A typical application, for neutrons, is illustrated in *Figure 6.21*. A pulsed-neutron generator is used, or a chopper is placed in front of a chemical neutron

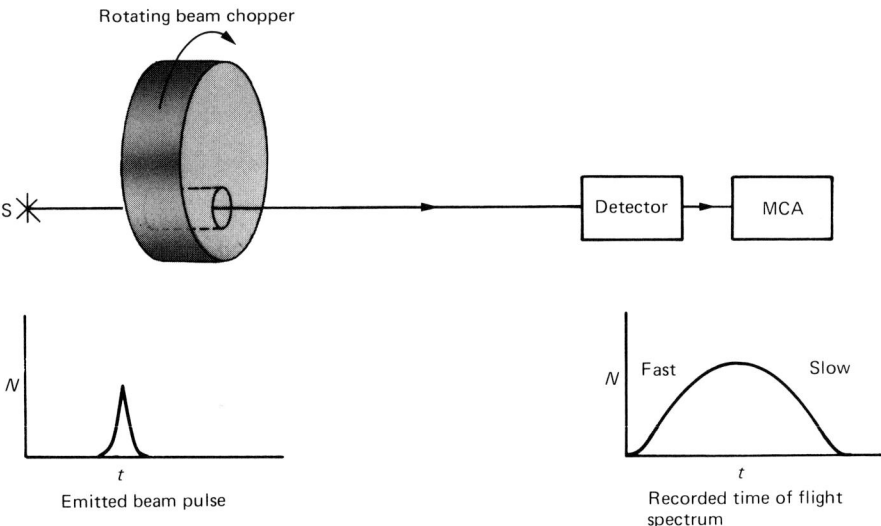

Figure 6.21 Time of flight spectrometry system for neutrons

source. As the pulse travels towards the detector, it spreads out. The fastest neutrons arrive first and the slowest arrive last. The detector output can be recorded by a multichannel analyser in a multiscaling mode triggered by each beam pulse to produce a time of flight spectrum. In this case, time is inversely proportional to velocity so the spectrum can be converted to a neutron energy spectrum. If necessary, the detector output can be single-channel-analysed to select a particular time of flight and record neutrons of a selected energy.

Time to amplitude conversion

Time to amplitude conversion is similar to the time of flight method but it does not require pulsed beams and applies to individual particles or photons. The time variable is the interval between a START and a STOP pulse generated by a single event in the detection system. These two pulses are fed into a time to amplitude (TAC) converter (or a time to height converter), which generates a single output pulse whose amplitude (or height) is proportional to the time interval between the input pulses. The TAC output pulses can be pulse-height-analysed to provide a time spectrum.

A typical application is to measure the position of an event within a detector, say, along the length of a single-wire gas counter. Each interaction produces a charge pulse at each of two electrodes. One of these pulses is taken out immediately, to provide the START pulse, but the other is delayed to form the STOP pulse. The time interval between these two pulses is proportional to the time taken for the STOP pulse to travel along the electrode so it is also proportional to the initial distance of the event from the output end of the electrode. A similar result can be obtained with an MCA, in the multiscaling mode, by using the START pulse to trigger the scaling process for each event.

BIBLIOGRAPHY

ANGER, H. O., 'Survey of Radioisotope Cameras', *ISA Trans.*, **5**, 311 (1966)
ATTIX, F. H. and ROESCH, W. C., *Radiation Dosimetry*, Academic Press, New York (1966)
BECK, R. N., 'Radioisotope Scanning Systems', *ISA Trans.*, **5**, 335 (1966)
BERGER, H., *Neutron Radiography*, Elsevier, Amsterdam (1965)
BROOKS, R. A. and DiCHIRO, G., 'Principles of Computer-Assisted Tomography (CAT) in Radiographic and Radioisotope Imaging', *Phys. Med. Biol.*, **21**, 689 (1976)
FAIRES, R. A., and PARKS, B. H., *Radioisotope Laboratory Techniques*, Butterworths, London (1973)
HAY, G. A. (Ed.), *Medical Images: Formation, Perception and Measurement*, The Institute of Physics, Bristol/John Wiley, London (1976)
McKAY, H. A. C., *Principles of Radiochemistry*, Butterworths, London (1971)
MEEK, G. A., *Practical Electron Microscopy for Biologists*, Wiley–Interscience, New York (1970)
MOODY, N. F., PAUL, W. and JOY, M. L. G., 'A Survey of Medical Gamma-Ray Cameras', *Proc. IEEE*, **58**, 217 (1970)
OESCHGER, H. and WAHLEN, M., 'Low Level Counting Techniques', *Ann. Rev. Nucl. Sci.*, **25**, 423 (1975)
ROGERS, A. W., *Techniques of Autoradiography*, Elsevier, Amsterdam (1973)
THORNBURN, C. C., *Isotopes and Radiation in Biology*, Butterworths, London (1972)
VENVERLOO, L. A. J., *Practical Measuring Techniques for Beta Radiation*, Macmillan, London (1971)

Chapter 7
Gas counters

7.1 INTRODUCTION

A gas counter consists of a *gas mixture* located between two or more electrodes. A potential difference is applied across the electrodes so that one, the *anode*, is positively charged and the other, the *cathode*, is negatively charged. The major component of the gas mixture is usually, but not always, a noble gas such as argon or xenon. When ionising radiation interacts with the gas it creates ion pairs, so that each detected particle or photon produces a number of electrons and an equal number of positive ions, which together constitute a charge pulse. The electrons are attracted to and collected at the anode while the positive ions are collected at the cathode, recording one event.

Within this framework there is room for considerable variety and there are many types of gas counter. The best known are the older, traditional gas counters, namely, the *ionisation chamber*, the *proportional counter* and the *Geiger–Müller counter*. Another well-established device is the *electrometer*. This is not a counter, in that it does not record individual events, but it does record integrated radiation intensity and will be discussed in this chapter as a special form of ionisation chamber.

In recent years, a new generation of gas counters has been developed. These are the *imaging gas counters*. They include the *spark chamber* and the *multiwire proportional counter*, which will be discussed in this chapter. From a functional point of view, however, they must be considered in relation to other imaging systems which are described in Chapters 10 and 11. The same is true of another imaging gas system, namely, the *ionography* process, which is more closely related to other radiographic systems and will be discussed in Chapter 10. It is, however, a form of ionisation chamber, although it is usually classified as an electrostatic imaging chamber.

In general, the output signal from a gas counter is one of three types. Charge pulses, read out as voltage pulses, may be used to detect individual events, the charge from many events may be accumulated, or integrated, to measure the total irradiation, or the charge may be allowed to leak away steadily as a current whose magnitude is proportional to count rate. It is also possible to observe light pulses, or scintillations, from de-exciting gas atoms, but these are usually in the ultra-violet wavelength range and difficult to make use of.

7.2 IONISATION CHAMBERS

An ionisation chamber, or ion chamber, is a gas counter in which all of the primary electrons and/or ions created by incident radiation are collected as a charge pulse or a current. Ideally, there should be absolutely no loss nor gain of charge in the detector so that the charge collected is exactly equal to the energy deposited, divided by the energy required to form an ion pair. Any device which has this property is an ion chamber.

The electrodes may be coaxial cylinders but they are more often parallel plates. The gas—electrode assembly is sealed in a chamber made of glass or some other non-porous material that does not outgas and contaminate the mixture. A standard arrangement, with electrical connections, is illustrated in *Figure 7.1*.

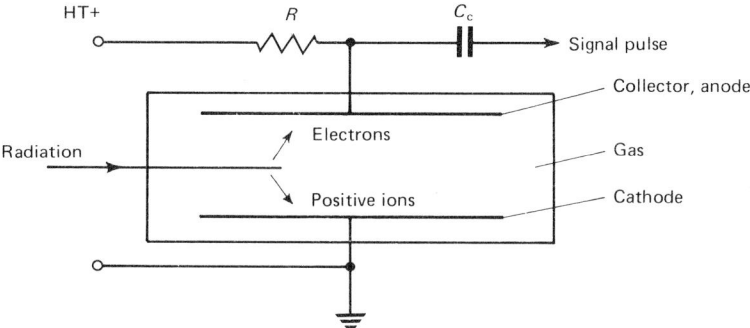

Figure 7.1 Basic structure of an ionisation chamber

With this geometry, the electric field, E, is uniform and constant throughout the sensitive volume between the electrodes, and given by the usual formula

$$E = V/d \tag{7.1}$$

where V is the applied potential difference, d is the electrode separation, and E is usually of the order of 100 V cm^{-1}.

The field direction, shown by the lines of force, is perpendicular to the electrodes except at the edges of the sensitive volume where the geometry is not that of a parallel-plate system. Ions and electrons created by interactions with incident radiation are attracted along these field lines to their respective electrodes. Electrons move towards the positive anode, with a net drift velocity which depends on the gas type and pressure but is usually of the order of 10^6 cm s^{-1}. Positive ions move more slowly in the opposite direction, with a drift velocity of the order of 10^3 cm s^{-1}. Both charge carriers suffer a large number of collisions on the way. As the electron cloud moves towards the anode it drives electrons in the anode back along the connecting circuit, causing a small decrease in the potential of that electrode, which reaches a maximum when the electron pulse is finally collected. More slowly, the positive ions produce a similar effect at the cathode. Both processes produce a small drop in the applied potential difference across the chamber to register the detection of one event.

The rate at which this voltage drop occurs depends on the rate at which the electrons and positive ions are collected. If, for example, a charged particle is incident along the median plane of a chamber with a 2-cm gap, then the electrons are collected, at a uniform rate, in about 1 μs and the ions are collected in about 1 ms. Thus the potential drops by a certain amount in 1 μs then by the same amount again in the following millisecond. The result is illustrated in *Figure 7.2*.

At the same time as it is being collected, the charge tends to leak away to earth. It is not held indefinitely by the electrodes. The leakage takes place mainly through the load resistance, R, and the high-tension supply. If these are very small resistances, the chamber operates in the current mode, leaking charge continuously. If R is extremely large, the chamber accumulates charge and its

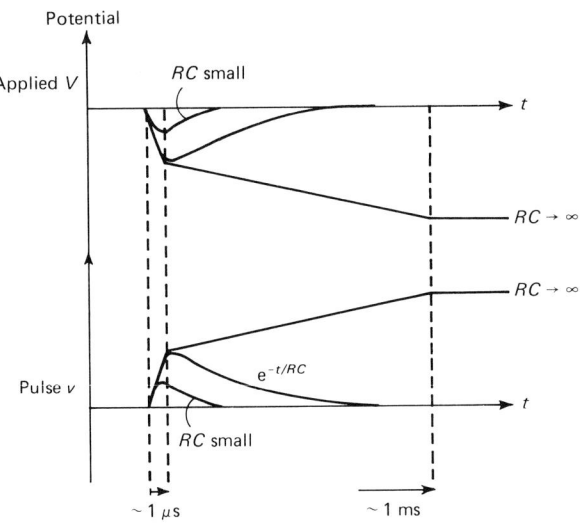

Figure 7.2 Effect of time constant, RC, on signal pulse shape

applied voltage continues to drop. In the pulsed mode, the charge is held for a short time, during which a small voltage drop is observed, then allowed to leak away, so that the original potential difference is restored. If the voltage across the detector is measured, then a small, transient pulse is observed when radiation is detected. This is usually read out through a coupling capacitance, C_c, which acts as a filter, blocking the d.c. voltage which would tend to overload the readout electronics, and passing only the a.c. signal pulse. As a result, a transient voltage pulse is obtained, as shown in *Figure 7.2*.

The pulse has an almost linear rise, determined by the charge collection rate, and an exponential decay, determined mainly by the value of the load resistance, R, and the detector capacitance, C. The pulse *rise time* is defined to be the time taken for the signal to increase from 10% to 90% of its maximum value, that is, its *pulse height*, v. The decay is given by the usual formula for transient voltage changes, as described in Chapter 12, i.e.

$$v(t) = v \cdot e^{-t/RC} \qquad (7.2)$$

where the product RC is known as the pulse *decay time constant* and is the time taken for the voltage to drop to $1/e$ of its maximum value, v.

The effects of different RC values are illustrated in *Figure 7.2*. If RC is infinite, there is no leakage. Charge is integrated at the electrodes, remaining there indefinitely, to form a voltage pulse in the form of a step function. If, on the other hand, RC is very small it may fail to integrate even the electron charge pulse. Ideally, it should be long enough to integrate the fast-electron pulse so that the total charge of at least one sign is measured. This produces a short pulse suitable for timing measurements and, by restoring the operating potential quickly, it reduces the chamber dead time. However, the pulse still has a long tail, since the positive ions have to be cleared from the chamber, so the total recovery time is about 1 ms.

If the decay time is long enough to integrate both negative and positive charge carriers, the dead time is increased to about 1 ms but the signal pulse height becomes proportional to the energy deposited in the chamber which is, therefore, an *energy spectrometer*. In fact, under these conditions, the chamber exhibits an almost unique property in that the charge pulse height is exactly 'equal' to the charge deposited, and, therefore, to the energy deposited. It is this feature, coupled with the possibility of using air as the chamber gas, that makes the detector so useful for the measurement of radiation exposure and dose. The number of ion pairs collected is $n = E/\epsilon$, where E is the energy deposited and ϵ is the mean energy required to form an ion pair. Hence, the total charge collected of each sign, q, is

$$q = ne = \frac{Ee}{\epsilon} \tag{7.3}$$

where e is the electronic charge, and the voltage pulse height, v, is

$$v = \frac{q}{C} = \frac{Ee}{\epsilon C} \tag{7.4}$$

Table 7.1 Properties of gases used in gas counters

Gas	Density at STP (mg cm^{-3})	Ionisation potential (eV)	ϵ (eV)	Excitation potential (eV)	Townsend coeff. α (cm^{-1})
He	0.179	24.5	42.3	20.9	0.0033
Ne	0.900	21.6	36.6	16.6	0.0032
Ar	1.784	15.8	26.4	11.6	0.0075
Kr	3.710	14.0	24.1	10.0	0.0067
Xe	5.851	12.1	22.0	8.4	0.0059
H_2	0.090	15.4	36.3	11.2	0.0034
N_2	1.251	15.5	36.0	6.1	0.0029
CO_2	1.977	14.0	33.0	10.0	0.0026
Air	1.205	15.0	34.0	6.0	0.0032

* Calculated for a typical proportional counter field, close to the anode wire, of 1.5×10^5 V cm^{-1}, and for gas at STP. Data from von Engel, A., *Handbuch der Physik*, Vol. 21, Springer-Verlag, Berlin, 504 (1956)

ϵ varies slightly with the type of gas and radiation involved (*Table 7.1*), but is generally of the order of 30 eV. Thus, if an energy of 1 MeV is deposited in the chamber, the number of ion pairs formed is

$$n = \frac{E}{\epsilon} = 3.3 \times 10^4$$

providing a charge pulse, of both signs, of

$$q = 2ne = 6.6 \times 10^4 \times 1.6 \times 10^{-19} = 1.06 \times 10^{-14} \text{ coulombs}$$

The capacitance of a parallel-plate electrode system, in air, is given by

$$C = \frac{8.85 \times 10^{-12} A}{d} \text{ farads} \tag{7.5}$$

where A is the plate area in square metres and d is the separation between the plates in metres. Thus, for a plate area of 100 cm and a separation of 1 cm, the detector capacitance is 8.85 pF and the maximum possible voltage pulse height, for a 1-MeV event, is given by

$$v = \frac{q}{C} = \frac{1.06 \times 10^{-14}}{8.85 \times 10^{-12}} = 1.2 \text{ mV}$$

Normally, the additional external load capacitance of connected cables and amplifiers reduces the signal to less than 1 mV. This result illustrates another characteristic property of the pulse ion chamber, namely, that output signals are very small and require amplification by factors of about 10^4 to be processed by standard nucleonic systems. Another way of demonstrating the problem is to compare signal size with the magnitude of preamplifier noise pulses. In the best, low-noise preamplifiers, these pulses are equivalent to about 500 ion pairs and indistinguishable, therefore, from signal pulses in which energies less than about 500 × 30 eV = 15 keV are deposited.

The energy resolution of an ion chamber is quite good. Being proportional to $\epsilon^{1/2}$ (equation 5.20), it is better than that of a scintillator and improved on only by a semiconductor detector. From equation 5.18, for a deposited energy of 1 MeV, the intrinsic energy resolution is calculated to be about 1.3%, FWHM. This assumes perfect ion chamber behaviour, with no charge loss or gain, and it ignores extrinsic effects, such as electronic noise. Generally, at 1 MeV, the energy resolution is about 3% and increases, as described by equation 5.18 and *Figure 5.8*, as energy decreases.

In order to maintain ion chamber conditions, both the field strength, E, and the gas mixture must be controlled. If the field is too strong and/or the mean free path of the electrons in the gas is too long (for example, if the pressure is too low), the electrons can be accelerated between collisions and gain sufficient energy to create further, secondary ion pairs, increasing the pulse size. If, on the other hand, the field is too weak, the electrons spend more time in passing gas atoms and colliding with them, so they are more likely to be captured, reducing the pulse height.

The effects of charge loss and gain can be shown in a graph of pulse height, v, against applied potential, V, as illustrated in *Figure 7.3*. This has a characteristic plateau over which v remains constant and the device behaves like an ion chamber. At lower potentials, v decreases owing to partial recombination and capture of

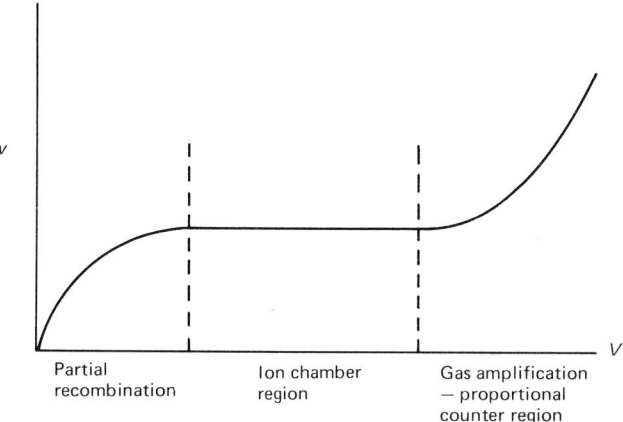

Figure 7.3 Graph of pulse height, v, against applied potential, V, for an ionisation chamber

electrons and, at high potentials, v increases owing to the formation of secondary ions. At this high-potential region, the chamber begins to operate like a proportional counter.

Grid ion chambers

The disadvantage of the spectrometric pulse ion chamber, described above, is the long dead time imposed by the need to integrate the positive ion pulse. This severely limits the high count rate capability of the detector. The obvious way to improve matters in this respect is to 'clip' the pulse by reducing the time constant, RC, so that only the electron pulse is integrated, but this leads to another complication.

Since the electrons are collected at a uniform drift velocity their collection time, and the rise time of the electron pulse, is proportional to the distance, x, travelled from the formation point to the anode. If this is clipped after a fixed time, the height reached is also proportional to x. In other words, pulse height is a function of position as well as energy, and is given by

$$v = \frac{q}{C} \cdot \frac{x}{d} \tag{7.6}$$

In order to use the chamber as a spectrometer, this position dependence must be eliminated. One method of doing so is to use a grid ion chamber, in which an open-mesh grid electrode is located very close to the anode and at a slightly lower potential. This screens the anode from charge movement in the bulk of

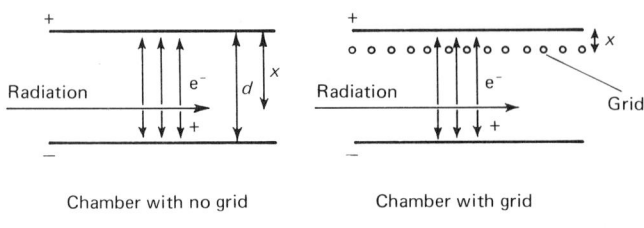

Figure 7.4 *Structure and function of a grid ionisation chamber*

the sensitive volume and registers a pulse only when the charge reaches the grid-anode region, so that this becomes the effective x value and is constant. A simple grid ion chamber is illustrated in *Figure 7.4*. Other devices may have two or more grids.

Guard ring chamber

The main reason for using plane-parallel electrodes is that the field, E, is uniform in magnitude and direction, so that controlled ion chamber conditions can be maintained throughout the sensitive volume. The only discrepancy occurs at the edges of the plates where the field lines bow outwards and the field strength gradually tails off to zero. In this region, electron and ion collection is incomplete.

The effect is overcome, as illustrated in *Figure 7.5*, by placing guard ring plates around the electrodes. These are co-planar with, and at the same potentials

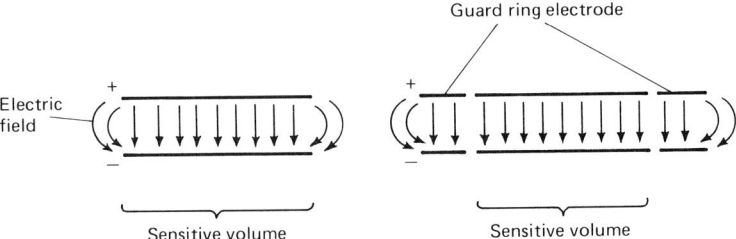

Figure 7.5 *Structure and function of a guard ring chamber*

as the electrodes, but they are not connected to the signal output system. They produce a uniform field throughout the sensitive volume between the signal plates and, incidentally, define the sensitive volume more precisely, enabling detection efficiency and, for example, radiation exposure per unit volume to be calculated accurately.

Integrating chambers

If the electrodes of an ion chamber are completely isolated from earth, its time constant is virtually infinite and charge deposited in the chamber remains on the electrodes, gradually cancelling the applied potential. Subsequent measurement of that potential, after a period of time, indicates the total charge deposited, with zero dead-time losses, and provides an indication of the radiation exposure or dose accumulated over that time. Such devices are used for radiation dosimetry. There are three basic types, as illustrated in *Figure 7.6*.

The first of these (*Figure 7.6a*) is an *air wall chamber*. It consists of an air-filled chamber with a guard ring and entry window which, between them, define a relatively small sensitive volume in the middle of the chamber. Secondary electrons, produced by incident radiation, may be scattered sideways into the air

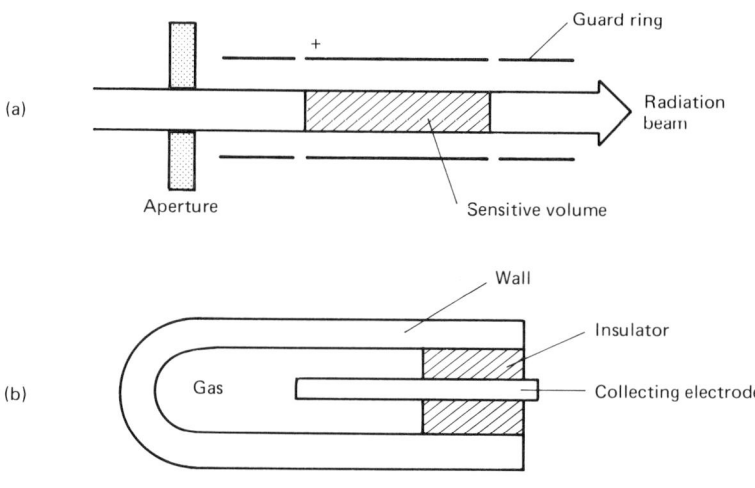

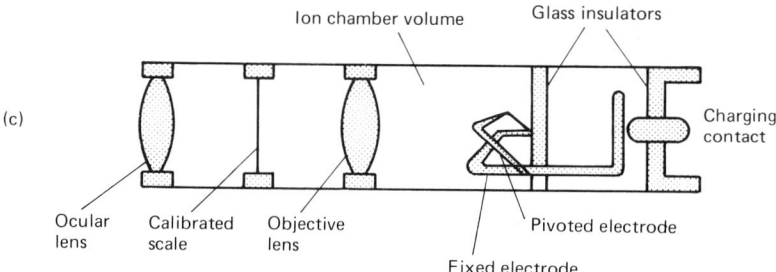

Figure 7.6 Integrating ionisation chambers. (a) Air wall chamber, (b) cavity chamber; (c) quartz fibre electrometer

volume between the collecting electrodes. Provided none of these reaches the electrodes, they deposit all of their ion pairs in the air and the charge collected corresponds to the full energy absorbed from the primary radiation. Electrons which are scattered backwards or forwards out of the sensitive volume lose some of their ion pairs to the guard ring, but this loss is compensated for by a gain of ion pairs from electrons scattered into the sensitive volume from the air between the guard rings. The compensation is virtually exact, that is, *electron equilibrium* is established, if all the electrons reaching the sensitive volume are scattered from air rather than some other medium which may have a different interaction cross-section.

In principle, the air surrounding the sensitive volume must have a thickness which everywhere exceeds the maximum range of the secondary electrons. In practice, this may be difficult for energetic primary radiation, and a smaller thickness can be accepted without much loss of accuracy. The chamber then measures the total charge deposited by primary radiation interacting with the sensitive volume and this, divided by the mass of air in that volume, gives the radiation exposure.

The chamber is bulky and inconvenient for personal or area dosimetry but it is used for absolute dosimetry and as a means of calibrating other systems. It should, of course, be operated under ideal ion chamber conditions in which ion pairs are neither lost nor gained. In air, some losses occur owing to electron attachment to electronegative molecules and a correction must be made for this effect.

The second device, shown in *Figure 7.6b*, is a *cavity chamber*, sometimes known as a *thimble chamber*. It consists of a central collecting electrode, a gas volume and an outer conducting wall which is the other electrode. The two electrodes are separated by a very good insulator such as amber, polystyrene or Perspex. The chamber may be charged to an operating voltage then allowed to discharge by the collection of ion pairs in the gas volume or it may be maintained at a fixed voltage and its charge or current output measured.

The chamber wall has a density that is of the order of 1000 times that of the gas but it is always constructed of a material whose photon mass—energy attenuation coefficient and electron mass stopping power are the same as those of the gas for the incident radiation energy. This is achieved by matching the effective Z numbers of these components. In an *air-equivalent* cavity chamber, the gas is air and the wall is a compound of low Z number materials, such as nylon, carbon and silica, lined with graphite to provide a conducting internal surface. In a *tissue-equivalent* chamber, materials with Z numbers similar to those of biological tissue are used. These include combinations such as ethylene—polyethylene and acetylene—polystyrene.

In both types of chamber, the result of matching the gas and wall materials is electron equilibrium. In effect, the air-equivalent wall is a condensed version of the air wall. Its equivalent thickness, in g cm^{-2}, should be the same as that of the air wall and exceed the maximum range of secondary electrons, but its physical dimensions are much smaller because of its greater density. The equilibrium condition follows from the fact that the same number of electrons are generated, *per unit mass*, in the gas and in the wall, and they have the same energy spectrum and angular distribution. Certainly, there are about 1000 times as many electrons produced in the wall, *per unit volume*, but the stopping power of the wall material is increased proportionately so that only as many electrons manage to leave it as enter it from the gas.

In operation, the cavity chamber is used to determine the absorbed dose, D, in some medium which is usually the liquid or solid medium of an irradiated specimen, biological tissue or phantom, but could be air. What the chamber actually measures is the total ionisation produced in the gas cavity. It is important that the chamber dimensions should be small compared with the mean electron range in the medium in which it is placed, in order to minimise the disturbance it introduces into the radiation levels in the medium. Larger chambers collect more charge and measure it more accurately, but their measurements must be corrected to account for their effects on the radiation environment.

If the radiation level in the gas cavity is the same as it would be in the medium displaced by it, then from equation 6.18, the absorbed dose, D, in the medium is proportional to the mass–energy attenuation coefficient, μ_{mE}, of the medium, and the absorbed dose, D_g, in the gas cavity is similarly proportional to its mass–energy attenuation coefficient, μ_{mEg}. If the ratio of these two coefficients, μ_{mE}/μ_{mEg}, is taken to be P, then D is equal to PD_g. From the definition of absorbed dose, D_g is equal to ϵJ, where ϵ is the mean energy required to form an ion pair and J is the number of ion pairs collected in the cavity per unit mass of gas, so the absorbed dose in the medium is given by

$$D = P\epsilon J \tag{7.7}$$

This is known as the *Bragg–Gray principle* and, since J is equal to the total charge collected divided by the mass of gas in the cavity, it relates absorbed dose to the charge collected.

The *electrometer* is derived from the classical electroscope. The device illustrated in *Figure 7.6c* is a quartz fibre electrometer, widely used as a personal dosemeter. It comprises two electrodes, one movable and hinged to the other, fixed electrode. They are charged to the same potential so that mutual repulsion drives the hinged electrode, which is usually made of platinised quartz, away from the fixed one. Electrons, or ions (depending on polarity), are collected at both electrodes, following the passage of ionising radiations through the sensitive volume, so the repulsive force decreases as the potential is gradually reduced and the quartz electrode moves closer to the fixed one. This movement can be viewed through transparent insulators against a scale calibrated in radiation dose or exposure to provide a continuous, integrated reading.

Current ion chambers

A quite different operational mode is obtained by removing the coupling capacitance, shown as C_c in *Figure 7.1*, and reducing the value of R. The effect of this modification is that charge is hardly integrated at all and leaks away through the low resistance of the external circuit as a continuous direct current. The magnitude of this current is measured by a d.c. amplifier and is proportional to the radiation flux through the chamber, provided the energy spectrum remains constant. Dead-time losses are virtually zero since the applied voltage never drops below its operational value, and the detector is widely used for measuring high radiation intensities. It is used, for example, to measure the beam current of an accelerator and radiation levels in a nuclear reactor.

Gas mixtures

In pulse ion chambers, especially when the electron pulse is being recorded, the major component of the gas mixture is usually argon, or one of the other noble gases: helium, neon, krypton or xenon. These are atomic gases, with stable electron shell configurations and exceptionally large binding energies. They are less likely than any other gases to capture electrons from a signal pulse or to be ionised by electrons in that pulse, so they facilitate the establishment of the ideal ion chamber conditions in which charge is neither lost nor gained from the signal pulse. Other, molecular gases, such as hydrogen, nitrogen, carbon dioxide (CO_2) and methane (CH_4), are slightly less satisfactory in this respect, and the highly electronegative gases and vapours, such as oxygen, water vapour and the halogens, must be excluded altogether. They have a strong *electron affinity*; that is, they tend to attach electrons, removing them from the signal pulse. This reduces the pulse height and may completely eliminate very small signals, reducing the detection efficiency of the detector.

A common practice is to add about 10% of CO_2 or CH_4 to argon. Such additives 'cool' the discharge and greatly increase the drift velocity of the electrons. When electrons collide with gas atoms they rebound in all directions and, in many cases, the applied field has to overcome backward momentum before drifting the electrons towards the anode. Because of their more numerous and accessible excited states, molecules undergo inelastic collisions more easily than noble-gas atoms do, so, when the mixture contains suitable molecules, the electrons tend to suffer inelastic collisions, losing some energy. They are then more easily turned towards the anode with the result that collection time is decreased, and count rate capability is increased. Too much molecular additive reduces pulse height and detection efficiency by capturing electrons.

The action of molecular and electronegative gases depends on the nature of the gas and the operating conditions. In general, however, the halogens, oxygen and water vapour must comprise less than 0.1% of the mixture in a pulse ion chamber. Consequently, gas purity of the order of 99.9% is required. A particular problem is desorption of impurity gases, especially oxygen and water vapour, from the chamber walls. This limits the useful lifetime of a sealed chamber. It can be minimised by using glass or metal walls or, alternatively, a throughflow system can be used in which the gas mixture is continuously flushed through the chamber.

In current and integrating chambers the gas mixture is much less critical and air, for example, can be used. The reason for this is that electrons captured by gas atoms to form negative ions are effectively lost from a fast pulse but they are eventually collected at the anode and are not lost permanently. Thus the negative charge pulse is not reduced in a current or integrating chamber, it is just extended over a longer time period. Besides air, used for radiation exposure measurements, roughly equal mixtures of argon and nitrogen, neon and hydrogen, and so on, are widely used.

7.3 PROPORTIONAL COUNTERS

In principle, the proportional counter is an extension of the ion chamber in which a higher field strength is used and internal pulse amplification takes place. The field is strong enough to accelerate electrons, between collisions with gas atoms, and to give them enough energy to produce secondary ion pairs. The secondary electrons are then accelerated to create further ion pairs, and so on. The net result is that each primary electron finally generates a large number, m, of electrons in a process known as a *Townsend avalanche*, as illustrated in *Figure 7.7*. The original charge deposited by the ionising radiation is multiplied by a factor, m, which is the *gas amplification factor* and may be as large as 10^5. Consequently, a large signal pulse of the order of 100 mV may be obtained.

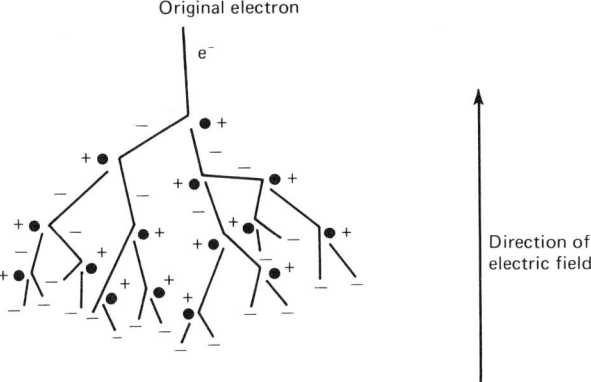

Figure 7.7 Townsend avalanche process

If n original ion pairs are formed in the gas, the signal pulse height, v, is given by

$$v = \frac{q}{C} = \frac{mne}{C} = \frac{meE}{C\epsilon} \qquad (7.8)$$

where E is the energy deposited as original ion pairs and C is the capacitance of the system; that is, of the detector and any external capacitance. The detector capacitance, C, is given by

$$C = \frac{5.56 \times 10^{-11} L}{\ln(r_2/r_1)} \text{ farads} \qquad (7.9)$$

where it has been assumed, for the moment, that cylindrical, coaxial electrodes of length L are being used. r_2 is the radius of the outer electrode and r_1 the radius of the inner one. As will be explained below, this is the standard geometry of a proportional counter and a Geiger–Müller counter. Normally, r_1 is much smaller than r_2 and may refer to an axial wire electrode.

Clearly, the pulse height is proportional to the number of ion pairs generated and, therefore, to the energy deposited, E, so the counter is an *energy spectrometer*. It is called a proportional counter because it differs from an ionisation chamber, in which pulse height is effectively equal to energy deposited, and a Geiger–Müller counter, whose response is independent of energy deposited. The intrinsic energy resolution is slightly poorer than that of an ionisation chamber, because of the additional statistical line broadening due to the gas amplification process, but extrinsic noise effects are relatively smaller because of the large signal pulses, so that net resolutions are usually of the order of 3%, like those of ion chambers.

In a uniform field, the current in a Townsend avalanche increases exponentially with distance, x, according to the formula

$$I = I_0 e^{\alpha x} \qquad (7.10)$$

where α is known as the *Townsend coefficient*.

Typically, the avalanche forms in about 10^{-7} s. The coefficient α depends on the field strength. It also depends on the mean free path of electrons in the gas and, therefore, on gas type and pressure. Some values of α are included in *Table 7.1* for a range of chamber gases. As in the case of ion chambers, noble gases are used so as to minimise recombinative losses, and a small percentage of gases or vapours, such as methane or ethanol, is added.

Electrode geometry

Unlike the ion chamber, however, the proportional counter cannot operate with plane-parallel electrodes. It has been pointed out that an ion chamber pulse height, in the absence of a grid, depends on the position of ion formation. In a proportional counter, the situation is much worse because gas amplification is also a function of the distance, x, of the original ion pair from the anode. From equation 7.10,

$$m = \frac{I}{I_0} = e^{\alpha x} \tag{7.11}$$

As a result, pulse height varies with formation position and all energy-proportionality is lost. The only solution, really, is to start all the avalanches at the same distance from the anode so that they all have the same length, x. This means that the field must exceed the critical value required for gas amplification only within a small, fixed distance from the anode. There is no reason why a grid should not be used to achieve this result but, since the proportional counter does not need a

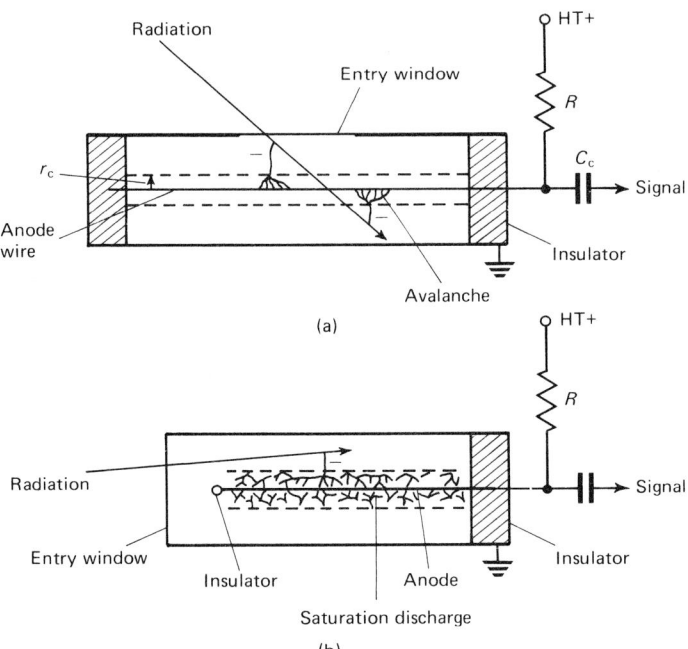

Figure 7.8 Structure of (a) a side window proportional counter and (b) an end window Geiger–Müller counter

uniform field, a more effective method is available: the chamber is constructed with cylindrical geometry.

A standard cylindrical counter is shown in *Figure 7.8*. The anode is a fine wire situated along the axis of a cylindrical cathode. Radiation usually enters through a thin side window of plastic, glass or metal foil. The electric field between the wire anode and the cathode decreases with radial distance, r, from the axis of the system according to the formula

$$E = \frac{V}{r \ln (r_2/r_1)} \qquad (7.12)$$

where V is the applied potential, r_2 is the radius of the cylinder and r_1 is the radius of the wire.

The field is strong enough to produce avalanches only within a certain *critical radius*, r_c, of the anode wire. r_c is very small (less than 1 mm), so the proportional region is confined to a narrow cylinder about the axial wire. Outside this region, ion chamber conditions prevail. Since this is the bulk of the sensitive volume, most interactions with incident radiations take place in this outer region. Primary electrons drift towards the anode and initiate avalanches only when they reach the critical radius. Virtually all avalanches start at $r = r_c$ and gas amplification is independent of the position of the primary ion pairs. In addition, the applied potential is not nearly as large as it would have to be to maintain proportional field conditions throughout the sensitive volume and, in fact, the voltages applied to proportional counters may be of similar magnitudes to those applied to ion chambers, although they are more often considerably larger.

Gas mixtures

Molecular gases can be used in proportional counters, as in ion chambers, but they tend to capture electrons and reduce gas amplification and pulse height. Gases such as H_2, CH_4, ^3He and BF_3 (boron trifluoride) are used for neutron detection but, when there is no particular reason to use such gases, the noble gases are preferred, usually with a small percentage of molecular additive. The most common mixtures are argon with 10% methane (CH_4), ethanol (C_2H_5OH), carbon dioxide (CO_2) or isobutane (C_4H_{10}). For such mixtures, gas amplifications in the range of 10^2 to 10^5 can be obtained, depending on the applied potential. In general, gains of more than 10^5 tend to saturate the counter so that pulses are of similar sizes, and energy dependence is lost. At this point the counter begins to function in the Geiger–Müller mode.

Small quantities of organic gases or vapours have two beneficial effects. As in ion chambers, they reduce the random electron movements by inelastic scattering, increase drift velocity and improve count rate capability and time resolution. As in Geiger–Müller counters, they act as quenching agents; that is, they limit the extent of the gas discharge and, in addition, prevent the tube from recycling (*see below*).

Quenching

In the avalanche discharge, excited atoms, as well as ions, are formed, and some of them are excited from the inner electron shells. They de-excite, emitting X-ray and ultra-violet photons which are capable of liberating outer electrons from the gas atoms and thereby creating further ion pairs by the *photoionisation* process. These may be some distance from the initial avalanche, spreading the discharge and continuing it indefinitely. In proportional counters, the process is terminated, or quenched, by the addition of a small quantity of *organic* gas or vapour. These have relatively large molecules which tend to absorb the fluorescent emissions of the noble-gas atoms. They also have smaller excitation potentials than the latter, so their de-excitation photons have insufficient energy to ionise the gas and propagate the discharge further. They reduce the mean free path of the noble-gas photons to about 1 mm and limit the extent of the discharge. For satisfactory *photon quenching*, the absorption spectrum of the quenching agent should match the emission spectrum of the noble gas. Methane and ethanol both satisfy this requirement and are widely used in proportional counters.

There is a second process, known as *recycling*, that may have to be quenched. In this, positive noble-gas ions bombard the cathode with sufficient energy to eject secondary electrons, which recycle the tube. Quenching agents with lower ionisation potentials than the gas atoms tend to lose electrons in collisions with these ions, becoming positive molecular ions themselves. Being heavier than the original ions, they approach the cathode more slowly, drawing electrons out by field emission. These cancel the ion and dissociate the molecule, dispersing energy without recycling the tube. Recycling is seldom a problem with proportional counters and the *ion quenching* effect, although present, is not really required.

Signal pulse

In a proportional counter, almost all of the ion pairs are generated inside the critical radius and very close to the anode wire. Electrons are collected long before the positive ions have moved away from the anode, which remains electrically neutral. The collected electrons are held in the wire by the nearby positive charge and it is only when these ions begin to move away that a negative

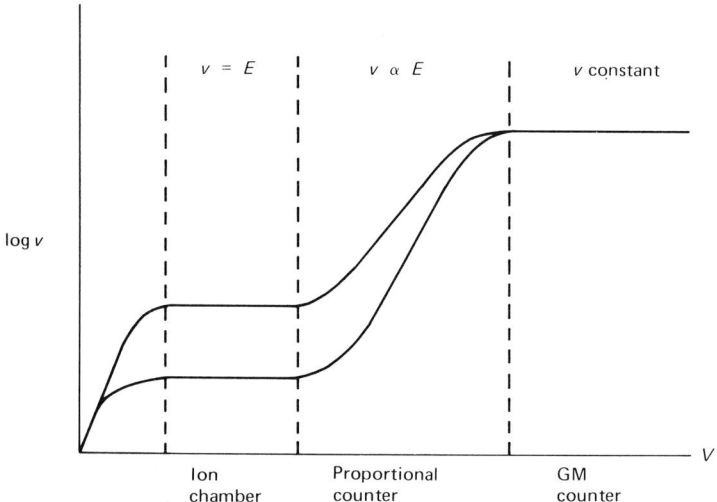

Figure 7.9 Pulse height as a function of applied voltage for gas counters, showing curves for two different radiation energies

charge pulse is observed by the signal processing circuit. Hence it is the movement of the positive ions that determines the pulse shape.

By comparison with the ion chamber, both electrons and positive ions have to move over a smaller distance and in a much larger field, so pulse rise times are faster and time resolution is of the order of 1 μs. For the same reason, dead time is less than that of the ion chamber and is of the order of 10^{-4} s when the counter is used for pulse height spectrometry. If, however, as in counting and timing measurements, one pulse is allowed to sit on the tail of a preceding one, dead time can be reduced to about 1 μs.

The pulse height, v, is proportional to energy, E, and to gas amplification, m. Since m is equal to $e^{\alpha x}$, log v is proportional to the Townsend coefficient, α, which in turn is proportional to the applied potential, V. Consequently, log v is proportional to V, as illustrated in *Figure 7.9*.

In passing, it is worth noting that excited atoms in the avalanche de-excite, emitting mainly ultra-violet radiation. Thus the proportional counter also behaves like a scintillation counter. The noble gases are inefficient scintillators but this is more than compensated for by gas amplification and the scintillating proportional counter is a real prospect.

7.4 GEIGER–MÜLLER COUNTERS

Like the proportional counter, the Geiger–Müller (GM) counter is constructed with cylindrical geometry. It has an outer cylindrical cathode with a thick axial wire, or cylindrical anode, and an end window to admit ionising radiation (*Figure 7.8*). The main difference is that compared with a proportional counter of the same size, the GM counter is operated at a higher applied potential. As a result, it produces larger discharges and may be more highly quenched.

With the increased applied potential, the critical radius, r_c, the avalanche size and the gas amplification factor are all increased. More excited atoms are produced and more photons penetrate some distance into the gas where they initiate *photoionisation* processes. Some *photon quenching* does take place but the amount of quenching agent present is not enough to terminate the discharge, which spreads throughout the tube, usually in about 10^{-7} s. The net result is that avalanches fill the entire axial volume within the critical radius, generating a *saturation discharge* which, for a given applied potential, is quite independent of the size of the original event.

This is the Geiger–Müller mode of operation and its output pulse heights are constant and entirely independent of the energy deposited by the ionising radiation. Thus, the GM counter is not an energy spectrometer, but it does have a high sensitivity to weakly ionising events and responds, for example, to a single ion pair. The output signal may be of the order of 1 V and easily processed, without further amplification.

The main function of a quenching gas, in a GM counter, is to prevent *recycling*. With the large potentials involved, positive ions reach the cathode in large numbers and with large energies that are sufficient to eject electrons, which recycle the counter. This mechanism must be quenched, either externally, using an electronic system, or internally, by means of gas additives which may be more electronegative than those used in ion chambers and proportional counters.

A number of external methods are available but the simplest approach is to make the load resistance, R, very large, of the order of 100 MΩ (10^8 Ω). When a large transient current, I, flows through the counter as a discharge, the same current flows through R generating a voltage drop, IR, across the resistor and drastically reducing the potential drop across the counter, so that Geiger–Müller conditions no longer prevail and the discharge terminates. The only disadvantage of this method is the long time constant of the external circuit, which may take up to 10 ms to re-establish Geiger–Müller conditions, producing a dead time of the same order of magnitude.

Internal *ion quenching* is accomplished by adding about 10% of an organic vapour, such as ethanol, or about 1% of a halogen, such as chlorine or bromine, to the noble gas in the chamber. The ionisation potentials of these quenching agents are substantially less than those of the noble gases so that in collisions between noble-gas ions and quenching-gas molecules the ionisation is transferred to the latter. When these reach the cathode, they are neutralised by electrons extracted, by field emission, from the cathode, but the electron energy is used up in dissociating the molecule instead of causing further ionisation.

Organic quenching agents remain dissociated and are gradually used up, imposing a useful lifetime, of the order of 10^8–10^9 counts, on the tube. The halogen molecules, on the other hand, reassociate and are not used up by the quenching process, but they are extremely reactive and tend to react with chamber components to produce impurities which limit tube lifetime, usually to about 10^{10}–10^{13} counts.

Signal pulse

The output pulse from a GM counter is independent of the energy deposited, and its height, v, is constant for a given applied potential. Pulse height, however, does increase slightly with applied potential because the critical radius, r_c, and the size of the discharge increase with potential.

The response of a GM counter can be compared with those of ion chambers and proportional counters by means of a graph of v against V, as illustrated in *Figure 7.9*. In the ion chamber, v is proportional to energy deposited, E, but independent of V. In the proportional counter, v increases with E and V. In the GM counter, v is independent of E and almost independent of V. Such a graph must be used with care, because it presupposes similar electrode geometries and gas mixtures. Ion chambers can be operated with cylindrical electrode geometry, but the range of operating voltage is considerably shortened. The use of highly quenching GM gas mixtures further shortens this plateau region to the extent that it may become unrecognisable. In addition, the ion chamber requires a low-noise preamplifier and a high-gain amplifier, neither of which is used with a GM counter, so it may be difficult to assess the absolute values of v, at the chamber, in each operating mode. Nevertheless, some resemblance of the graph can be obtained, as V is increased from zero, to confirm that a particular counter is, finally, operating as a GM counter.

A more standard method of finding the optimum voltage range for a GM counter is to produce a graph of count rate, N, against applied potential, V. This is known as the *characteristic curve* of the counter and is illustrated in *Figure 7.10*. At the low-voltage end of the curve, count rate falls off to zero

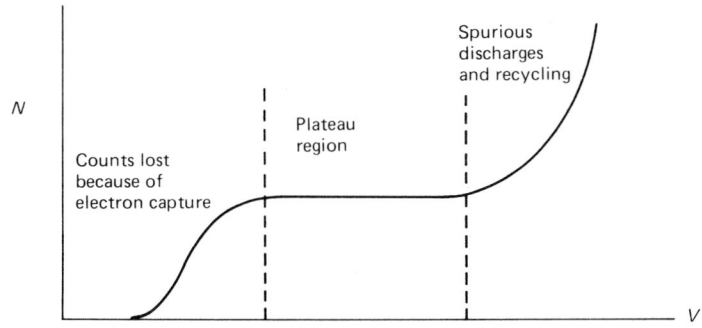

Figure 7.10 Characteristic curve of a Geiger–Müller counter

because the field is not strong enough to attract all of the primary electrons into the avalanche region before they are captured by the quenching-gas mixture. At higher potentials, a plateau is formed; that is, there is a range of voltage over which the count rate remains fairly constant. The counter should be operated somewhere in the middle of this range. In organically quenched tubes, this plateau is quite flat as virtually all primary electrons reach the avalanche region and all events are recorded. In halogen-quenched tubes, the highly electronegative gas stops some electrons before they reach the critical radius, so some events fail to trigger the tube and detection efficiency is reduced by a few percent. In this case, the critical radius increases with applied potential, increasing the probability that the primary electrons will reach the avalanche region and the characteristic plateau shows a small gradient. At the highest potentials, recycling begins to take place and count rate increases rapidly. Continued operation at such voltages overheats and damages the counter. Typical operating voltages are 600 V for a halogen-quenched counter and 1000 V for an organically quenched counter.

7.5 SPARK CHAMBERS

The spark chamber is a direct extension of the Geiger–Müller counter in which the applied potential is increased to the point where it is almost, but not quite, sufficient to cause a spark breakdown between the electrodes. When ionising radiation passes through the gas mixture, any ion pairs deposited are sufficient to initiate a spark discharge which records the event. As in the GM counter, it is a saturation response and does not measure the energy deposited but it does measure the position of the event, so the spark chamber is an *imaging gas counter*. For this reason, it is widely used in high-energy and nuclear physics as a tracking device and an alternative to the bubble chamber. It has also been used for bidimensional imaging and should be compared to the scintillation camera and radiographic systems.

A spark chamber is illustrated in *Figure 7.11*. It consists of two plane-parallel electrodes, separated by a gap of about 1 cm and containing a mixture of a noble gas, such as argon, plus about 10% of a quenching agent, such as methane or ethanol. This cannot be said to be a typical arrangement, because spark chambers vary considerably in design, but it serves to illustrate the detection mechanism.

The electric field is not only strong enough to produce Townsend avalanches, but uniform, because of the plane electrode geometry, so avalanches occur anywhere in the gas and not just near to the anode as in a proportional or GM counter. As shown in *Figure 7.12*, the ion pairs deposited by incident radiation generate an avalanche which grows rapidly. Electrons, with their higher mobility, surge to the head of the avalanche leaving positive ions behind. This polarisation produces a space charge effect with a reverse field inside and alongside the

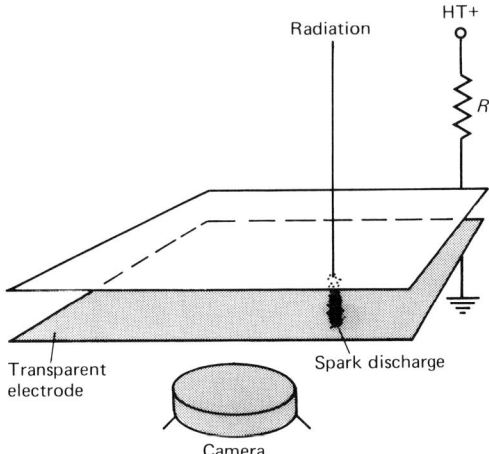

Figure 7.11 *Visual spark chamber operated in the d.c. mode*

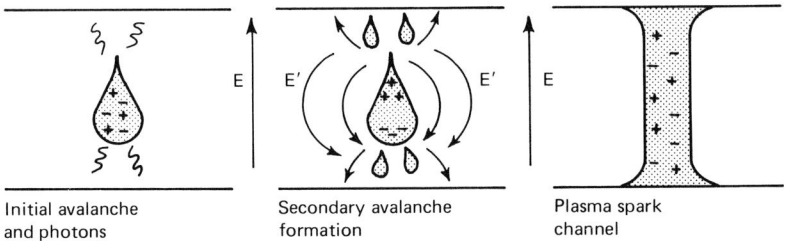

Figure 7.12 *Development of a spark discharge. The net field at any point is the algebraic sum of the applied field, E, and the induced field, E'*

avalanche, slowing it down, and an increased field strength at each end of it. Between these ends and the electrodes, intensive *photoionisation* takes place, causing further avalanching in these regions. These secondary avalanches link with the initial avalanche before it has time to spread and the result is a highly localised charge plasma connecting the electrodes. This has a low resistance, of the order of 1 Ω, and spark breakdown occurs along the channel provided. Typically the whole process is established in about 10^{-8} s. The difficult part is terminating, or quenching, the discharge.

Quenching

There are two forms of quenching. One is *photon quenching*, by the molecular gas or vapour in the gas mixture, and the other is *external quenching*, by means of a large, external resistive load or by pulsing the applied potential.

Photon quenching depends on the addition of about 10% of an organic quenching agent, such as ethanol or methane, to the noble gas. If less than this amount is used, the discharge becomes non-localised and the chamber tends to be more unstable against spurious discharges. If more is used, the slightly electronegative vapour tends to attach electrons, slowing down the initial avalanche, and higher applied potentials must be used to overcome this effect. Unfortunately, the spark discharge dissociates the quenching agent, which is used up rapidly and also forms carbonaceous radicals that eventually deposit on the electrodes. As a result, the gas mixture may have to be continuously refreshed and inconvenient *throughflow gas systems* are required. Although it is possible to purify and recycle gas mixtures, this feature makes it quite expensive to use high Z number gases, such as krypton and xenon. In addition, the spark discharge etches the cathode by positive-ion bombardment. Over a period of time, this ruins the electrode and further pollutes the gas mixture with metallic ions.

External quenching is required because once the spark is initiated it is not possible to terminate it by internal ion quenching. Two methods are used, leading to two types of spark chamber. Once can be described as operating in a d.c. mode and the other in an a.c. mode. In the former (*Figure 7.11*), a continuous, d.c. potential, of the order of several kilovolts, is applied to the chamber through a large resistive load of about 40 MΩ. This is similar to the method used to quench a GM counter externally. When a spark occurs, current flows through the load, developing a large potential drop across it and reducing the chamber potential to below its sparking point. The time constant for this circuit is very long, so

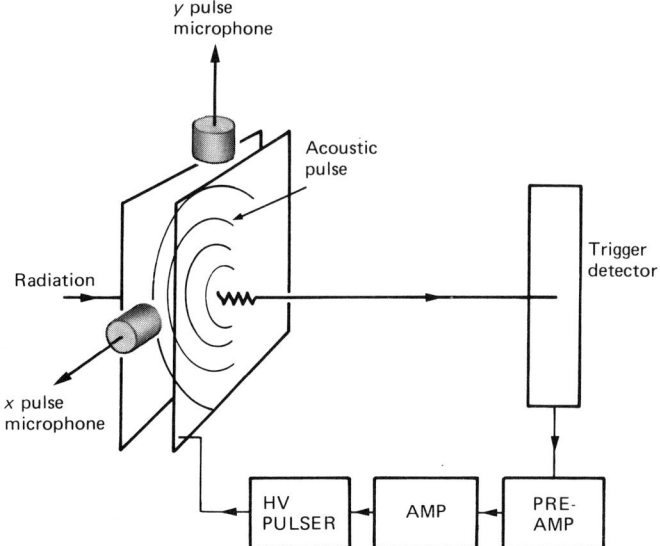

Figure 7.13 *Acoustic spark chamber operated in the triggered, a.c. mode*

although the spark develops in about 20 ns it requires up to 10 ms to extinguish. Thus the chamber dead time is very long and count rate capability is quite poor.

The alternative method of external quenching is to apply the high voltage in the a.c. mode, that is, as a pulse. This must reach several kilovolts in a few tens of nanoseconds and it must be applied while the ion pairs liberated by incident radiation are still present in the gas mixture. One way of achieving this coincidence is to employ a *trigger system*, as illustrated in *Figure 7.13*. In this arrangement, a separate trigger detector is employed to detect ionising radiation that has passed through the spark chamber. The output signal from the trigger detector is amplified and used to trigger a high-voltage pulse that is applied to the spark chamber while the ions are still present. Normally, this must be within a microsecond of the ionising event. The spark discharge takes place along the ion trail then extinguishes when the pulse terminates. Residual ions are swept away by a small d.c. potential applied continuously to the chamber to produce a *clearing field*.

The triggered system is analogous to a triggered bubble chamber and, like the latter, has been widely used in high-energy physics for particle tracking. The need for a second detector and a trigger system may be regarded as a nuisance, but in high-energy physics it is an advantage, since the trigger detector can be used as a particle identification system, to identify specific radiations so that only these are imaged. In medical applications, it is inconvenient because these mainly involve photon imaging and each photon interacts usefully in the spark chamber or the trigger detector, but not both. Although internally triggered chambers have been developed, most of the systems used for medical imaging have been operated in the d.c. mode. Compared to the triggered, a.c. mode, this is more prone to spurious discharge and capable of a lower count rate.

Signal output

A spark discharge generates visible light, sound, heat, electric current and a magnetic field, so it is virtually impossible to avoid recording it — as operators of nearby electronic equipment will verify. All of these effects can be used to count and locate the sparks, giving rise to a variety of chamber designs. Further diversity derives from a wide range of electrode configurations. These may be solid metal electrodes, deposits of transparent conductors such as tin oxide (SnO_2) or metallic films on transparent substrates, such as glass, or they may be wire mesh or parallel wire arrays. With the solid electrodes, the sparks must be photographed, or otherwise recorded, through the interelectrode gap. This approach is frequently used in particle tracking measurements. With the other systems, the sparks can be photographed through the electrodes, as must be the case when photon distributions are being imaged visually. Usually, the chamber is examined by a camera whose shutter is left open for the duration of the exposure to obtain a time-integrated picture of the radiation distribution (*Figure 7.11*).

The system shown in *Figure 7.12* is an acoustic chamber. The sound pulse generated by the spark travels at a uniform velocity through the gas mixture to be recorded by two or more microphones. The time delay between the application of the high-voltage pulse and the reception of the acoustic pulse measures the distance of the spark from each microphone and, therefore, its spatial coordinates.

Stereophonic recordings with three microphones generate three-dimensional localisations, as do stereophotographic recordings with three cameras.

With wire electrodes it is possible to extract electrical readout signals that are directly computer-compatible. A typical assembly is shown in *Figure 7.14*. It consists of two orthogonal wire systems. The spark generates a signal in one wire of each electrode array and the positions of these two wires indicate the x and y coordinates of the event. These data can be read out and displayed on a cathode ray oscilloscope or processed via an on-line computer.

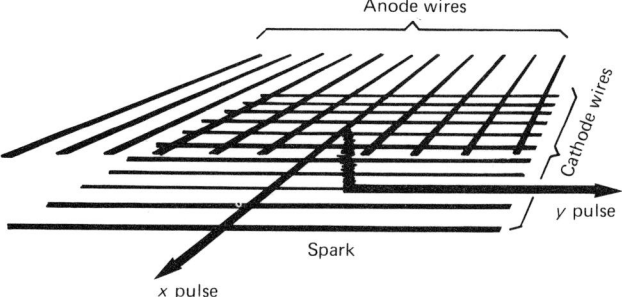

Figure 7.14 Multiwire spark chamber

The *position resolution* of a spark chamber depends on the readout mechanism and on the nature of the ionising radiations. With *charged particles*, the intrinsic, track resolution is of the order of the width of the spark, which is a fraction of a millimetre. With *photons*, the intrinsic resolution is much poorer because of the effects of *oblique incidence* and *photoelectron range*. If the chamber is used in radiography, or as a gamma camera with a pinhole aperture, photons pass through the gas at an oblique angle and can interact anywhere along that path. Consequently, there is some lateral uncertainty with regard to the exact x and y coordinates of an event viewed through one electrode. In addition, the photoelectron produced in an interaction can be emitted in any direction and generate the spark at any point along its path, introducing a further uncertainty as to the position of the initiating event. In argon, at STP the range of a 30-keV photon is about 7 mm and the net, FWHM spatial resolution is of the same order of magnitude. At higher photon energies the situation is even worse. Resolution is improved by using high gas pressures and high Z number gases to reduce the photoelectron range, but it does not generally equal that of a scintillation camera and is much poorer than that of any of the radiographic detectors.

Apart from its relatively poor spatial resolution, the spark chamber has another major deficiency, in terms of radiographic applications, namely, its poor count rate capability. It could be used as a photon camera, especially for low-energy photons, but it has few advantages and several disadvantages in this application, compared with a multiwire proportional counter, which measures photon energy, has less need of a throughflow gas system, and has a higher count rate capability.

7.6 MULTIWIRE PROPORTIONAL COUNTERS

The multiwire proportional counter, or MWPC, was developed by Charpak in 1968, and such chambers are sometimes described as *Charpak counters*. The

MWPC differs from the traditional proportional counter in that its anode is a plane array of many wires instead of a single wire. The field geometry about each wire and the gas discharge mechanism are the same as in a single-wire chamber, and the output signal from each wire is a voltage pulse whose height is proportional to the energy deposited in the chamber. In addition, the location of an ionising interaction is recorded as the position of the affected wire, so the MWPC, like the spark chamber, is an *imaging gas counter*, as well as an *energy spectrometer*.

A typical MWPC is illustrated in *Figure 7.15*. It has two cathodes, each comprising a set of parallel wires, and these are mutually orthogonal, that is, at right

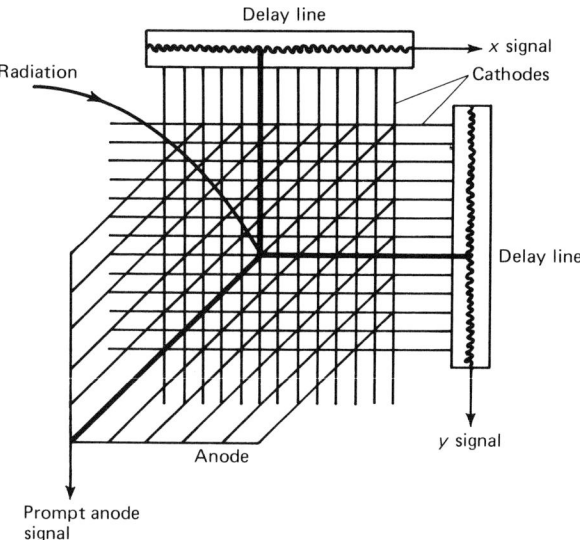

Figure 7.15 Multiwire proportional counter with delay line readout

angles to each other, so that one measures the x coordinate and the other the y coordinate of an ionising event. Between these two, and usually orientated at 45° to each cathode, is an array of anode wires. A Townsend avalanche takes place very close to an anode wire, which collects electrons and measures pulse height. Positive ions are collected at both cathodes, which record the position coordinates of the discharge as the array positions of the two wires from which voltage pulses are derived.

In principle, the simplest method of extracting position information is to connect each cathode wire to its own amplifier and readout electronics but, with 100 or 1000 wires per cathode, this can be an expensive and difficult undertaking, even with modern integrated circuitry. In practice, the preferred method is the delay line readout shown in *Figure 7.16* as a block diagram. In this system, a prompt, or START, pulse is derived from the anode. The cathode pulses are passed along a continuous delay line, where they are delayed, typically by about 5 ns mm^{-1}, so the total delay time measures the position of the activated wire. Each ionising event in the chamber generates one prompt, anode pulse and two cathode pulses. All three pulses are amplified and passed on to time-to-amplitude (TAC) units. Each TAC unit measures the time interval between the anode START pulse and a cathode STOP pulse and outputs a single voltage pulse whose height

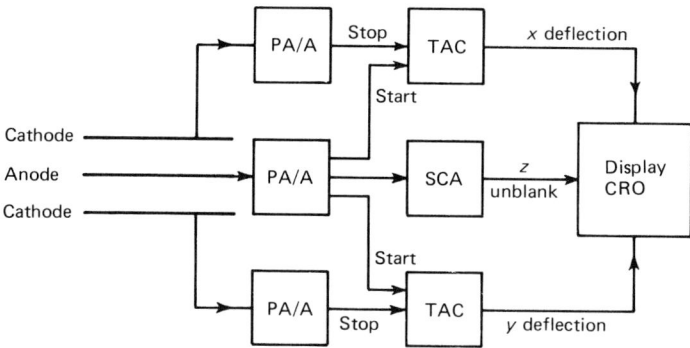

Figure 7.16 Delay line readout electronics. PA/A, preamplifier and amplifier; TAC, time-to-amplitude converter; SCA, single-channel analyser

is proportional to the time interval and, therefore, to the x or y coordinate of the event. The two TAC pulses are used to deflect the beam of a display oscilloscope which writes one point at the appropriate x–y position and, over a period of time, produces a bidimensional image of radiation distribution in the chamber. Energy selection can be included by single-channel-analysing (SCA) the height of the prompt signal and using this output to unblank the display pulses. If required, a time-to-digital conversion (TDC) can replace the TAC units to digitise the data for storage and analysis in an on-line computer.

Being a gas counter, the MWPC has a poorer *detection efficiency* than other detectors. In high-energy physics, it is preferred to a bubble chamber when stopping power is less important than data-processing capability. The bubble chamber does not produce computer-compatible output signals. In medical and other photon-imaging applications, it has to compete with the established scintillation gamma and positron cameras, which have high efficiencies.

The detection efficiency of an MWPC can be improved by the addition of a *drift region* to form what is one type of *drift chamber*. The drift region consists of a large gas volume interposed between the chamber entry window and the electrode assembly. An electric field is maintained across this region sufficient to generate ion chamber conditions so that any electrons formed in it, by ionisation events, are drifted along the field lines to the electrodes where they initiate proportional counter responses. The effect is to increase the gas volume and, therefore, the stopping power of the chamber. Further increases can be achieved by using high Z number gases (such as xenon), high gas pressures, and high-density photoelectron converters in the chamber entry window. These measures can lead to a chamber efficiency which approaches that of a sodium iodide scintillator.

The *spatial resolution* of an MWPC is affected by oblique incidence of radiation and photoelectron range, as in a spark chamber. It can be improved by employing high-density gas mixtures, and, if special signal-processing methods are used, may be smaller than the wire spacing, but is normally of the order of 1 cm.

The *gas mixtures* used in MWPC can be the same as those used in single-wire chambers but, in general, larger signal pulses and, therefore, larger gas amplifications are required. For this reason, gas mixtures known as *magic mixtures* have been developed. They provide gains of up to 10^8. One such mixture comprises argon with 16% of isobutane (C_4H_{10}) and less than 0.3% of Freon-13Bl (CF_3Br).

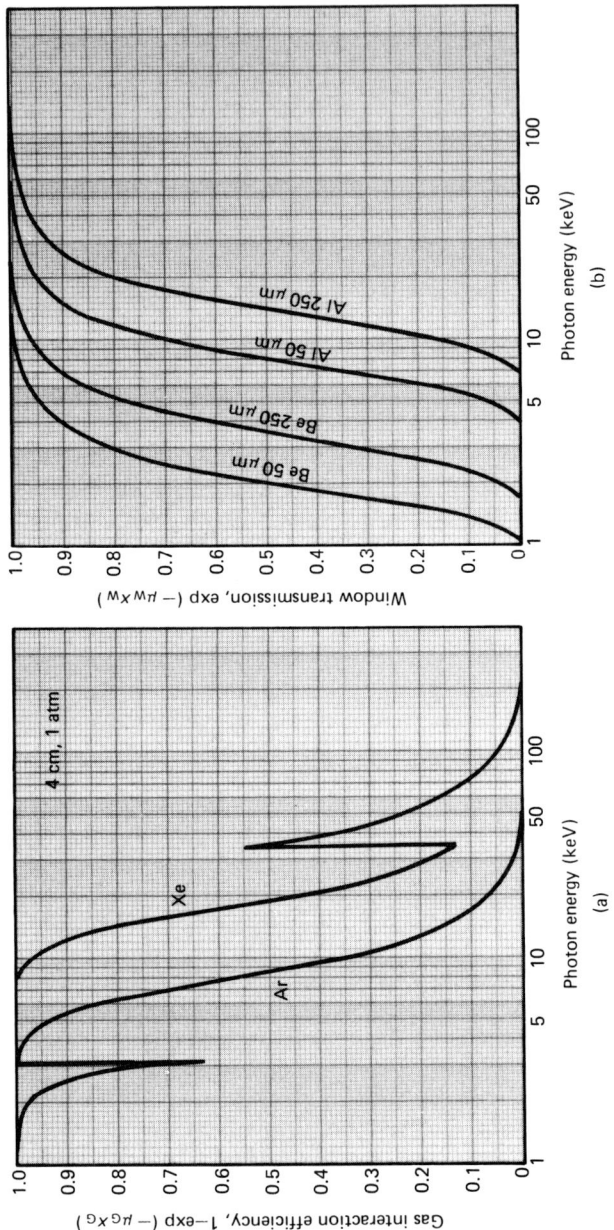

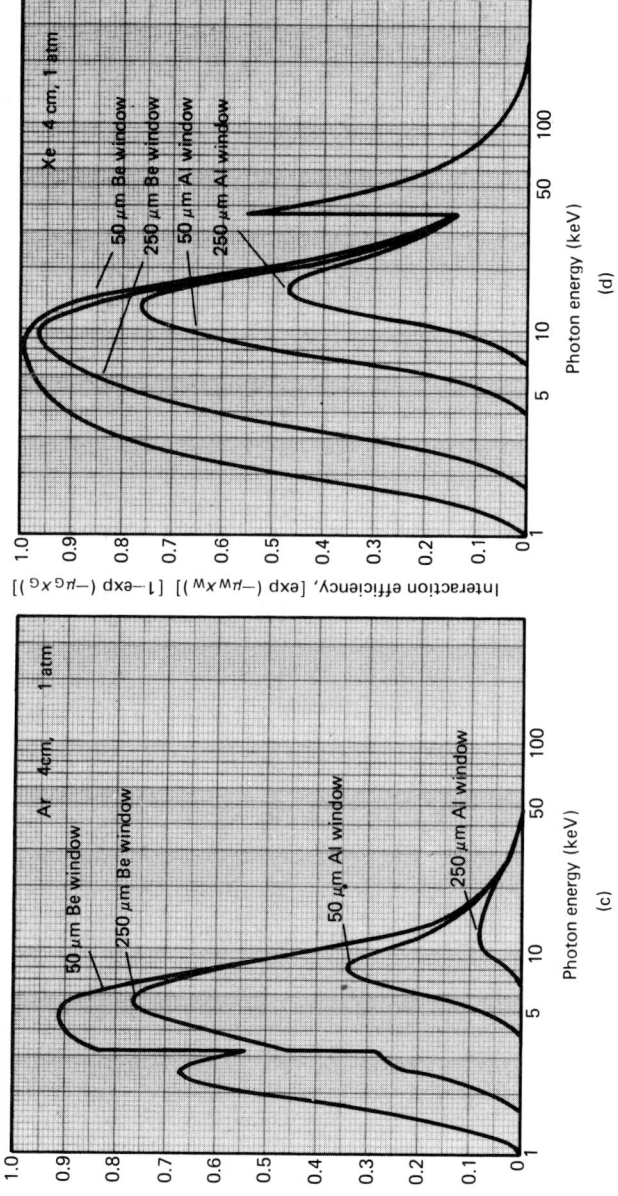

Figure 7.17 Interaction efficiency of gas counters for X-ray photons, excluding the effects of photoelectron conversion in the chamber walls and window. (a) Attenuation of 4 cm, 1 atm argon and xenon; (b) transmission of Be, Al windows; (c) sensitive range of a 4 cm, 1 atm argon chamber with various windows; (d) sensitive range of a 4 cm, 1 atm xenon chamber with various windows

7.7 FUNCTIONS

Counting

The dead time of a proportional counter used for non-spectrometric work can be less than 1 μs, so it can handle count rates of up to about 10^6 s^{-1}. The dead time of a GM counter is of the order of a few hundred microseconds so its detection efficiency falls off at count rates of the order of 10^3 counts s^{-1} and it is incapable of measuring count rates much higher than this. It is useful for measuring low activity and background counts because of its high sensitivity to weakly ionising events.

For *charged particles*, the interaction efficiency, R, is about 100% except in halogen-quenched tubes, in which it is a few per cent less. There are two other factors which strongly affect detection efficiency. First of all, the mean free path of the particle in the gas must not exceed the dimensions of the chamber, otherwise no ion pairs may be formed. This means that the specific ionisation must substantially exceed one ion pair per path length in the chamber. By this criterion, beta particles with the minimum ionising energy, of about 1 MeV, require considerably more than about 1 mm of argon at STP, a condition that is easily satisfied. The second factor is the entry window, which must be thin

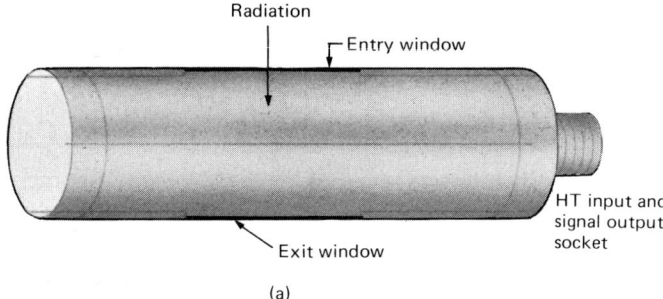

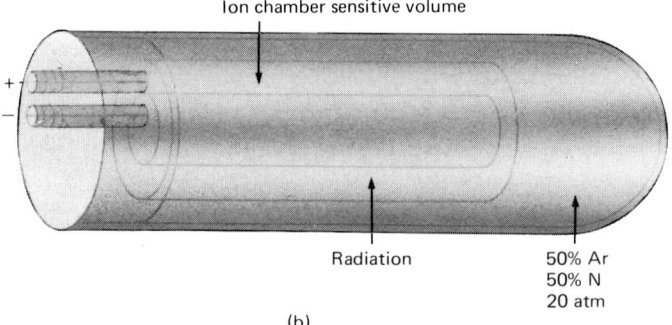

Figure 7.18 Gas counters for photon detection. (a) X-ray proportional counter; (b) cylindrical ion chamber for gamma radiation

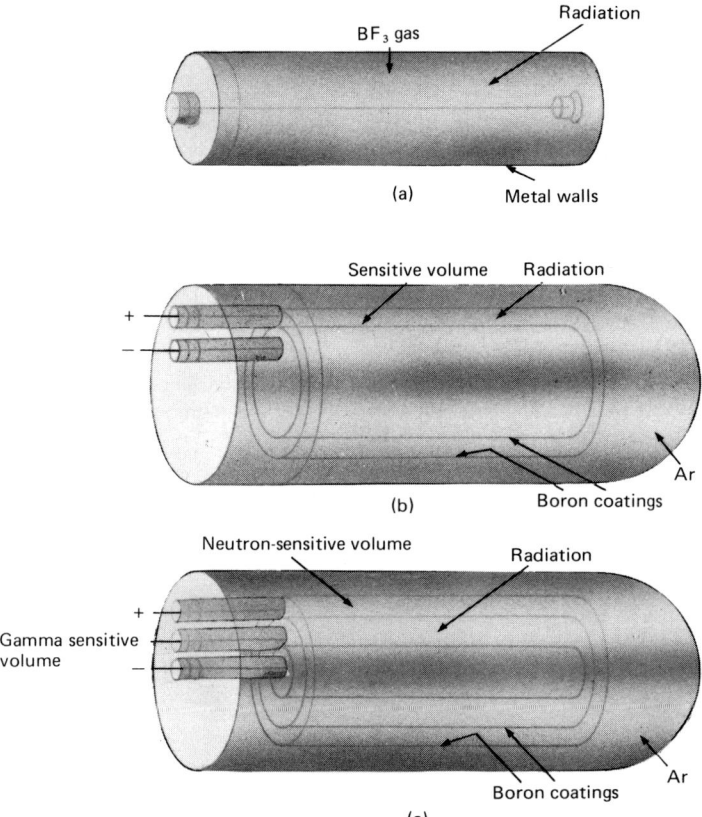

Figure 7.19 Gas counters for thermal-neutron detection. (a) BF_3 proportional counter; (b) cylindrical, d.c. ion chamber; (c) compensated, d.c. ion chamber

enough to allow the charged particle to reach the gas. Thin windows of plastic or mica can be used. Thicknesses of a few mg cm^{-2} allow alpha particles into the chamber, since the range of a 5-MeV alpha particle in a low-Z material is of the order of 5 mg cm^{-2}.

For *photons*, the interaction probability is a function of attenuation coefficient and falls off rapidly with increasing energy. As shown in *Figure 7.17*, the efficiency of 4 cm of argon at atmospheric pressure is less than 1% for energies greater than about 40 keV. The use of xenon, with a Z number of 54, extends the useful range to about 150 keV and, at a few atmospheres of pressure, the range can be extended further to about 250 keV. Higher pressures cannot be used for X-rays because very thin entry windows have to be employed in order to transmit the low-energy photons. Normally, a few mg cm^{-2} of beryllium or aluminium are used and the devices are standard detectors, in the form of proportional counters, for X-rays. For gamma rays, very high pressures, of the order of 20 atm, are used and thick entry windows can be tolerated. The detection efficiency of such counters is a few per cent, half of which is due to photoelectric conversion in the chamber walls and electrodes. These are operated as current ion chambers, in high-activity areas such as reactors, or as GM counters for low count rates. Typical X-ray and gamma counters are illustrated in *Figure 7.18*. The latter is an ion chamber constructed with cylindrical geometry in which the outer electrode is the anode, to prevent avalanche formation.

For *thermal neutrons*, current ion chambers lined with ^{10}B or ^{235}U are used for reactor flux monitoring. These can cope with high count rates and are less damaged by radiation than some other detectors. For small neutron fluxes, proportional counters filled with boron trifluoride (BF_3) are standard. They too have metal walls to contain the extremely toxic gas filling. ^{3}He fillings are also used and spherical casings may be employed to measure fluxes without imposing any directional preference. Another standard device is the gamma-compensated thermal-neutron detector. This is a cylindrical anode, current ion chamber containing two sensitive volumes — one boron-loaded to be neutron-sensitive as well as gamma-sensitive, and the other sensitive only to photons. Subtraction of one output from the other gives the neutron flux.

For *fast neutrons*, gas counters filled with H_2, CH_4, or other hydrogenous gases, can be used, but the most common approach is to locate a thermal-neutron detector behind a moderator which thermalises the fast neutrons. In reactors, graphite 'thermal columns' are used. In dosimetry, polyethylene moderators are used.

Some typical neutron detectors are illustrated in *Figure 7.19*.

For counting *liquid samples*, a number of gas counters are available. One of these, known as the *Veall counter*, is illustrated in *Figure 7.20*. Essentially, this

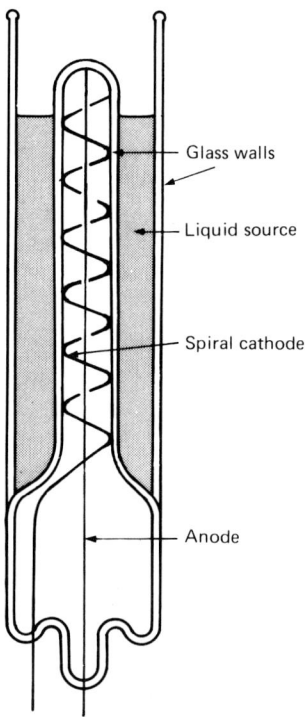

Figure 7.20 Veall counter

comprises a GM counter built into a test tube so that the radioactive liquid in the test tube surrounds the GM counter. This provides almost 2π geometry and relatively good detection efficiency. The device can be used as a test tube, to perform any desired chemical reaction, then plugged into a suitable counting circuit.

Pulse height spectrometry

Ionisation chambers are used for alpha and beta ray spectrometry. They have the advantage over some other systems that the source can be located inside the chamber, eliminating any energy loss which would derive from an entry window. Throughflow gas systems are required and pulse ion chambers, with low-noise preamplifiers, are used. Throughflow gas systems may present some problems and require some special expertise, however, so such detectors are mainly used in research work.

Proportional counters are widely used for X-ray spectrometry and are much less expensive than the alternative lithium-drifted silicon detectors. Although they have poorer energy resolution and rather smaller detection efficiencies, they are good enough for many routine purposes. X-ray pulse height spectra for ^{55}Fe 5.9-keV photons are illustrated in *Figure 7.21* to show the energy resolutions

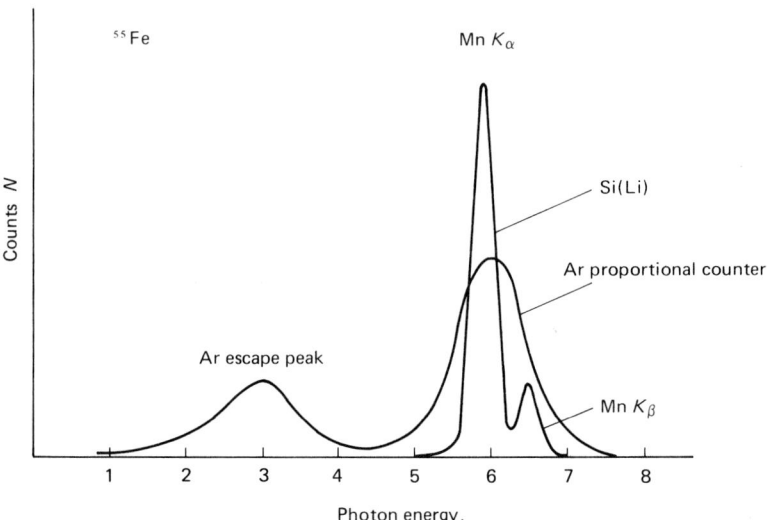

Figure 7.21 X-ray pulse height spectra for ^{55}Fe

obtainable with gas proportional counters and lithium-drifted silicon detectors. The latter can resolve K_α and K_β peaks, the gas counter cannot. The 3-keV peak in the gas counter spectrum is formed by the escape of argon K X-rays.

Hydrogen-filled proportional counters are used for fast-neutron spectrometry and ^{3}He-filled chambers are used for low-energy neutrons.

One of the problems of pulse height spectrometry, with both photons and neutrons, is conversion in the walls, electrodes and windows of the chamber.

Photoelectrons and neutron reaction products originating in these materials lose some energy before reaching the sensitive gas volume so pulse heights are randomly reduced and energy resolution suffers. For this reason, sensitive gases rather than liners are used in neutron spectrometers, and high-Z, high-pressure gases together with thin entry and exit windows are used in X-ray spectrometers.

Dosimetry

Radiation dosimetry is probably the major application of traditional gas counters. Certainly they are used for this purpose more than any other detectors except in personal dosimetry, where other systems are available.

The air wall ionisation chamber provides a direct measurement of radiation exposure in air and is the most accurate means of doing so for low photon energies. At higher energies, the low stopping power of air is insufficient to scatter as many photoelectrons into the sensitive volume as are scattered out of it, so a tissue-equivalent material is used to maintain electron equilibrium in the cavity chamber. Both systems are used for accurate dose and exposure measurements, when integrated values are required.

For area monitoring, dose rate measurements are provided by ion chambers or GM counters operating in the current mode. These have to be calibrated, usually in mR h^{-1}, since it is exposure which they actually measure. The ion chamber is capable of measuring energy deposited and can be calibrated in mrad h^{-1}, but neither can be constructed of tissue-equivalent material to measure biological dose in mrem h^{-1}. This is a general problem in radiation dosimetry. All that can be done with these gas counters is to use variable window thicknesses, that is, removable caps, to distinguish between beta and gamma radiation. X-ray dosimetry is always difficult since most windows are fairly opaque, and most GM counters are insensitive to photons with less than about 10 keV of energy.

Neutron dosimetry is effectively carried out with boron-filled chambers, with or without moderators. In either case, the energy and the RBE (see p. 151) of the radiations are known so the instrument can be calibrated in millirems per hour.

Integrating chambers such as the quartz fibre electrometer are widely used as personal monitors. They are small, robust, inexpensive, reusable and provide a continuous, integrated reading so it is immediately clear when the user has reached the limit of exposure. They are, however, less accurate than film badges and less able to distinguish between different types of radiation, so they measure exposure rather than biological dose. They are also quite insensitive to low-energy photons and to neutrons and should not be used when such radiations are the major hazard.

Imaging

Spark chambers and MWPC are extensively used in high-energy and cosmic-ray physics applications as tracking systems. Normally the objective is to identify particles that are the products of nuclear reactions, which may occur inside or outside the detector. In the former case, the target nuclei must be contained in the detector — for example, in the electrode material. Particle identification is carried out by measuring track parameters such as range, ion density, delta ray density, multiple scattering and curvature in a magnetic field. To obtain this information, the tracks must be recorded in three dimensions, then measured and analysed. A typical arrangement consists of a series of stacked chambers, through which the ionising radiations pass in turn, or, in the case of the spark chamber, a multielectrode chamber.

An example of the latter is shown in *Figure 7.22*. It has a large number of thick lead electrodes, alternately charged positively and negatively by the high-voltage pulse, so that each interelectrode space is effectively a single chamber.

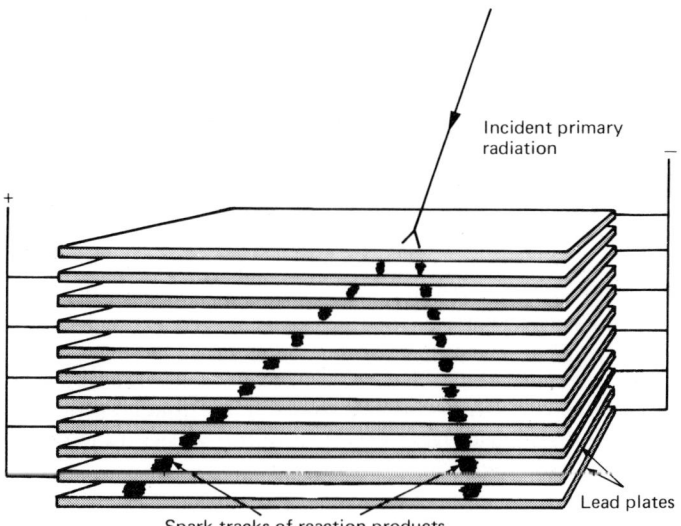

Figure 7.22 Multiplate visual spark chamber

Tracks are recorded by stereophotography. The advantage of this type of chamber is the extremely high stopping power of the lead electrodes for penetrating radiations such as neutrinos and high-energy photons, so it is popular in gamma ray astronomy as well as high-energy physics. The disadvantage is its visual output signal, which is not immediately computer-compatible, so data reduction can be a tedious process, as with the bubble chamber and the nuclear emulsion discussed in Chapter 11.

A similar geometry is provided by a stacked array of MWPC. In this case dense target materials cannot be incorporated in the chamber electrodes but they can be inserted between chambers to produce the same effect. This type of assembly is more difficult to construct but it offers energy-measuring capability and electrical output signals which are computer-compatible. On-line computer data reduction and analysis represent a great advantage over other tracking devices.

Besides spark chambers and MWPC, *streamer chambers* and *drift chambers* are used in tracking systems. The streamer chamber is similar to the spark chamber except that its high-voltage pulse is of much shorter duration and spark discharges are not allowed to grow and link the electrodes. Ionising particles leave tracks of small, luminous sparks. Drift chambers may be of the type described above or they may be arranged to measure the time taken for ions or electrons to drift to the collecting electrode. Measurement of this time provides information regarding the distance of the initiating event from the collector and determines position in that direction.

Imaging gas counters have also been used for X-ray and neutron *radiography* and as gamma and positron *cameras*. At present, they do not compete very successfully against established systems and most of the work being carried out is of an investigative nature. It is unlikely that they will challenge films and fluorescent screens as X-radiographic detectors because of their relatively poor spatial resolution and count rate capability, but they show some promise for neutron radiography, being able to incorporate expensive target materials such as gadolinium. For radionuclide imaging, the MWPC, at least, shows more promise. It compares favourably with a scintillation camera for X-ray imaging, for example, of ^{125}I thyroid concentrations, and with high-density photon converters may be suitable for gamma and positron cameras. It has a poorer detection efficiency than a scintillator but may have better spatial resolution and a larger interactive area. It should also be considerably less expensive.

One area in which spark chambers have become established is *radiochromatography*, for which commercial systems are available. They are particularly sensitive to low-energy beta radiations, from ^{3}H and ^{14}C, for example, and offer spatial resolutions that are quite adequate for thin film and paper chromatography. Spark distributions are usually recorded on Polaroid film.

Computed tomography is another application to which gas counters are well suited, and commercial systems, based on arrays of xenon-filled counters, are available. At photon energies greater than 100 keV they are not quite as efficient as scintillators but they can be used in large arrays to reduce scanning time. A multiwire chamber should be even better in this respect.

Timing

The time resolution of a gas counter is not the best available but it is good enough for many applications described in Chapter 6. Proportional counters and GM counters can measure time to within a microsecond or so and are used whenever their other properties meet the experimental specifications. They are usually excluded from photon coincidence counting by their low detection efficiencies but they are adequate for counting charged particles. Their advantage is that they are relatively inexpensive and simple to operate. Typical applications include cosmic ray telescopes and anticoincidence shields for low-level counting. They are also suitable for thermal-neutron timing.

BIBLIOGRAPHY

ALLKOFER, O. C., *Spark Chambers*, Verlag Karl Thiemig, Munich (1969)
RAETHER, H., *Electron Avalanches and Breakdown in Gases*, Butterworths, London (1964)
RICE-EVANS, P., *Spark, Streamer, Proportional and Drift Chambers*, Richelieu Press, London (1974)
SNELL, A. H. (Ed.), *Nuclear Instruments and their Uses*, Vol. 1, John Wiley, New York (1962)

Chapter 8
Solid-state detectors

8.1 INTRODUCTION

The original idea behind the solid-state detector was, simply, to replace the gas of a gas counter with a solid. Two major advantages can be expected of such a substitution. First of all, the stopping power of a solid is much greater than that of a gas, so the *detection efficiency* should be improved. Secondly, the energy required to form a single ion pair in a solid is equal to the forbidden energy gap between the valence and conduction energy bands, and usually much smaller than the ionisation potential of an isolated gas atom. In both cases, the mean energy, ϵ, deposited per ion pair formed is about twice that required to ionise a single atom so, in general, ϵ is smaller in a solid, more ion pairs are formed and *energy resolution* should be improved. Typically, the energy gap in a solid insulator is of the order of 7–10 eV so that the mean energy required to form an ion pair is about 15 eV, roughly half that required for a gas.

Since a potential difference must be applied across the chamber, without generating a current, the solid has to be a good insulator. Furthermore, for any particular applied potential, the electric field generated in a solid is much weaker than would be obtained in a gas. The field is reduced by a factor K, the *dielectric constant*, which, for example, is in the region of 5 to 10 for glass and may be much larger in other insulators. As a result, extremely large potential differences would have to be applied across a solid chamber to maintain field strengths similar to those used in gas counters. Proportional and Geiger–Müller conditions are difficult to maintain and the only practical proposition is an ionisation chamber.

Solid-state, parallel-plate ion chambers have been constructed. They are usually referred to as *conduction counters* but, since crystal insulators have been preferred to amorphous materials, they are also known as *crystal detectors*. In practice, however, they have not been very successful and are not widely used at present. The main problem is the *trapping* of signal pulse charges by impurity ions fixed in the solid matrix. Since they are different from the atoms of the host material, impurity atoms do not have the same energy level structure and their allowed energy states tend to fall somewhere in the forbidden energy gap between the valence electron band and the conduction band of the host. Being relatively isolated they also form discrete levels rather than a band structure, as illustrated in *Figure 8.1*. If these intermediate, impurity levels are occupied by valence electrons, they are easily ionised, since only a small thermal or electrical energy is required to raise their electrons to the conduction band of adjacent host atoms. Consequently, they tend to form positive ions, at fixed positions,

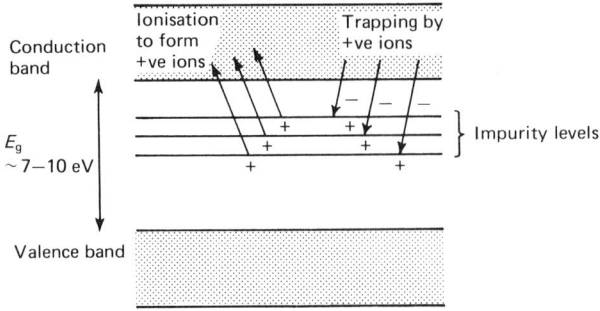

Figure 8.1 Energy level structure of an extrinsic insulator

which easily capture or trap electrons from a signal pulse. The effect is similar to that of electronegative components in a gas mixture, except that a neutral atom is formed, permanently or temporarily, and the charge is completely lost from the signal. Other types of impurity can trap positive charge carriers and crystal defects can have the same effects.

If an attempt is made to improve charge collection efficiency by increasing the applied potential, the normal result is that electrons are excited to the conduction band from the valence band of the host atoms and a leakage current flows across the chamber. The signals cannot be observed against this background. In effect, it is not possible to maintain true ion chamber conditions in a solid dielectric unless the material is absolutely pure and perfect. Such *intrinsic* materials do not exist. In practice, all have some impurities as well as crystal defects and are *extrinsic*, but some materials can be prepared to very high degrees of purity. Of these, diamond is probably the best available, at present, but high-purity diamond is a very expensive gemstone and not readily obtainable in large sizes. Silicon and germanium can also be produced to high specifications but these are semiconductors, with small energy gaps of the order of 1 eV, and prone to current leakage at very low applied potentials.

There are, however, two less demanding alternatives to the conduction counter and both of them have led to useful detector systems. The first of these is the *semiconductor counter*, in which a semiconducting material is constructed in the form of a diode rectifier so that a reasonably large potential can be applied, in one direction, without discharging current. The other approach has led to a variety of devices which have no single name and have to be referred to as *integrating solid-state devices*. In these, no attempt is made to read out individual signal pulses but, instead, the charge carriers are allowed to be trapped at impurity centres. Over a period of time these processes alter the properties of the material to produce observable effects which measure integrated radiation count or dose. These are analogous to the integrating ion chambers and widely used as radiation dosemeters.

8.2 SEMICONDUCTOR COUNTERS

At present, the only semiconductors that can be obtained with sufficient purity and perfection to be used as radiation detectors are silicon and germanium, and this restriction is probably the major disadvantage of the semiconductor counter.

Group	II	III	IV	V	VI
	$_4$Be	$_5$B	$_6$C	$_7$N	$_8$O
	$_{12}$Mg	$_{13}$Al	$_{14}$Si	$_{15}$P	$_{16}$S
	$_{30}$Zn	$_{31}$Ga	$_{32}$Ge	$_{33}$As	$_{34}$Se
	$_{48}$Cd	$_{49}$In	$_{50}$Sn	$_{51}$Sb	$_{52}$Te
	$_{80}$Hg	$_{81}$Tl	$_{82}$Pb	$_{83}$Bi	$_{84}$Po

Figure 8.2 Section of the Periodic Table containing silicon and germanium

Silicon and germanium belong to Group IV of the Periodic Table, along with carbon, tin and lead (*Figure 8.2*). They are tetravalent and form covalently bonded crystals with the same energy band structure and crystal structure as diamond (carbon).

The energy band structure of intrinsic material is illustrated in *Figure 8.3*. Isolated atoms of these Group IV elements have outer-electron configurations in which two electrons occupy the s state and two occupy the p state. Both states are represented by sharply defined energy levels. In carbon, silicon, germanium, tin and lead the energy shells involved are, respectively, L, M, N, O and P, so the states are, respectively, $2s$ and $2p$, $3s$ and $3p$, $4s$ and $4p$, $5s$ and $5p$, and $6s$ and $6p$. In all cases, the s state is full but only two out of six allowed p states are occupied.

As atoms are brought close together to form a solid, a number of changes take place. The four outer electrons form attractive covalent bonds which pull the atoms closer together; that is, electron energy decreases as separation decreases. This effect is balanced by the mutual repulsion of atomic nuclei so that electron energy, shown by the level structure, begins to increase if separation decreases beyond a certain point. This minimum energy point is the natural separation of atoms in solid crystals, at a given temperature. Because of the additional electron shielding, this separation increases from carbon, through silicon, germanium and tin, to lead.

While the energy levels are changing they are also broadening into a number of bands. The s state splits into two bands, each containing one electron state per atom. The p state also divides into two bands, each containing three atomic states. In addition, the lower p band and the upper s band cross over, so that, in a solid, the four lowest states are one of the s and three of the p states. These

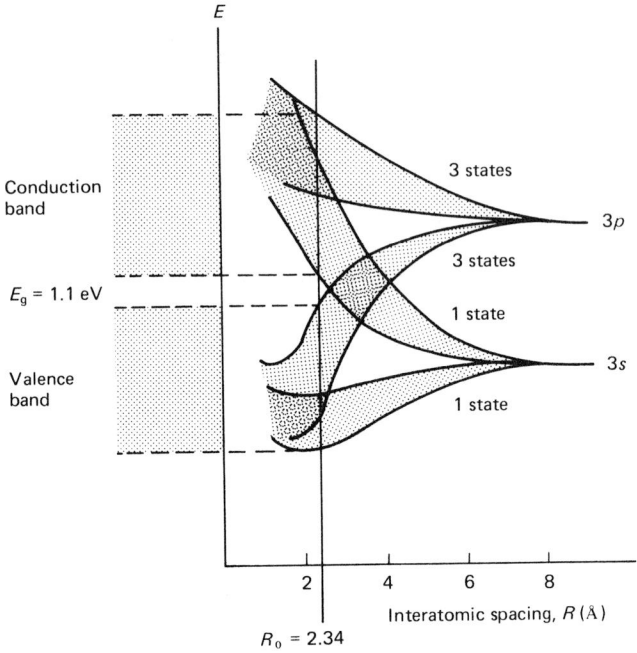

Figure 8.3 Band structure of silicon. For germanium, $E_g = 0.67\,eV$

are occupied by the valence electrons, while the other four, vacant states become the conduction band. The forbidden energy gap, E_g, between these bands decreases as the atomic separation increases. It is about 7 eV for diamond, which is an insulator, 1.12 eV and 0.67 eV for the semiconductors, silicon and germanium, 0.1 eV for grey tin (formed at low temperatures from normal tin), which is virtually a conductor, and zero for lead, which is a conductor.

Besides silicon and germanium, there are several semiconductors which could produce useful radiation detectors. The most promising are the covalently bonded binary compounds of Group III–V and Group II–VI elements. These include gallium arsenide (GaAs) and cadmium telluride (CdTe) which, at present, appear most likely to meet high purity requirements in the near future. Some properties of these and other semiconductors are listed in *Table 8.1*.

Table 8.1 Properties of Si, Ge and other semiconductors

Semiconductor	Atomic number	Energy gap, room temp. (eV)	Electron mobility ($cm^2\,V^{-1}\,s^{-1}$)	Hole mobility ($cm^2\,V^{-1}\,s^{-1}$)
Silicon	14	1.12	1500	500
Germanium	32	0.67	3800	1800
Gallium arsenide	31, 33	1.43	8500	420
Cadmium telluride	48, 34	1.5	~100	~100
Indium antimonide	49, 51	0.17	7800	750
Gallium antimonide	31, 51	0.67	4000	1400

Electrical properties

The intrinsic energy level structure of a semiconductor is shown in *Figure 8.4*. At absolute zero temperature, all the electrons, outside the core shells, are in the valence band, which is full. Since there are no vacant energy states in the valence band, to which other electrons may move, no electron movement can take place

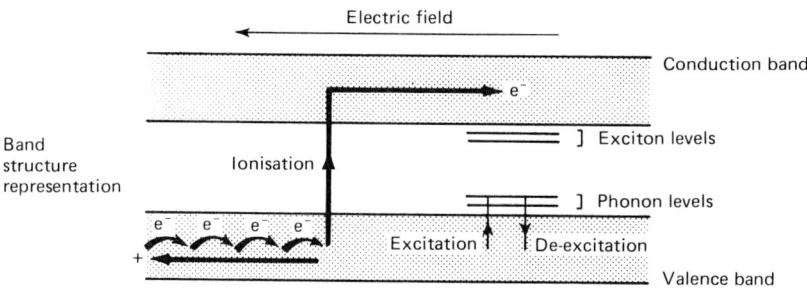

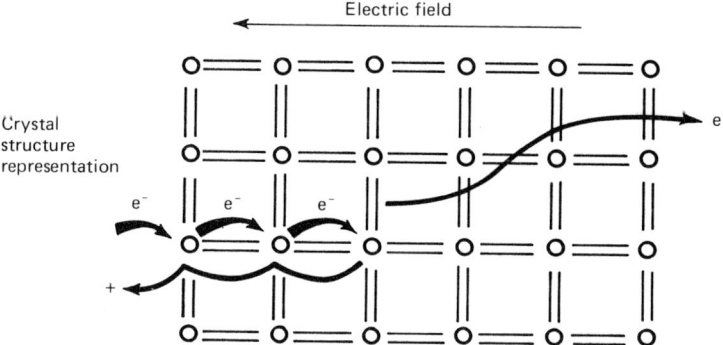

Figure 8.4 Electrical conduction in an intrinsic semiconductor

and no current can flow unless electrons are somehow given sufficient energy to reach the conduction band. This energy can be supplied by ionising radiation or a strong electric field. It can also be supplied as heat, so that at room temperature some electrons are raised to the conduction band, leaving vacancies in the valence band. If a small electric field is applied the electrons easily move through the conduction band, where there are many free states, to form an electric current. As these are removed at one end of the semiconductor they are replaced, at the other end, by other electrons from the external circuit to maintain the net electrical neutrality of the crystal and a continuous current flows. This current increases as temperature increases. It is also greater for small energy gaps than for larger ones.

Raising an electron to the conduction band creates an ion pair and leaves a vacant energy state in the valence band. This is associated with a positive ion which is fixed in the crystal lattice and cannot be collected as an ion current. It does, however, provide a vacant state into which a nearby valence electron may move, driven by an applied electric field. This in turn leaves another vacancy which is filled by another electron, and so on. Vacancies are successively filled by electrons in such a way that they appear to move, like positive charges, in the opposite direction. In effect, these vacancies are *positive charge carriers* and are known as *positive holes*. They move in the field direction with a velocity which, because of the momentary delays before electron jumps, is about three times smaller than that of the *negative charge carriers*; that is, the *electrons* in the conduction band. As for the electrons, a continuous discharge takes place because, at one end of the crystal, the hole is filled by an electron from the cathode and, at the other end, it gives an electron to the anode.

The same process can be illustrated in terms of the physical structure, instead of the energy band structure of the material, as shown in *Figure 8.4*. The crystal structure can be represented by a two-dimensional projection in which each atom is connected to four neighbours by eight lines, denoting the valence electrons. If an electron is raised to the conduction band it ceases to appear as a line and is drawn as a free particle which is swept away by an applied field. The residual positive hole is successively infilled by valence electrons and effectively moves in the opposite direction.

Raising an electron to the conduction band is equivalent to the atomic *ionisation* process and requires a minimum energy equal to the forbidden energy gap, E_g. *Excitation* processes can also take place in which electrons are raised to intermediate levels in the energy gap. In intrinsic crystals, there are two types of intermediate level. Lying above the valence band are the restrahlen, or *phonon*, levels corresponding to the vibrational energy states of crystal ions. Just below the conduction band are the *exciton* levels corresponding to processes in which electrons are given enough energy to move through the crystal but are always associated with positive holes. Both systems are electrically neutral, so no net charge movement takes place. Energy is dissipated by thermal collisions and de-excitation can restore the original configuration. Thus, although the energy required to form one ion pair is E_g, much of the energy deposited in the crystal is 'wasted' on excitation, so the mean energy deposited per ion pair formed, ϵ, is, as for gases, much larger than E_g. In silicon, E_g is about 1.1 eV and ϵ is about 3.6 eV. In germanium, E_g is about 0.7 eV and ϵ is about 2.8 eV.

The electrical resistivity of a semiconductor is determined by the number of charge carriers present. In an intrinsic material the number of electrons, n_e, and the number of positive holes, n_p, must be equal and are usually denoted by n_i. The resistivity, ρ_i, is then given by

$$\rho_i = \frac{1}{n_i e (\mu_p + \mu_n)} \tag{8.1}$$

where e is the electronic charge and μ_p and μ_n are the mobilities, that is, the velocities per unit applied field, of the positive holes and the electrons, respectively.

Resistivity is the resistance, R, of unit length and unit cross-sectional area of a material. It may be expressed in units of ohm centimetres (Ω cm) and is given by

$$\rho = \frac{RA}{l} \tag{8.2}$$

where A is the cross-sectional area and l the length of material.

The calculated values of n_i for silicon and germanium at room temperature are, respectively, 1.5×10^{10} cm^{-3} and 2.4×10^{13} cm^{-3}. These values lead to useful estimates of intrinsic resistivities that can be compared with measured values in order to determine the impurity content of real crystals. For silicon, ρ_i has the value 2.3×10^5 Ω cm and for germanium, 47 Ω cm. In general, the resistivities of conductors are of the order of 10^{-6} Ω cm while those of insulators tend to exceed 10^7 Ω cm.

Extrinsic semiconductors

Typically, the resistivities of the best semiconductors available are an order of magnitude smaller than the values expected for intrinsic crystals. This means that approximately ten times as many charge carriers are provided, by impurities and defects, as are generated intrinsically, and leakage currents are correspondingly larger. There are two types of impurity, namely, pentavalent and trivalent atoms, and two types of crystal defect, namely, interstitial atoms (that is, extra atoms) and lattice vacancies (missing atoms). These produce two types of extrinsic semiconductor.

If *pentavalent* atoms of, for example, phosphorus (P), arsenic (As) or antimony (Sb) (*Figure 8.2*) replace the tetravalent atoms of silicon or germanium, each atom provides one extra electron which does not form a covalent bond. Instead, these electrons occupy discrete energy levels just below the conduction band of the crystal — since the energy level of a Group V element is slightly lower than the corresponding level of a Group IV element. A similar effect is provided by *interstitial atoms*, whose *p*-shell electrons tend to be surplus and occupy levels closer to the conduction band than to the valence band. In both cases, these electrons are easily thermally excited to the conduction band, even at low temperatures, to generate a number n_e of electrons, or *negative charge carriers*, that are free to move through the crystal, and an equal number of residual positive ions that are fixed in the lattice.

If *trivalent* atoms, such as boron (B), aluminium (Al), gallium (Ga) or indium (In) (*Figure 8.2*), replace the tetravalent host atoms, or if *lattice vacancies* are present, then a number n_p of covalent bonds are missing. These vacancies easily attract valence electrons so that the impurities or vacancies form negative ions with discrete energy levels just above the valence band — since the energy levels of the extra electrons are slightly higher than those of the silicon or germanium atoms. The net result is the formation of n_p negative ions, fixed in the lattice, and an equal number of positive holes, or *positive charge carriers*, which are free to move through the crystal by means of the successive infilling process described above.

Natural materials tend to have equal numbers of each type of impurity and defect, but near-intrinsic crystals can be doped with one particular type of impurity to create a large majority of that type. Pentavalent impurities, and crystals doped with them, are known as *n-type*. They have a surplus of negative charge carriers and positive ions which produce *donor levels* just below the conduction band — so called because they donate electrons to the conduction band. Trivalent impurities, and crystals doped with them, are known as *p-type*. They have a majority of positive charge carriers and fixed negative ions which have accepted electrons from the valence band and are known as *acceptor* levels.

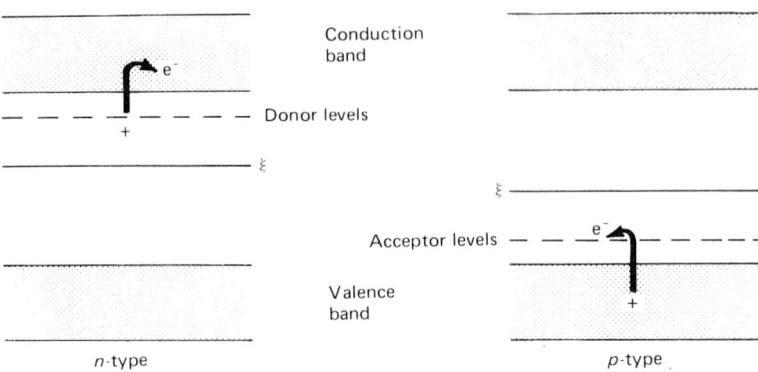

Figure 8.5 Energy band structure of extrinsic semiconductors

These energy level structures are illustrated in *Figure 8.5*. Also shown is the Fermi energy, ξ. This is the mean electrochemical energy of the crystal. In an intrinsic material it is in the middle of the forbidden energy gap, because there are as many charge carriers above it as below it. In *n*-type material it is displaced towards the conduction band and in *p*-type material it is closer to the valence band.

Junction diode detectors

A junction diode is made by doping one end of a single high-purity crystal with *n*-type, and the other end with *p*-type impurities. The resulting energy band structure and charge distributions are shown in *Figure 8.6*. At the junction between the two regions, the mobile charge carriers are attracted towards each other. Electrons cancel positive holes, in effect by falling into them, to create a narrow region in which there are no charge carriers. This region is known as the *depletion layer* or the *intrinsic (i) layer* and is the sensitive volume of a junction diode detector.

Since the fixed impurity ions do not move, this process leaves a surplus of positive charge on the *n*-type side of the depletion layer and a surplus of negative charge on the other side. Although the crystal as a whole remains electrically neutral there is a net charge redistribution which creates a *built-in potential*. The polarity of this potential difference is such as to drive positive charge carriers back into the *p*-type region and negative charge carriers back into the *n*-type region, so it restricts the thickness of the depletion layer, usually to a few tens of micrometres, and stabilises the diode structure.

The band structure diagram, which is, after all, a graph of electron energies, shows the same effect. In order to move from *n*-type to *p*-type material, electrons have to be given additional energy equal to the built-in potential. This is true of conduction electrons as well as valence electrons, so positive holes moving through the valence band encounter the same barrier. The Fermi energy, being the mean of the whole crystal, remains constant.

If an external potential difference is applied across the crystal in opposition to the built-in potential, it can give electrons sufficient energy to overcome the barrier and move into the *p*-type region and, then, to the external circuit. Positive holes move in the opposite direction. The arrangement is known as *forward bias* and it generates current across the diode. If, on the other hand, the external, bias potential is in the same direction as the built-in potential it reinforces the latter, driving negative and positive charge carriers further apart and widening the depletion layer, so that no current flows. This is a *reverse bias* and it creates the conditions necessary for the semiconductor to act as a radiation detector by

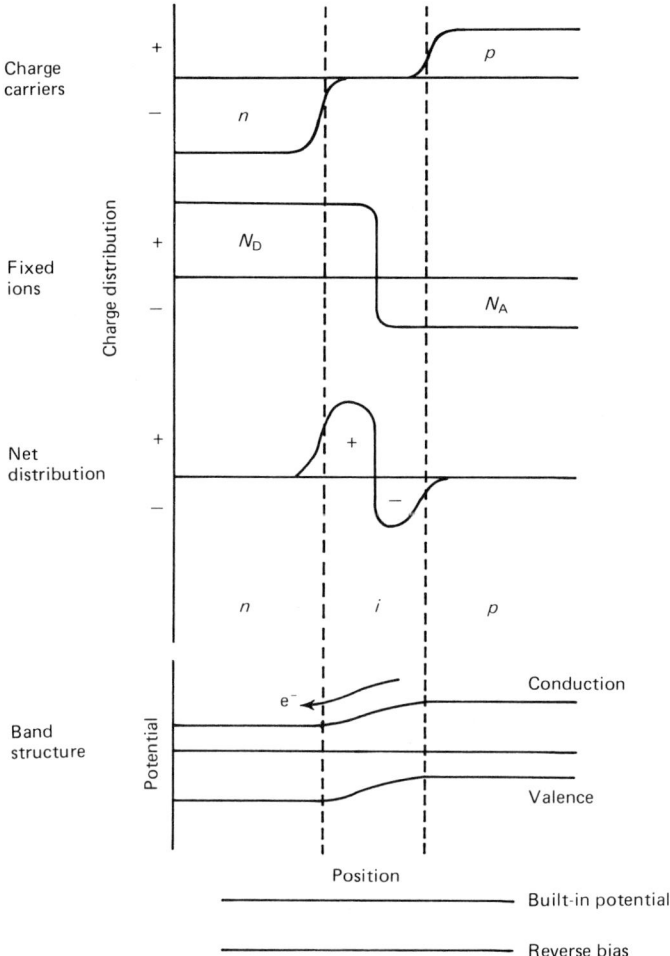

Figure 8.6 *Charge and potential energy distribution in a junction diode*

providing a state in which a potential difference can be applied without discharging current. The device is a diode *rectifier*, since it passes current in one direction only, and is usually described as a $p-n$ or $n-p$ junction.

Under reverse bias, the depletion layer behaves like an insulator but the n- and p-type regions, heavily doped with charge carriers, are more like conductors, and act like the electrodes of a gas counter. Almost all of the applied potential extends across the depletion layer, which acts as the sensitive volume. The

218 Solid-state detectors

thickness of this region increases with the purity of the material and, therefore, with its resistivity. It also increases with applied potential. Most diodes have depletion layers of a few hundred micrometres and bias voltages of less than 100 V but they can be obtained with thicknesses of up to 5 mm to be operated at over 1000 V.

A planar diode detector is illustrated in *Figure 8.7*. When ionising radiations interact with the depletion layer they create ion pairs, in the form of electron–positive hole pairs, by raising valence electrons to the conduction band. These

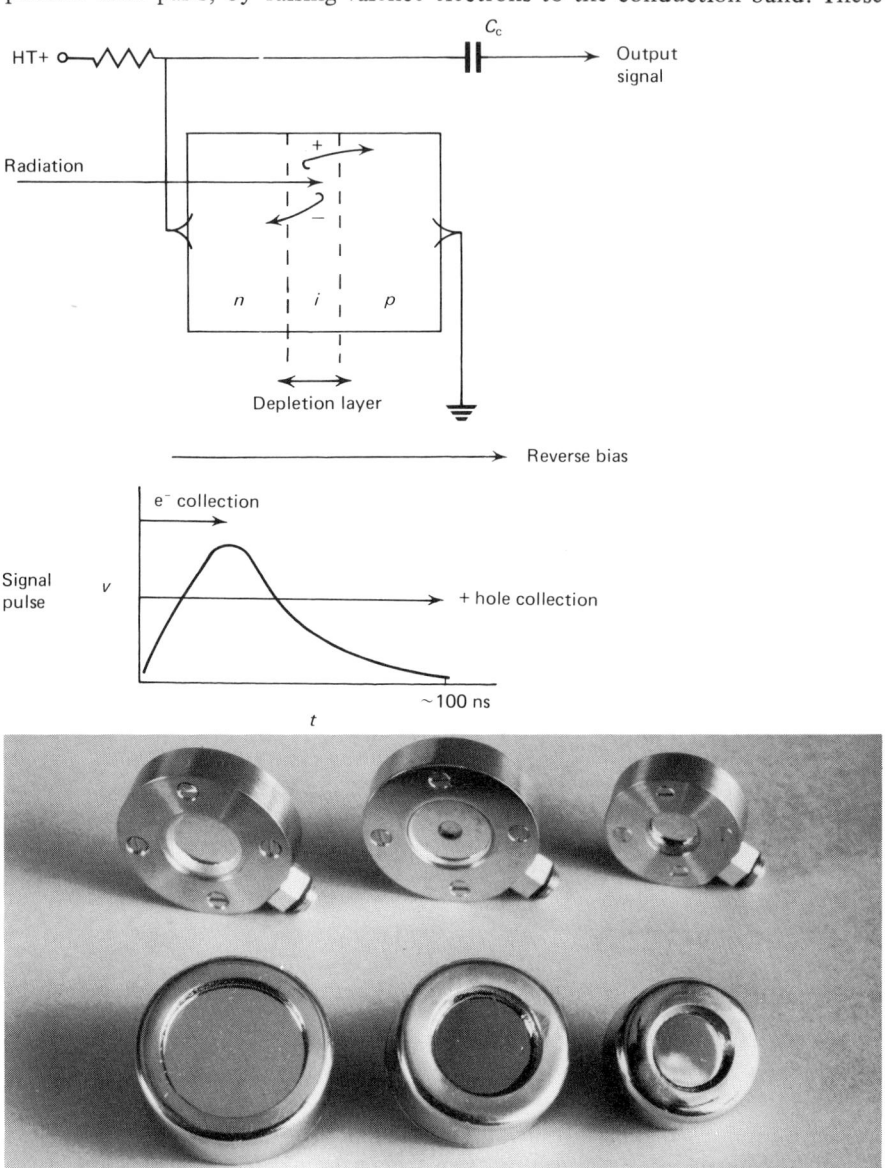

Figure 8.7 Surface barrier semiconductor detectors. (Courtesy of PGI International)

charge carriers are collected at their respective electrodes as a charge signal pulse which is integrated, as for an ion chamber, at the detector, and load capacitance forming a negative voltage pulse.

The mobility of electrons in silicon is about 1350 cm s^{-1} per unit field. For an applied potential of 1000 V over a depletion layer of 0.5 cm (that is, a field strength of 2000 V cm^{-1}), the electron drift velocity is about 2.7×10^6 cm s^{-1}, so the pulse rise time is less than 0.2 μs, even for this exceptionally thick detector. The mobility of positive holes is about one-third that of the electrons, so they are collected in about 0.6 μs in a thick detector. As a result the pulse from a semiconductor tends to be much faster, and to have a much shorter tail than that of a gas counter, which can extend to 100 μs. For a 100-μm depletion layer, a typical pulse width is 100 ns.

It is possible to collect almost 100% of the charge carriers, so the charge pulse, q, is given, to a good approximation, by

$$q = ne = \frac{Ee}{\epsilon} \tag{8.3}$$

where n is the number of primary ion pairs formed, E is the energy deposited by radiation, ϵ is the mean energy required to form an ion pair and q refers, of course, to charge of one sign only.

If this charge is integrated, a voltage pulse is obtained whose height, v, is given by

$$v = \frac{q}{C} = \frac{Ee}{\epsilon C} \tag{8.4}$$

where C is the total integrating capacitance of the detector and any external capacitance.

These formulae are identical to those obtained for an ion chamber (equations 7.3 and 7.4). They show that the charge produced, q, and the signal pulse height, v, are proportional to the energy deposited, E, so the semiconductor is an *energy spectrometer*. Values of ϵ for silicon and germanium are 3.6 eV and 2.8 eV, respectively, which is about one-tenth of the energy required to form an ion pair in a gas. Consequently, ten times as many ion pairs are formed and the FWHM energy resolution, given by equation 5.18, is three times smaller, usually of the order of 1% at reasonably high energies. The energy resolution of a semiconductor is generally better than that of any other counter. This is its major advantage — probably its only advantage — over other types of detector. The fact that it is very widely used is evidence of the importance of this parameter.

The capacitance of a plane electrode system containing a medium of dielectric constant K is given by

$$C = \frac{8.85 \times 10^{-12} AK}{d} \quad \text{farads} \tag{8.5}$$

where A is the electrode area in square metres, d is the separation (that is, the depletion depth) in metres, and K is equal to 12 for silicon and 16 for germanium.

For example, a silicon detector of area 5 cm^2 and depletion depth 1 mm has a capacitance of about 50 pF. An energy of 1 MeV absorbed in the depletion layer should produce a signal height of about 2 mV, which is similar to that expected of an ion chamber.

Although it has excellent energy resolution, the junction diode has a number of disadvantages and problems compared with other types of detector. These relate to leakage currents, trapping, detection efficiency and signal formation.

Leakage current is produced by minority charge carriers, that is, by p-type impurities in the n-type region and n-type impurities in the p-type region. These set up a background diode structure for which the applied voltage is a forward bias, so that a small leakage current is generated. Variations in this current are integrated as small voltage pulses which add randomly to the signal pulses, increasing their FWHM pulse height spread and increasing the energy resolution. To minimise this effect, very pure crystals are required and it is for this reason that only silicon and germanium can be used, at present. Impurity leakage can be very bad at the edges of the diode if they are exposed to the atmosphere. Oxidation and other processes increase the impurity level to such an extent that the junction diode is effectively short-circuited.

A second source of leakage current is the thermal excitation of electrons from the valence to the conduction band, to produce intrinsic charge carriers which counteract the reverse bias. At room temperature, this is tolerable in silicon, for which E_g is 1.1 eV, but for the best possible energy resolution the detector should be cooled. For germanium, E_g is about 0.67 eV and thermal ionisation is quite considerable at room temperature. Germanium semiconductors have to be cooled to about 90 K to be useful although, in practice, the convenient 77 K of liquid nitrogen is used. The detector has to be mounted on the end of a conducting rod, the other end of which is immersed in liquid nitrogen in a cryostat (*Figure 8.8*). To minimise thermal absorption, the diode is encapsulated with insulation, which adds considerably to the entry window for the radiation, so that only penetrating radiations can be detected. In addition, the liquid nitrogen must be replenished regularly and the cryostat is bulky and inconvenient. For these reasons, germanium is not widely used for simple junction diode counters. They are usually made of silicon.

Trapping is also caused by impurity atoms, but impurities whose electron energy states lie closer to the middle of the energy gap than the donor and acceptor levels. Like the latter, they form ions by losing electrons to the conduction band or gaining them from the valence band and they also capture signal charge carriers from these bands. Unlike the donor and acceptor ions, they tend to hold on to these charge carriers, either temporarily, in the *trapping* process, or more permanently, in the *recombination* process. In either case, charge carriers are removed from the signal pulse which is, therefore, severely attenuated. Again, the only solution to this problem is to use very high-purity materials, limiting the range of choice to silicon and germanium.

Detection efficiency, or, more accurately, *interaction efficiency*, is close to 100% for charged particles, but so is that of the gas counters. For gamma rays, efficiency is very poor because of the small sensitive volumes and the low Z number (14) of silicon. This is unfortunate in view of the fact that one of the major reasons for using a solid-state material is to improve detection efficiency. Junction diodes can be used for alpha, beta and X-ray detection but for gamma rays, higher Z numbers are required and thicker depletion layers must be achieved. This means that germanium ($Z = 32$) must be used, with all its disadvantages, and intrinsic material, or its equivalent, must be produced. The latter requirement has led to the development of compensated diodes, as described later.

Signal formation is complicated by the fact that the detector capacitance

Figure 8.8 Ge(Li) detectors with cryostats and preamplifiers. The detectors are mounted in various positions. (Courtesy of Canberra Instruments Ltd)

varies with depletion layer thickness and therefore with the bias voltage, so that the voltage pulse height, v, also depends on the bias voltage. The signal pulse itself reduces the bias voltage across the diode and the net result is that, if the signal is integrated at the detector, there is some instability in the operating conditions and the pulse height, adding to the energy resolution. This problem is overcome by removing the signal from the diode as a charge pulse, so that it does not affect the bias voltage, and integrating it in a *charge-sensitive preamplifier*. This has a large capacitative feedback loop which integrates the charge without changing in value and presents a large, stable input capacitance to the charge pulse. In addition, of course, it must have a very low noise level in order to realise the energy-resolving capability of the semiconductor detector. Normally, preamplifiers based on field effect transistors (FET) generate the smallest noise pulses.

Construction

Basically, there are three ways of making a semiconductor junction diode. The products are known as *surface barrier, diffused-junction* and *ion-implanted* diodes. All of them have planar geometry, with narrow depletion layers, effectively forming thin slabs. The most effective orientation of such a geometry, in terms of detection efficiency, is with the flat surface normal to the direction of incident radiations — this subtends the largest solid angle at the source. Consequently, the entry windows of these diodes have to be through the n- or the p-type layer and one of the major objectives is to make this layer as thin as possible.

Surface barrier detectors are made by oxidising one surface of a cleaned, etched slice of n-type silicon. The oxidised layer forms a very thin p-type layer which is the entry window. This is then protected by a thin metal deposit, usually gold. Metallic electrical contacts are diffused into each side and the edge surfaces are encapsulated to prevent deterioration and leakage currents. The gold deposit also prevents the silicon from being exposed to ambient light. Silicon is photosensitive and irradiation by light generates a considerable leakage current by the production of intrinsic charge carriers. Typically, the total entry window thickness is less than the equivalent of 0.5 μm of silicon. A surface barrier detector is shown in *Figure 8.7*.

Diffused-junction diodes are made by diffusing phosphorus into p-type silicon to form the n-type region. There are various methods of carrying out this process but they all tend to produce relatively thick, n-type entry windows, typically equivalent to about 1 μm of silicon. For this reason, diffused junctions are less widely used than surface barrier devices.

Ion implantation is a process in which the silicon surface is bombarded with ions accelerated to energies of the order of hundreds of kilo-electronvolts. The technique requires light ions so, typically, n-type material is bombarded with boron ions. These cancel out the existing n-type impurities and create an excess of p-type ions to a depth which depends on the beam energy. This approach requires an accelerator and produces some radiation damage to the crystal lattice but it appears to result in a p-type region of very uniform and controllable depth and ion density.

Lithium-drifted diodes

The excellent energy resolution of the semiconductor diode detector makes it very attractive for gamma ray spectrometry but the small sensitive volume rules it out for this purpose. Since intrinsic crystals are not readily available and larger bias potentials cannot be used to increase depletion layer thickness without causing electrical breakdown, the only alternative is *compensation*. In this process, impurity ions of the opposite type to those already present are introduced. The charge carriers of each type annihilate and the electric fields of the two ion types cancel out to produce the equivalent of intrinsic material, which has a high resistivity so that large bias voltages can be applied and thick depletion layers produced.

There are at least three ways of making these compensated diodes. Ion implantation has been used but causes radiation damage which is not easily annealed out of thick regions. Irradiation of silicon with thermal neutrons converts silicon

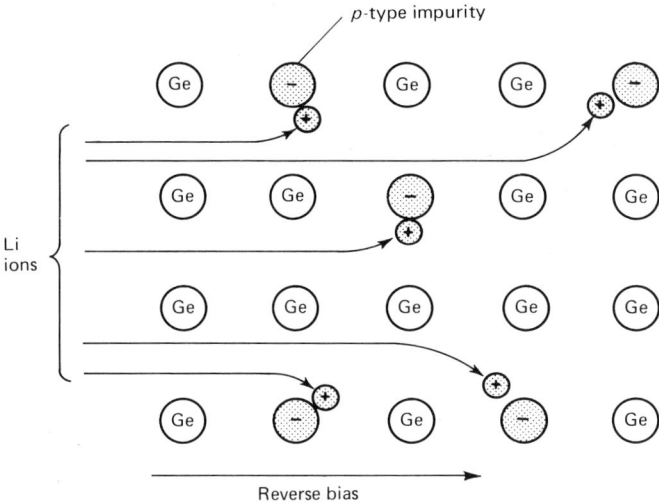

Figure 8.9 Lithium ions drifted into p-type material

nuclei to phosphorus, compensating p-type crystals but, again, the radiation damage is considerable. In general, semiconductors should not be exposed to neutrons, for this reason. The third alternative is lithium drifting.

In this method, lithium ions ($_3^7\text{Li}^+$) are drifted into the lattice of a p-type crystal of silicon or germanium, usually from a gas environment or a surface coating. Being very small ions, they move interstitially, that is, between the lattice ions, behaving like heavy, positive electrons. Effectively, they are n-type *dopants*. Under a reverse bias they are attracted deep into the p-type material and tend to form neutral pairs with the negative acceptor ions (*Figure 8.9*), establishing a very thick and uniformly compensated depletion layer. Such devices are usually referred to as p–i–n (or n–i–p) junction diodes or as Si(Li) and Ge(Li) detectors.

Planar diodes (*Figure 8.10*) can be prepared from thick slices of crystal. The surface through which the lithium ions are diffused is overcompensated to form

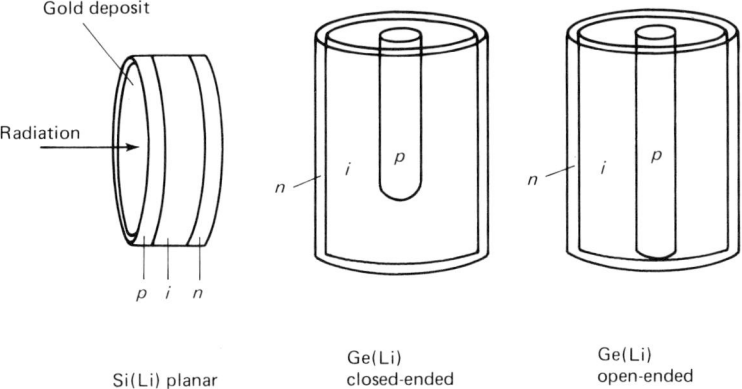

Figure 8.10 Si(Li) and Ge(Li) detector geometries

an *n*-type layer. The other surface is etched down to the intrinsic region, then a *p*-type surface barrier layer is formed to provide a thin entry window. These diodes can have depletion depths of up to 15 mm and sensitive volumes in excess of 15 cm^3. They are used for weakly penetrating radiations and are excellent for alpha, beta and X-ray spectrometry. Since germanium is not required for these applications, they are usually made of silicon.

For gamma ray detection, germanium is used and even larger sensitive volumes are produced. These are obtained by diffusing lithium ions into a *p*-type cylindrical (or trapezoidal) crystal so that the residual *p*-type material lies along the axis and the excess lithium ions form a surface *n*-type region. The result is a cylindrical detector (*Figure 8.10*) with a very large intrinsic volume of up to about 150 cm^3 and a claimed detection efficiency of up to 30% of that of a 3 in × 3 in (762 × 762 mm) sodium iodide scintillation counter.

As mentioned above, germanium detectors have to be kept cold to reduce thermal leakage current. This requirement is even more important for Ge(Li) detectors. If they are allowed to remain at room temperature for more than a few minutes, the lithium ions diffuse through the crystal. A similar catastrophe is produced, of course, if the bias voltage, typically 2 or 3 kV, is applied in the wrong direction, creating forward bias conditions. In both cases, the result is inconvenient and expensive since the detector has to be replaced, or redrifted by the manufacturers.

The polarity of the bias voltage should be permanently fixed and not reversible by the users. The detector is encapsulated and mounted on a conducting 'cold finger' which is cooled by liquid nitrogen in a cryostat (*Figure 8.8*). The liquid nitrogen must be regularly topped up and, needless to say, must be readily available. Thus, a Ge(Li) detector is permanently associated with a large, non-portable cryostat, which is somewhat inconvenient. To reduce noise levels further, the preamplifier is also mounted on the cryostat.

The crystal can be prepared and mounted in its encapsulation in more than one way. As illustrated in *Figure 8.10*, it can be open-ended, with the axial *p*-type region extending to both of the flat surfaces, or it can be closed-ended, with a region of intrinsic, and some *n*-type, material at one end. In the latter case, it can be mounted with the closed end or the open end facing the radiation source and acting as the entry window. The closed end presents a thicker entry window, placing a low energy limit of about 30 keV on the detectable photon energy. The open-ended window is usually transparent down to less than 10 keV, but it has a dead, *p*-type layer in the most sensitive, axial position so detection efficiency is reduced and statistical variations in sensitivity are increased, leading to poorer energy resolution.

8.3 INTEGRATING SOLID-STATE DEVICES

A number of solid-state materials can be used to measure integrated counts. Mostly, they are used to record radiation dose, both for personnel monitoring and in experiments involving specimen irradiation. Radiations induce changes in the physical properties of these materials and these changes are indicative of the accumulated dose. In general, irradiation excites electrons from the valence band to the conduction band and they are subsequently trapped at impurity centres. In this case, the impurities behave like *activators*. They are specifically selected and added to the material in order to produce a particular effect. The effect determines the way in which output signal formation takes place and provides the basis for classifying different types of device. At present, three processes are important, namely, coloration, radiophotoluminescence (RPL) and thermoluminescence.

Coloration

Impurity ions tend to form vacant, sharply defined energy states between the conduction band and the valence band of the host material. In an insulator, they provide the means of absorbing visible-light photons so that an otherwise transparent, colourless material becomes coloured. The visible-light photons are absorbed when they have sufficient energy to excite electrons from the valence band to the impurity levels. For example, the presence of different impurities in crystalline Al_2O_3 converts it from colourless corundum to blue sapphire, green emerald or red ruby.

A number of glasses, plastics and dyes change colour when irradiated. The incident radiations have sufficient energy to excite electrons from the valence band to the conduction band. The electrons are then trapped at the activator centres so that energy levels which were previously able to absorb visible-light photons are now full and unable to do so (*Figure 8.11*). As these levels are progressively filled the colour of the material changes. There are other mechanisms but basically they all involve the redistribution of the electron population of interband energy states leading to changes in light transmission.

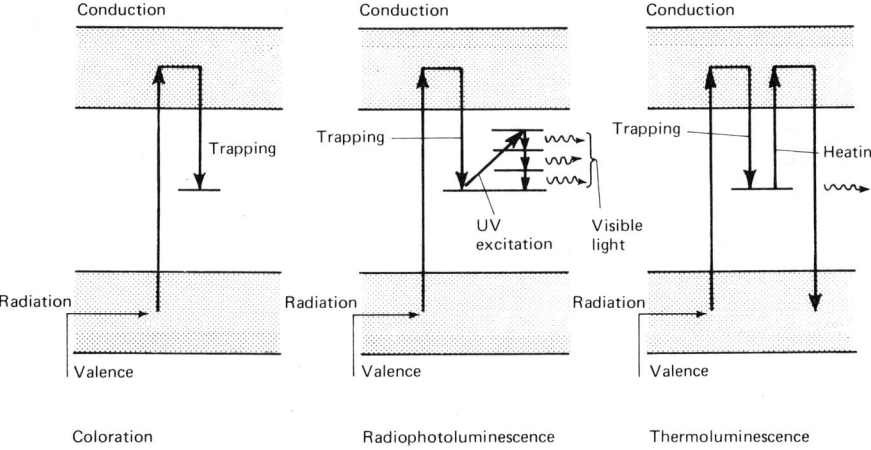

Figure 8.11 Signal formation in solid-state devices

In most materials the colour changes tend to fade as trapped electrons are released by thermal excitation. On the other hand, heating can also be used to erase readings so that the devices can be reused. Glasses, such as silver-activated phosphate glass and cobalt-activated borosilicate glass, have been used. Their main disadvantage is that their range of sensitivity tends to be in the region of 10^3–10^5 rad, which is too high for personnel dosimetry.

Radiophotoluminescence

In radiophotoluminescence (RPL), the trapped electrons are released from the impurity levels by irradiation with ultra-violet light. This raises the electrons to higher energy states, within the energy gap, from which they de-excite in a cascade process, eventually returning to the original trapping level (*Figure 8.11*). The absorbed dose is proportional to the intensity of the light emitted in the luminescence decay process and the reading can be erased by heating.

At present, glasses are most widely used as RPL devices and can be prepared to diverse specifications: for example, they can include high Z number elements for photon detection and lithium or boron for thermal neutrons. Phosphate glasses are preferred, activated with various mixtures of silver, aluminium and lithium. They can register doses of less than 50 mrad and are suitable for personnel monitoring, although glass is less tissue-equivalent than some other materials.

Thermoluminescence

In radiothermoluminescence, to give the process its full name, electrons are again raised to the conduction band by incident radiation, then trapped at activator levels in the forbidden energy gap. They are released by heating the material, and excited up to the conduction band. Enough of them de-excite to the valence band, missing the trapping levels, to generate a measurable light output which is proportional to the absorbed dose. In this case, data readout also erases the system so that it can be used again.

Most solids exhibit some thermoluminescent properties and some of them are suitable for thermoluminescent dosimetry (TLD). These include lithium fluoride (LiF), calcium fluoride (CaF_2), calcium sulphate ($CaSO_4$) and aluminophosphate glass, all activated with manganese. Of these, lithium fluoride is the most popular, because its response is almost independent of photon energy from about 10 keV upwards, and it forms the basis of commercially available TLD systems. It is also sensitive to thermal neutrons, because of its lithium content, and can register doses as small as 20 mrad.

More generally, the advantages of TLD systems are the availability of a wide range of materials and dose ranges, to conform to a variety of environmental and experimental conditions. Thus, they can be selected to perform a highly specific function. Some, such as LiF, are very sensitive to neutrons, while others, such as CaF_2 and aluminophosphate glass, are insensitive to neutrons. Composite devices can be constructed to be selective. They also vary in range of sensitivity. For LiF, the useful exposure range is about 20 mR to 5×10^4 R, while for $CaSO_4$, it is about 50 R to 10^4 R. They are all accurate to within about 3% for high exposure and to within about 15% for low exposures.

8.4 FUNCTIONS

Counting

For *charged particles*, the interaction efficiency, R, of a surface barrier detector is about 100%, provided two conditions are met. The first of these is that the entry window must be thin enough to transmit the particles with sufficient residual energy to register in the detector. The second is that the intrinsic region must be thick enough to absorb as much energy as is required to generate a signal pulse larger than noise pulses. Whether or not these conditions are satisfied depends on the specific energy loss of the radiations and on the stopping power, hence the dimensions, of the detector. This information can be derived from graphs of range against radiation energy, such as are presented in *Figure 8.12*. For a silicon surface barrier detector, the entry window may be equivalent to about 0.5 μm of silicon (0.12 mg cm^{-2}). To take the least penetrating radiation as an example, the range of a 5-MeV alpha particle, in silicon, is about 24 μm (5.5 mg cm^{-2}), so that in general this condition is easily satisfied.

To ensure that the second condition is satisfied, reference must be made to the manufacturer's data, which will quote noise levels for the detector and the preamplifier, usually in energy-equivalent units of kilo-electronvolts. If, for

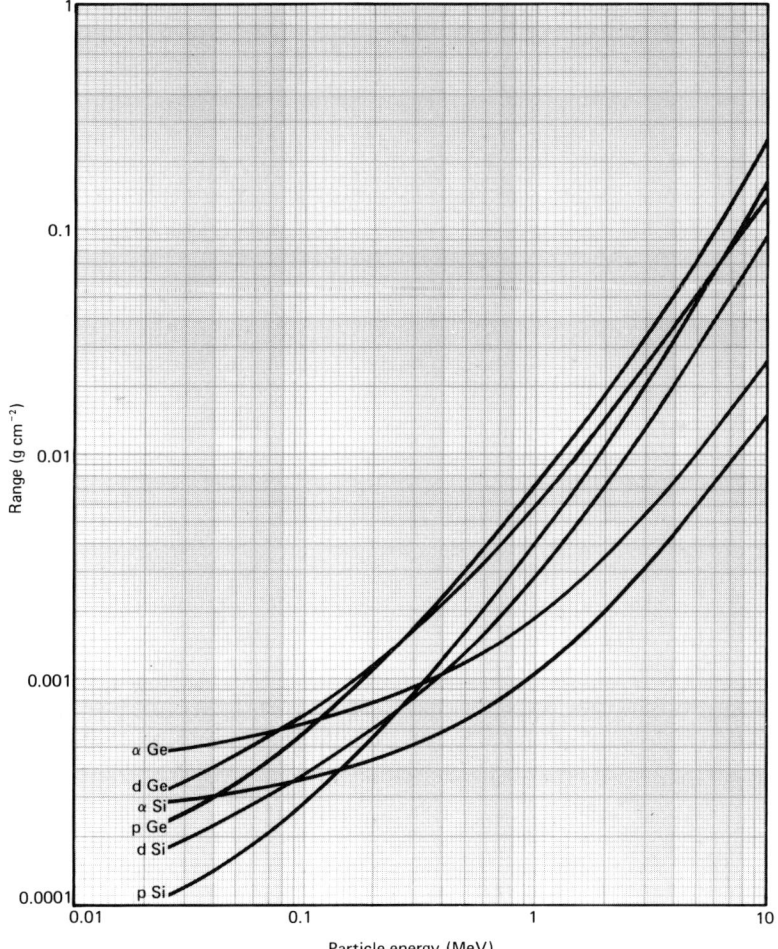

Figure 8.12 Range–energy graphs for charged particles in silicon and germanium. Si: 1 g cm^{-2} = 0.429 cm; Ge: 1 g cm^{-2} = 0.187 cm

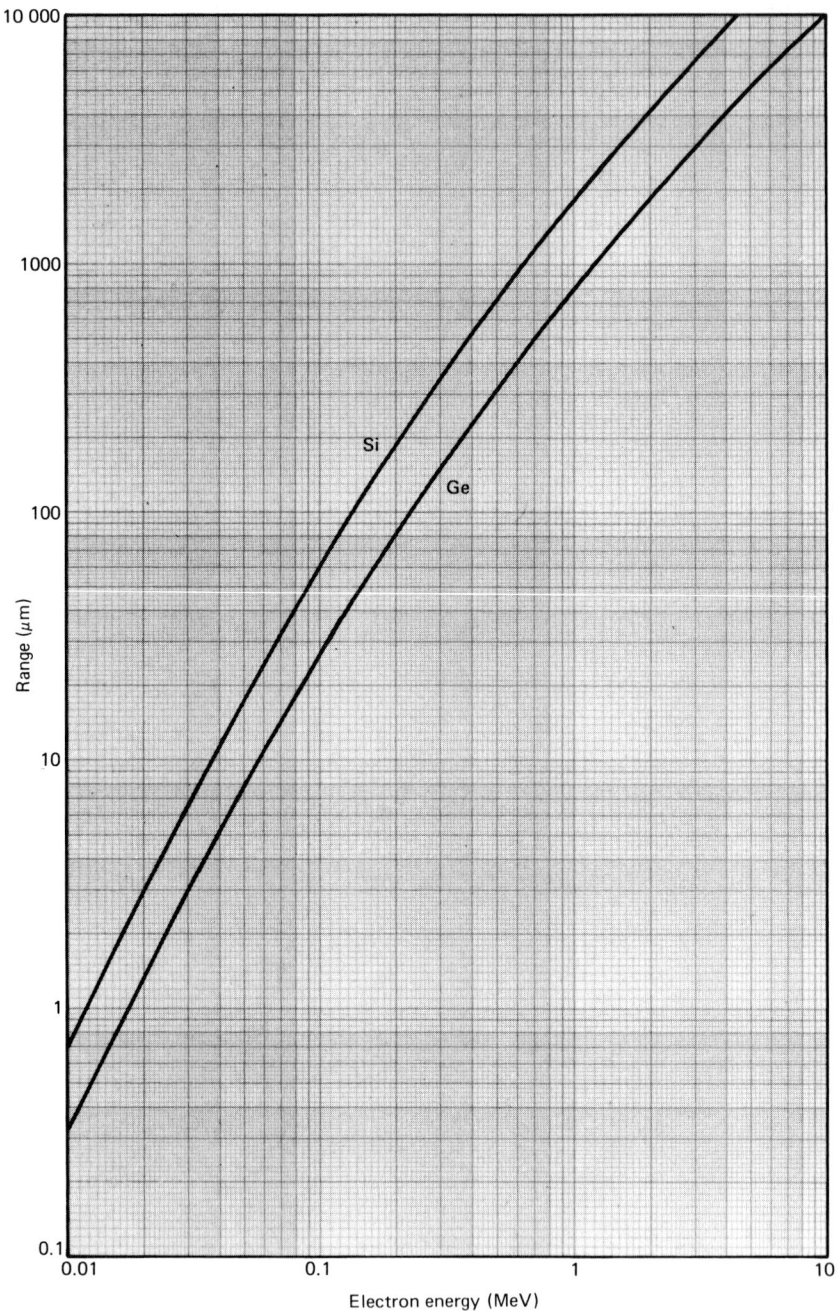

Figure 8.13 Range–energy graphs for electrons in silicon and germanium

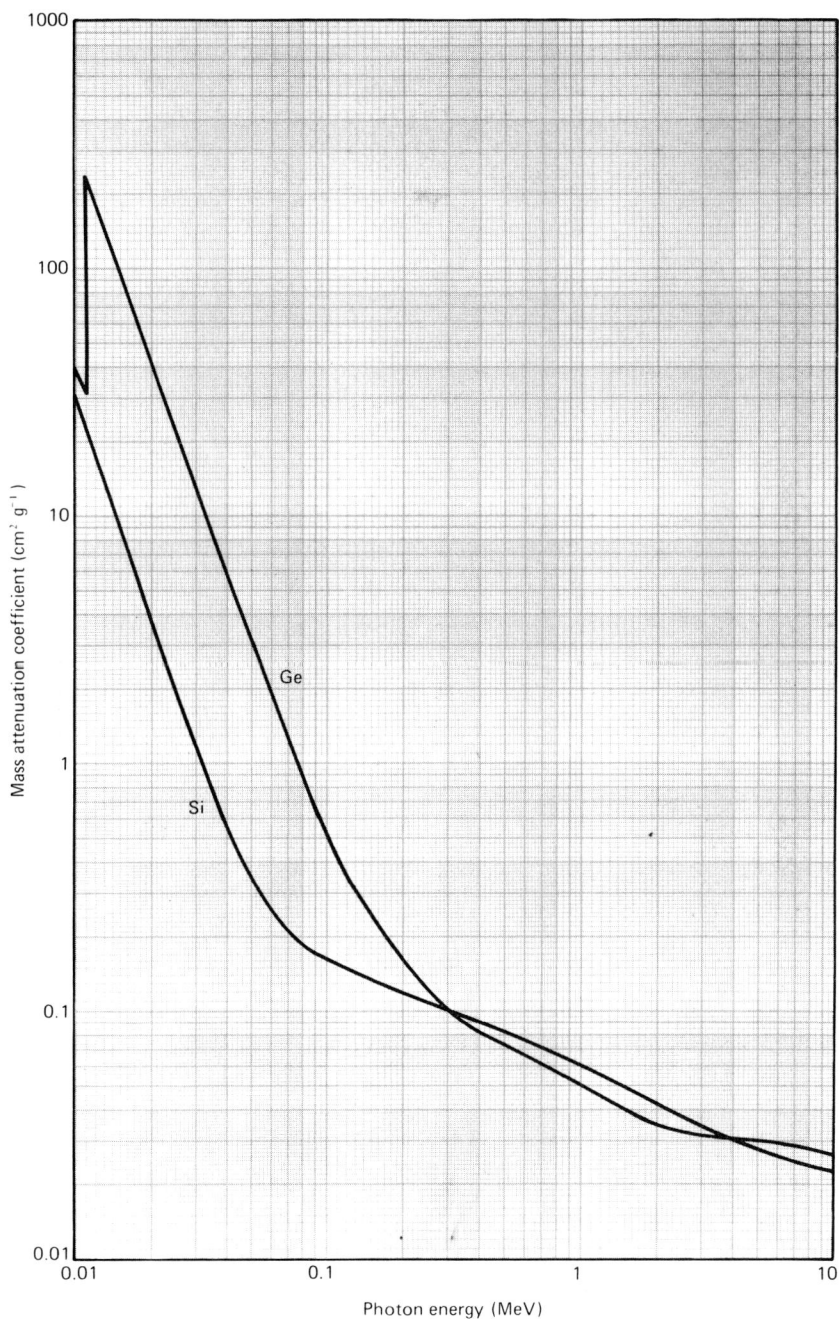

Figure 8.14 Photon attenuation coefficients for silicon and germanium

example, the total is 100 keV, then the depletion layer must be thick enough to absorb more than this amount of energy. From *Figure 8.13* it can be seen that a beta particle entering the depletion layer with 1 MeV (range 420 mg cm^{-2}, or 1.8 mm) will lose 100 keV in about 60 mg cm^{-2}, or 25 µm, so this condition too is easily satisfied, even for beta particles. If the particle has to be completely stopped in the sensitive volume, then of course its thickness must exceed the residual range. For the 1-MeV beta, this would require a depletion depth greater than about 1.8 mm of silicon.

For *X-rays*, thin entry windows and the thickest possible intrinsic regions are required. The useful energy range is limited at its lower end by attenuation in the window material, and at its upper end by the interaction efficiency, $R = 1 - e^{-\mu x}$ (equation 5.7), of the intrinsic region. These limits can be determined and the interaction efficiency can be calculated, as a function of photon energy, by reference to graphs of attenuation coefficients against energy (*Figure 8.14*) and the manufacturer's data on the relevant thicknesses.

If this calculation is performed for Si(Li) detectors, the results are as shown in *Figure 8.15*. For a 1-cm depletion depth, the interaction efficiency is greater than 10% for the range 0.8 keV to 90 keV, and is 100% for the range 6 keV to

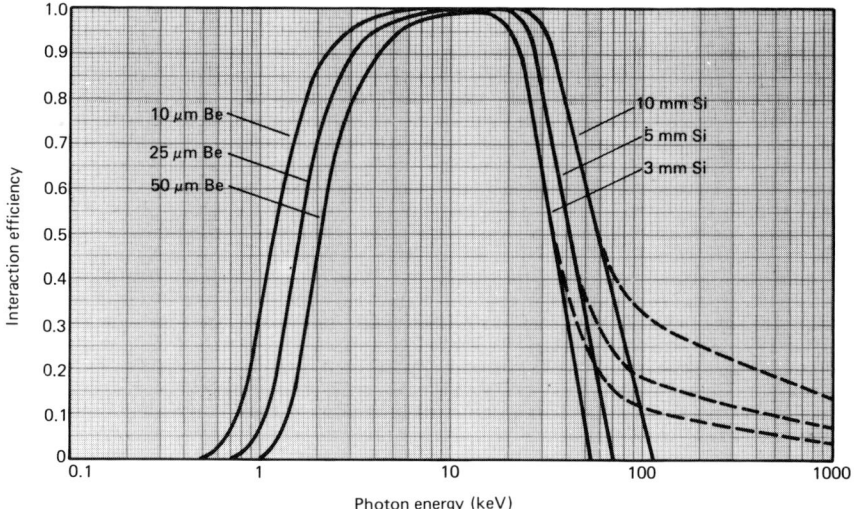

Figure 8.15 *Interaction efficiency of a Si(Li) detector as a function of photon energy, for various Be window thicknesses. The dashed lines include Compton scattering events which do not lead to a photopeak*

10 keV. Ordinary surface barrier devices have thinner entry windows and depletion layers, so they tend to cover a lower energy range. For the opposite reasons, pure germanium detectors can be applied to a higher energy range, but the advantage of a higher Z number tends to be offset by the need for a cryostat.

For *gamma rays*, Ge(Li) detectors have to be used. The detection efficiency of a Ge(Li) detector, for photons, is usually quoted as a percentage of the efficiency of a 3 in × 3 in (762 mm × 762 mm) NaI(Tl) detector — that is, a thallium-activated sodium iodide scintillation counter with a cylindrical crystal of diameter 3 in and height 3 in. The reasons for this convention are largely historical but it is a convenient practice because it allows a single figure to be quoted. The ratio of detection efficiencies does vary with energy but less rapidly than the efficiency of either one of them. NaI(Tl) is one of the most efficient gamma ray detectors available, a 3 in × 3 in crystal is quite a large one, and standard data are available for the absolute detection efficiencies of these crystals.

Typically, a Ge(Li) detector has a sensitive volume in the region of 25 cm^3 to 50 cm^3 and a detection efficiency of about 10% of a 3 in × 3 in NaI(Tl) detector. Larger efficiencies can be obtained but are unnecessary for most purposes. Usually, the efficiency ratio is stated for a particular photon energy, such as the 1.33-MeV ^{60}Co line.

For *neutrons,* Si(Li) detectors can be used. Thermal neutrons are detected by means of nuclear reactions with the lithium nuclei. This tends to use up the lithium but a more damaging effect is produced by the reaction products, which knock silicon ions out of their lattice positions, creating crystal defects which impair the properties of the detector. Similar radiation damage is caused by other radiations although they tend to be less penetrating or less likely to eject silicon ions. In general, solid-state devices should not be exposed to high levels of radiation, especially neutrons. The numbers of total counts that can be accepted before radiation damage becomes severe are shown in *Table 8.2.*

Table 8.2 Radiation damage in semiconductors

Type of detector	Radiation (particles cm^{-2}) required to produce significant deterioration in performance			
	Electrons	Fast neutrons	Protons	Alpha particles
Surface barrier	10^{13}	10^{12}	10^{10}	10^9
Diffused junction	10^{13}	10^{12}	10^{10}	10^9
Si(Li)	10^{12}	10^{11}	10^8	
Ge(Li)		10^8-10^9		

For *low-level counting,* semiconductor counters are particularly useful, especially for photon sources. Although the detection efficiency of these devices is much smaller than that of a sodium iodide scintillator, they have excellent energy resolution. This enables the semiconductor to isolate the activity being investigated from general background radiation and more than compensates for the relatively small detection efficiency. A Ge(Li) detector, with an anticoincidence shield or an 'old steel' shield (*see* p. 141), is a standard arrangement for low-level counting and probably the best general-purpose system available.

Pulse height spectrometry

The most accurate method of measuring the energy of charged particles, and especially high-energy particles, is by deflection in a magnetic spectrometer. For photons with energies less than about 300 keV, the most accurate method is by crystal diffraction. Both of these systems involve large, expensive systems designed to perform only the spectrometric function and they require relatively intense beams of radiation. For general purposes and for photons with more than about 300 keV, the semiconductor counters are the best energy spectrometers available.

The intrinsic, FWHM energy resolution, dE/E, is proportional to $(F/\bar{n})^{1/2}$ (equation 5.19). In a semiconductor, the mean energy required to form an ion pair is less than in any other material so the mean number of ion pairs, $\bar{n}$, is larger than in any other absorption detector. The result is improved by Fano factors, F, which are of the order of 0.1 to 0.3 for germanium and 0.1 to 0.2 for silicon. Consequently, FWHM resolutions of the order of, or less than, 1% are obtainable for reasonably high radiation energies. As pointed out in Chapter 5, energy resolution deteriorates at very low energies (*Figure 5.8*) for all types of detector.

This intrinsic resolution is of the same order of magnitude as that generated by the preamplifier, so preamplifier noise must also be minimised if the excellent resolution is to be realised in practice. Besides being charge-sensitive, they must be low-noise devices and, usually, based on field effect transistors (FET). If the detector is cooled, then the preamplifier is also cooled to reduce thermal noise still further.

The charge pulse from the detector is integrated into a voltage signal pulse at the preamplifier feedback capacitance and at the input capacitance; that is, of the detector itself and of the connecting cable between it and the preamplifier. Preamplifier noise is also affected by, and increases with, the input capacitance, so the connecting cable should be as short as possible or, better still, the preamplifier should be mounted directly on the detector. Since input capacitance will vary from one situation to another, the preamplifier noise level is always quoted by the manufacturer as a function of input capacitance. Normally, it is given as line broadening, in terms of FWHM in kilo-electronvolts. Typically, it is about 1.5 keV for zero input capacitance and rises to about 3 keV for 100 pF. The total FWHM resolution, squared, is the sum of this figure, squared, and the intrinsic resolution, squared; that is, it is obtained by adding the separate contributions in quadrature (equation 5.23).

The spectrometric capabilities of silicon detectors are illustrated in *Figure 8.16*.

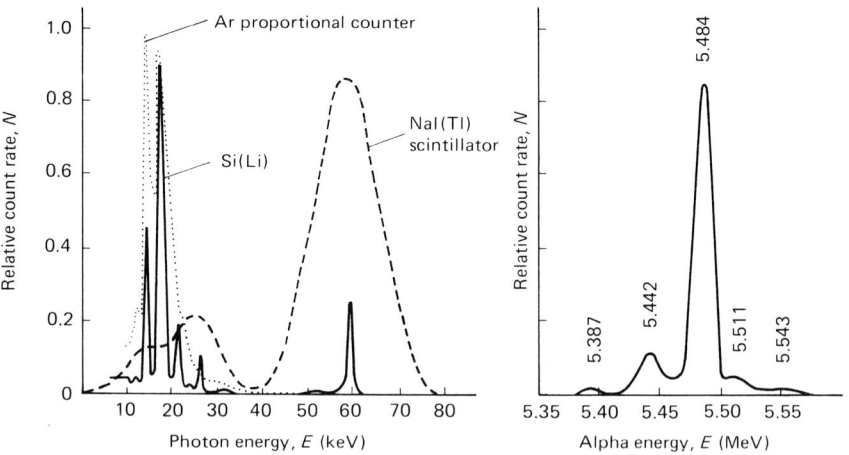

Figure 8.16 Pulse height spectra obtained with Si(Li) detectors. (a) X-ray spectra showing the effects of detector energy resolution and interaction efficiency; (b) alpha particle spectrum, both of ^{241}Am

This shows the alpha particle spectrum of ^{241}Am with sufficient resolution to identify fine-structure lines corresponding to the excited states of the daughter nuclide, ^{237}Np. It also shows the photon spectrum from the same source and clearly resolves the 59.5-keV gamma ray line (isomeric nuclear decay of ^{237}Np following the emission, from ^{241}Am, of a 5.486-MeV alpha) from the L X-rays of neptunium. The alpha spectrum was obtained with a surface barrier detector and the X-ray spectrum was obtained with a Si(Li) detector but both spectra could be observed, simultaneously, with the same silicon detector.

Ge(Li) gamma ray spectra for ^{137}Cs, ^{60}Co, ^{133}Ba and ^{207}Bi are shown in *Figure 8.17* and, in one case, compared with spectra produced by Si(Li) and NaI(Tl) detectors. Because of its very low Z number (14) the silicon produces almost no photopeaks at gamma ray energies, and is not used for gamma ray spectrometry. The sodium iodide is more efficient than the germanium but the

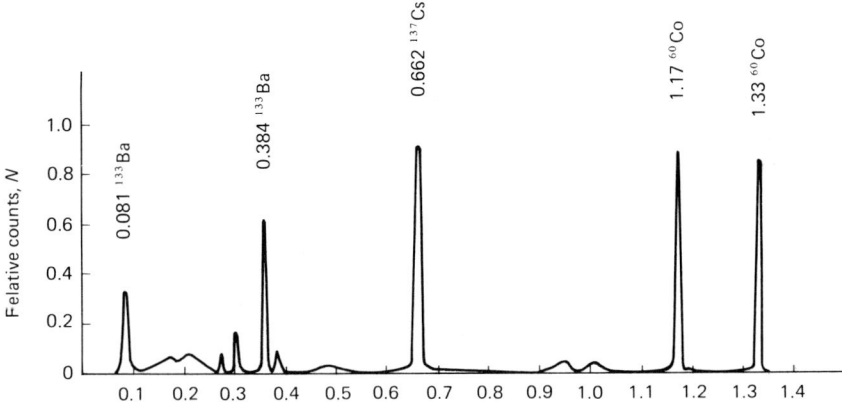

(a)

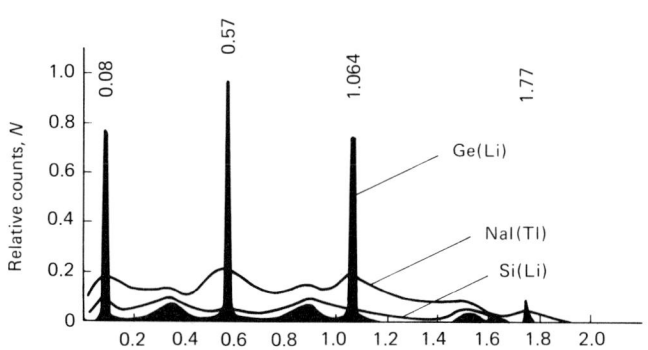

(b)

Figure 8.17 (a) Pulse height spectrum for a mixture of ^{133}Ba, ^{137}Cs and ^{60}Co, obtained with a Ge(Li) detector; (b) pulse height spectra for ^{207}Bi obtained with different detectors

photopeaks of the latter are much more pronounced. Although they contain fewer counts they are spread out over fewer channels so they are relatively taller.

Because of the thick entry windows of the Ge(Li) detector, the internal-conversion electrons and the beta particles, from the ^{137}Cs and ^{60}Co sources, are not detected. In this particular case, the same is true of the 30–36-keV Cs X-rays from the ^{133}Ba source, but the higher energy of the 72–87-keV Pb X-rays from the ^{207}Bi source is sufficient to produce a response.

One of the most important features of a Ge(Li) pulse height spectrum is the fact that the photopeaks are not only very narrow but also much taller than their corresponding Compton edges and backscatter peaks. Typically, peak-to-Compton ratios exceed 20. This makes it possible to record a large number of photopeaks in one spectrum without loss of clarity. In neutron activation analysis, for example, the irradiation of a specimen can produce a large number of active nuclides, all emitting photons. As another example, a radioactive source containing a long-lived parent nuclide and a large number of daughter products in a chain also generates a large number of photon energies. Before the development of the Ge(Li) detector, these emissions could be studied only by chemically separating the active constituents. Now, chemical separation is often unnecessary.

Dosimetry

Conduction counters can be used as current-generating dose rate monitors but leakage currents have so far prevented the development of a very successful device. The integrating detectors, however, have been incorporated into personal dosemeters and some systems are commercially available. RPL and TLD systems are the most promising, with the latter, based on lithium fluoride, showing the best results at present.

These solid-state devices have high stopping powers so they can be made quite small and are ideal in this respect for personnel dosimetry. They are good for all types of radiations, including neutrons, and show a linear response to photon energies down to relatively low values. In general, because of their high stopping power, they are more accurate than ionisation chambers and they have the advantage over film badges that they are reusable. Their main disadvantages relate to the high cost of the devices themselves, and to the relative complexity of the processing technique. The initial cost of installing a complete monitoring system is higher than that of the alternative systems. If processing is carried out *in situ*, the results can be obtained quickly but readout is not immediate by personal inspection, as with a gas electrometer.

Imaging

Ge(Li) detectors are used in gamma scanning techniques when energy resolution is particularly important: for example, to eliminate background radiations or to identify two or more isotopes. In general, however, the energy resolution of a scintillation counter is good enough and, because of its higher detection efficiency, it is preferred to the germanium detector.

In cosmic ray and nuclear physics, two-dimensional arrays of semiconductor diodes have been used to measure the spatial distributions of radiations, again when energy resolution has been of primary importance to identify types of radiation. The spatial resolution of such an array is limited to the size of each diode, however, and not very good. At present, this approach has found no wider applications.

A promising development is the semiconductor triode illustrated in *Figure 8.18*. This is a *position-sensitive detector*; that is, it has intrinsic spatial resolution. It operates on the principle of *charge division*. Incident radiation creates ion pairs

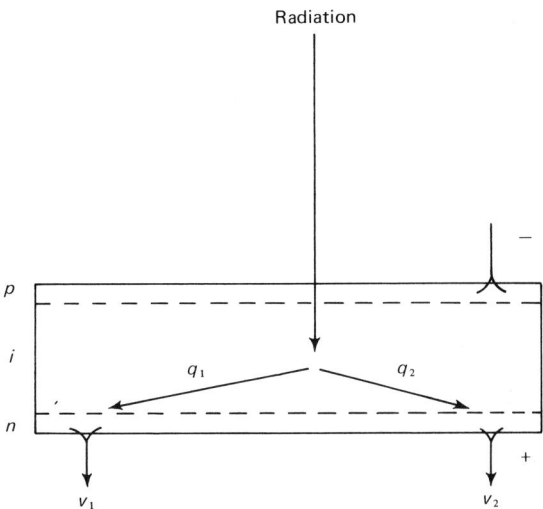

Figure 8.18 Semiconductor triode detector

at some point within the intrinsic region. Positive holes are collected at the single negative electrode and provide a measure of the energy deposited. Electrons are collected at two widely separated positive electrodes. The electron pulse behaves like all charge pulses and obeys Kirchhoff's law governing the division of current along two parallel resistances, R_1 and R_2. In this case, these are the resistances of the intrinsic material between the point of incidence and the collecting electrodes, so that R_1 is proportional to the distance to electrode 1 and R_2 is proportional to the distance to electrode 2. Thus, a charge pulse, q, is collected at one electrode, giving an output signal v_1 that is inversely proportional to R_1. Similarly, the signal at the other electrode is inversely proportional to R_2. The ratio v_1/v_2 can be measured to determine the ratio R_1/R_2, and therefore the distance of the point of incidence from one electrode.

At present, position-sensitive detectors of up to 5 cm in length are available with a claimed spatial resolution of 0.1 mm. If larger devices could be made and bidimensional output developed, they could be extremely useful imaging detectors, with the best energy resolution available and a reasonable detection efficiency. Unfortunately, this poses some very difficult technical problems, in growing large crystals.

Timing

With a time resolution that can be less than 1 ns, the semiconductor counter is suitable for a wide range of timing functions. As with gas ionisation chambers, however, the pulse rise time depends on the position of ion formation, so the time resolution deteriorates with depletion depth. Thin surface barrier detectors are very fast but do not have the stopping power and, therefore, the detection efficiency required for coincidence and anticoincidence counting techniques. In general, a plastic scintillator is more convenient and less expensive.

Ge(Li) detectors have time resolutions of up to about 10 ns but they have two advantages over the plastic scintillators, namely, better resolution and a higher Z number and, consequently, better efficiency for gamma rays. They are used whenever these properties are required. Their advantage is a relatively small size which means that they cannot, for example, form anticoincidence shields. A typical compromise is to use a Ge(Li) detector with a plastic scintillator as an anticoincidence shield.

There is one other point to note when using Ge(Li) detectors for timing measurements. The charge-sensitive preamplifier, with a large input capacitance, is not ideal for fast pulse output. It is standard practice to provide two preamplifier circuits, one for energy spectrometry and the other for timing. Signals are taken from the preamplifier output sockets according to which of these two functions is the more important.

BIBLIOGRAPHY

BERTOLINI, G. and COCHE, A., *Semiconductor Detectors,* North-Holland, Amsterdam (1968)
CAMERON, J. R., SUNTHARALINGAM, N. and KENNEDY, G. N., *Thermoluminescent Dosimetry,* University of Wisconsin Press, Madison (1968)
DEARNALEY, G. and NORTHROP, D. C., *Semiconductor Counters for Nuclear Radiations,* E. & F. N. Spon, London, (1966)
International Atomic Energy Agency, Technical Report No. 109, *Personnel Dosimetry Systems for External Radiation Exposures,* Vienna (1970)

Chapter 9
Scintillation counters

9.1 BASIC SYSTEMS

Most radiation detectors, including gas counters and semiconductor detectors, utilise the ion pairs generated by incident radiation. Scintillators, like Čerenkov counters and some solid-state devices, depend upon the production of excited atoms. In any material, an incident particle or photon leaves a number of electrons in excited energy states from which they tend to decay, emitting photons of ultra-violet or visible light in the form of a light flash, or *scintillation*. It would appear, therefore, that all materials are capable of such a response and all can be used as scintillators. In practice, however, a number of conditions must be satisfied if the material is to be used as a radiation detector.

In the first place, the scintillation process must be immediate and not delayed until after the triggering event. Thus, the materials must *fluoresce* rather than phosphoresce. Scintillators are properly described as 'fluors', not 'phosphors', although the latter term is widely used. The second condition is that the de-excitation process must tend to be *radiative* rather than non-radiative so that a large fraction of the available energy is converted into light rather than heat. Otherwise the material is a very inefficient scintillator. Thirdly, the material must be *transparent* to its own scintillations – an apparently limiting condition in view of the fact that emission and absorption spectra are usually similar.

For these reasons, not all materials are useful scintillators and there is a considerable diversity of efficiency among those that are. Nevertheless, there is a wide variety of scintillating materials, including inorganic and organic crystals, plastics, glasses, liquids and gases. They can be in almost any size or shape and can incorporate almost any material. Consequently, the main advantage of the scintillation counter is *versatility*, in sharp contrast to the semiconductor detector, for example, which must be fabricated from either silicon or germanium, to a limited range of specifications. Because of the variety of scintillators available, this is probably the most widely used type of detector, for general purposes. For the same reason, it is difficult to describe one basic system and the versions discussed below must be regarded as examples rather than as standards.

Before that discussion, however, one point should be clarified. A fluorescent screen is a scintillator, but not a scintillation counter. Certainly, Rutherford, in his alpha-scattering experiments, used a cadmium sulphide scintillator, in the form of a fluorescent screen, and was just able to observe, and count, individual scintillations. Most scintillations are much weaker and cannot be detected with the naked eye, so some form of photosensitive device must be used. Normally, this is a *photomultiplier* tube, which not only counts events but also measures their relative intensities, so that a scintillation counter is usually a scintillator–photomultiplier assembly. Fluorescent screens will be discussed in Chapter 10, as will screened-film systems, in which scintillations are effectively recorded on film.

In general, a scintillation counter comprises three elements: a *scintillator*, an *optical coupling system* and a *photomultiplier tube*. Two common arrangements are illustrated in *Figure 9.1*, for a solid and a liquid scintillator.

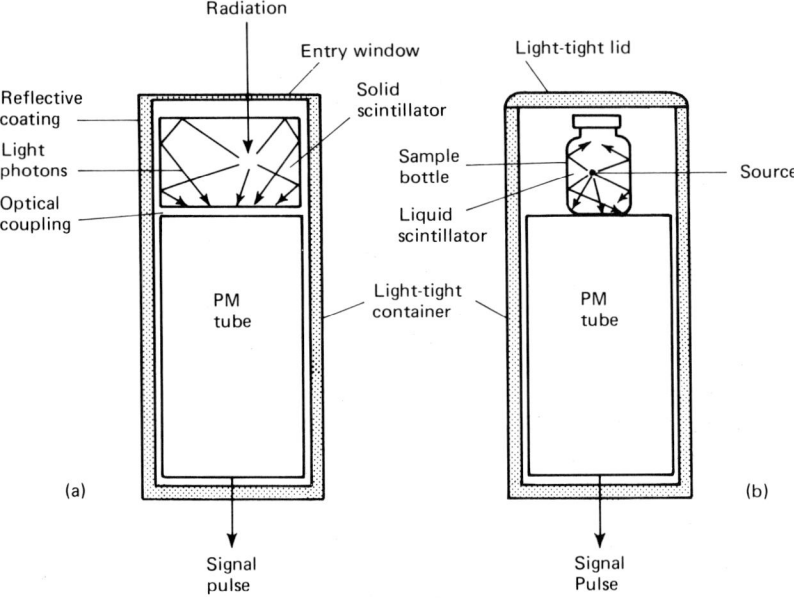

Figure 9.1 Scintillation counter assemblies. (a) Solid scintillation counter for external source; (b) liquid scintillation counter for internal source

The scintillator

The scintillator is the sensitive volume of the detector. Each particle or photon which interacts with this volume generates a single scintillation pulse. This is a very weak pulse, typically comprising fewer than 1000 photons, so it has to be viewed by a highly sensitive photomultiplier (PM) tube and the whole assembly must be enclosed in a light-tight container to isolate the scintillation from ambient light. The smallest pinhole in this container is sufficient to admit enough light to swamp the signal and, probably, overload the photomultiplier. With a newly assembled counter, the photomultiplier should be activated slowly; that is, the high-tension voltage applied to it should be increased gradually to observe the effects of light leakage before the tube is fully operational.

Liquid scintillators usually contain the source of radiations but, for other types, the source is external and an entry window must be provided. In order to maximise detection efficiency, especially for low-energy and weakly penetrating radiations, this window must be as thin as possible, but a lower limit is set by the need for optical opacity. For X-rays, photons with energies down to about 10 keV can be detected with a 0.025-mm (6.7 mg cm^{-2}) Al window and this limit can be extended to about 3 keV with a 0.2-mm (37 mg cm^{-2}) Be window.

The intensity of the scintillation and, therefore, the amplitude of the signal pulse is proportional to the energy deposited in the scintillator, so the counter can be used as an energy spectrometer. Energy resolution is determined by statistical fluctuations in the signal pulse height, as given by equations 5.19 and 5.20. Unlike the gas counters and the semiconductor detectors, however, the net resolution does not depend only on the mean energy, ϵ, required to form an ion pair. The initial signal has to pass through a number of process stages, each of which adds a statistical contribution and degrades the resolution. The first of these occurs in the scintillator with the conversion of deposited energy into light. If the *conversion efficiency* is defined to be the mean fraction, $\bar{f}$, of energy which is converted into light, then the FWHM intrinsic resolution is related to the mean energy, $\bar{w}$, required to produce one photon, rather than the mean energy, ϵ, required to form an ion pair and equation 5.20 can be rewritten as

$$\text{FWHM} \frac{dE}{E} = \frac{2.35(\bar{w})^{1/2}}{E^{1/2}} \tag{9.1}$$

where

$$\bar{w} = \epsilon/\bar{f} \tag{9.2}$$

Because the final signal height depends on other processing stages, absolute values of conversion efficiency are not easily obtained and it has become standard practice to describe a scintillator's performance in terms of its *light output* or *scintillation efficiency*. This is intrinsic signal height as a percentage of that of a thallium-activated sodium iodide scintillator used under similar conditions. NaI(Tl) scintillators are among the most efficient available and their conversion efficiencies are of the order of 13%.

Besides determining the intrinsic energy resolution, scintillation efficiency also affects detection efficiency by defining the minimum size of the signal pulse. Low-energy radiations are not detected unless the scintillation efficiency is sufficient to provide signals which are larger than noise effects. This is an important factor because photomultiplier noise is considerable and, along with the relatively thick entry window, tends to limit the low-energy response to events depositing more than about 10–20 keV in the scintillator. For higher energies, however, the detection efficiency is determined mainly by the size, density and interaction efficiency of the scintillator. Very large sizes can be obtained. Sodium iodide crystals of diameter 15 in and depth 15 in (3800 × 3800 mm) are available and the size of a liquid or plastic scintillator is limited only by the need to extract the light pulse. In addition, interaction efficiency can be maximised by a suitable choice from a variety of materials. Sodium iodide, with an effective Z number of 50, and the other inorganic crystals are very good for gamma rays, plastics are useful for fast neutrons and loaded plastics and glasses are excellent for thermal neutrons. For general purposes, scintillators are the most efficient detectors available and it is for this reason that the 3 in (diameter) by 3 in (depth) (762 × 762 mm) NaI(Tl) crystal is used as a reference standard when specifying the detection efficiency of other detectors, especially semiconductors.

Two other properties are important. One is the *spectral quality* of the scintillation; that is, the range of photon wavelengths produced. This has to match the transmission properties of the scintillator, the optical coupling system and, especially, the entry window of the photomultiplier as well as the spectral sensitivity of the photomultiplier. Any mismatch means a reduced signal height and poorer statistics. Most photomultipliers respond best to visible light. Some, with quartz or silica windows, can accept short-wavelength UV, but in general the scintillator output must be located in the visible-light range.

The other property is the signal *pulse shape*; that is, its time characteristics. Most scintillators have fast decay times so they impose very small intrinsic dead times and can cope with high count rates. Plastics and liquids are especially fast and are widely used for timing measurements. In all cases, the basic requirement is that the scintillation should have a major fluorescence component and a minimal amount of phosphorescence.

Optical coupling

In the scintillation process, light is emitted isotropically and somehow has to be channelled towards the photomultiplier. Any loss at this stage reduces the signal pulse height, decreases the low-energy sensitivity and degrades the energy resolution. The transfer is effected by means of an optical coupling system, which may vary from virtually nothing to a highly sophisticated arrangement. In the solid counter illustrated in *Figure 9.1*, for example, it comprises a number of standard features. All the surfaces of the scintillator are highly polished, except the one adjacent to the photomultiplier, to encourage total internal reflection at these surfaces. In addition, reflective materials, such as aluminium oxide (Al_2O_3)

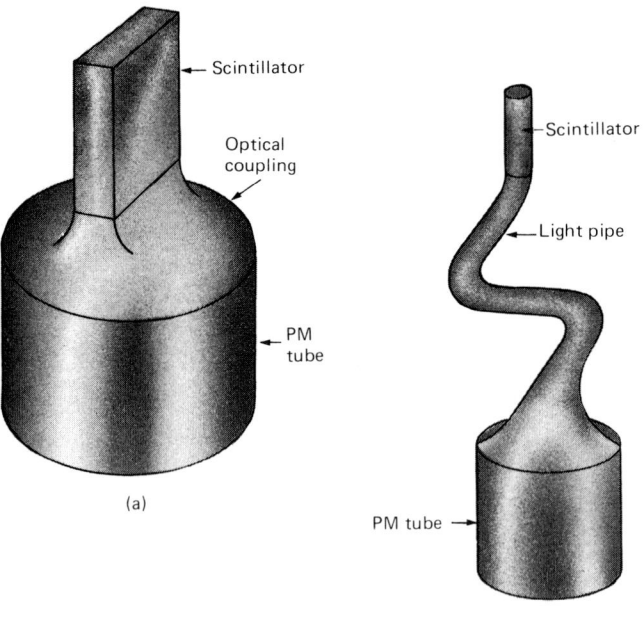

Figure 9.2 Optical coupling systems. (a) Scintillator of different shape from PM tube; (b) scintillator remote from PM tube

or magnesium oxide (MgO), may be packed around the polished surfaces to reflect more light back into the scintillator. The remaining surface is unpolished to avoid internal reflection, and transmission through this surface is increased by filling the small air gap between it and the glass window of the PM tube with an optical cement or grease. This should have a refractive index intermediate between that of the scintillator and the glass window to reduce reflections at all the surfaces involved. Proprietary cements and greases are available but, if necessary, any transparent medium is better than none at all.

With liquid scintillators it is not generally possible to incorporate all of these features. In the simple arrangement shown in *Figure 9.1*, the sample bottle should be coupled to the PM tube with optical grease but in some cases, as for example when the scintillator is examined by two photomultipliers counting in coincidence, even this degree of coupling has to be lost.

On the other hand, more sophisticated coupling systems can be employed to suit the experimental conditions. Mainly, these involve the addition of shaped pieces of Perspex (lucite) or glass to connect the scintillator and the photomultiplier. They are used, for example, to match different shapes or sizes of scintillator and photomultiplier or to connect a photomultiplier with a remote scintillator which may be in a magnetic field or in a specimen under investigation. In the latter case, fibre optics systems are widely used. Both arrangements are illustrated in *Figure 9.2*. The coupling material is shaped to transfer light pulses by total internal reflection and is sometimes known as a 'light pipe'.

The photomultiplier

The photomultiplier tube is a special type of electronic valve. It has a negative *photocathode*, a series of intermediate electrodes, called *dynodes*, at successively higher (i.e. more positive) potentials, and a positive collector *anode*. These are enclosed in an evacuated glass envelope, as illustrated in *Figure 9.3*.

The tubes used in radiation detectors have end-windows of glass or, if UV light has to be transmitted, of silica. The photocathode is a layer of semitransparent photosensitive material deposited on the inside surface of this window. Scintillation photons pass through the window to produce electrons by photoelectric interactions in the photocathode. These photoelectrons are accelerated towards and focused onto the first dynode by the electric field between cathode and dynode. This dynode, like all the others, is coated with a material, such as Cs–Sb or Ag–Mg, that emits a number of secondary electrons for each incident electron, effectively converting electric field energy into electrons. The process is repeated at each successive dynode, so that the electron pulse is amplified as it travels through the tube. If the gain at each dynode stage is p, and there are n stages, then the net gain, G, is given by

$$G = p^n \tag{9.3}$$

Typically, p is of the order of 5 and there may be 8 to 14 dynode stages giving total gains in the region of 5^8 to 5^{14}; that is, roughly 10^6 to 10^9. Although electron transfer is not 100%, as has been assumed, gains of this magnitude are easily realised so the photomultiplier acts as a high-gain pulse amplifier. It is also a linear amplifier for, provided the interdynode potential differences remain

242 *Scintillation counters*

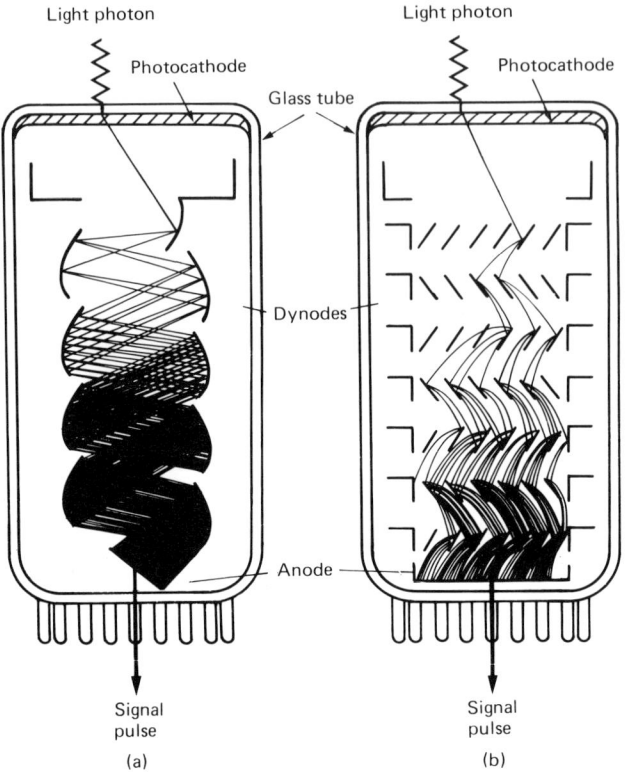

Figure 9.3 Photomultiplier tubes. (a) Focused dynodes; (b) venetian-blind dynodes

constant, p is constant and so is G. Thus, the proportional relationship between pulse height and the original energy deposited in the scintillator is maintained and *energy spectrometry* is possible. The final electron charge pulse, for each event in the scintillator, is integrated at the anode into a voltage pulse whose height may be of the order of 1 V, so further amplification is often unnecessary.

In operation, a single HT potential is applied to the anode and the successively lower dynode potentials are taken from a potential divider system comprising a series chain of resistors. This is known as a *dynode chain*. It is usually located outside the tube and connected to the dynodes via a number of pins in the tube base (although this is not shown in *Figure 9.3*). Normally, interelectrode potential differences of about 100 to 150 V are required, so the total applied potential is in the region of about 600 to 2000 V, depending on the number of dynodes. This HT potential must be highly stabilised since each stage gain, p, depends on potential difference and is raised to the power n to determine overall gain. Slight variations in applied potential are sufficient to alter the signal pulse height and degrade energy resolution.

As illustrated in *Figure 9.3*, there are two types of dynode structure. *Focused dynodes* are curved to focus the electron beam pulse from one dynode to the next, while *venetian-blind dynodes* are sets of parallel inclined planes through which the electrons pass in a more random cascade process. Focused dynodes are more efficient at transferring electrons and provide slightly higher gains but they are more susceptible to variations in applied potential. The major difference between the two systems is that the focused dynodes transfer electrons more quickly, producing pulse rise times of the order of 2 ns compared with about 10 ns in a venetian-blind system. Consequently, focused dynodes are preferred for timing measurements.

Count rate is also affected by applied potential. If this is too low, some pulses are lost owing to inefficient transfer, especially at the cathode—dynode stage.

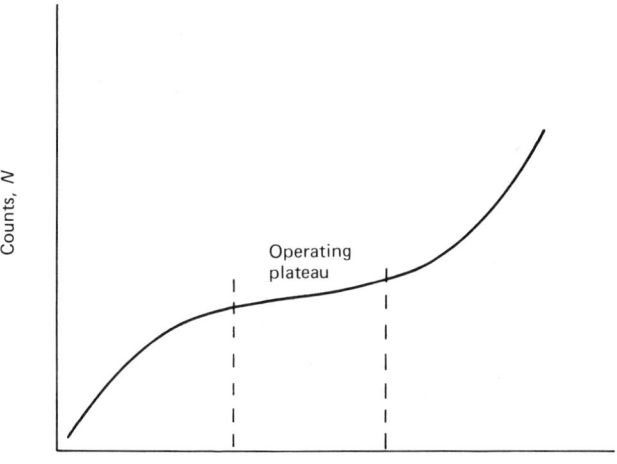

Figure 9.4 Characteristic curve of a scintillation counter

At higher potentials, spurious discharges may take place, registering as counts and, incidentally, tending to damage the PM tube by passing too much current through it. A *characteristic curve* should be obtained for a scintillation counter as for the gas counters. As shown in *Figure 9.4*, this has a rather poorly defined plateau region and it is in this region that the counter should be operated.

Since the photomultiplier is an integral part of the scintillation counter, its properties are as important as those of the scintillator itself. Elementary properties, such as gain and size, must be considered. The gain and, therefore, the number of dynodes must meet scintillator output and electronic input requirements while the size should match that of the scintillator or the optical coupling system. At a more technical level, the major properties may be described as *spectral response*, *noise*, and *time characteristics*.

Spectral response is the product of *window transmission* and *photocathode quantum efficiency*. Both are functions of photon wavelength. Window transmission is the fraction, or percentage, of photons of a given wavelength which are passed through the entry window. Quantum efficiency is the fraction, or percentage, of photons which produce photoelectrons in the photocathode; that is, the number of photoelectrons per photon of a given wavelength. Some typical spectral response curves, for a number of window and photocathode materials, are shown in *Figure 9.5*. These show the short-wavelength extension provided by fused silica windows and they also show the relatively low efficiency of the

244 Scintillation counters

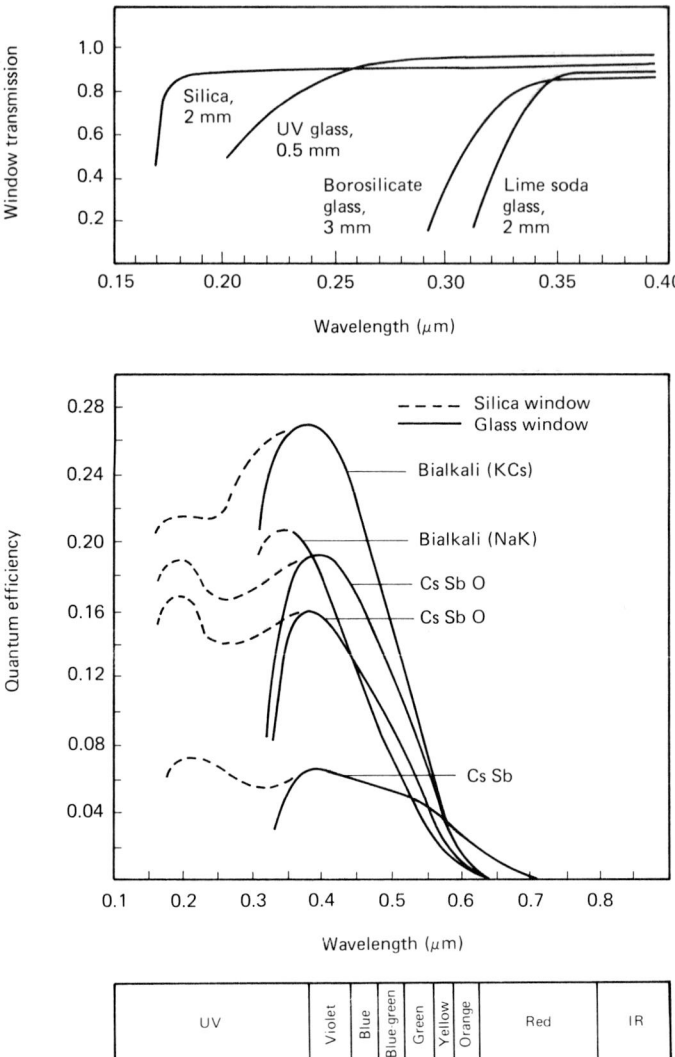

Figure 9.5 Spectral response functions for some photomultiplier tubes

whole conversion process. In general, fewer than 25% of incident scintillation photons yield photoelectrons. This indicates that energy resolution statistics are seriously worsened at this stage and low-energy detection efficiency may suffer. It emphasises the need to match the spectral response function of the photomultiplier to the spectral quality of the scintillation.

Electronic noise is the other major contributor to line broadening. When a photomultiplier is operated in complete darkness, a small *dark current* continues to flow through it and random variations in this current are amplified and recorded as noise pulses. Dark current is mainly due to thermally emitted electrons from the photocathode and to photoelectrons generated by radiations from residual radioactive materials, such as ^{40}K, in the glass window. It can be reduced, by cooling the tube down to about $-40°C$ and by using low-activity quartz windows, but not eliminated.

In general, the total contribution of the photomultiplier, via spectral response statistics and electronic noise, is about equal to the intrinsic resolution of the scintillator and its optical coupling system. These add in quadrature, being statistical uncertainties, so that a typical overall resolution of 10% may be produced by 7% resolutions in each of the scintillator and the PM tube, i.e.

$$\frac{dE}{E} = (0.07^2 + 0.07^2)^{1/2}$$

$$= (0.01)^{1/2}$$

$$= 0.1$$

The *time characteristics* of a photomultiplier are usually very good; that is, it is a very fast amplifier. For timing measurements, time resolution is the important parameter and is defined by the rise time introduced by the tube. For focused dynode tubes this is of the order of 2 ns. Counting dead time is dependent on the electron transit time through the tube but, again, this is very fast, usually about 30 ns, and is seldom an experimental limitation.

Apart from the properties described above, there are other factors which may affect the performance of the photomultiplier. It will not function properly in magnetic and electric fields since these deflect electrons from their design trajectories. High-temperature environments are inimical in that they increase dark current and degrade energy resolution. High-radiation environments also cause problems since they produce counts, due to fluorescence and Čerenkov interactions, in the tube. Neutrons, of course, activate the tube to generate an increase in dark current which persists until the active isotopes have decayed back to stable ones. Finally, visible light has a similar effect on the photocathode. If a tube has been exposed to light and is then sealed in its light-tight container, its photocathode will continue to phosphoresce for some time. It should not be used as a radiation detector until this activity has subsided (a process which normally takes less than half an hour).

9.2 INORGANIC SCINTILLATORS

The properties of some inorganic scintillators are summarised in *Table 9.1*. There are three main groups: crystals, glasses and gases. Of these, the crystal scintillators are undoubtedly the most widely used. They are mainly alkali halides and, compared to other scintillators, they have high Z numbers, high densities and high light outputs, so they are generally the most efficient gamma ray detectors available. Because of their large light outputs they have relatively good signal-to-noise ratios and comparatively good energy resolutions, and they can be made very large and still yield usable scintillation pulses, making them even more efficient photon detectors. Sodium iodide crystals of diameter up to 15 in and depth 15 in (3810 × 3810 mm) can be obtained.

Table 9.1 Properties of some inorganic scintillators

Scintillator	Density (g cm^{-3})	Light output* (% of NaI(Tl))	Decay constant † (ns)	Wavelength of maximum emission (Å)
NaI(Tl)	3.67	100	300	4200‡
NaI at 77 K	3.67	190	60	3030‡
CsI(Tl)	4.51	40	1100	5800
CsI at 77 K	4.51	220	600	4000
LiI(Eu)	4.06	33	1200	4750‡
CaI$_2$(Tl)	3.96	110	1000	4000‡
KI(Tl)	3.13	25	1000	4100
Glass	2.7	10	75	4000
Argon	1.8 × 10^{-3}	4	20	2500
Xenon	5.9 × 10^{-3}	~20	20	3300

* Absolute values can be deduced from the fact that the conversion efficiency of NaI(Tl) is 13%.
† Decay constants depend on the nature of the radiation. Data are for gamma radiation.
‡ Hygroscopic.

There are two kinds of inorganic crystal scintillator: pure or *intrinsic* crystals and doped, *extrinsic* crystals. Although these are identical as far as interactions with ionising radiations are concerned they differ in respect of their scintillation processes and, therefore, with regard to their properties as scintillators. For present purposes, they can be considered as separate categories.

Intrinsic crystals

The scintillation mechanism in intrinsic crystals can be described by reference to the usual band structure diagrams for solid-state materials, as illustrated in *Figure 9.6*. Incident radiation both ionises and excites valence band electrons. Ionisation liberates them from their ionic bonds and raises them to the conduction band, leaving positive holes in the valence band. Excitation forms a weakly coupled electron–hole pair; that is, an *exciton* whose electron energy state is just below the conduction band.

If the valence and conduction bands are drawn as graphs of potential energy against distance, R, of the electron from its parent atomic nucleus, they both form potential wells. Each has a minimum-energy position corresponding to an equilibrium electron position analogous to the Bohr orbits of isolated atoms. Because they are more tightly bound, valence band electrons have minimum-energy positions closer to the nucleus, as shown in *Figure 9.6*.

In the ionisation process, electrons are mainly raised from the valence band minimum to the same radial position in the conduction band. This is known as the Franck–Condon principle. As a result, ionised electrons have some excess energy in the conduction band. Provided the temperature of the crystal is

Inorganic scintillators

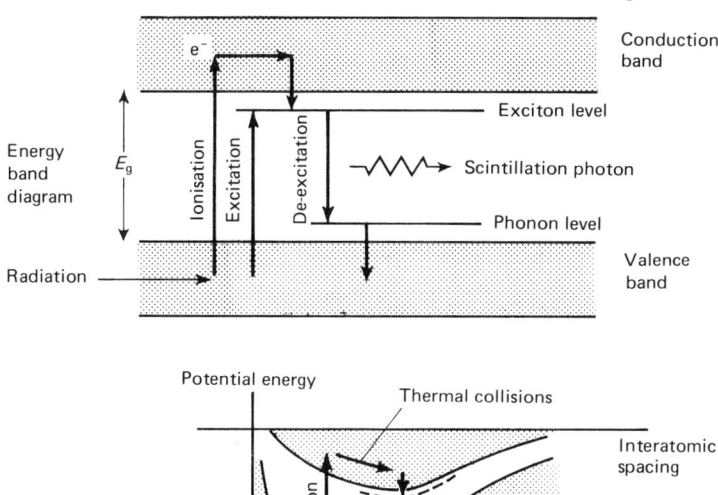

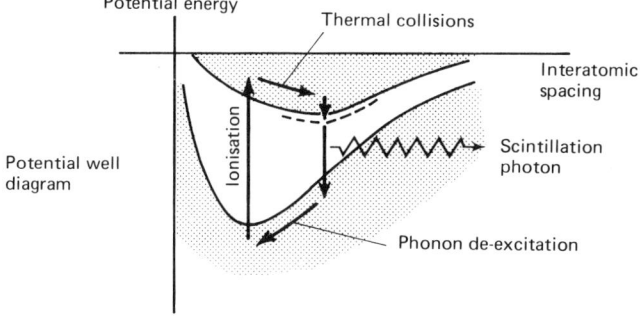

Figure 9.6 Scintillation process in intrinsic inorganic crystals

sufficiently low, they lose this energy in thermal collisions, moving down the energy scale and outwards in the configuration scale, to the equilibrium radius. Since the electrons are not in an electric field, they drift about in the crystal and are easily captured by positive ions. This is the electron–hole recombination process in which the electron drops down to the valence band, emitting a photon of electromagnetic radiation. It leaves the electron with excess energy in a vibrational, *phonon* state, from which it eventually decays, by thermal energy loss, to the valence band minimum.

Because of the relative displacement of the conduction and valence band minima, recombination energy is less than ionisation energy, emission and absorption spectra are different and the crystal is transparent to its own scintillations. Exciton decay also results in photon energies that are too small to produce ionisation, so the overall effect is a very efficient scintillation process, typically with a conversion efficiency of about 25%.

There is, however, one important condition to be met. Electrons ionised into the conduction band must be able to lose energy in thermal collisions. They can do so only if the crystal is cooled. Normally, intrinsic crystals are operated at a temperature of 77 K, obtained using liquid nitrogen. At room temperature, they are quite inefficient because emission and absorption spectra are too similar.

There is a further complication. In insulators, the forbidden energy gap, E_g, associated with ionisation is quite large, in the region of 5 to 10 eV, for example. Recombination involves a smaller transition but it still generates relatively high-energy photons in the UV wavelength band. Sodium iodide, for example, produces approximately 4-eV photons — which, from equation 1.22, is equivalent to a wavelength of about 300 nm. As shown in *Figure 9.5*, these are UV photons, requiring special optical coupling systems and photomultiplier windows.

Pure crystals, such as NaI and CsI, are just about the most efficient scintillators available but, because of the low temperature and UV transmission requirements, they are not widely used. They have another advantage over extrinsic crystals in that the scintillation process is a very fast one so the pulse decay time is relatively short and they can be used for very high count rate measurements.

Extrinsic crystals

Extrinsic crystals are formed by doping intrinsic materials with small quantities of heavy-metal impurities, such as thallium (Tl) and europium (Eu). These are known as *activators* and are specified, in parentheses, after the name of the host crystal. For example, thallium-activated sodium iodide is written as NaI(Tl).

The scintillation process begins, as it does in the intrinsic crystals, with the production of ion pairs and excitons, some of which decay, emitting UV photons. At this stage, however, the process is strongly affected by the presence of activator impurities. The energy level structure of an activator centre is illustrated in *Figure 9.7*. There is a ground state in the lower half of the forbidden energy

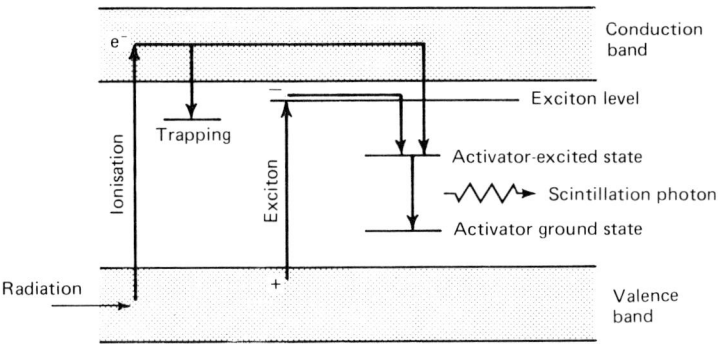

Figure 9.7 Scintillation process in extrinsic inorganic crystals

gap and a first excited state in the top half. Normally, both states are unoccupied. An electron in the ground state is easily captured by a positive hole in the valence band leaving the activator as a positive ion. Naturally, this attracts and captures electrons moving freely in the conduction band. It also tends to capture excitons and UV photons. In all cases, the result is usually an excited activator atom; that is, an atom with an electron in the excited energy state, from which it de-excites to the ground state, emitting a photon of electromagnetic radiation. Since this transition is much smaller than the forbidden energy gap, the crystal is transparent to its own emissions, which are usually in the visible-light range.

The activator centres perform two major functions. First of all, they capture and utilise electrons with excess thermal energy in the conduction band, so the crystal can operate at room temperature. Secondly, they act as *wavelength shifters*, converting UV light into longer-wavelength visible light. Neither of these functions is carried out with 100% efficiency, so the scintillation efficiency of an activated crystal is usually less than half that of the pure crystal at liquid nitrogen temperature. In addition, the extra stage in the scintillation process makes it longer, and scintillation decay times are usually doubled. Nevertheless, extrinsic crystals are efficient enough and fast enough for most purposes and are very widely used. NaI(Tl) is probably the most common scintillator and is the standard with which other detectors are compared.

Crystal scintillators may contain other types of impurity ion which form energy levels just below the conduction band. These tend to capture electrons from the conduction band and delay the decay of the scintillation pulse. They are known as *trapping centres* and increase the phosphorescence component of the scintillation. If they capture electrons for very long periods of time, they reduce the light output and act as *quenching agents*.

Sodium iodide, and most of the alkali halides, have only one disadvantage – they are *hygroscopic*. Such crystals deteriorate on exposure to moisture and have to be permanently sealed in airtight casings. These have glass end-windows to transmit scintillations to the photomultipliers. KI(Tl) is non-hygroscopic and CsI(Tl) is only very slightly hygroscopic.

Glasses

Glass is an amorphous material comprising small regions of local crystal-type structure randomly orientated with respect to each other. The scintillation process within a glass is similar to that in a crystal except that the more random structure attenuates the scintillation pulse, and scintillation efficiency is not very high.

In addition, glasses have lower densities and Z numbers than the crystals so they are less efficient photon detectors. On the other hand, they have a number of advantages and it is likely that their usage will increase. They can be made to a wide range of size, shape and composition specifications and they can tolerate extremely severe environments.

At present, their main application is to neutron detection and neutron radiography. Cerium-activated lithium silicate glass, loaded with a few per cent of lithium, is used for this purpose. Leaded glass is likely to be increasingly used for gamma ray detection and large-area imaging detectors can be constructed.

Gases

The only gases which scintillate are nitrogen and the noble gases. The process involves direct atomic excitation and de-excitation, so emission spectra match absorption spectra and scintillation efficiencies tend to be low, usually a few per cent of that of NaI(Tl). Also, because of the closed-shell structure of these gases, the scintillation is in the UV wavelength region, ranging from 250 nm for argon to 390 nm for helium. Thus the photomultiplier window, the optical coupling system and the gas container must all be transparent to UV light and the photocathode sensitivity must extend down to these wavelengths. Alternatively, wavelength shifters may be added to the gas to absorb UV photons and re-emit visible-light photons. A variety of such substances are available, usually in the form of solids evaporated onto the internal surface of the gas container. They are the same substances which are used in liquid scintillators and will be discussed, later, in that context.

Because of these operational problems, gas scintillators have not been widely used. They have an additional disadvantage in their low stopping power, especially for gamma rays. This can be overcome by using liquid and solid forms of the noble gas but, in general, the expense and practical inconvenience of such an approach outweigh the advantages to be obtained. Gas scintillators do have very fast pulse times of the order of 20 ns or less, but liquid and plastic scintillators are even faster.

There is, however, one way in which a gas scintillator can be used to advantage, namely, when it is the sensitive volume of a proportional counter. In this case, it forms a hybrid detector known as a 'scintillating proportional counter'. Light output is strongly and adversely affected by quenching agents such as methane but this is more than compensated for by the gas amplification, which produces many more excited atoms, and, therefore, scintillation photons, per event than a standard scintillator. The light output of such a detector is still in the UV wavelength range but it exceeds that of NaI(Tl). As a result, the energy resolution of a scintillating proportional counter is better than that of an equivalent standard scintillator, and may be better than that of a standard proportional counter.

9.3 ORGANIC SCINTILLATORS

There are three types of organic scintillator: crystal, liquid and plastic. Compared with the inorganic crystals, they have much lower densities and Z numbers, so they are usually very poor gamma ray detectors, but the liquids and plastics have specific advantages and are probably as widely used as NaI(Tl). Liquid scintillators are inexpensive and can incorporate various materials including radioactive sources such as ^{3}H and ^{14}C. Plastic scintillators are easily loaded with specific materials, such as boron or lithium for neutron counting, and can be manufactured in almost any shape or size. In addition, all the organic scintillators are extremely fast, especially liquids and plastics, and are widely used in timing measurements.

Organic crystals

Most of the organic crystal scintillators are aromatic hydrocarbons based on benzene ring structures. The most widely used are anthracene ($C_{14}H_{10}$), *trans*-stilbene ($C_{14}H_{12}$) and quaterphenyl ($C_{24}H_{18}$). Their light outputs are about half that of NaI(Tl) and in the visible-light range. As with all organic scintillators, their mean Z numbers are close to those of carbon and hydrogen, and too low to detect gamma rays, but the hydrogen content makes them useful as fast-neutron detectors.

The scintillation process is a simple one of direct excitation and de-excitation of the π electrons. These are less tightly bound than the σ electrons, which form the benzene rings, and they are responsible for all scintillations. As illustrated in *Figure 9.8*, incident radiation excites a π electron from its ground state to an excited state. The ground state has a minimum potential energy at a smaller radial coordinate than that of the excited state so the process leaves the excited electron, with excess energy, in one of the higher vibrational sub-states. It loses energy by thermal interactions, moving down the energy scale and outwards radially nearer to the minimum-energy position of the excited state. It then de-excites, from this position, to the ground state. In many respects, the process

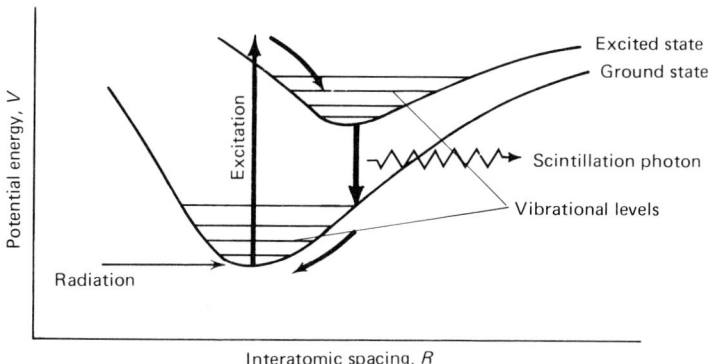

Figure 9.8 Scintillation process in organic crystals

is similar to that described for an intrinsic inorganic crystal, except that molecular levels are involved instead of crystalline bands. The result is the same. Emission spectra do not match absorption spectra and the crystal is transparent to its own scintillations.

Organic liquids

A variety of substances can be used as, or contained in, a liquid organic scintillator and the range of applications of such detectors is correspondingly wide. In high-energy and nuclear physics they provide large volumes and can incorporate target materials for the study of specific nuclear reactions. In addition, because of their bulk and fast response, they are used as anticoincidence counting shields.

Their major applications, however, are probably in the biomedical and biochemical fields, where they are used for the assay of radioactively labelled compounds. For this purpose, their main advantage is their ability to contain the radioactive material, which is often organic and usually a low-energy beta emitter. Such emissions do not easily penetrate detector entry windows, and the only efficient way to count them is to include the source material in the sensitive volume of a detector. Also, large numbers of samples may have to be counted, so the scintillator should be relatively inexpensive. Normally, the scintillator is purchased in bulk and dispensed into sample bottles to which the source materials are added. These are placed in contact with a photomultiplier tube to be counted, as illustrated in *Figure 9.1*.

The simplest system is a *binary liquid scintillator*, which consists of an organic scintillator, known as the *activator* or the *primary solute*, dissolved in an *organic solvent*. Typically, concentrations are less than 10 g l^{-1}. Some of the substances used are listed in *Table 9.2*. Primary solutes include the compounds known as PPO, PBD and TP, which are described in *Table 9.2*, and popular solvents are toluene, xylene and 1,4-dioxan. The main requirements of the solute are that it should be an efficient scintillator and soluble in the solvent used. The solvent should be transparent to the scintillations of the solute and should be able to dissolve not only the solute but also a wide range of source compounds and other substances that may have to be incorporated in the scintillator.

Table 9.2 Properties of some organic scintillators

Scintillator–activator–wavelength shifter	Density (g cm^{-3})	Light output (% of NaI(Tl))	Decay constant (ns)	Wavelength of max. emission (Å)
Crystals				
Anthracene	1.25	40	30	4400
trans-Stilbene	1.16	20	6	4100
Quaterphenyl		36	4	4400
Plastics				
Polystyrene–tetraphenylbutadiene	~1.0	14	5	4500
Polyvinyltoluene–*p*-terphenyl–*p,p'*-diphenylstilbene	~1.0	19	3	3800
Polyvinyltoluene–*p*-terphenyl–tetraphenylbutadiene	~1.0	18	4	4450
Liquids				
Toluene–PPO	0.88	20	3.8	3650
Toluene–*p*-terphenyl–POPOP	0.88	24	3.5	4250
Toluene–PBD	0.88	26	3.0	3650
Xylene–PBD	0.89	28	3.0	3650
1,4-Dioxan–PPO–POPOP	1.04	26	3.8	4250
Toluene–PPO–POPOP	0.88	30	3.7	4250
Xylene–naphthalene–POPOP	0.87	31	3.7	4250

PPO   = 2,5-diphenyloxazole, $C_{15}H_{11}NO$
POPOP = 1,4-di-(2-(5-phenyloxazolyl))-benzene, $C_{24}H_{16}N_2O_2$
PBD   = 2-phenyl-5-(4-biphenylyl)-1,3,4-oxadiazole, $C_{20}H_{22}N_2O$

The scintillation process is a complex one and reference should be made to one of the specialised works listed at the end of this chapter for a full treatment. Basically, the beta particles, or other incident radiations, interact with the solvent molecules, causing ionisation and excitation. The ionised molecules tend to recombine immediately, forming more excited molecules. These excited molecules diffuse through the medium and transfer their energy from one to another in the usual collision processes. Thus, excitation energy diffuses rapidly throughout the liquid volume. The primary solute molecules, which have lower excitation potentials, absorb this energy and de-excite, in the same way as an organic crystal, to produce a scintillation pulse.

The peak wavelengths of these scintillations are usually just inside the UV range at about 370 nm, presenting a problem in terms of light transfer and photocathode sensitivity. Special silica-windowed PM tubes may be used, along with similar windows in the base of the sample bottles or, more conveniently, a *wavelength shifter*, in the form of a *secondary solute*, may be added to the liquid scintillator. This has an even lower excitation potential than the primary solute. It absorbs the emissions of the latter and re-emits them in the visible-light range. The most popular wavelength shifter is known as POPOP. Others include dimethyl-POPOP and bis-MSB.

Another process which may take place in a liquid scintillator is *quenching*. This has the effect of reducing the light output and, therefore, the signal pulse height. It may eliminate weak signals completely, and this is usually the case for tritium beta particles.

There are three types of quenching. The most common of these is known as *impurity quenching* (or primary or chemical quenching). It is due to the presence of impurity materials that interfere with the energy transfer from the solvent to the primary solute, absorbing excitation energy and de-exciting non-radiatively. It increases rapidly with the amount of impurity present and is minimised by purification. Dissolved oxygen has this effect and is removed by bubbling nitrogen through the liquid. Source compounds may also introduce impurity quenching.

A second type of quenching is known as *colour quenching* (or secondary quenching) and is caused by the presence of coloured impurities which absorb light and interfere with the light collection process. This effect can be caused by source compounds and is reduced by bleaching these.

The third type of quenching is *radiation quenching* and is due to the fact that energy transfer is sensitive to the intensity of ionisation and excitation produced by the incident radiation. In beta counting this is not generally an important factor

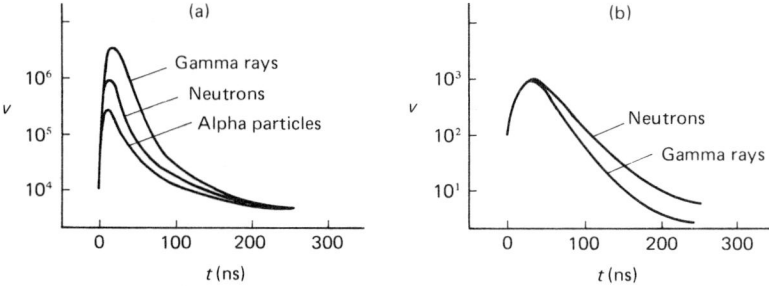

Figure 9.9 Pulse shape discrimination with organic scintillators.
(a) Stilbene; (b) quaterphenyl

but it is used to advantage in the technique known as *pulse shape discrimination*. Since different types of radiation have different specific energy losses, dE/dx, they generate different amounts of radiation quenching. As a result, signal pulse shapes and heights depend on radiation type, as illustrated in *Figure 9.9*. Electronic circuits can distinguish between the different shapes and identify types of radiation. All organic scintillators exhibit this property.

Plastics

A plastic scintillator is a solid organic solution and its properties and scintillation process are essentially the same as those of a liquid organic scintillator. In particular, it has a light output similar to that of the latter, it has the same fast energy transfer mechanism and, therefore, a short pulse decay time of a few nanoseconds, and is easily loaded with specific target elements. It has the disadvantage that, unless it is made up by the user, it does not incorporate source material. If its surface is cleaned with an organic solvent, it is virtually windowless (apart from being in a light-tight enclosure) but source materials placed in contact with it absorb some of their own radiations so it can never achieve the detection efficiency of a liquid system for weakly penetrating radiations. On the other hand, plastic is easily handled and it can be shaped to meet any requirements. For this reason, it is widely used in nuclear physics applications, where large or exotically shaped scintillators are required, especially for timing measurements and neutron detection. Like liquids, plastics have a low Z number and a low density and are not very efficient for gamma ray detection, producing almost no photopeaks for photon energies greater than a few tens of kilo-electronvolts.

The most common primary solute is p-terphenyl (TP) at a concentration of about 30–40 g l^{-1}. Solvents used include polyvinyltoluene and polystyrene. Wavelength shifters, such as p,p'-diphenylstilbene and tetraphenylbutadiene are used. Since plastics tend to absorb atmospheric and other materials the surface is likely to become heavily quenched.

9.4 FUNCTIONS

Counting

For *charged particles*, the interaction efficiency of a scintillator is usually 100% but detection efficiency may be much smaller because of absorption in the optically opaque entry window. To admit and detect 5-MeV alphas, window thickness must be less than about 4 mg cm^{-2} of low-Z material. Thin polymer foils are available to this specification. They are used on alpha radiation monitors, but they are fragile and have to be protected by a rigid grid. Betas are more easily admitted but still require thin windows. Because of the continuous energy spectrum and the range straggling effect, detection efficiency decreases gradually with increasing window thickness.

X-ray photons are also strongly absorbed by the entry window. As mentioned above, 10-keV photons can be substantially transmitted through a 6.7 mg cm^{-2} Al window, while 3-keV photons penetrate a 37 mg cm^{-2} Be window. In practice, noise pulses from the photomultiplier may be of a similar size to the signal pulses, increasing these limits slightly. To achieve reasonable signal-to-noise ratios, photon energies usually have to exceed 10 keV. For lower energies, proportional counters or Si(Li) detectors are preferred. To illustrate the point, pulse height spectra for the 6-keV (Mn K X-rays) photons from ^{55}Fe, obtained with a Be-windowed NaI(Tl) scintillator and an argon proportional counter, are shown in *Figure 9.10*.

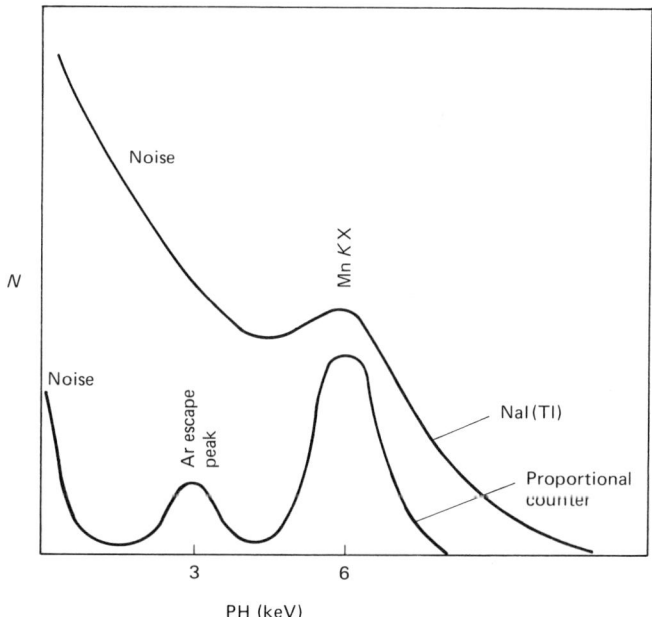

Figure 9.10 Pulse height spectra of ^{55}Fe obtained with a NaI(Tl) scintillator and an argon proportional counter

For *gamma rays*, the high Z number, high-density inorganic crystals are ideal, and, with thick entry windows, they have a built-in discrimination against charged particles. NaI(Tl) has an effective Z number of 50 and a density of 3.67 g cm^{-3}. The detection efficiency, η, for a point source at a distance h along the axis from a 3 in × 3 in (762 × 762 mm) NaI(Tl) crystal has been described in *Figure 5.3*. This shows that the interaction efficiency is about 100% for energies up to about 0.1 MeV. The attenuation coefficient, μ, as a function of photon energy, is described in *Figure 9.11*.

Fast neutrons are detected by the hydrogenous organic scintillators. Plastic scintillators are, probably, the most widely used for this purpose. The detection mechanism depends on the ionisation and excitation produced by knock-on protons; that is, hydrogen nuclei. For thermal neutrons, plastics or glasses loaded with lithium or boron are almost 100% efficient. The only problem, with regard to detecting fast or thermal neutrons, is *radiation damage*, which affects all solid-state materials. In high-level radiation environments, such as those near the core of a nuclear reactor or some accelerators, gas counters are preferred to scintillators.

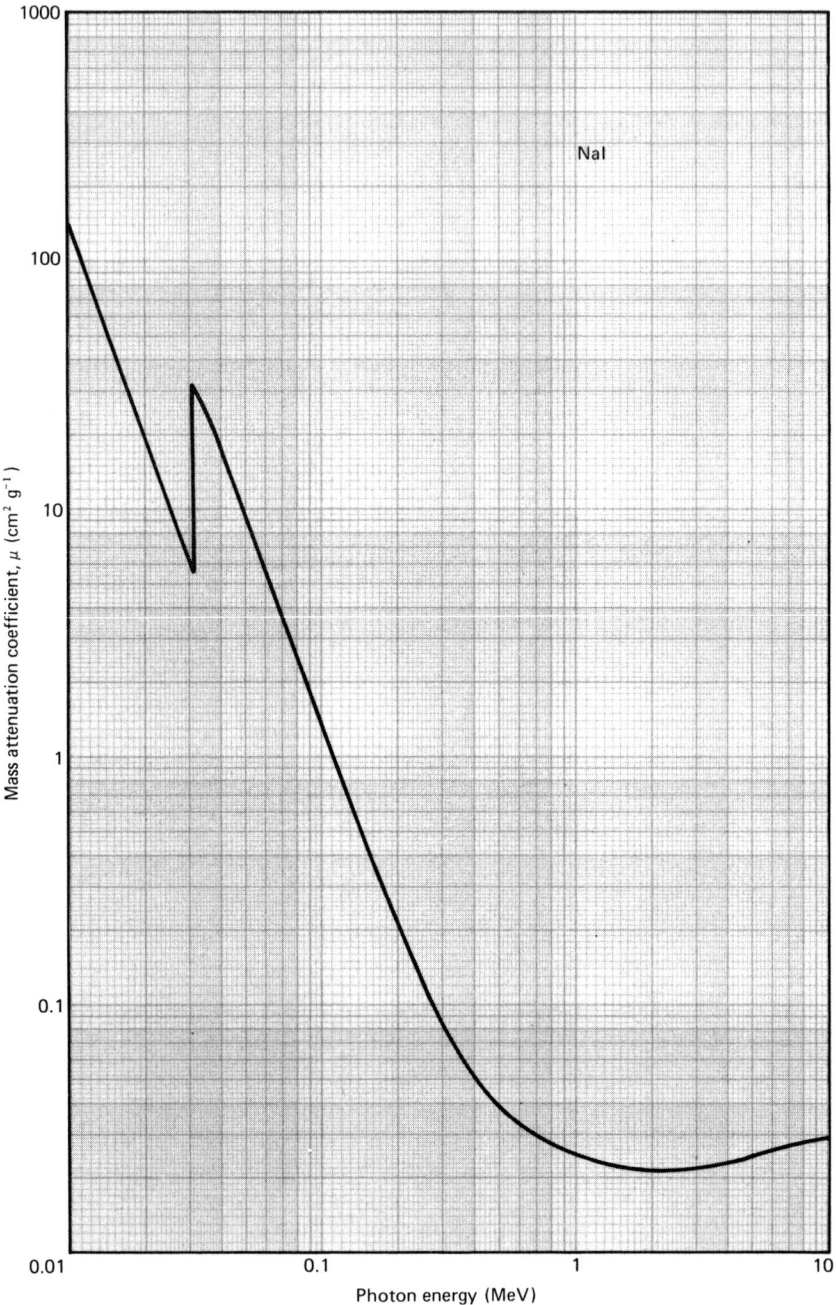

Figure 9.11 Mass attenuation coefficient as a function of photon energy for sodium iodide

A useful detector is the *Hornyak button*, and its derivatives, designed to detect fast neutrons in the presence of gamma rays. A typical device is illustrated in *Figure 9.12*. It comprises concentric and alternating cylinders of ZnS scintillator and non-scintillating plastic. The photons interact with the high Z number ZnS but, since this is opaque to its own scintillations, these interactions are not recorded. The neutrons interact with the plastic, producing knock-on protons which generate scintillations on the surfaces of the ZnS layers. These are recorded. If the plastic is loaded with lithium, the device functions as a thermal-neutron

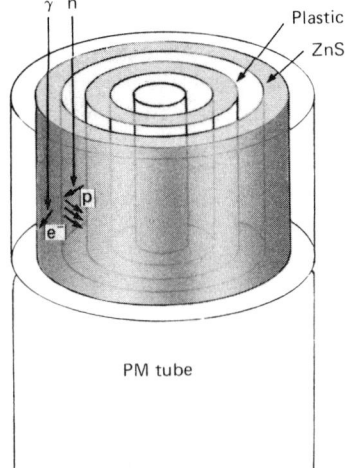

Figure 9.12 Structure of a Hornyak button neutron detector

detector. Fast-neutron interaction efficiencies are of the order of 5%; thermal-neutron efficiencies usually exceed 50%. In addition, because the protons are knocked predominantly forwards, the ZnS–plastic matrix can be arranged to measure the direction of the neutrons.

In *low-level counting* measurements, a scintillation counter offers the advantage of high detection efficiency. Since it detects source and background radiations with equal efficiency (unless these are subject to discrimination), the signal-to-background ratio is not improved but the counting statistics are. A better arrangement is to use a large liquid or plastic scintillator as an anti-coincidence counting shield surrounding the source, and another detector, which may be a scintillator or, better, a semiconductor. The system, illustrated in *Figure 9.13*, records events which are seen by the inner detector but not the outer one. It is used to eliminate cosmic ray counts, which register in both detectors (for example, in determining ^{14}C activity in carbon dating techniques).

Liquid samples can be counted in two ways. If they emit gamma radiation, the simplest procedure is to place the sample, in a test tube, in a well-shaped inorganic scintillator, as shown in *Figure 9.14*. This provides almost 4π geometry and discriminates against beta particles, by absorption in the test tube wall and the scintillator entry window. If the sample is a low-energy beta emitter or, less commonly, an X-ray emitter, liquid scintillation counting is the standard approach.

In *liquid scintillation counting*, the source is usually a labelled organic compound which can be dissolved in an organic liquid scintillator. Typically, the method is used for ^{3}H (emitting betas with a maximum energy of 18 keV and

258 Scintillation counters

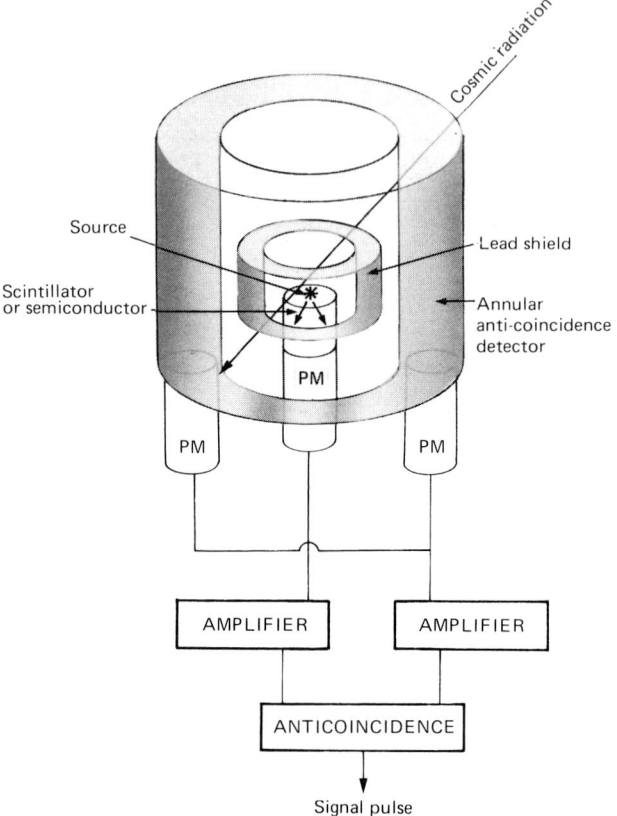

Figure 9.13 *Anticoincidence counting system for low-level counting*

a mean energy of 6 keV) and other beta emitters such as ^{14}C (maximum energy, 156 keV) and ^{35}S (maximum energy, 168 keV). In the simplest arrangement, the source material is dissolved in a liquid scintillator, in a sample bottle, which is then presented in a light-tight container to a photomultiplier tube for counting. In a more sophisticated arrangement, already described in Chapter 6 (*Figure 6.20*), the sample is examined by two photomultipliers counting in coincidence. Scintillations are observed simultaneously by both photomultipliers and counted as true coincidences. Apart from a relatively small number of random coincidences, noise pulses are not coincident and are rejected by the circuit. This method greatly improves the signal-to-noise ratio and allows weak events to be recorded. An estimate of random noise coincidences can be made by switching a 1-μs delay into one line. This delays one set of pulses, making scintillation signals out of coincidence so that only the random coincidences are counted.

In a second refinement, the signals from one of the PM tubes are pulse-height-analysed to measure energy. Beta particles have continuous energy spectra and, if two or more are present, they overlap to some extent. Nevertheless, the technique is usually good enough to resolve two or three different energy bands and to identify the counts from two or three different sources. Fixed single channels

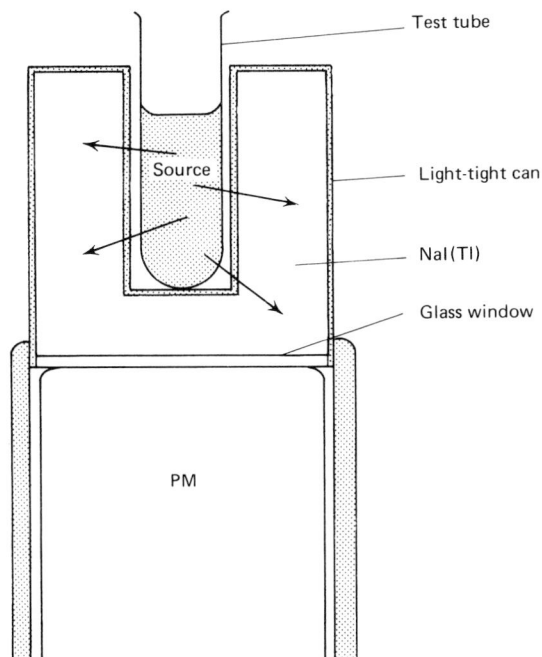

Figure 9.14 Liquid sample counting system using a well-shaped crystal

Figure 9.15 Automatic sample changer for liquid scintillation counting. (Courtesy of Beckman Ltd.)

260 Scintillation counters

can be used, corresponding to particular sources, such as ^{3}H, ^{14}C and ^{35}S, or these can be varied to accept any pulse height band.

Finally, a third extension is provided when, as is often the case, large numbers of samples have to be assayed. This is known as an *automatic sample changer*. It is a mechanism which automatically presents each of a large number (for example, 200) of samples to the counting head. It can often be programmed to perform a specific routine and may include some computerised data reduction to make it a fully automatic facility. A typical system is shown in *Figure 9.15*.

One of the problems of liquid scintillation counting is the fact that the detector efficiency is extremely sensitive to the presence of impurities and of source material, both of which may have a quenching effect. It is normal practice to include standard samples of known activities to assess these effects and calibrate the system.

Pulse height spectrometry

The total energy resolution, dE, of a scintillation counter is the sum, in quadrature, of a number of components due to the various statistical effects involved in the whole detection process. These include ion and excited-atom formation, light transfer, conversion to photoelectrons in the photocathode, secondary emission of electrons at the dynodes and fluctuations in tube dark current. Hence, total linewidth is given by an equation of the form

$$dE^2 = dE_1^2 + dE_2^2 + dE_3^2 + dE_4^2 + dE_5^2 \tag{9.4}$$

and is relatively large. FWHM resolution is usually in the region of 8% to 15%.

As a rule, scintillation counters are used for energy spectrometry only when detection efficiency, or some other property, is more important than energy

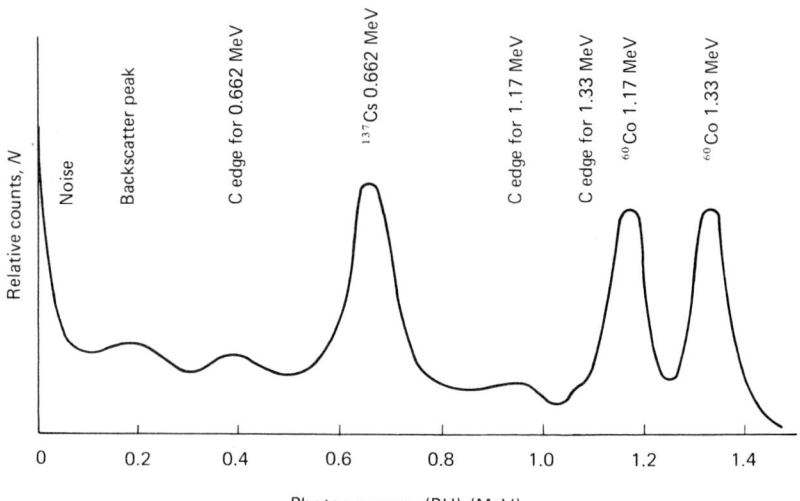

Figure 9.16 *Gamma ray spectra obtained with a NaI(Tl) scintillation counter for a mixture of ^{137}Cs and ^{60}Co*

resolution. Otherwise semiconductor detectors or, less frequently, gas counters are used.

In gamma ray spectrometry, a scintillator may be used for identified sources when it is known that its resolution is good enough to isolate the photopeaks of interest. It is especially useful for weak or transient sources for which counting statistics have to be optimised. As an example, many of the products of neutron activation analysis are very short-lived and their half-lives are more accurately determined with a large scintillator than with a less efficient semiconductor. Some typical gamma ray spectra are shown in *Figure 9.16*.

Similar conditions apply to charged-particle spectrometry, and scintillators are used for low, and rapidly varying radiation intensities. In addition they have two other advantages. First of all, they have high stopping powers and are capable, for example, of measuring the energies of high-energy particles to which other detectors are transparent. Secondly, they can be made very thin to serve as dE/dx spectrometers. Typical particle identification systems employ a thin dE/dx scintillator followed by a stopping, E, detector as described in Chapter 5 (*Figure 5.14*) and are widely used in nuclear physics applications.

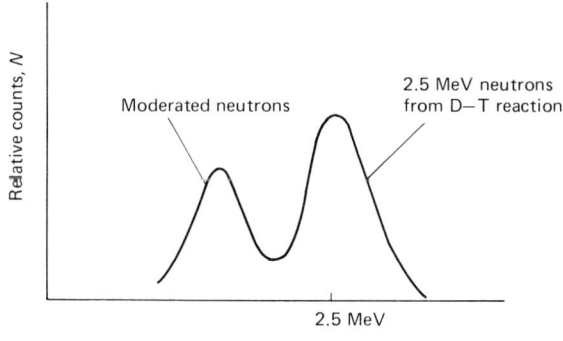

Figure 9.17 Neutron spectrum obtained with a $^6LiI(Eu)$ scintillation counter

The detection efficiency of plastic scintillators for fast neutrons is exceptionally high and these detectors are very good for neutron spectrometry. A further advantage is their relative insensitivity to gamma rays. Organic crystals and LiI(Eu) crystals are also used for neutron spectrometry and *Figure 9.17* shows the spectrum obtained with the latter for $^3H(p, n)^3He$ neutrons. The low-energy peak is produced by neutrons which have been moderated by scattering interactions in environmental materials.

Dosimetry

Scintillators can be used for all types of dosimetry except personnel monitoring. Their major advantages are sensitivity, the ability to select different types of radiation and measure their energies and the relative tissue equivalence of some of the organic scintillators. Against this, they require stabilised high-voltage supplies and photomultiplier tubes, so they tend to be bulkier and more expensive than alternative systems.

Scintillation counters are widely used for *whole-body monitoring*, mainly because of their high sensitivities but also because of their abilities to select and identify radiations and their source isotopes. Generally, they can detect smaller quantities of source material than any other system. Various configurations have been used, including single NaI(Tl) crystals, multiple arrays of NaI(Tl) crystals, plastics and liquid scintillators. One arrangement is illustrated in *Figure 9.18*.

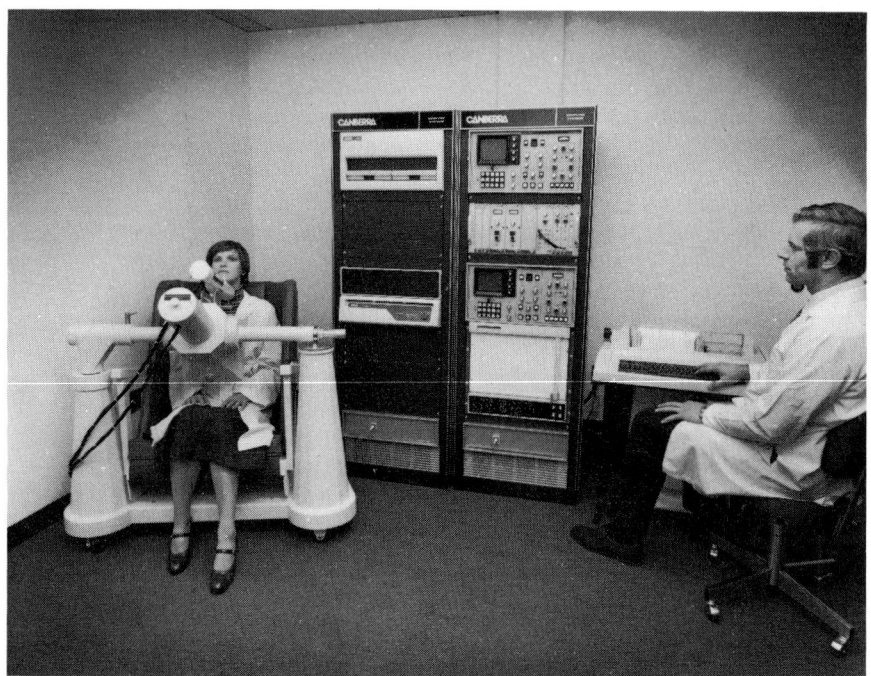

Figure 9.18 Body burden analysis system. (Courtesy of Canberra Instruments Ltd)

Area monitors are frequently based on scintillation counters, again because of their sensitivity but also because of their selectivity. NaI(Tl) is used for gamma monitoring, plastics for fast neutrons and loaded plastics or liquids for thermal neutrons. For alpha monitoring, NaI(Tl) or plastic scintillators are used with very thin (less than 0.1 µm), optically opaque, aluminised plastic windows.

Imaging

At present, most of the detectors used in gamma cameras, isotope scanners and computed-tomography systems are scintillation counters and almost all of these employ NaI(Tl) crystals. Their main advantages are high gamma ray detection efficiencies, required to minimise radiation dose to the patient while generating satisfactory image statistics, and computer-compatible output signals, which facilitate image processing and medical diagnosis. Other advantages include reasonable count rate capabilities, energy sensitivity and infinite latitude. A high count rate capability allows short exposure times to be used, reducing the probability that patient movement will blur the image. Energy sensitivity eliminates background and scattered radiations to produce a clearer image. Infinite latitude means that, unlike a film, the detector does not saturate after a fixed number of counts but can count indefinitely. Thus, there is no need to assess exposure and detector response in advance of the measurement and the final image can be presented in terms of any set of intensity ranges. Each range is given a colour or a grey scale so that relative intensities are clearly defined.

Various *gamma camera* systems have been developed. The majority employ a single, large-area NaI(Tl) crystal, typically 15 in × ½ in thick (3810 × 127 mm). This is examined by a hexagonal array of photomultiplier tubes, as illustrated in *Figure 9.19*. In modern scintillation cameras, 37 tubes are used. The light from

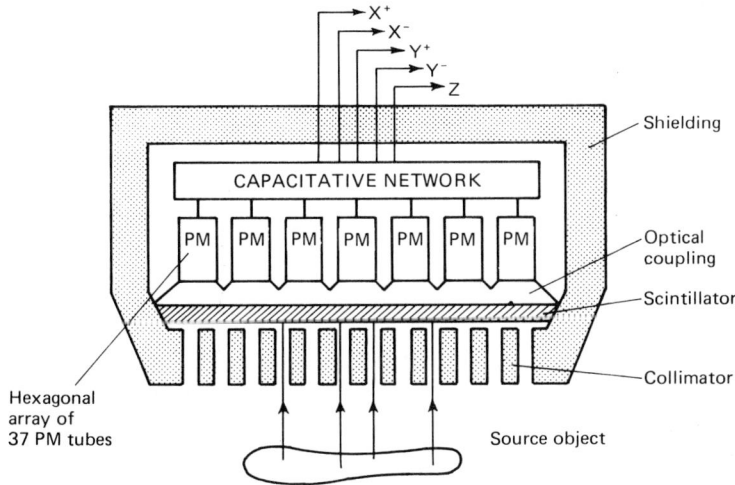

Figure 9.19 Scintillation counter gamma camera system

each scintillation is divided among the tubes according to their proximities to the scintillation point. The voltage pulses from the PM tubes are taken out through a network of capacitances in such a way that five output signals are obtained. One of these, the Z output, is proportional to the total scintillation intensity and therefore to the energy deposited by the event. An X^+ and an X^- signal are derived, effectively from opposite ends of the array and the difference in their magnitudes measures the x coordinate of the event. This is divided by the Z signal to provide an energy-independent measure of position. The Y^+ and Y^- signals similarly generate the y coordinate of the event. *Positron cameras* operate in the same way except that their output signals must be in coincidence with those of the focal point detector. In both systems the readout mechanism is a major factor in determining spatial resolution.

There are two other major contributions to spatial resolution. One of these is due to the collimator hole diameter and, in order to maximise detection efficiency, this has to be quite large. In general, a spatial resolution in the region of 5 to 10 mm (FWHM) is regarded as a satisfactory compromise between resolution and sensitivity. The other contribution is the electronic signal-to-noise ratio, which affects the readout process. This limits the low-energy response of the camera, introducing an intrinsic resolution of about 6 mm at 140 keV and 1.4 cm at 60 keV.

Scanners also employ NaI(Tl) crystals, but these can be smaller than the ones used in cameras. The early devices had single-hole collimators and scanned the patient in a rectilinear pattern. In modern systems, detection efficiency and scan time are improved by using multihole collimators focused onto one point in the patient and by using several collimator–scintillator arrangements simultaneously. One such system is illustrated in *Figure 9.20*. The increased detection efficiency

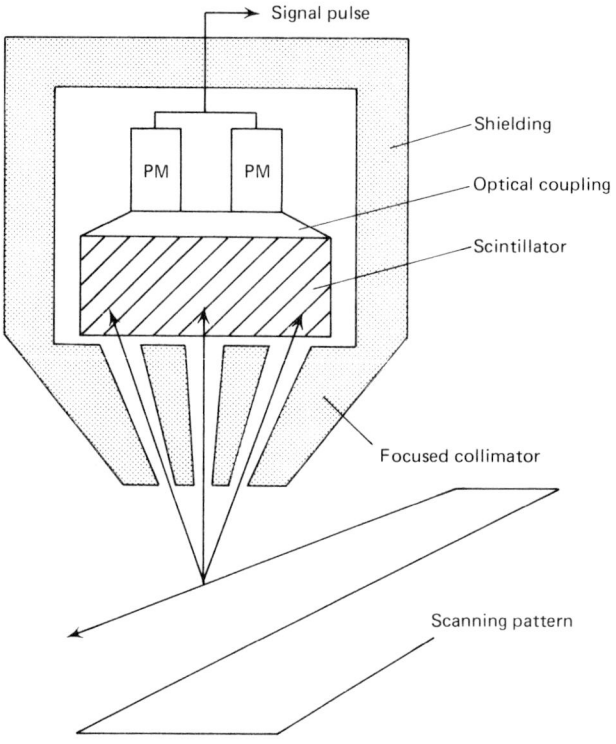

Figure 9.20 Scintillation counter gamma scanner system

allows the scanner to produce better spatial resolution than the camera but patient movement must be restricted during the scanning process. *Transverse scanning* or *tomographic scanning*, to image a narrow cross-section of the labelled organ, can also be carried out with scintillation counters.

Most of the *computerised axial tomography* or *computed tomography* (CT) systems follow the example of the original EMI scanner and use NaI(Tl) crystals. The EMI whole-body scanner is shown in *Figure 9.21* along with a typical scan. In order to minimise radiation dose and scan time, this device incorporates a narrow, fan-shaped X-ray beam incident on an array of 30 detectors.

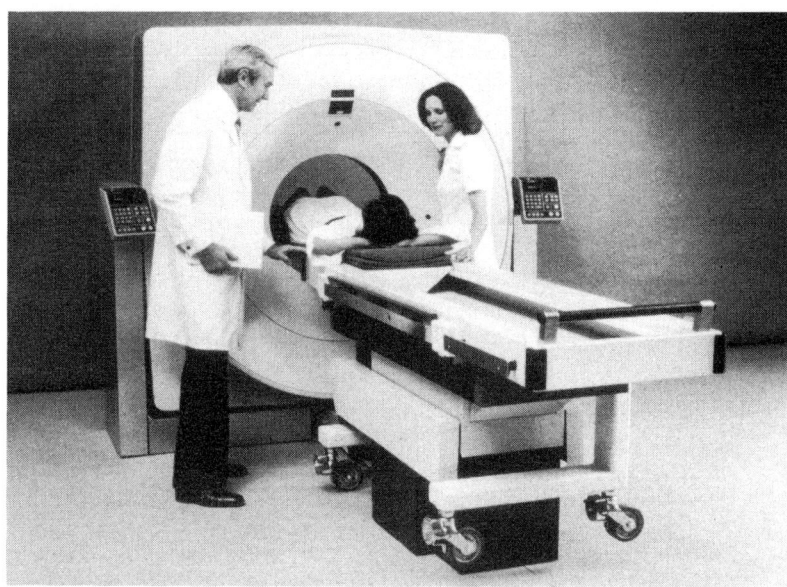

(a)

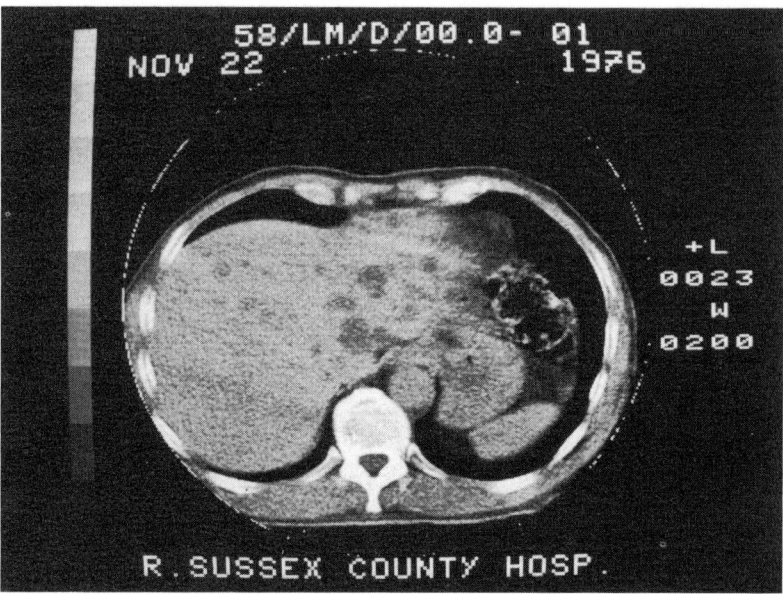

(b)

Figure 9.21 (a) Computed tomography system — the EMI scanner 7070 (courtesy of EMI Medical Ltd). (b) EMI scan showing a liver riddled with translucent rounded areas due to multiple congenital cysts (courtesy of the Royal Sussex Hospital)

Timing

Organic scintillators are among the fastest detectors available and are used for timing measurements whenever their other properties, such as detection efficiency, are suitable. Their applications in anticoincidence counting have been described above. They are also used in the other timing techniques described in Chapter 6, including time-of-flight measurements and the determination of time relationships between events in nuclear reactions.

NaI(Tl) counters are used for photon timing experiments when detection efficiency is the primary consideration. This applies particularly to the use of anticoincidence shields to eliminate Compton scattering events, to the coincidence counting of annihilation photons in a positron camera, to the measurement of short half-lives in, for example, neutron-activated isotopes and to time-of-flight counting.

BIBLIOGRAPHY

BELL, C. G. and HAYES, F. N., *Liquid Scintillation Counting*, Pergamon Press, Oxford (1958)

BIRKS, J. B., *The Theory and Practice of Scintillation Counting*, Pergamon Press, Oxford (1964)

CROOK, M. A., JOHNSON, P. and SCALES, B., *Liquid Scintillation Counting*, Vol. 2, Heyden, London (1971)

DYER, A., *Liquid Scintillation Counting*, Vol. 1, Heyden, London (1971)

Chapter 10
Visual imaging systems

10.1 FILMS

The film, or *photographic emulsion*, is one of the most widely used types of radiation detector, with major applications in radiography, crystallography, radiation dosimetry, autoradiography, electron microscopy and nuclear physics. It is almost the oldest form of detector, in view of the fact that it was the fogging of film by the emissions from uranium ore which led to the discovery of radioactivity by Becquerel, in 1896.

A typical modern film consists of a layer of emulsion, usually between 10 and 25 μm thick, coated onto one or both sides of a transparent plastic base. The base provides structural support for the emulsion, which is the sensitive volume of the film. The active ingredients of the emulsion are crystalline *grains* of silver bromide, each containing something of the order of 10^{10} atoms and held in suspension in a colloidal gelatin medium. In ordinary films, that is, those used for photography, radiography, and so on, but not nuclear track measurement, the silver bromide forms about 40% of the total mass of the emulsion. The grains are closely packed and their diameters range from about 0.3 μm in slow, or insensitive film, to about 2 μm in the faster, more sensitive film used for imaging ionising radiations. The gelatin contains some absorbed water; glycerine, to improve its flexibility; and a number of additives, to assist radiation detection, film processing and other effects. Its functions are to support the silver bromide grains, to allow access to those grains by processing chemicals and to transfer energy from incident radiations to the grains. In addition, there is a thin protective layer of gelatin, about 0.5 μm thick, on the outside surface of each emulsion.

The primary function of the ordinary emulsion is the bidimensional imaging of incident radiations. There is another type of emulsion, known as a *nuclear emulsion*, which is used in nuclear and high-energy physics to record tracks of individual particles. This is much thicker than an ordinary emulsion, in order to provide high stopping power for energetic particles. Its silver bromide grains make up about 80% of the emulsion mass, but they are smaller and more widely dispersed than in ordinary emulsions. This structure leads to better spatial resolution for track localisation and it also makes the nuclear emulsion relatively insensitive to light.

Ordinary films are extremely sensitive to visible light, except at the red end of the spectrum, and, to be used as radiation detectors, they have to be enclosed in light-tight containers. These may be paper or cardboard *envelopes* or more substantial *cassettes* made of aluminium or some other relatively radiotransparent material. The complete system may be *screened* or *unscreened*. In a detector

268 Visual imaging systems

using unscreened film, the emulsions form the sensitive volume and the other materials, including container, excess gelatin and film base, collectively form an entry window. In the alternative, screened system, thin layers of scintillating material, in the form of fluorescent screens, are inserted in front of and behind the film, usually as integral components of the cassette. These provide additional, and more efficient sensitive volumes. They greatly increase the sensitivity of the detector, although there is some deterioration in image quality because of the increased lateral spread of each event. The light flashes in the screens are detected in the emulsion along with intrinsic interactions. Both systems are illustrated in *Figure 10.1*.

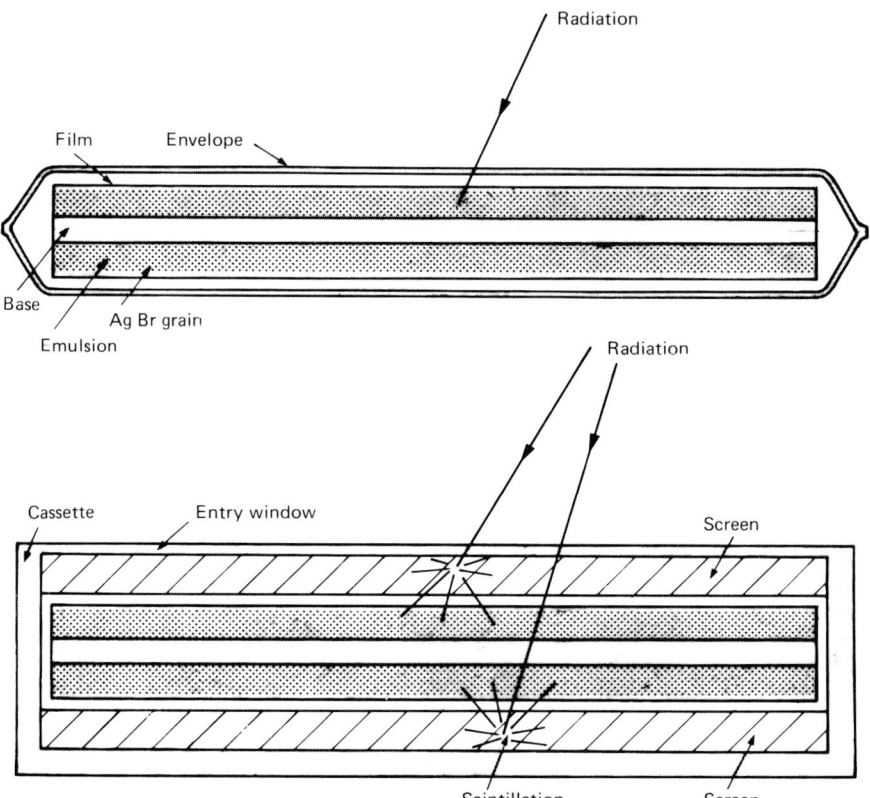

Figure 10.1 Film detection systems

Screens are particularly useful for imaging X-rays and neutrons. For X-rays, high Z number scintillators are the most effective, while plastics doped with lithium or boron, for example, are suitable for neutrons. The detection of thermal neutrons and of gamma rays can also be enhanced by using *converters* instead of screens. These are thin metal foils in which incident radiations generate reaction products that are detected by the film. For gamma rays, high Z number metals are employed and photoelectrons register in the film. For neutrons, metals such as gadolinium, dysprosium and indium provide high capture cross-sections and suitable reaction products. In both cases, there is an optimum converter thickness, below which the efficiency is reduced and above which too much absorption takes place in the foil. The ideal thickness is equal to the range of the photoelectrons or the reaction products.

Image formation

Silver bromide is an ionic crystal formed by the donation of one electron from each silver atom to each bromine atom to establish effectively closed shell structures in both. Overall, the matrix has net electrical neutrality and has the energy band structure illustrated in *Figure 10.2*. The conduction band is the

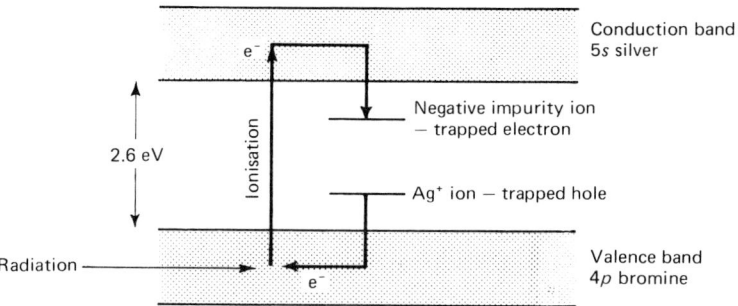

Figure 10.2 Band structure diagram of a silver bromide grain

empty 5s state of silver, the valence band is the full 4p state of bromine, and there is an energy gap of 2.6 eV between the bands. Incident radiation ionises the silver bromide either directly or indirectly, following initial interactions in the emulsion which release energetic electrons, to form electron–hole pairs. The electrons and positive holes then diffuse through the silver bromide grain, preferably with minimal recombination. The electrons are eventually trapped at crystal defects and impurity atoms, such as gold and sulphur, incorporated specifically for that purpose, while the positive holes tend to be trapped at interstitial silver atoms to form positive silver ions. It is these silver ions which record the positions of incident radiations. They form a *latent image* that can be developed into a visible image by chemical processing.

If a silver bromide grain is left with more than about four silver ions, then the whole grain becomes developable. This has two important consequences. First of all, it means that there is a gain of about 10^{10} atoms for every four ion pairs created. Since ionising radiations are capable of producing four ion pairs in single interactions, each interaction is amplified, in visual terms, by a factor of about 10^9 and the size of each event, that is, the spatial resolution, is the size of one grain, or about 1 μm. The second consequence is a minimum energy threshold below which no image is formed. Photons of red and orange light have considerably less than 2.6 eV, so that most films can safely be exposed to red light, in photographic darkrooms, without effect. In fact, because of the distribution of energy between excitation and ionisation, the mean energy deposited per ion pair formed is about 5 eV. In addition, there is the statistical effect of random energy distribution, such that many grains receive more energy, and form more ion pairs than they require. As a result, the average developable grain has about ten ion pairs corresponding to an absorbed energy of about 50 eV.

It should be noted that this energy can be supplied in ways other than by intentional irradiation. Static electricity, mechanical pressures, bending forces and chemical reactions can produce localised effects known as *artefacts*. A more uniform exposure to chemicals and to cosmic and other background radiations results in an even distribution of ionised grains known as *fogging*. Normally, both effects should be avoided or, at least, minimised. Because of background radiations and chemicals, fogging tends always to be present and to increase with the age of the film.

The first stage of the chemical processing is the *development* stage in which the film, still maintained in a dark environment, is placed in a solution of developer. This solution penetrates through the emulsion to the silver bromide grains where it reduces silver ions to atomic silver. The process affects ionised grains more quickly than the rest so that, if terminated at the appropriate time, it converts ionised grains into black deposits of silver and the latent image into a visible image. The time required for development depends on the nature and purity of the developer as well as its temperature, but is typically about four minutes. Developer solutions contain many components, including a reducing agent, which is usually an aromatic organic compound; alkali and buffer materials to maintain a constant pH, which affects development time; sulphites, to prevent oxidation by air; and other additives. A number of proprietary solutions are available and should be used as recommended by the manufacturers of the film.

Following development, the film is transferred to an acid stop bath for about 30 seconds. This is a weak acid such as acetic acid, and stops the development stage by lowering the pH value of the emulsion. The next stage involves *fixing* the emulsion and can be carried out in a normal, ambient light environment. The film is placed in a solution of fixer, such as sodium thiosulphate, or 'hypo', which dissolves out all the undeveloped silver bromide leaving only the visible image formed by the silver deposits. It also hardens the emulsion to make it more permanent and less susceptible to chemical interaction with its environment. For a really durable record, fixing should continue for about 30 minutes but less permanent effects can be produced in a shorter time. The minimum time is that required for all the excess silver bromide to be removed and is usually a few minutes. The process can be observed visually and terminated when the image becomes clear and sharp. After the fixing process, the film is washed free of chemicals in clean running water, for about one hour, then allowed to dry slowly at room temperature.

The end product is a *negative* image in which the darker areas denote the regions of higher incident radiation intensity. In general, this is quite satisfactory for ionising radiations, since there is no psychological need to associate light areas with high irradiation, but in photography it means that bright parts of the object come out as dark parts of the image. In photography, a *positive* print is made from the negative by shining light through it onto a photosensitive paper to reverse the light–shade distribution. The paper is coated with a silver bromide or chloride emulsion which is then developed and fixed as for a film.

Automatic processing

There are two forms of automatic film system. One is widely used in radiography, to provide negative transparencies or prints, while the other, the Polaroid Land system, is employed in photography to produce 'instant' positive prints. Both utilise special emulsions which contain silver bromide grains and a developer. After exposure to radiation, the emulsion is activated by a strongly alkaline solution, which penetrates the emulsion and allows the developer to convert the latent image into a visible, negative image.

In the radiographic system, the emulsion may be coated onto a transparent base or a paper base. The latter has the advantage that the final print can be viewed directly, without the need for transillumination. After exposure, the film or paper is transferred, in a darkroom, to a processing machine. Here it is passed through a series of rollers to make uniform contact, first with the activator solution, then with a stabiliser solution. In the latter, the unexposed silver bromide is converted to a colourless and virtually light-insensitive compound. The whole process takes less than about ten seconds, after which the transparency or print can be viewed. For complete permanency, it can be fixed in the usual way.

In the Polaroid Land system, the activator is contained in a pod which is broken as the film is pulled out of the camera, between rollers. At the same time, the film is brought into close contact with a sensitive paper. The activator develops the exposed film to form a negative image and releases the unexposed silver bromide. This diffuses out of the film and onto the adjacent surface of the sensitive paper where it is fixed to existing silver atoms to form a positive print.

10.2 PROPERTIES OF FILMS

The most important property of a film, and its major advantage over other imaging systems, is its excellent intrinsic spatial resolution. The position of an incident particle or photon is recorded as being somewhere within the silver bromide grain rendered developable, so it is defined to within the diameter of one grain. Thus, spatial resolution is of the order of 1 μm. In screened film, the resolution is somewhat larger, for, although the grain size in a fluorescent screen is also of the order of a micrometre, there is some increase in the size of the light pulse as it spreads outwards towards the film. A second advantage of the film system is that it is relatively inexpensive and uncomplicated, especially with regard to data readout. It provides a direct visual record with no need for electronic signal processing. Both properties have been influential in establishing the film as a standard radiation imaging system and the latter in particular has been a major factor in its adoption for use in personnel dosimetry.

As for the disadvantages, the film is a thin detector, which saturates after a certain amount of irradiation, so its detection efficiency is quite small and non-linear. It has no intrinsic ability to measure the energy or the type of incident radiation, nor the time of incidence. In addition, it does not satisfy the increasing need for computer-compatible signal output. These intrinsic disabilities can be overcome, to some extent, with external assistance. For example, gamma- or neutron-sensitive screens can be used to increase sensitivity to those radiations, cameras provide some degree of time measurement, defined by the shutter speed, cine cameras allow dynamic studies to be undertaken, and densitometers can be employed to convert visual data into electrical signals. It is even possible to measure energy, to some extent, by inserting appropriate filters in front of the film. Despite these arrangements, however, films do not compare favourably with some other detectors in these respects.

Detection efficiency

The efficiency with which a film responds to incident radiations can be measured and defined in one of two ways. The first of these utilises the concept of absolute *detection efficiency*, η, or the relative *interaction efficiency*, R, defined in Chapter 5. The latter is more often referred to as the *quantum efficiency* of the film and is the fraction of incident radiations that produce an interaction in its sensitive volume. This parameter is useful for comparing the response of a film with that of another detector, but it depends on the measurement of the number of interactions that have taken place. If it is assumed that one interaction produces one developable grain, then the number of grains rendered developable, N, is the recorded number of counts.

In practice, however, N is not directly measurable. The response of a film is usually described in terms of degree of blackening, or *optical density*, d, rather than in integrated counts. For this reason, the more common method of representing film response is by its *sensitivity* or *film speed*. These specify a quantity of radiation required to produce a given optical density.

Optical density, averaged over a small area of film, can be measured by means of a densitometer in which the area is illuminated by visible light of intensity I_0 photons (or any other units) per cm^2 and transmits an intensity I photons per cm^2. For light photons, absorption is exponential, so that these intensities are related by the expression

$$I = I_0 e^{-Na} \tag{10.1}$$

where N is the number of developed, opaque grains per cm^2 and a is the average cross-sectional area of one grain, in cm^2.

The product Na is the total obscured area per cm^2. It may be greater than 1.0 because the grains may overlap each other. In this case, some light is transmitted through the film by being scattered, in the emulsion, round the grains. Optical density is proportional to Na but is usually defined in terms of the measurable quantities I and I_0, as

$$d = \log(I_0/I) \tag{10.2}$$

which, from equation 10.1, reduces to

$$d = (\log e)Na = 0.43 Na \tag{10.3}$$

which shows the relationship between optical density, d, and the total recorded counts, N.

There is, however, a further complication. Unlike an electronic device, a film has no dead time or coincidence counting losses. Nevertheless, it does not record all the interactions that take place in its sensitive volume. The reason for this effect is, essentially, position-coincidence losses. A silver bromide grain, having recorded one count, becomes insensitive to further interactions so that the number of sensitive grains decreases with continued exposure to radiations, as does the data recording efficiency, until both reach zero when all grains have become developable and the film is saturated. If the total number of silver bromide grains per cm^2 is N_0 and the number of grains per cm^2 rendered developable at a given time during an exposure is N, then the number of remaining sensitive grains per cm² is $N_0 - N$. A further exposure of $d\Phi$ particles or photons per cm² produces an increase dN in the number of developable grains, given by

$$dN = \sigma(N_0 - N) \cdot d\Phi \tag{10.4}$$

where σ is the interaction cross-section per grain, describing the probability that an interaction will take place and produce a developable grain. If, as is approximately the case, an incident particle interacts with only one grain and one such interaction makes the grain developable, then σ is equal to a, the projected grain area. Equation 10.4 is a differential equation whose solution gives the number of developable grains, or the recorded counts, N, as a function of the total, integrated radiation intensity, Φ per cm². Φ is known as the *particle fluence*, and the solution of equation 10.4 is (assuming $\sigma = a$)

$$N = N_0(1 - e^{-a\Phi}) \tag{10.5}$$

From equation 10.3, optical density is given by

$$d = 0.43 N_0 a (1 - e^{-a\Phi}) \tag{10.6}$$

so that both N and d increase exponentially with Φ, rather than linearly as would be the case for an electronic counter.

The response of a film to different quantities of radiation can be described by equation 10.6, or by the graph of d against Φ illustrated in *Figure 10.3*, but

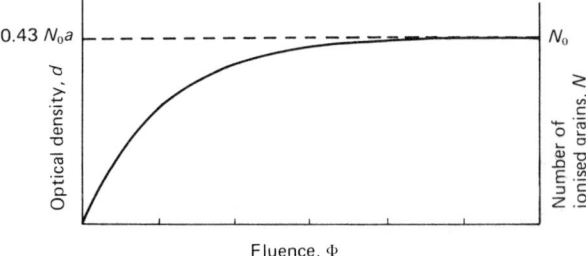

Figure 10.3 Graph of optical density and the number of ionised grains against particle fluence

this representation is inconvenient in some respects. First, the applications of films to radiography and dosimetry make it more useful to define radiation quantity in units of exposure dose, X, rather than particle fluence, Φ. For charged particles, absorbed dose, D, is more appropriate. Secondly, it is necessary to define the contrast of an image by comparing the optical densities for a range of exposures within the image. This should be the gradient of the d–X graph, but the gradient does not have a constant value. For this reason it is standard practice to plot a graph of optical density, d, against the logarithm of the exposure. Such a graph is known as the *characteristic curve* of the film. It has a linear region over which d is proportional to $\log X$ and the gradient of d against $\log X$ is constant. Typical characteristic curves are shown in *Figure 10.4*.

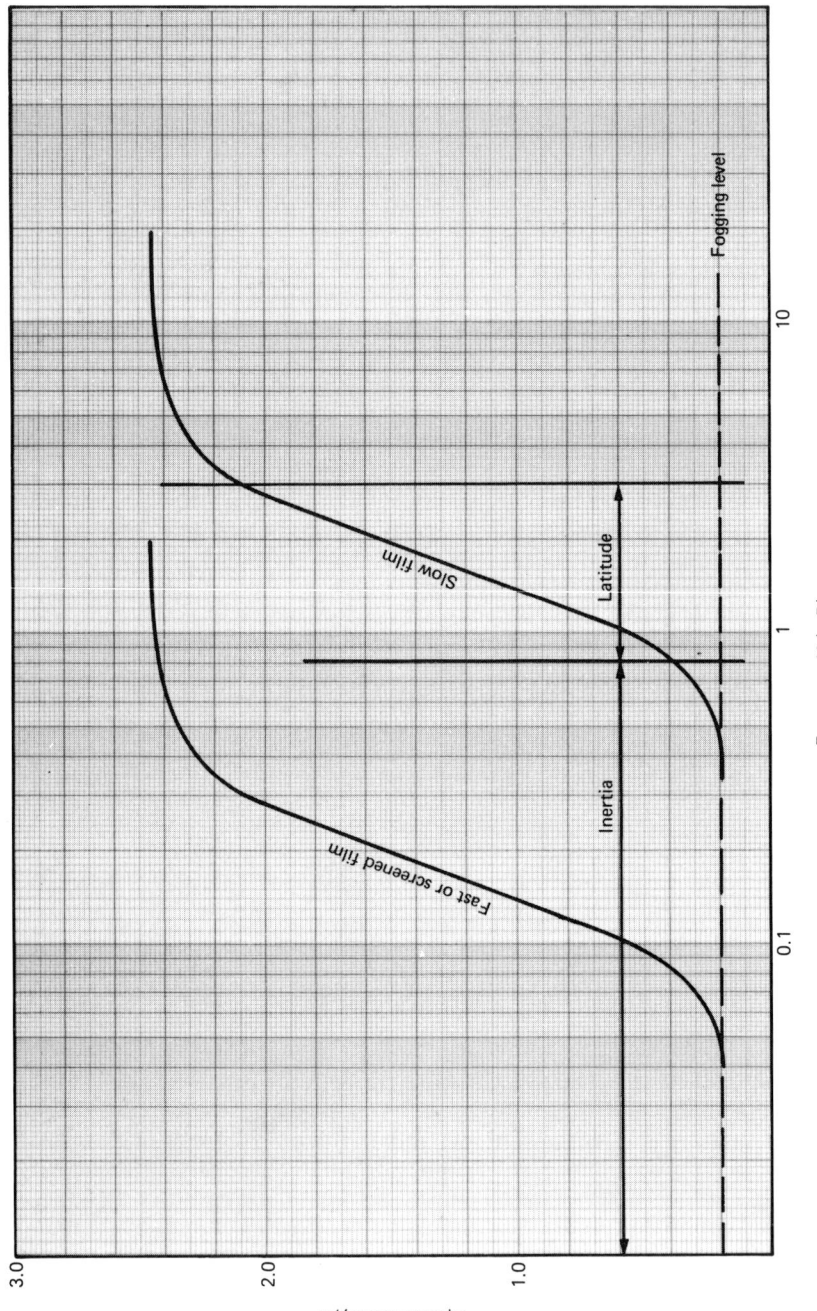

Figure 10.4 Characteristic curves of films

Properties of films 275

The useful range of exposure is defined by the length of the linear part of the characteristic curve and known as the *latitude* of the film. For unscreened film, this region may extend from about $X = 0.8$ mR to $X = 3$ mR, or from $d = 0.3$ to $d = 2.0$. The limited latitude of a film is a major disadvantage and one that is familiar to all photographers. It means that exposure must be predetermined and controlled to ensure that the optical densities of all parts of the image lie within the latitude range. If the film is overexposed, the image density range encroaches on the top, flat end of the curve where there is no density variation with exposure and no contrast in high-density areas of the image. If the film is underexposed, contrast is lost at the low-exposure areas of the image. At extremely high exposures, the optical density begins to decrease with further exposure because of radiation damage to the silver bromide crystals. This effect is known as *solarisation*. At the other end of the curve, d flattens out to a threshold level of about 0.1 or 0.2. This optical density is uniformly distributed across the film and is due to *fogging*.

The *contrast*, γ, of a film determines the range of optical density generated, within an image, from a given range of exposure. If d_1 and d_2 are two values of optical density within the linear part of the characteristic curve, and X_1 and X_2 are their corresponding exposures, then the contrast is given by

$$\gamma = \frac{d_2 - d_1}{\log X_2 - \log X_1} \tag{10.7}$$

Two other useful parameters are the *inertia* and the *film speed* or *sensitivity*. The inertia may be defined as the minimum exposure required to reach the linear part of the characteristic curve, or as the exposure value to which the linear part extrapolates. It describes the minimum exposure that should be present in an image. Film speed and sensitivity are related to inertia and defined as the reciprocal of exposure required to produce a particular density, usually at the bottom of the linear part of the characteristic curve. In photography, exposure, X, is measured in metre-candle-seconds for an optical density of 0.1 above the fogging level. There are two common systems:

$$\text{ASA/BSI:} \quad \text{Speed} = \frac{0.8}{X}$$

$$\text{DIN:} \quad \text{Speed} = 10 \log \left(\frac{1}{X}\right) \tag{10.8}$$

For X-ray films, exposure is measured in röntgens for an optical density of 0.3 and given by

$$\text{Sensitivity} = \frac{1}{X} \ \text{R}^{-1} \tag{10.9}$$

At the relatively low exposures for which film sensitivities are defined, equation 10.6 reduces to

$$d = 0.43 N_0 a^2 \Phi \tag{10.10}$$

This shows that, for a fixed value of d, the reciprocal of the fluence, Φ, and, therefore, the film sensitivity, $1/X$, is proportional to N_0 and a^2. In photographic emulsions, film speed increases with the number of grains per cm^2 and with grain size. For the more penetrating ionising radiations, it also increases with emulsion thickness, since N_0 increases with emulsion thickness, and is largest for double emulsions. Screened films are even faster because each light flash in the fluorescent screen affects a large number of grains in the emulsion. In this case, the optical density, given by equation 10.6, must be multiplied by a *screen factor* equal to the average number of grains affected by each scintillation. Typically, this factor ranges from about 10, for high-resolution screens, to about 100, for high-efficiency screens.

Fast films, and especially screened films, provide a greater amount of blackening than slow films, for a given exposure and for a given interaction efficiency. They also tend to produce more contrast in an image exposure range. This is a useful feature in X-radiography involving biological tissue, where low exposures are preferred and small ranges of transmitted exposures are obtained. In other radiographic techniques, the range of Z numbers and densities present in the specimen is usually larger so that less contrasting, slower films are required to reproduce structural details. Slow films offer better spatial resolution than fast films and produce better image quality.

According to the theory developed so far, optical density depends on total exposure and not on exposure time. This hypothesis is known as the *reciprocity law*. It suggests that if, as is usually the case, exposure rate per second is constant, then optical density is proportional to exposure rate and to exposure time. Thus, for example, doubling the exposure rate and halving the exposure time results in the same optical density. The law is not accurate for visible light, nor very high radiation exposures, but it is reliable for most applications involving ionising radiations. It is particularly useful in X-radiography, where exposure rate is proportional to machine current, for a fixed operating voltage, and easily adjustable. It does not apply to screened film, which depends on the detection of visible light from the screen.

Another assumption that is implicit in the above discussion is that the characteristic curve and the sensitivity of a film are constant properties, independent of the nature or the energy of incident radiations. This is not true. Exposure is a measure of a quantity of electromagnetic radiation defined in terms of its effect on air. Its effect on a silver bromide emulsion may be quite different. In fact, optical density is more closely related to the energy deposited in the silver bromide and, therefore, to the absorbed dose, than to the exposure. In other words, sensitivity can be regarded as the reciprocal of the exposure required to deposit a specified absorbed dose, D, where D is the dose, in rads, required to produce an optical density, d, of 0.3.

For *photons*, the relationship between dose and exposure is given by equation 6.21 and leads to the formula

$$\text{Sensitivity} = \frac{1}{X} = \frac{0.89}{D} \times \frac{\mu_{mE}}{\mu_{mE,\,air}} \tag{10.11}$$

where μ_{mE} and $\mu_{mE,\,air}$ are the mass–energy attenuation coefficients, or mass absorption coefficients for silver bromide and air, respectively.

The ratio of these absorption coefficients and the relative sensitivity of a typical film, as a function of photon energy, are shown in *Figure 10.5*. The two curves are similar, both showing maxima at about 40 keV, although the latter is slightly flatter. This is due to the fact that many photon interactions in the emulsion produce energetic electrons which reach, and ionise, silver bromide grains. The effective absorption coefficient, μ_{mE}, is partly that of the emulsion as well as the silver bromide and more similar, therefore, to that of air. For a screened film, sensitivity is proportional to $\mu_{mE}/\mu_{mE,\,air}$, where μ_{mE} is the absorption coefficient for the screen. For plastic screens, the component Z numbers and the absorption coefficient are very similar to those of air, so sensitivity is almost constant and independent of photon energy. For metal converters, the sensitivity is a maximum near the K absorption edge of the metal employed.

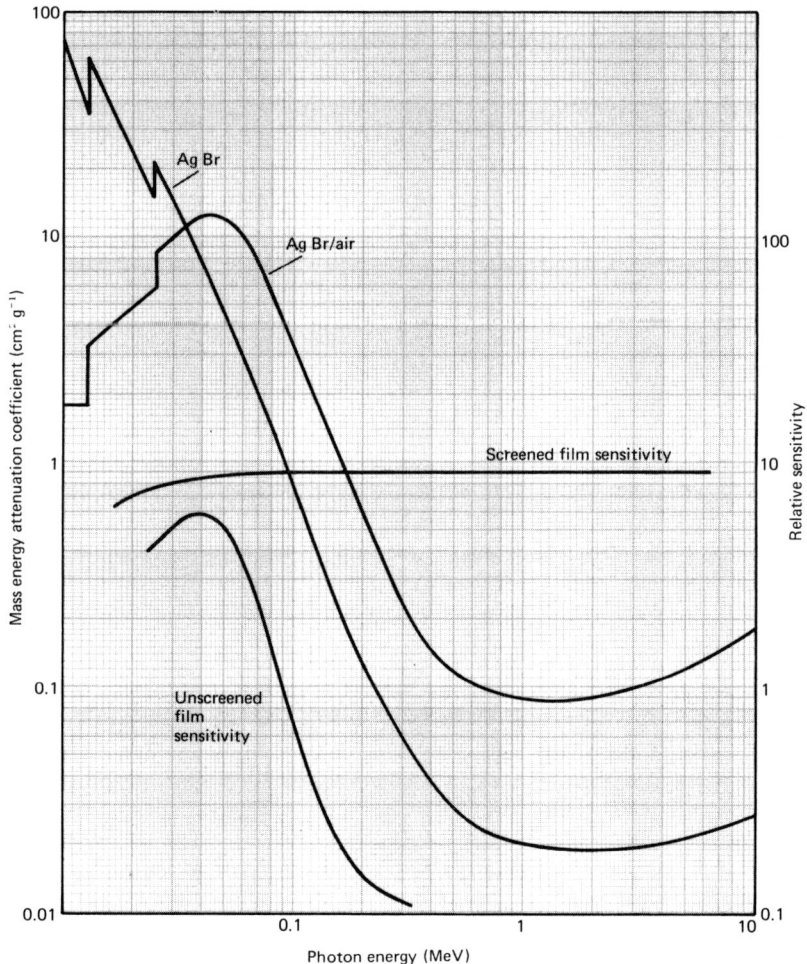

Figure 10.5 Mass energy attenuation coefficient for silver bromide and its ratio to that of air, along with typical relative sensitivities of screened and unscreened film

Sensitivity to charged particles may be defined as the reciprocal of the absorbed dose, D, required to produce an optical density $d = 0.3$. Thus, as shown by equation 6.18, sensitivity should decrease with increasing LET, or stopping power, dE/dx. This effect is observed at high energies, but at low energies the finite particle range is the dominant factor, as illustrated for electrons in *Figure 10.6*. At low energies, sensitivity increases with energy, as range and penetration into the emulsion increase and more silver bromide grains become accessible. The thickness of an ordinary emulsion is about 4 mg cm^{-2} and about equal to the range of a 50-keV electron.

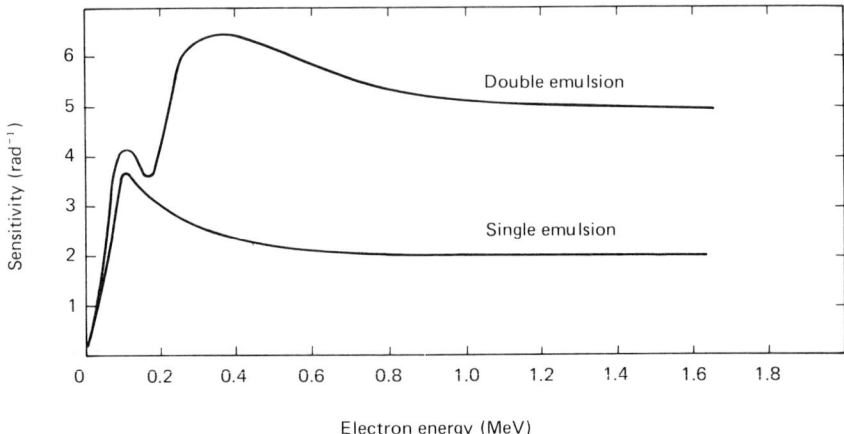

Figure 10.6 Film sensitivity as a function of electron energy

Sensitivity increases to a maximum at about 100 keV, where electron range and dE/dx are optimal, then decreases with increasing dE/dx. Since dE/dx is fairly constant at these energies, the decrease is slight. For double emulsions, two peaks are observed, the second representing optimum penetration of the film base and the second emulsion.

The absolute detection efficiency of a film may be measured in terms of its *recorded interaction efficiency*, R_D, given by

$$R_D = n/\Phi \qquad (10.12)$$

where n is the number of recorded interactions per cm^2 and Φ is the integrated radiation fluence per cm^2.

If one recorded interaction corresponds to one developable silver bromide grain, then n is equal to N, the number of developable grains produced, per cm^2. In practice, however, an incident particle or photon usually ionises several grains so N is much larger than n, and does not provide a useful measure of interaction efficiency. In fact, interaction efficiency is not a particularly appropriate parameter since image density is determined by N and a; that is, by film speed, rather than by the number of recorded events.

For *photons*, the instantaneous recorded interaction efficiency, at the beginning of an exposure, is given by the usual formula, $1 - e^{-\mu x}$, where μ is the mass attenuation coefficient and x the equivalent thickness of silver bromide. In a 5 mg cm^{-2} emulsion containing 40% silver bromide, this is about 2.5% for 40-keV photons. As the exposure continues and the number of sensitive grains decreases, the recorded interaction efficiency also decreases. For a typical image, with an average optical density, d, equal to 1.0, the integrated value of R_D is about 1.5%.

For *charged particles* with sufficient range to penetrate to all the silver bromide grains, R_D is of the order of 100%.

10.3 FUNCTIONS

Dosimetry

The *film badge* is the most widely used form of personnel dosimeter. It consists of a small piece of film in a light-tight package, contained in a *film badge holder*, and it is sensitive to beta, gamma, X- and thermal-neutron radiations. Special badges are available for fast-neutron dosimetry.

For personnel dosimetry, the film has a number of advantages over alternative systems. It is small, inexpensive, reliable and easily available. It provides a permanent record of exposure, needs no special expertise or equipment to use, and is quite easily processed. By itself, a film does have a number of disadvantages. The main problem is the fact that its sensitivity depends on the nature and the energy of incident radiations so there is no unambiguous relationship between optical density and absorbed dose or exposure. In other words, it is not completely tissue-equivalent. A second disadvantage is the limited latitude. It is insensitive at very low exposures and saturates at very high exposures, losing all sensitivity to further irradiation. A third disadvantage is that the information regarding exposure is received retrospectively and not, as it should be, during the exposure. There is little that can be done to overcome the last of these, apart from employing area monitors when large exposures are expected, but the film badge is designed to reduce the effects of the first two problems.

One type of film badge is illustrated in *Figure 10.7*. The film can be selected from a range of available sensitivities to suit the user requirements. As a safeguard against unexpectedly high exposures, it may have two emulsions, one fast and the other slow. The fast emulsion is sufficient for most purposes but, if it saturates, the slow emulsion records higher exposures.

The holder is designed to identify different types and energies of incident radiation by selective absorption and conversion in a number of filters. In the example shown, the holder is made of 300 mg cm^{-2} plastic which stops beta radiation. The 50 mg cm^{-2} window admits beta particles with energies greater than about 0.5 MeV and the open window admits all beta radiation. Any film blackening behind these regions, in excess of that recorded in the rest of the film, identifies beta radiation and provides an approximate measure of beta energy. Soft X-rays penetrate these windows and the 300 mg cm^{-2} plastic, but not the metal filters. Penetration of the Dural filter increases with photon energy and is almost 100% at about 150 keV. At higher photon energies, penetration through the lead filters increases and becomes about 100% at gamma ray energies approaching 1 MeV. The cadmium is a converter for thermal neutrons, producing gamma rays and internal-conversion electrons which are detected in the film. The tin filter is designed to provide the same gamma ray attenuation as in the cadmium, so that any excessive blackening behind the latter can be ascribed, unambiguously, to thermal neutrons. The indium strip provides an immediate response to very

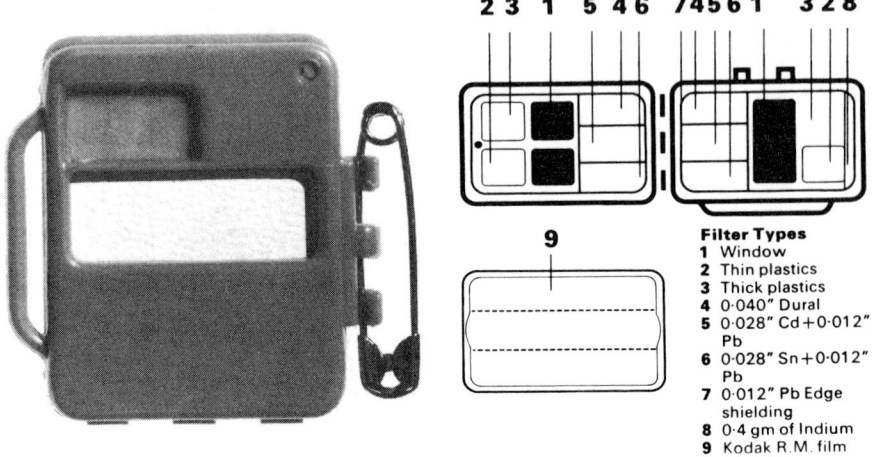

Figure 10.7 A film badge for measuring the level of radiation exposure to which a person may be exposed. This contains a photographic film which is affected by radiation in the same way such film is affected by light. Film badges are issued to all people who may be exposed to radiation during the course of their work. (Courtesy of AERE Harwell and British Nuclear Fuels Ltd)

high thermal-neutron exposure. The induced radioactivity can be monitored with a Geiger–Müller counter. Fast neutrons can be monitored, although not very conveniently, by means of proton recoil tracks from plastic inserts or in special badges with thick plastic film bases.

Imaging

Film systems are recognised as standard equipment for a number of bidimensional imaging processes, including X-radiography, X-ray crystallography, autoradiography, electron microscopy and neutron radiography. They are also used, in the form of nuclear emulsions, for particle tracking measurements in nuclear and high-energy physics, although they are no longer the most popular choice for this application.

In all of these applications, films have two major advantages over other detection systems. First, they have an intrinsic spatial resolution of a micrometre or less, which is as good as that of the human eye and is not bettered by any other detector. Secondly, they are convenient; that is, they are inexpensive, commercially available, simple to use and easy to process. Compared with electronic counters, they provide an integrated record with zero dead-time losses so they are particularly suitable for high count rate, short time exposure applications such as X-radiography.

On the other hand, films do have a number of disadvantages. One of these is the limited latitude, which produces a non-linear response and makes it necessary to predict and control exposures. A second problem is the low interaction efficiency for photons. This is accepted in X-radiography and it is less important in industrial radiography, where large doses to the specimen can be tolerated, but it means that films are not suitable for isotope imaging, as gamma cameras.

Another disadvantage is the lack of a computer-compatible output signal. This precludes the use of films in the increasing number of applications, such as computerised tomography and nuclear track recording, in which data processing is a major element. Finally, films do not identify radiation type, nor do they measure energy, so they are not selective and are unable, for example, to distinguish between signal and background radiations.

In practice, spatial resolution is determined by extrinsic as well as intrinsic factors and, with films, the former are usually the larger. In X-radiography,

Figure 10.8 Typical images obtained with film of a leaf: (a) Optical photograph of the underside of the leaf. (b) X-radiograph at 15 kVp. (c) Autoradiograph (plant grown in ^{32}P). (d) Scanning electron micrograph of underside of the leaf, mag. ×200. (Courtesy of M. N. Richardson, North East London Polytechnic)

spatial resolution depends on the source spot size and on the source—sample geometry. For most applications, an image resolution of about ten lines per millimetre is regarded as satisfactory but better resolution can be achieved if required. In X-ray crystallography, the uncertainty in diffraction angle again depends on source spot size and source—crystal—film geometry. It is equal to the angle subtended by the source spot at the sample. In electron microscopy, very good detector resolution is required to avoid degrading the resolution of the electron optics. In transmission electron microscopy, electrons are imaged directly on film, while in scanning electron microscopy, the beam is usually incident on a fluorescent screen, which is photographed. Spatial resolution is limited to about 5 to 10 μm by electron scattering in the emulsion.

The other applications have a common geometry as far as the imaging process is concerned. In all cases, the source is effectively a thin layer in close contact with the film and spatial resolution is determined by the lateral spread of radiations from each point in the source. In autoradiography, the source is the labelled specimen. If it is labelled with tritium, then the short range of the betas emitted enables a resolution of about 2 μm to be obtained. For more energetic betas, from ^{14}C and ^{35}S, 100 μm can be achieved. For photons, the resolution obtained in autoradiography is usually inadequate. In gamma radiography, metallic converters provide photoelectrons whose lateral spread in the film depends on their range. The same is true of the converters used for neutron radiography. Gadolinium, for example, produces photons and internal-conversion electrons, which are imaged in the film.

Some examples of film images are shown in *Figure 10.8*. These illustrate the image quality that can be obtained with films.

10.4 FLUORESCENT SCREENS

As mentioned at the beginning of this chapter, the film is almost the oldest type of system used to detect ionising radiations. In fact, the oldest system of all is probably the barium platino-cyanide fluorescent screen used by Röntgen in 1895 to detect his newly discovered X-rays.

The fluorescent screen is a thin scintillator of large area. Its response to individual radiations is very weak, as is the case for all scintillators, so that without external assistance, single events are not easily visible. In addition, it is not generally used to count events and should be regarded as a scintillation detector rather than a scintillation counter. It is used for the bidimensional imaging of incident radiations.

In principle, any thin scintillator can be regarded as a fluorescent screen, and, for example, plastic screens loaded with suitable target isotopes are used to image thermal neutrons. In practice, however, most screens have a granular, crystalline structure. These provide the largest and fastest light output and, since each scintillation is mainly contained within one grain, they offer the best spatial resolution. Grain size and intrinsic spatial resolution are of the order of 1 μm, as for films.

If the incident radiation beam is sufficiently intense, the fluorescent image may be bright enough to be viewed directly, as is the case, for example, in a cathode ray oscilloscope or a television receiver. For optimum efficiency, the spectral output of the fluor should match the sensitivity of the eye, which is at its maximum in the blue-green wavelength range. Zinc cadmium sulphide, for example, meets this requirement and is widely used in X-radiography, or *fluoroscopy*, as it is often called. In such applications, the screen must be viewed through a thick glass window to avoid overexposure on the part of the observer. Alternatively, in *fluoroscopy*, the screen may be photographed.

There are other methods of viewing and recording fluorescent images indirectly. One of these, involving the combination of screen and film, has been described above. The others provide image intensification so that less intense beams of radiation may be employed. These include the use of *image intensifiers, channel plates* and *television cameras*.

An arrangement based on an *image intensifier* is illustrated in *Figure 10.9*. Various types of image intensifier are available but in principle they are all similar to the scintillator–photomultiplier assembly used in scintillation counters.

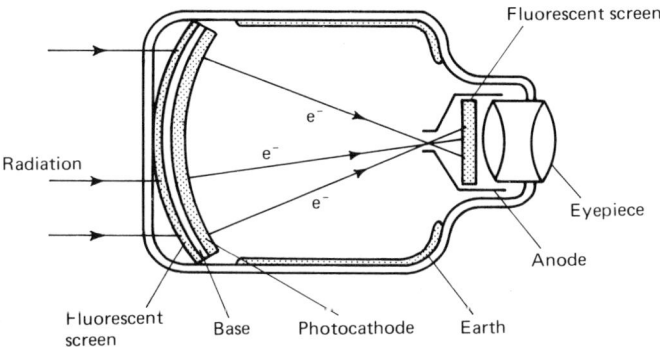

Figure 10.9 The structure of an image intensifier

In the example shown, the fluor is mounted on a thin transparent base in an evacuated tube. Scintillations are transmitted through the base to a photoemissive coating. This is a photocathode, from which photoelectrons are accelerated along a focused, radial electric field towards an anode. They pass through an aperture in the anode to be imaged on a second fluorescent screen which is viewed by the observer, or photographed. The acceleration process provides the electrons with additional energy so that the image on the viewing screen is brighter than that on the primary screen, by a factor of up to 10^3. Other types of image intensifier employ intermediate electrodes, at which secondary electron emission takes place, to increase the gain. Further increases in gain can be achieved by coupling two or more intensifiers in series. The secondary screen of one is connected to the primary screen of the next intensifier by a fibre-optic coupling system, or a lens.

Multistage image intensifiers can provide a net gain of the order of 10^6, limited by dark current and other statistical effects which reduce image quality. Their disadvantages are that they are bulky and require very high applied potentials. They also suffer from some image distortion because of edge effects in the focusing electric fields.

An alternative to the image intensifier is the *channel plate* system illustrated in *Figure 10.10*. This consists of a glass disc, typically about 0.5 mm thick, perforated with a large number of holes, each with a diameter of about 10 μm.

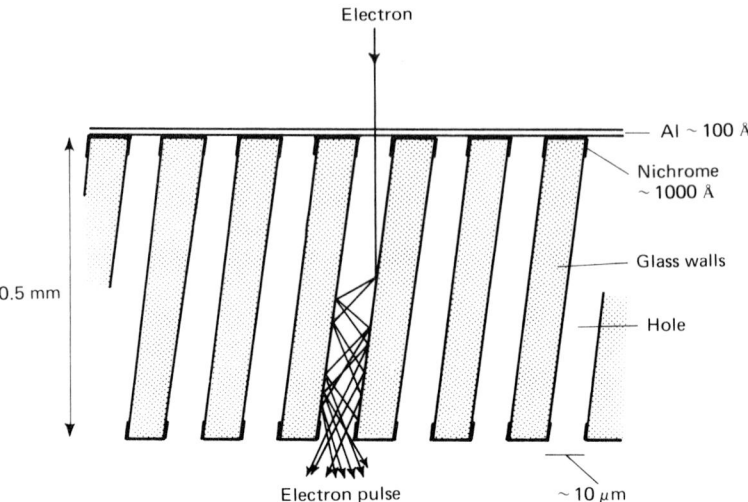

Figure 10.10 A cross-section of the structure of a channel plate

Each surface of the disc is coated with a conductor, which does not block the holes, and a potential difference is applied across the disc. The internal surfaces of the holes are coated with a sensitive layer which produces secondary electrons. The holes are angled slightly so that all incoming electrons must strike the emissive surfaces. Secondary electrons are accelerated along the holes by the applied potential to produce a net electron gain of about 10^3, and a channel plate, in a single image intensifier, between the photocathode and the electron focusing field, results in a total gain of the order of 10^6.

The screens can also be examined by *television cameras*. These too come in a variety of types offering large image gains of up to about 10^5. In conjunction with television receivers, they allow considerable control of image density, contrast and magnification. They can provide videotape recordings and they can be operated remotely, so that the screens can be located in areas of high radiation intensity. Consequently, they represent a convenient method of monitoring accelerator beam profiles and reactor neutron fluxes.

In general, fluorescent screens offer no improvement over films, in terms of spatial resolution or detection efficiency. They do, however, have higher 'film speeds' and they generate immediate output signals. This latter property is a necessity for reactor or accelerator flux monitoring and is useful in dynamic radiography, in which the effect of patient or specimen movement can be studied. They are also used in mass miniature radiography, in which a fluorescent screen is photographed, without intensification, by an ordinary camera. This technique results in the patient receiving a relatively high dose, but it is convenient for large-scale screening operations such as are carried out by chest X-ray units.

10.5 XEROGRAPHY

The Xerox process has been used for some time as a means of copying documents; that is, of imaging patterns of visible light. Recently, it has been developed as a method of detecting and imaging ionising radiations, especially X-rays, and a commercial system for *xeroradiography* is now available.

The detector is a photoconductive layer of amorphous, or vitreous selenium, mounted on a conducting plate in a light-tight cassette. The selenium has a discontinuous energy band structure characteristic of an amorphous material. Its energy gap is about 2.3 eV, so electrons are easily excited, by visible light, from the valence band to the conduction band, forming loosely associated electron–hole pairs. It is most sensitive to blue light but it also responds to X-rays, for which the mean energy required to form an ion pair is about 7 eV.

Before exposure to radiation, the selenium is given a uniform electrostatic charge, usually by passing the plate beneath a fine wire maintained at a high electric potential so that a corona discharge takes place between the wire and the plate. The process is completed, and the plate is transferred to the cassette, in light-tight conditions. When the plate, in its cassette, is exposed to X-rays, ion pairs are formed in proportion to local radiation intensity. As illustrated in *Figure 10.11*, these reduce the resistivity of the selenium, allowing the surface charge to leak away to the backing plate, so that the final charge distribution forms a latent image of the incident radiation pattern.

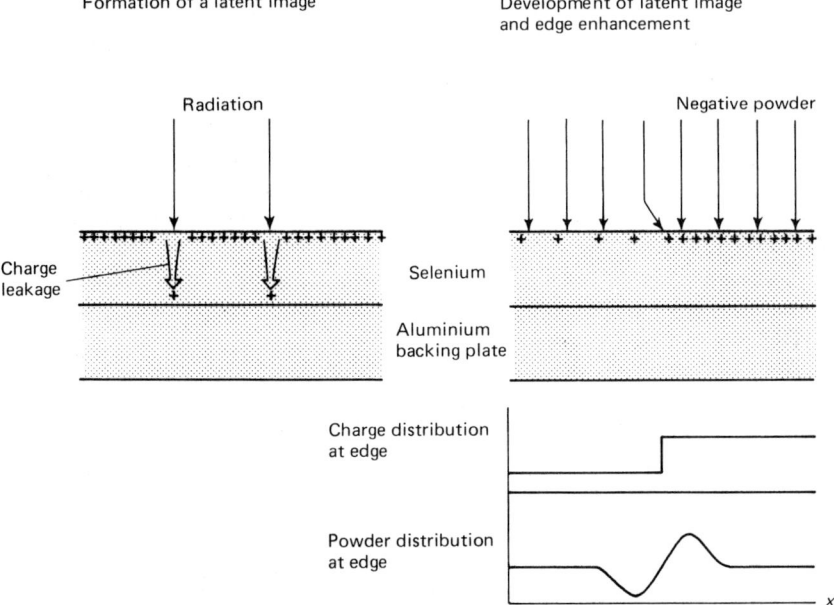

Figure 10.11 The xerographic process: formation and development of the latent image

The latent image can be developed by spraying the plate with a pigmented, charged powder. This adheres to the selenium surface, in proportion to the charge density, to form a visible image, which is then transferred to a specially prepared paper to yield a permanent copy of the image.

In the early stages of the development process, the charged powder settles almost uniformly on the latent charge image. There is very little discrimination between regions of high and low charge density; that is, as much powder is deposited on one as on the other. Half-tone reproduction is not only poor but almost non-existent. Continued development overcomes the problem, as the charge capacity of each image region saturates and, eventually, powder density matches charge density. There is, however, some advantage in terminating development at an early stage. Half-tone reproduction is very poor but an interesting effect, known as *edge contrast*, or *edge enhancement*, is produced instead. This results in an image whose structural lines and edges are emphasised at the expense of half-tone quality and the product is more like a line drawing than a radiograph.

The cause of edge enhancement is illustrated in *Figure 10.11*, which shows the electrostatic field configuration near a discontinuity in the latent image charge distribution. The field lines are uniformly parallel except in the region close to the discontinuity, where they deflect towards the area of higher charge density. Charged powder particles experience no lateral force unless they are incident within this edge region, in which case they are deflected towards the area of higher charge density. Thus all the lines or edges within the image are

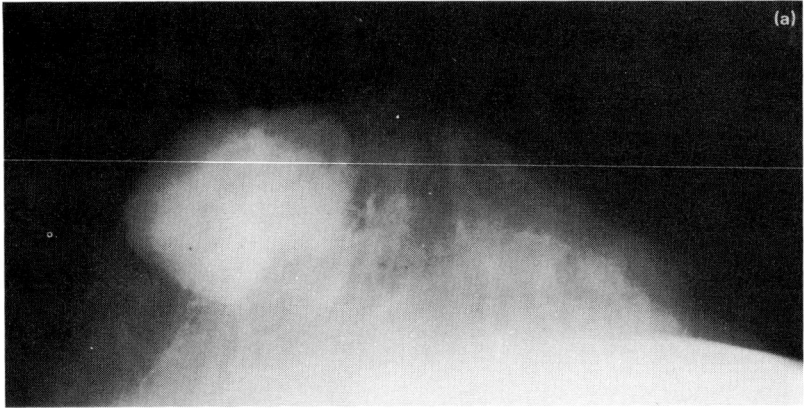

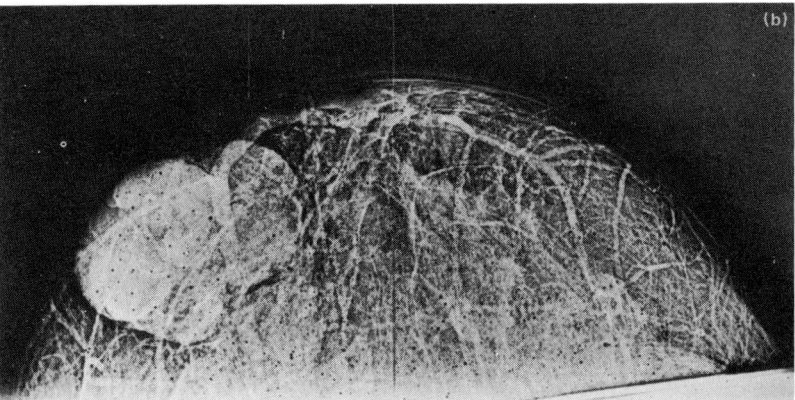

Figure 10.12 (a) A film radiograph and (b) a xeroradiograph showing a breast tumour. The comparison is unfair since more can be seen in looking through the film than is apparent on the print but it does illustrate the effect of edge contrast. (Courtesy of Prof. J. W. Boag)

prominently marked by narrow, high-density powder deposits, while other regions remain uniformly developed.

The effect of edge contrast is illustrated in *Figure 10.12*, which compares a radiograph of a breast taken by xeroradiography with one obtained by means of standard film technique. Since the latter should be viewed by transillumination for optimum effect, the comparison is not entirely fair, but it does demonstrate the specific nature of edge enhancement. It is particularly useful for revealing fine, overlapping structures in soft tissue where there is little variation in density and Z number. For this reason, xeroradiography is widely used for mammography.

Xerography is still under development, to some extent, so it is difficult to make categoric statements regarding its merits as a radiation detection system. At present, it provides spatial resolution as good as that of a film. Its sensitivity is slightly better than that of unscreened film but poorer than that of screened film. In some ways, it is more convenient than film techniques. In the commercial system, marketed by the Xerox Corporation, the plate is cleaned, charged and sealed in its cassette in a *conditioner* unit, then, after exposure, it is developed and copied in a second, *processor* unit. The process is dry, quick and easily controlled. Positive or negative prints can be obtained and the degree of edge enhancement adjusted to meet user requirements.

10.6 IONOGRAPHY

Ionography, like xerography, is an electrostatic imaging process. It is also known as *electron radiography* (ERG). It is similar to the xerography process except for the fact that the sensitive volume is a thick volume of xenon gas, at several atmospheres' pressure, instead of a relatively thin layer of selenium. The main reason for this replacement is to improve the sensitivity of the system to the extent that it may be superior to that of screened film.

The process is based, therefore, on an integrating, imaging ionisation chamber. In order to realise the potentially high stopping power of the xenon gas, the chamber has to be quite thick; that is, with an interelectrode spacing of the order of 1 cm, or more. In principle, detection efficiency can be increased almost indefinitely by simply increasing the thickness of the sensitive volume. In practice, with plane-parallel electrodes, spatial resolution deteriorates with increasing thickness because of the smearing effects of obliquely incident radiations. This is a general problem with all thick, planar imaging devices used with a diverging radiation beam as, for example, in radiography. It is illustrated in *Figure 10.13*. A photon entering the chamber at an angle θ to the normal through the front electrode may interact anywhere along its path in the sensitive volume. If d is the thickness of this volume, the projection of this interaction will be, randomly, anywhere within a lateral spread of $d \tan \theta$. Thus, each event is recorded with an uncertainty of $d \tan \theta$ as to its point incidence.

In ionography, this difficulty is overcome by constructing a chamber with spherical electrodes whose centres of curvature are located at the focal spot of the radiations. The chamber is illustrated in *Figure 10.14*. It is a complicated arrangement but all the radiations do pass normally through the chamber electrode along the radial field. Since the photoelectrons and the positive ions released in the gas also diffuse radially along the field lines there is very little spatial smearing due to oblique incidence.

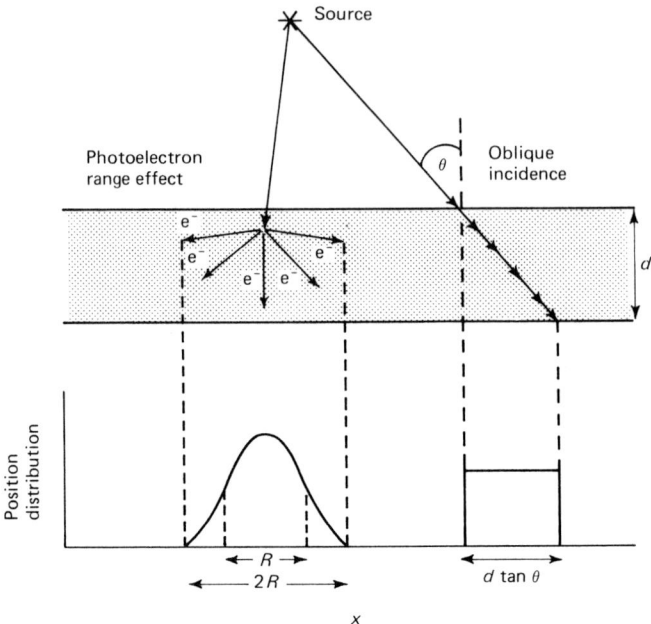

Figure 10.13 The effects of photoelectron range and oblique incidence on the spatial resolution of an imaging gas counter

A second, major contribution to spatial resolution is not eliminated by the spherical electrode geometry. This, too, is illustrated in *Figure 10.13*. A photoelectron can be emitted in any direction from the point of interaction, producing ion pairs all along its path. In the worst case, in which the photoelectron is emitted at right angles to the direction of the initiating photon, the recorded position of the event is smeared over a distance equal to the range, R, of the photoelectron. In high-density, high Z number materials, such as silver bromide and sodium iodide, this effect is not usually a major problem but in gases, electron range is quite large and increases with energy, as given in Appendix 4. For example, in argon at STP, a 40-keV photon ejects a photoelectron from the atomic K shell with a net energy of about 37 keV and a range of about 14 mm. In xenon, the photoelectron would have a net energy of about 5.4 keV and a range of less than 0.3 mm, so high Z number gases offer considerable improvements in spatial resolution. High pressures provide further improvement but, in general, ionography and other techniques employing gas counters are restricted by this factor to relatively low-energy photons; that is, to X-rays.

The chamber illustrated in *Figure 10.14* has spherical electrodes, separated by a gap of 15 mm, and contains xenon at a pressure of 10 atm. Incident photons interact with the gas to produce electrons and positive ions by the usual processes, and these are directed, by the field, towards their respective electrodes. One of these is covered by a thin plastic foil which, being an insulator, collects and integrates the charge to form a latent image. The foil can be removed from the chamber and developed by exposure to charged powder or liquid toner, as in the xerography process, to provide a final visual image of the charge distribution.

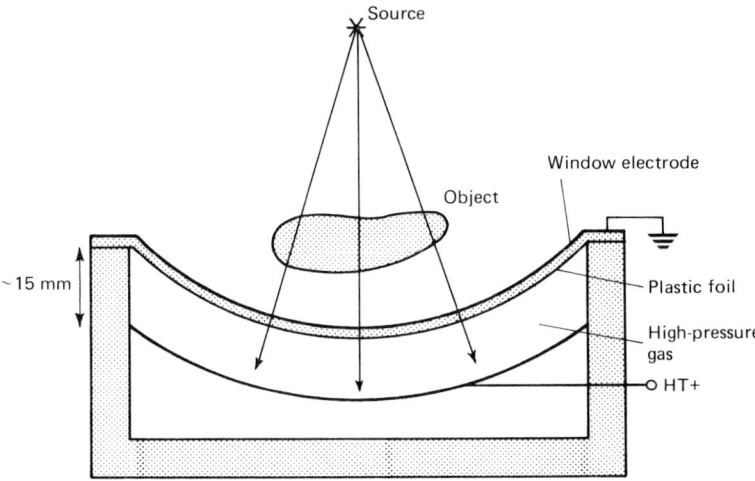

Figure 10.14 *An ionographic chamber with spherical electrodes*

Normally, the foil is stretched over the cathode to collect the positive ions. These, being less easily deflected than the electrons, produce better spatial resolution.

This type of system is capable of resolving about 10 lines per millimetre with a detection efficiency which for clinical X-ray sources is superior to that of screened film. Edge contrast is also available, although less readily than in a xerographic system, and development can be automated and completed in about a minute or less. An ionograph (or electron radiograph) is illustrated in *Figure 10.15*.

Recent improvements to the prototype system described above have included the replacement of xenon by the less expensive Freon 13B1 and the development of a closed system in which the plastic foil is mounted on the outside of the chamber and induced charges form the latent image. In addition, there is some possibility of using liquid instead of gas to improve photon stopping power. If this modification is operationally successful, the chamber should be applicable to the imaging of gamma rays in nuclear medicine.

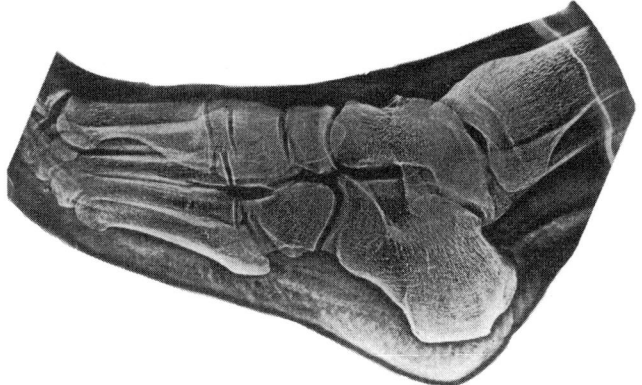

Figure 10.15 *An ionograph of a foot. (Courtesy of Prof. J. W. Boag, from Brit. J. Radiol.* **49**, *755 (1976)*

BIBLIOGRAPHY

BOAG, J. W., 'Xeroradiography', *Phys. Med. Biol.,* **18**, 3 (1973)
HERTZ, R. R., *The Photographic Action of Ionising Radiation,* Wiley–Interscience, New York (1969)
JOHNS, H. E., 'New Methods of Imaging in Diagnostic Radiology', *Brit. J. Radiol.,* **49**, 745 (1976)

Chapter 11
High-energy particle detectors

11.1 GENERAL PRINCIPLES

There is no formal definition of a high-energy particle but, as a rough guide, it can be regarded as a particle with an energy greater than about 100 MeV. At such an energy, the associated de Broglie wavelength is very small so that, when the particle interacts with a nucleus, the reaction tends to be highly localised, usually to a single nucleon which may be considered to be a free, target particle. High-energy reactions differ from lower-energy nuclear reactions in a number of respects. For example, large numbers of nucleons are, typically, emitted from the nucleus, which may itself be split into two or more residual, product nuclei. In a low-energy nuclear reaction the product nucleus is only slightly smaller than the original target nucleus. Also, the particle–nucleon reaction often produces new fundamental particles, including pions and muons, which may decay radio actively or initiate secondary reactions to generate further products. These, in fact, are the characteristics that distinguish high-energy particles from low-energy ones.

In high-energy physics, high-energy particles are obtained from accelerators and used to study nuclear forces and reactions. The aim is to establish fundamental laws of nature. The same is true of cosmic-ray and space physics, in which high-energy particles of extraterrestrial origin are studied. As a spin-off from these fundamental studies, high-energy particles have found at least one application – in nuclear medicine, where they are used in radiation therapy. At present, pions are mainly used for this purpose. Compared with gamma radiation, they, and other charged particles, have the advantages of an energy-determined, finite range and a tendency, illustrated by the Bragg curve (Chapter 4), to produce more ionisation towards the ends of their paths. As a result, they deposit relatively large radiation doses in the target region rather than in the intervening tissue between that and the source.

In general, high-energy particles have two characteristic properties which necessitate special detection systems. First, they are exceptionally penetrating and secondly, they readily undergo nuclear reactions. Consequently, absorption spectrometry is not usually the best means of identifying and measuring the energies of high-energy particles. Large sodium iodide scintillators are capable of stopping protons with energies of about 200 MeV, but they are less effective with higher energies and lighter particles, they do not identify incident and product radiations, and they do not distinguish between ionisation interactions and nuclear reactions. In fact, the most common situation is one in which the incident particles are not, themselves, under investigation but it is the products of specific nuclear reactions that have to be detected and identified. Ideally, the detector should be one that contains the reaction and separately records all the products.

Apart from the Čerenkov detector, which is a special case, the most widely used high-energy particle detectors are those with intrinsic, three-dimensional spatial resolution sufficient to record particle tracks. The track parameters, such as specific ionisation, delta ray density and multiple scattering, help to identify particle type, and nuclear reactions are easily recognised. Energy can be measured by using range—energy formulae or, more accurately, by deflecting charged particles in a magnetic field. The field may be outside the detector, in which case the exit positions of the particles are measured. In other words, the detector is located on the focal plane of a momentum-analysing magnet to measure the radial positions of emerging particles. Alternatively, the detector itself may be between the poles of a magnet so that the field is within its sensitive volume and appropriately curved particle tracks are recorded. In both cases, a considerable amount of data reduction is required to identify particles and reaction products; that is, to determine their masses, charges and energies.

11.2 NUCLEAR EMULSIONS

Nuclear emulsions are similar to the photographic emulsions described in Chapter 10 except that they are much thicker, to provide additional stopping power, and they have smaller silver bromide grains, to resolve details of individual particle tracks. The emulsion thicknesses range from a few micrometres to about 1 mm, grain sizes are typically in the region of 0.1 to 0.3 μm, and grain spacing may be as large as 2 μm. Composition and density vary from one type of emulsion to another and, since the emulsion is designed to absorb water in the processing stages, they also vary with humidity. Normally, the silver halide is about 50% to 85% of the total mass and density ranges from 2 to 4 g cm^{-3}.

The emulsions are available as photographic plates, with glass backings, or in the form of thin, unbacked sheets, or pellicules. The latter can be stacked together to increase the total thickness and stopping power. They are separated and mounted on glass plates for processing.

The processing stage is difficult because the thickness of the emulsion inhibits the rapid and uniform access of the processing solutions. It is also critically important since it is absolute rather than relative information that is to be derived from the developed image. Underdevelopment, overdevelopment and uneven development must all be avoided if the final image is to present an accurate record of the number of grains ionised and rendered developable by the incident radiation.

Typically, an exposed emulsion is soaked in distilled water for about 2 hours and cooled to about 5 °C then transferred to a similarly cooled developer solution and left for a further period of about 2 hours. It is removed from the developer and allowed to dry at room temperature for another 2 hours then cooled again and placed in an acetic acid stop bath for 2 hours. Washing and fixing are similarly carried out at low temperature and over long periods of time so that the whole emulsion is uniformly and accurately processed. Unavoidably, however, large quantities of silver bromide are removed from the emulsion in the fixing process, so it shrinks, usually by a factor of 2 or 3. This must be accounted for if particle ranges are to be measured.

When a charged particle passes through an emulsion, it ionises and renders developable a proportion of the grains along its path. These are developed to form a visible track that provides various data relating to the mass, charge and energy of the particle, as follows:

(1) The *position* of the event is useful if the emulsion is used as a focal-plane detector for a momentum-analysing magnet. This measures the radius of curvature, ρ, of the particle's path in the magnetic field and, with the value of the magnetic induction, B, yields the momentum, p, of the particle. For non-relativistic particles, $p^2 = 2Em$, so the product of energy and mass is obtained. If the particle is magnetically deflected inside the emulsion, then the *curvature* of its path gives the same information.

(2) The *range* of the particle can be measured if it is stopped in the emulsion. Range depends on the energy, mass and Z number of the particle, as given by equations 4.27 to 4.29 and illustrated, for a typical nuclear emulsion, in *Figure 11.1*.

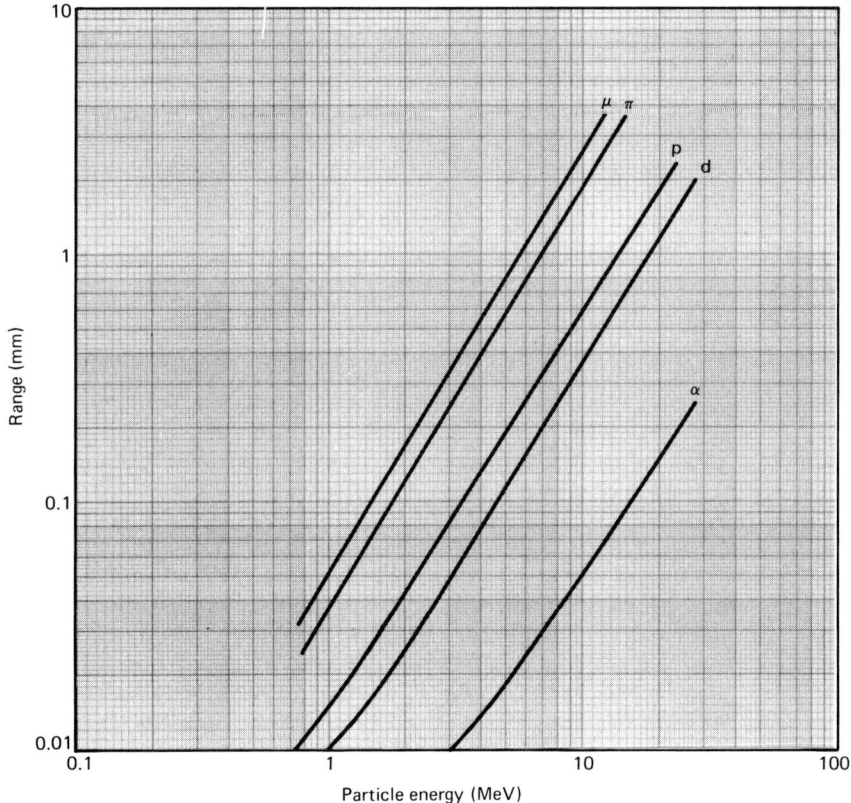

Figure 11.1 *Typical range–energy curves for a nuclear emulsion*

(3) The *grain density* can be determined by counting the number of developed grains per unit path length. This too is related to particle energy, mass and Z number, and is given by equation 4.25, the Bethe–Bloch formula.
(4) The *delta ray density*, that is, the number of delta rays formed per unit path length, can also be observed. This provides another function of particle energy, mass and Z number.
(5) The *scattering* of the particle, as it passes from one grain to another, can be measured and specified as the mean scattering angle. This increases along the track; that is, with decreasing residual range. It also decreases with particle mass and Z number.

In most cases, it is sufficient to measure only three of these track parameters in order to determine energy, mass and Z number (or charge). In effect, they describe three simultaneous equations in three unknown variables. The measurement process, however, is something of a problem. The most direct method involves the systematic scanning of the emulsion with a high-powered optical microscope. This can be a tedious and time-consuming operation, notoriously prone to human error. Automatic scanning systems have been developed to eliminate this source of inaccuracy. Basically, these are densitometer-type systems in which a light beam is transmitted through the emulsion and its output signal is digitised and passed to an on-line computer for data reduction.

Nuclear emulsions are used in nuclear physics research as focal-plane detectors. They have zero dead-time losses and they are inexpensive and reliable. They are also used in cosmic-ray studies, where they have the additional advantage of light weight and are easily carried by balloons. In general, however, they have a number of disadvantages compared with alternative systems, and their use has been declining for some time. The major disadvantages are the lengthy and complicated processing required to produce viable tracks and the equally complicated data reduction process.

11.3 CLOUD CHAMBERS

There are two types of cloud chamber, known as *expansion* and *diffusion* chambers respectively. The former is the original version invented in 1912 by Wilson. The latter is a more recent development, but neither is used very much nowadays. At present the cloud chamber is of historical rather than practical value but it does have some intrinsic interest because it is based on a unique operating process and has led directly to the development of the more widely used bubble chamber. It also enabled the introduction of new particle detection methods, including triggering techniques and stereophotography.

In the *expansion chamber* a mixture of gas, such as air or argon, and a vapour, such as that of ethanol, is contained in a cylindrical chamber with a glass top, for viewing into the chamber, and side entry ports, to admit ionising radiation. The chamber contains, or has direct access to, a reservoir of liquid ethanol, so the vapour is in equilibrium with the liquid and is at saturation vapour pressure. It is sensitised to ionising radiation by suddenly and adiabatically (without heat loss) increasing the volume of the gas–vapour mixture. This can be done by withdrawing a piston, or a diaphragm, as illustrated in *Figure 11.2*. The increase in volume produces a decrease in pressure and temperature according to the usual gas laws. It also leaves the mixture in a supersaturated condition; that is, with more vapour than it can hold in the prevailing conditions.

At this point, some of the vapour condenses. It condenses on the liquid surface, on the internal chamber surfaces, on dust particles and on ions, if they are present, but not with equal preference. Condensation appears more readily on dust particles than on the flatter surfaces of the chamber and liquid, and on charged particles rather than uncharged particles. In the absence of any ionising radiation, vapour condenses, like raindrops, on dust particles. If, however, a charged particle has just passed through the chamber, leaving a trail of ion pairs, then vapour condenses mainly on the ions to form a track of droplets that can be viewed and photographed immediately. Shortly afterwards, the droplets diffuse away from the track and begin to evaporate as the chamber warms up to room temperature. The chamber is then recompressed in readiness to record a further event.

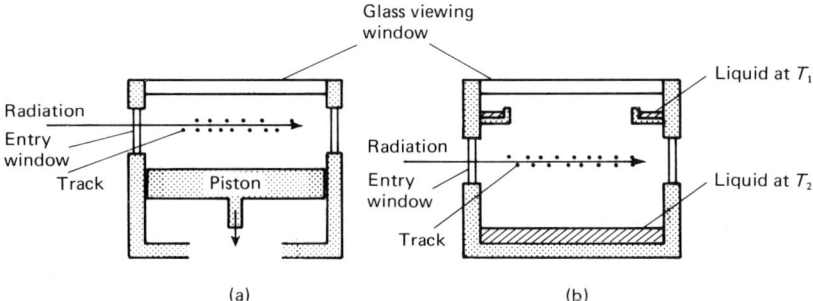

Figure 11.2 Cloud chamber systems. (a) Expansion chamber; (b) diffusion chamber

The chamber has to be expanded immediately after the passage of ionising radiation in order to record a track. It can be expanded at regular intervals in the hope of finding random coincidences with interesting events, but this is an unsatisfactory procedure. A better approach is to use a *triggering* technique in which a second detector is placed behind the cloud chamber to detect particles that have passed through the chamber. This trigger detector does not have to be a tracking device but it is very useful if it is an energy spectrometer or particle-identifying system, so that particularly interesting radiations can be *selected*. When the trigger detector records an interesting event its output signal causes the chamber to expand and record the track of the particle that has just passed through. The arrangement is illustrated in *Figure 11.3*. It is used with a number of other detectors, including bubble chambers and spark chambers. Although it is something of a nuisance that two detectors are required to obtain a result from one, it is a great advantage that, between them, they provide a selective tracking system. There is, of course, no reason why the trigger detector should not be located in front of the chamber except that the latter, being a gas counter, is likely to be the more transparent.

Besides expanding the chamber, the trigger pulse also switches on flashlights and cameras to photograph the event. Three or more cameras can be used to obtain a three-dimensional reconstruction of the track. This technique is known as *stereophotography* and it too has been applied to bubble chambers and spark chambers.

The chamber is sensitive only for about 0.1 s and, because of the time required to reset it, which is typically of the order of 100 s, it has a very long dead time. If used at the end of an accelerator, it expensively ignores much of the beam live time. The *diffusion chamber*, also illustrated in *Figure 11.2*, was developed to overcome this problem. It contains two liquid reservoirs, one at the top, which is usually heated, and one at the base, which is cooled. Vapour is produced by evaporation at the top reservoir, to form an unsaturated mixture, then diffuses down through the temperature gradient to become supersaturated and condense on the base reservoir. Between these extremes there is always a layer with the right amount of supersaturation to record ion tracks. This is the sensitive volume of the chamber. It has zero dead time and requires no triggering. As in the

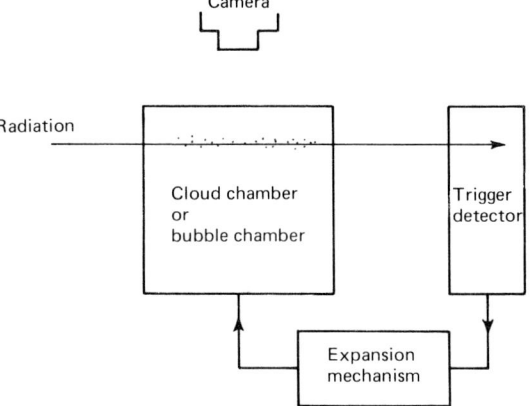

Figure 11.3 Trigger system for a cloud chamber or bubble chamber

expansion chamber, residual ions can be removed by the application of a continuous electric field called a *clearing field*.

Cloud chambers are not used as focal-plane detectors, but they are operated in magnetic fields to produce curved particle tracks. Compared with nuclear emulsions, they have some advantages. They can be operated at various pressures and incorporate various gases, vapours and additives to provide suitable stopping power, path lengths and reactions. Generally, paths are longer in cloud chambers than in films and they are more easily photographed, so data reduction is facilitated, but although the liquid droplets can be very small, they are not sufficiently static to be photographed through a microscope, so some spatial resolution is lost. The expansion chamber has the considerable disadvantage of a very long dead time but, against this, it can be operated in the triggered mode. This allows it to be selective and greatly facilitates the data reduction process. The bubble chamber, however, shares this advantage over the emulsion and has the added attraction of a denser medium with more stopping power, so it is usually preferred to the cloud chamber.

11.4 BUBBLE CHAMBERS

The bubble chamber was originally developed by Glaser, in 1952. It is a direct extension of the cloud chamber in which the gas—vapour mixture is replaced by a liquid, so its stopping power and its total reaction cross-section are both increased by factors of about 10^3. The liquid is volatile and is maintained at a temperature and pressure just below its boiling point. It readily becomes superheated and vaporises, forming small bubbles at the chamber surfaces, dirt particles and ions created along the paths of incident and reaction product particles. As with cloud chambers, the process is more likely to occur at ions than at dirt particles but, in this case, the preference is due to microscopic, localised heating produced by the excessive kinetic energy of the ions.

Corresponding to the diffusion and expansion types of cloud chamber, there are two types of bubble chamber: those which operate *continuously*, and those which are *triggered*. In the continuous mode, the chamber liquid is very close to its boiling point, so that the small amount of additional energy supplied by ionising radiation is sufficient to induce local boiling. This demands accurate control of temperature and pressure, and, because the chamber is so close to boiling, it tends to do so spuriously at internal chamber edges and dirt particles. To minimise these effects, the chamber must be constructed very carefully under clean conditions. These requirements impose a practical limit on the chamber size, so that continuously operated chambers are usually relatively small and produce track images with poor signal-to-background ratios.

The chamber can be triggered by detectors placed in front of it to respond to incident particles, or by detectors located to identify reaction products leaving it. In the first case, the chamber records all reactions. In the second case, it records only those reactions which generate the specific products. For example, the chamber may be surrounded by muon detection systems in order to record only those reactions in which muons are produced. In both cases, the output signal from the trigger detector activates a piston or diaphragm in the chamber, causing it to release the pressure and initiate the boiling process by reducing the boiling point. A typical arrangement is described in *Figure 11.4*.

Provided the expansion is triggered immediately before or after, that is, within a few nanoseconds of the formation of ion pairs, bubbles are created preferentially along the tracks of ionising particles involved in the reaction. The bubbles grow

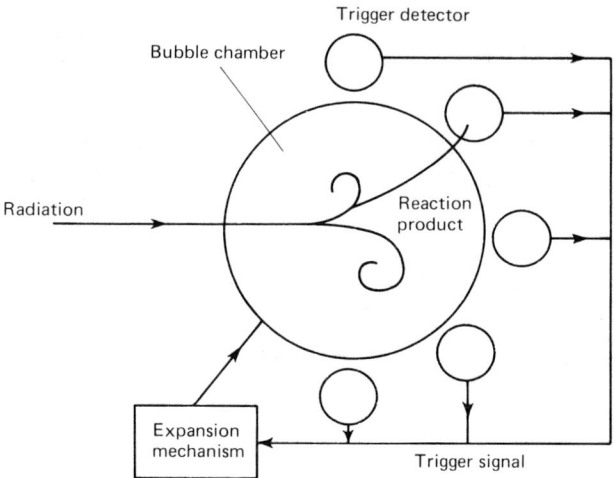

Figure 11.4 Bubble chamber triggered by reaction products

298 High-energy particle detectors

to a useful size in a few milliseconds, when they are illuminated by flashlights and stereophotographed to obtain a three-dimensional record. The chamber is then recompressed, ready for further action and the whole cycle can be completed in about 20 ms. Time resolution is about 1 ms and spatial resolution, determined by the bubble size, is usually of the order of 100 μm.

The operating conditions of the triggered chamber are less critical, so that, besides being more selective, it is more stable than the continuously sensitive chamber. It has a better signal-to-background ratio and can be constructed to much larger specifications. Chambers of several metres' diameter are currently in use and produce clear, detailed images such as that shown in *Figure 11.5*.

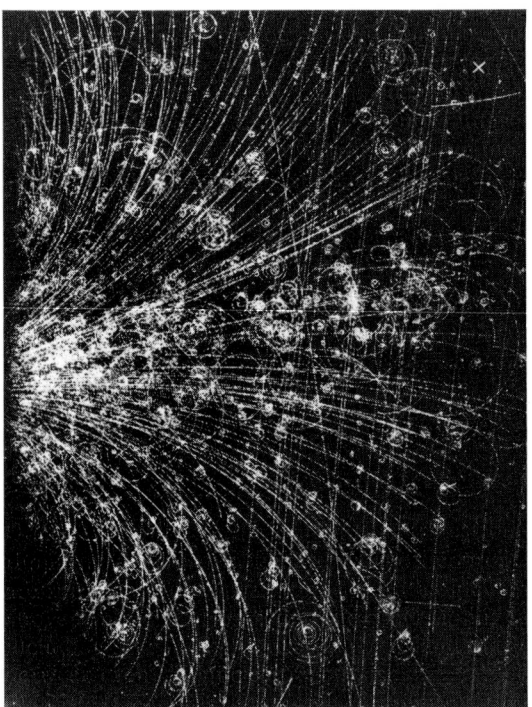

Figure 11.5 A cosmic ray shower in the 2-m hydrogen bubble chamber at CERN. It is interpreted as being caused by a 100-GeV particle interacting with the piston of the chamber. (PHOTOCERN)

Because of their size and clarity, they provide excellent stopping power and reaction probability for high-energy particles and they are preferred for many high-energy physics experiments.

As an additional advantage, a variety of liquids and additives can be used to present specific target nuclei and to study particular types of nuclear reaction. Liquid hydrogen is widely used because it unambiguously produces reactions with protons. Other liquids include deuterium, helium, propane, Freon and liquid xenon. Typical operating parameters for these liquids are listed in *Table 11.1*. Also, strong magnetic fields can be applied across the chamber to produce tightly spiralling particle tracks characteristic of bubble chamber photographs. These assist in the identification of reaction products.

Table 11.1   Bubble chamber liquids and operating parameters

Liquid	Temperature (°C)	Pressure (atm)	Density (g cm^{-3})
Hydrogen	−246	5	0.06
Deuterium	−240	7	0.13
Helium	−270	1	0.12
Propane	58	21	0.43
Freon (CF$_3$Br)	30	18	1.5
Xenon	−19	26	2.2

In common with the nuclear emulsion and the cloud chamber, the bubble chamber has one major disadvantage in that its data readout is visual and not immediately computer-compatible. Output signals cannot be read directly into an on-line computer. In view of the substantial amount of data reduction and data analysis that has to be undertaken, this is a serious drawback and has encouraged the development of electronic imaging systems, such as the spark chamber and the multiwire proportional counter, which do have this capability.

Normally, data processing involves three successive stages: scanning, measurement and analysis. In the first stage, photographs are scanned for interesting events and the positions of these are recorded with respect to fiducial points marked on the chamber. In the second stage, selected events are tracked through the photographic projections, track parameters are measured against calibrated reticules and these data are recorded on punched cards or some other form suitable for input to a computer. The final stage involves computer analysis to determine the reaction parameters. Although much of the process can now be carried out by automatic systems, it remains a complex undertaking.

11.5 ČERENKOV COUNTER

Čerenkov radiation is electromagnetic radiation, usually in the ultra-violet and visible-light wavelength range, generated when a charged particle travels through a dielectric medium faster than the velocity of light *in the medium*. It is also produced by neutrons, which have no charge but do have magnetic dipole moments, but in this case is usually much too weak to be of any practical value. The radiation is emitted as a cone of light projected forwards from the path of the particle but, of course, travelling more slowly than the particle. It is analogous to the bow wave produced by a boat moving through water faster than surface water waves, and to the sonic boom generated by aircraft exceeding the velocity of sound waves in air.

300 High-energy particle detectors

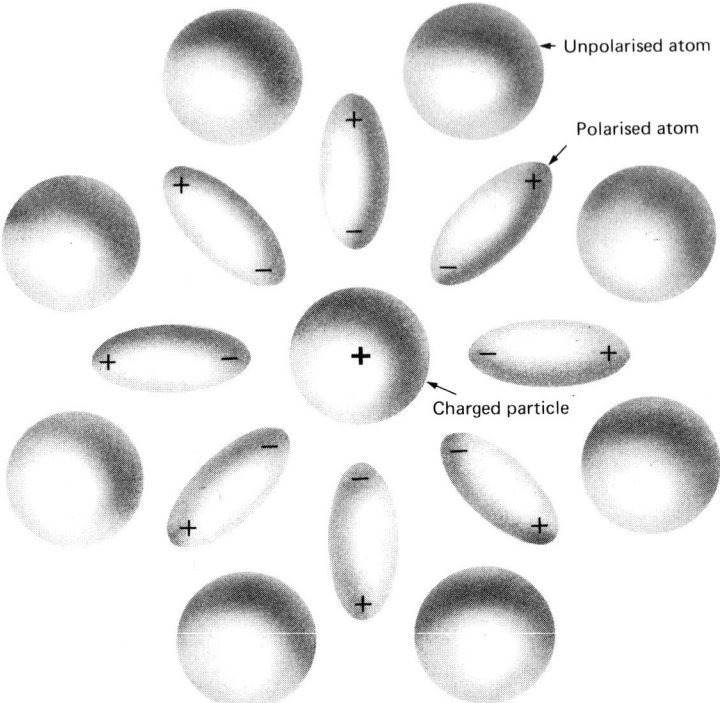

Figure 11.6 The effect of a stationary charge on a dielectric medium

The production of Čerenkov radiation can be illustrated by reference to *Figures 11.6* and *11.7*. *Figure 11.6* shows a stationary charged particle in a dielectric medium. Its electrostatic field polarises the surrounding atoms. A positively charged particle, for example, attracts atomic electrons and repels nuclei so that the atoms from electric dipoles aligned radially about the particle. The atoms are then in a weakly excited state and, if the particle could be removed suddenly, they would de-excite, emitting low-energy photons. Since the field strength decreases with distance from the particle and is further attenuated by the dielectric medium, the effect is limited to atoms very close to the particle.

Figures 11.7a and *11.7b* are drawn to a much smaller scale, in which the polarised atoms are effectively confined to the particle track. They show a particle moving through the medium with a velocity, v, that in one case is less than and in the other case is greater than the velocity, V, of light waves in the medium. Atoms in the wake of the particle have been excited and have subsequently de-excited emitting UV and visible light. Each point along the track is stimulated in turn, and can be regarded as an isotropic source of spherical light waves. In optics, this representation is known as the construction of Huygens secondary wavelets and all the points on the same wavelet surface have the same phase.

When v is less than V, these wavelets are travelling in different directions with different phase relationships so they are not coherent; that is, they interfere

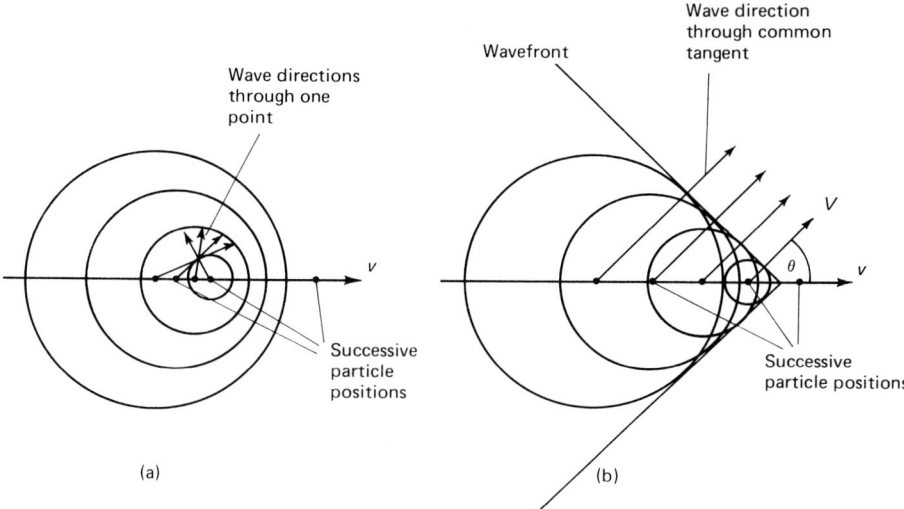

Figure 11.7 The production of Čerenkov radiation. (a) Particle velocity v less than the wave velocity V; (b) particle velocity v greater than the wave velocity V

with each other to produce cancellation and no net optical effect. When v is greater than V, however, there is a common tangential surface to all of the wavelets, on which they have the same phase and reinforce each other to produce a coherent wavefront travelling forwards at an angle θ to the direction of the particle. This is Čerenkov radiation and it is projected forwards as a cone of light with a half-angle θ.

From *Figure 11.7b*, it can be seen that the angle θ is given by

$$\cos \theta = \frac{V}{v} \qquad (11.1)$$

If the refractive index of the medium is n, then V is equal to c/n, where c is the velocity of light in a vacuum and equal to 3×10^8 m s^{-1}, and equation 11.1 can be written as

$$\cos \theta = \frac{c}{vn} \qquad (11.2)$$

Thus, if a charged particle has sufficient energy to exceed the velocity of light in a transparent medium it emits a cone of light whose half-angle, θ, is related to its velocity and, therefore, to its energy. Measurement of θ determines energy. The threshold velocity, above which Čerenkov radiation is emitted, depends only on the refractive index of the medium. For glass and Perspex, n is about 1.5, so the threshold velocity is $0.67c$. Since energy depends (relativistically, of course) also on the mass of the particle, the corresponding threshold energies vary from one type of particle to another. These energies can be calculated from equation 1.13. For a medium of refractive index 1.5, they are as follows:

Alpha particle:	1600 MeV
Proton:	320 MeV
Pion:	48 MeV
Muon:	36 MeV
Electron or beta particle:	0.175 MeV

Čerenkov radiation is a useful means of detecting and measuring the energy of charged particles. From the above data, it can be seen that it is mainly limited to high-energy particles but there is at least one notable exception in that beta particles with only about 175 keV can be detected. This raises the interesting possibility of detecting the beta particles from radionuclides by means of Čerenkov radiation. They can indeed be detected in this way, and easily distinguished from other radioactive emissions which do not have enough energy to generate Čerenkov radiation, but there are two complications. First, the betas have to complete with electrons and muons in cosmic rays which also produce Čerenkov radiation, and secondly, the intensity of the Čerenkov radiation depends on the energy of the particle beam, which is usually quite low for radioactive emissions.

The intensity of Čerenkov radiation can be expressed as the number of photons, dN, emitted in a given wavelength range in a path length dx. In a wavelength range of 3500 to 5000 Å, it is given by the approximate formula

$$dN/dx = 40 \sin^2\theta \text{ (photons cm}^{-1}) \tag{11.3}$$

For example, in glass, with a refractive index of 1.5,

for $\theta = 5°$, $dN/dx = 0.4$ photons mm^{-1}
$\theta = 10°$, $dN/dx = 1.6$ photons mm^{-1}
$\theta = 48°$, $dN/dx = 22$ photons mm^{-1}

It is therefore a very weak effect and increases with θ; that is, with particle energy. The radiation can be detected with a photomultiplier, but on average this requires about 10 photons to produce one photoelectron, so it is clear that Čerenkov radiation is of significant practical value only for energetic beams of radiation. These are the norm in high-energy physics, for which the process provides a number of standard detection systems.

There are many ways in which Čerenkov radiation can be utilised and many different detection systems have evolved, but generally they fall into one of two types. The first type can be described as a *threshold* detector. It merely detects particles with more than the appropriate threshold energy. The second type is sometimes known as a *differential* detector. It measures the emission angle, θ, and, therefore, the energy of the particle. The following are examples of these two types of Čerenkov counter.

Threshold detectors

The simplest threshold detector is a dielectric medium, called a radiator, examined by a photomultiplier tube, as illustrated in *Figure 11.8*. Any charged particle with more than the threshold energy for the radiator used generates a pulse of Čerenkov light which is recorded by the photomultiplier. The threshold energy can be varied to some extent by using materials of different refractive indices. Gases, for example, have refractive indices of the order of 1.2, and this varies with pressure; water has a refractive index of 1.33, and Perspex and most glasses have refractive indices in the region of 1.5. The only restriction on the type of medium used is that it must not be a scintillator, otherwise Čerenkov radiations are completely swamped by scintillations, which are produced by all types of incident radiation.

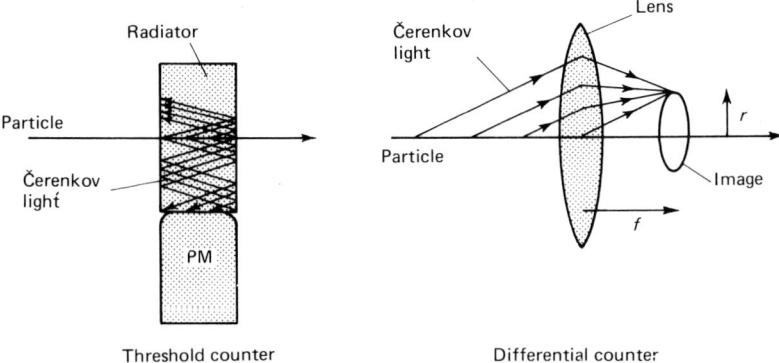

Figure 11.8 Two types of Čerenkov counter

Because they are highly selective, threshold Čerenkov counters are widely used as trigger detectors. Since they are also very transparent, they can be located in front of the main detector. They are also used in coincidence and anticoincidence counting systems, for the same reasons. As an example, consider a 1-GeV beam of protons and π^+ particles incident on a water radiator. The threshold velocity for Čerenkov radiation is $0.75c$, the proton velocity is $0.73c$ and the pion velocity is $0.99c$, so only the pions are detected. If the Čerenkov counter is connected in coincidence with a second detector, such as a scintillator, which detects both types of particle, then the pions are counted. If the two detectors are connected in anticoincidence, the protons are counted.

One of the major advantages of the Čerenkov counter for these and other timing measurements is that it is very fast, probably the fastest of all detectors. It has a time resolution that is usually better than 1 ns.

Differential detectors

Many ingenious arrangements have been devised to measure the half-angle, θ, of Čerenkov light and determine the velocity and energy of incident particles. One of these differential, or focusing, detectors is illustrated in *Figure 11.8*. In this, the radiator is a gas medium and the particle beam is incident along the axis of a converging lens. The Čerenkov light is focused into a ring, of radius r, about the focal point of the lens. From the diagram, it can be seen that θ is given by the relationship, $\tan \theta = r/f$.

Energy resolution depends on the accuracy with which θ can be measured. This in turn depends on the width and divergence of the radiation beam, optical dispersion and aberration, diffraction, particle scattering and the statistics of photon collection. For particles with much more than threshold energies it can be of the order of 0.5%, so energy can be measured with great accuracy.

Finally, it may be noted that Čerenkov counters can be used to detect gamma rays by means of the radiation produced by photoelectrons and positrons. It is a relatively weak effect because of the low interaction probability of gamma rays with matter and the difficulty of obtaining high Z number, transparent materials. Since photoelectrons are emitted in all directions with respect to the direction of an incident photon beam, it is not generally possible to measure energy by measuring the Čerenkov angle, but photon energy can be determined from the intensity of the Čerenkov radiation. From equation 11.3, this depends on θ and, therefore, on photon energy. However, energy resolution is not usually very good and is of the order of 30%.

11.6 IMAGING GAS COUNTERS

One of the disadvantages shared by the nuclear emulsion, the cloud chamber and the bubble chamber is that they produce visual output signals that are not directly compatible with on-line computer analysis. The bubble chamber, in particular, is widely used in high-energy physics but there are many applications in which there is a need for an electronic system, with intrinsic spatial resolution, capable of on-line computer analysis and control. This need has been met by the development of the imaging gas counters.

These include the *spark chamber*, the *streamer chamber*, the *multiwire proportional counter* (**MWPC**) and the *drift chamber*. From a fundamental point of view, they are gas counters and have been described, in that context, in Chapter 7. From a functional point of view, they should be compared with the detectors considered in this chapter, since they are widely used as particle-tracking devices in high-energy physics. Some of them are also used for photon imaging and should, therefore, be regarded as alternatives to the visual imaging systems discussed in Chapter 10.

BIBLIOGRAPHY

BARKAS, W., *Nuclear Research Emulsions*, Academic Press, New York (1963, 1973)
BRADNER, H., 'Bubble Chambers', *Ann. Rev. Nucl. Sci.*, **10**, 109 (1960)
HENDERSON, C., *Cloud and Bubble Chambers*, Methuen, London (1970)
JELLEY, J. V., *Čerenkov Radiation and its Applications*, Pergamon, Oxford (1958)
LITT, J. and MEUNIER, R., 'Čerenkov Counter Technique in High-energy Physics', *Ann. Rev. Nucl. Sci.*, **23**, 1 (1973)
SHUTT, R. P. (Ed), *Bubble and Spark Chambers*, Vols 1 and 2, Academic Press, London (1967)

Chapter 12
Nuclear electronics

12.1 PULSES

Nuclear electronics, or nucleonics, is a specialised branch of electronics devoted to the processing of electrical signals from those radiation detectors which provide them. Mostly, the signals take the form of charge or voltage *pulses*, so a nucleonic system usually involves *pulse electronics* and its most important property is its *transient* behaviour; that is, its effect on pulse shape as a function of time.

A nucleonic system, like other electronic systems, can be regarded as an assembly of separate *units*, each performing a well-defined function. An amplifier, for example, is a unit whose function is to alter the size of a signal. The system can be represented by means of a *block diagram* in which the units are shown as labelled boxes interconnected by lines denoting signal pathways. Each unit consists of a number of *elements* or *components*, such as transistors, valves, resistances, capacitances and inductances, which make up an electronic *circuit*. A detailed representation of a unit, or the whole system, showing its structural components, is known as a *circuit diagram*.

These concepts are illustrated in *Figure 12.1*, which describes a typical nucleonic unit. As it happens, the figure refers to a preamplifier, but it demonstrates principles that are more generally applicable to other units. There are three diagrams. The first one is a circuit diagram, depicting all the components, the second is a block diagram, showing no internal structure, and the third is a physical diagram, showing what the unit actually looks like. The present treatment will be based on block diagrams but some elementary circuits will be employed to illustrate fundamental processes.

Electronic circuit analysis depends on the calculation of electric potentials (often described as voltages) at key points in the circuit and, especially, at the input and output ends of the units, for which purpose block diagrams are sufficient. These are normally measured with respect to zero, or earth potential. so they are really potential differences. There are two types of potential, namely d.c. or constant potentials, V, and a.c. or varying potentials, v. The d.c. potentials are of secondary interest. They include voltage *supplies* to transistors and valves as well as the voltages applied, through nucleonic units, to detectors. The latter are known as *high-tension* (HT) or *bias* voltages, although this terminology is not entirely satisfactory in that discriminator settings are also referred to as bias voltages. The a.c. voltages include the signal voltages, which are of primary interest and can often be considered in isolation from the d.c. voltages, when the latter merely provide suitable operating conditions. A signal voltage, v, is a function of time and should be expressed as $v(t)$. In the present discussion the symbol v will refer to the *pulse amplitude*; that is, the *pulse height* (PH).

Circuit diagram

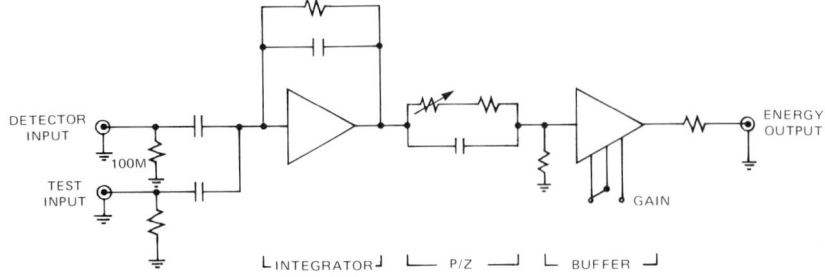

Block diagram

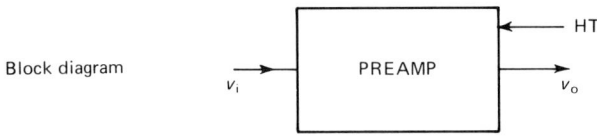

Physical diagram

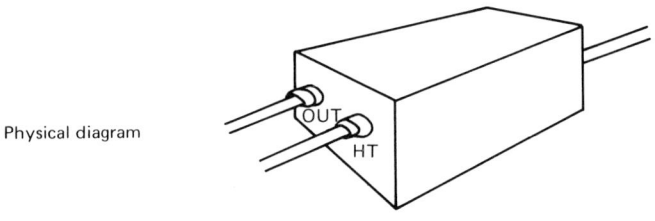

Figure 12.1 Three different representations of a preamplifier. (Photograph courtesy of Canberra Instruments Ltd)

A nucleonic unit has an input signal voltage of amplitude v_i and an output signal voltage of amplitude v_o. In the circuit diagram of *Figure 12.1*, these are shown as potential differences across the input and output ends of the circuit. In the block diagram, the earth points are not shown but are assumed to be present. In the physical diagram, the signal is, in this particular case, a potential difference between the axial wire and the outer, earthed sheath of a coaxial cable.

The most widespread branch of electronics is probably audioelectronics, in which signals are continuously varying voltages with regular, periodic waveforms. They convey information via their frequencies and their amplitudes. Frequency corresponds to the pitch of a sound and amplitude relates to its intensity. In nuclear electronics, each signal is a single voltage pulse (although it may start off as a charge pulse) produced by the detection of one particle or photon; that is, by one event. Pulse *frequency* is not regular, but random because of the random arrival of radiations in the detector, and corresponds to *count rate*. Since pulses may arrive almost simultaneously or very far apart, the system has to cope with a very wide band of signal frequencies, typically ranging from more than 1 MHz down to zero. In general electronic terms, this requirement is for a very *large bandwidth*.

When pulse *amplitude* is considered it becomes possible to define two types of pulse. There are known as *analog* (or linear) pulses and *digital* (or logic) pulses, and are illustrated in *Figure 12.2*. Digital pulses are all of the same

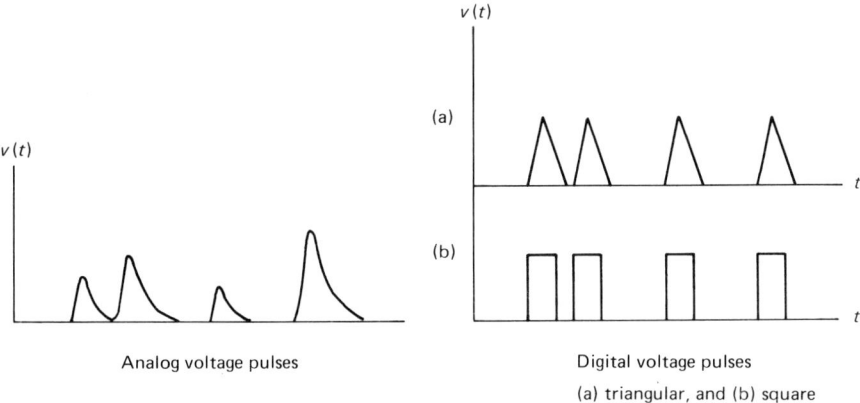

Figure 12.2 Analog and digital pulses

amplitude. They convey information regarding count rate and the times of events. Analog pulses contain additional information in that their amplitudes, or heights, are proportional to some other property, which may be the energy deposited by the incident radiation. Also, the *shape* of an analog pulse may contain further information relating, for example, to the specific ionisation, dN/dx, of incident radiation. Thus, analog pulses are more informative but, because of their varying size and shape, they are more difficult to process, so they are invariably converted to digital pulses as soon as their useful additional data have been extracted. Digital pulses can have any convenient shape. They may be square, or triangular, as illustrated in *Figure 12.2*.

Signal pulses

A nucleonic pulse originates as a *transient* voltage change, $v(t)$, across the output end of a radiation detector, as a result of charge deposited in the detector by the passage of ionising radiation. The process can be described by considering, first of all, transient effects in R–C circuits. *Figure 12.3* shows a circuit comprising a

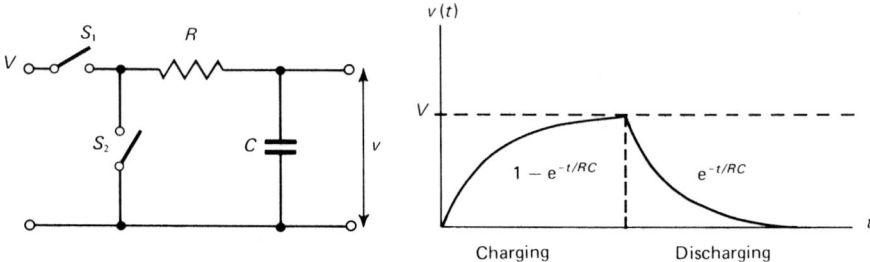

Figure 12.3 Transient effects in an R–C circuit

resistance, R, and a capacitance, C, connected in series. If a d.c. voltage, V, is suddenly applied across this combination, by closing switch S_1, the transient behaviour of the circuit is such that the voltage across C, $v(t)$, increases exponentially to a maximum value, $v = V$. Thus, the capacitance charges up according to the formula

$$v(t) = v(1 - e^{-t/RC}) \tag{12.1}$$

The product RC is known as the *time constant* of the circuit and is the time taken for the voltage to reach $1/e$ of its final value. Obviously, the smaller the value of RC the faster is the rate at which the capacitance is charged.

If the voltage supply is shorted out of the circuit, by closing switch S_2, the capacitance discharges through R. The voltage across C falls exponentially to zero, again with a time constant RC, and is given by

$$v(t) = v e^{-t/RC} \tag{12.2}$$

Any radiation detector that generates a charge pulse, such as a gas counter, a scintillator or a semiconductor counter, can be represented by a circuit diagram similar to *Figure 12.3*. As shown in *Figure 12.4*, a voltage, V, is applied, through a load resistance R, to a detector with a capacitance C, so that, in the absence of an ionising event, the voltage across the detector is equal to V. If, then, a charge q is collected at one of the electrodes, the voltage across them falls by an amount v, given by

$$v = \frac{q}{C} \tag{12.3}$$

Following this event, the voltage recovers exponentially to its original value and the net effect is that a small, negative voltage pulse is superimposed on the much larger, positive d.c. potential. The process can be explained by the fact that the HT supply recharges the detector through the load resistance, R, and its

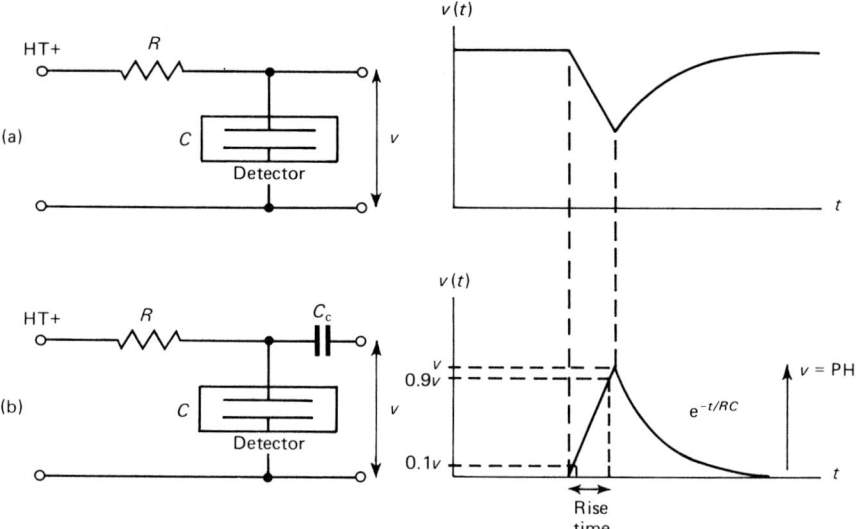

Figure 12.4 Typical detector circuits and output voltage pulses. (a) d.c. coupled and (b) a.c. coupled

own negligible internal resistance. Alternatively, by considering only the a.c. voltages, it can be suggested that the charge dissipates by leakage round the same external circuit. In either case, charge moves through a series $R-C$ circuit.

Normally, it is desirable to isolate the signal pulse from the much larger d.c. potential in order to avoid overloading subsequent units. This is done by reading the signal out through a *coupling capacitance*, C_c, which transmits only a.c. voltages. The signal pulse then has the form illustrated in *Figure 12.4*. It increases linearly, because of the uniform rate of charge collection, to a maximum value, v, which is the pulse height. The *rise time* is defined to be the time taken for the voltage to increase from 10% to 90% of its maximum value and depends on the charge collection mechanism within the detector. The pulse then decays exponentially with a *decay time constant* equal to RC, which is the time taken

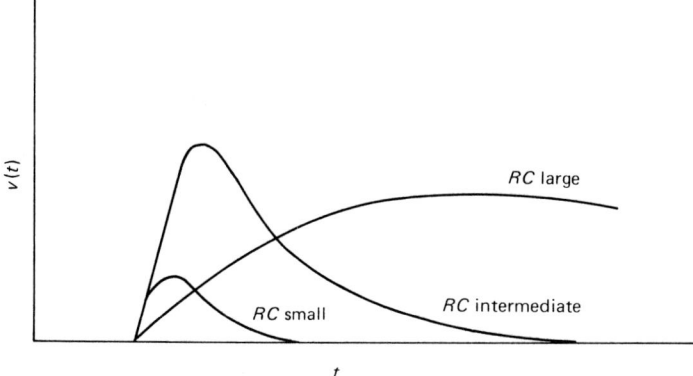

Figure 12.5 The effect of time constant on pulse shape

for it to decrease to 1/e of its maximum value. If the product RC is very small, the charge may leak away through the resistance R and the HT supply before it is completely collected, or integrated, on the detector electrodes. In this case, as shown in *Figure 12.5*, v does not reach the value given by equation 12.3 and is not necessarily proportional to the charge deposited. If, on the other hand, RC is very large, v does reach this value but retains it for a long time during which a second event may be detected. Thus, long time constants are ideal for pulse height spectrometry but short time constants reduce the dead-time counting losses and improve counting efficiency. In practice, a compromise is effected, or a system is designed specifically for spectrometry or for timing and high count rate measurements.

The above theory applies both to spectrometric detectors, which generate analog pulses, and to other detectors — such as GM counters — that produce digital pulses. In the latter case, however, the charge collected and therefore the pulse height, v, is not proportional to energy deposited, so that all pulses have the same height. An externally quenched GM counter and a d.c. spark chamber have very large load resistances, of the order of 10 to 100 MΩ, to slow down the supply of current to the chamber and extinguish the discharge. Consequently, they also have long recovery times and cannot cope with high count rates.

Other pulses

A nucleonic system has to process the signal pulses from a radiation detector. In doing so, it utilises pulses, of various types, produced by *pulse generators* and it minimises the effects of undesirable pulses due to *electronic noise* and *electrical interference*. Thus other pulses, besides signal pulses, play a part in nuclear electronics, in some cases constructively and in others destructively.

In the first category are the *sinusoidal pulses* obtained from an *oscillator*. These can be used in the form of continuous, *bipolar* pulses, alternating from positive to negative polarity, or they can be rectified to produce a series of positive or negative pulses as shown in *Figure 12.6*. With easily adjustable frequencies and amplitudes, they may be employed as clock pulses, to measure time intervals in discrete, i.e. digitised, increments, or they may be used to simulate detector pulses for circuit testing. Another useful type of pulse is the *sawtooth* or *ramp* voltage pulse, as generated, for example, by a gas thyratron valve which charges up slowly then suddenly discharges to produce a series of triangular pulses with long rise times and almost zero decay times. These are used as the time bases of cathode ray oscilloscopes and other automatic scanning operations requiring a series of uniform voltage increases from zero to some maximum value.

Another, more important type of pulse generator is the *multivibrator*, which generates *square* pulses — if such a term can be applied to a shape measured in volts, in one dimension, and time interval, in the other. There are three different types of multivibrator. An *astable*, or *free-running multivibrator* produces a continuous series of square, digital pulses which, like those of the sinusoidal oscillator, have adjustable size and frequency and may be used as clock pulses or simulated detector pulses. A *monostable multivibrator* has one stable state. When it receives an input pulse big enough to trigger it, it is displaced to an unstable state from which it returns in a short, controllable time, to produce a square pulse whose height and duration are independent of the trigger pulse. Thus, it can be used to convert analog pulses to digital pulses. The third type of multivibrator is a *bistable multivibrator*, sometimes known as a *flip-flop circuit*. It has two stable states. An input pulse displaces it from one state to the other and a second input pulse returns it to its original state so two input pulses generate one, digital output pulse, as shown in *Figure 12.6*. This is the essential property of a *binary counting* device and the flip-flop circuit, in the form of an integrated circuit (IC), is used in nucleonic *scalers*, to count events, as well as in *digital computers*.

Of the undesirable pulses present in a nucleonic system, those presenting the greatest problem are the random *noise pulses*. These have many distinct sources, both in the detector and in electronic units, but they are all basically due to statistical fluctuation in current and voltage. They are small triangular pulses, of varying height, and very similar to small signal pulses. They can be mistaken

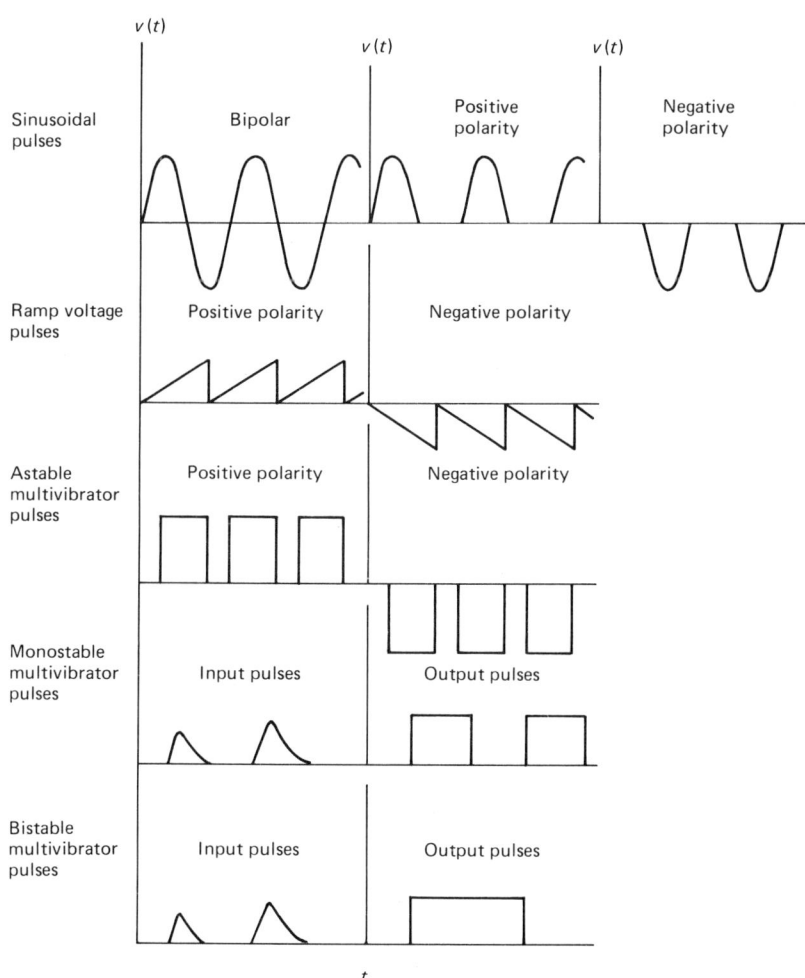

Figure 12.6 Voltage signals obtained from different types of generator

for detector signal pulses, they occupy system live time, increasing dead-time counting losses, and, by adding randomly to signal pulses, they increase pulse height fluctuation and energy resolution. They are, of course, most troublesome when they are produced in the detector or at the first stage in the electronic processing, since they are then amplified along with the signal pulses.

There are three ways of dealing with noise pulses. A simple but sometimes impracticable method is to cool the detector and first-stage electronics, since a major component of electronic noise is due to thermal processes. This approach is used with photomultipliers and some semiconductor detectors. An alternative is to eliminate them by means of a low-level discriminator, which transmits only pulses whose heights exceed a preset value. This is a standard practice when signal pulses are much larger than noise pulses, but it does not reduce dead-time losses in the detector nor does it remove the effects of signal–noise coincidence on energy resolution. The third method is a standard approach for all situations and involves a combination of pulse-shaping and amplification, usually in a *preamplifier*. It will be described in a later section. A major function of a preamplifier is to improve the *signal-to-noise ratio*.

Electrical interference can also result in pulses which conceal weak signal pulses, increase dead-time losses and affect pulse height resolution. *Radiated electrical interference*, sometimes known as radiofrequency (RF) pick-up, is produced by fluctuating currents in nearby electrical equipment. These generate electromagnetic radiations which are received by the detector circuit in the same way as a radio aerial receives broadcast radio waves. The effect is especially severe if large currents are nearby, but, in all save the most extreme circumstances, it is easily eliminated by enclosing all parts of the circuit in earthed containers. These act as screens, conducting induced signals to earth. When a signal has to be transferred from one unit to another, outside the screened container, similarly screened coaxial cable is used. This has a central signal wire surrounded by an insulating material which, in turn, is enclosed in an earthed conducting sheath which acts as the screen.

Conducted electrical interference involves the transmission of pulses from one piece of equipment to another, the detector circuit, via the earth line of the mains power supplies. This can be a severe problem. It is dealt with by establishing separate and substantial earth points to conductors which penetrate down to the underground water table or by using a filter circuit at the mains electrical supply.

Pulse shaping

The shape of a signal pulse is one of its most important properties. It can affect the dead time, energy resolution, time resolution, signal-to-noise ratio and other parameters of a detection system. It is easily altered, however, both by design and otherwise, by pulse shaping mechanisms which occur at various stages during signal processing. Some knowledge of the causes and effects of pulse shaping is useful in purchasing a nucleonic system, setting it up, optimising it and testing it.

The subject can be introduced by considering the simple *CR* and *RC* circuits shown in *Figure 12.7*. The former is a *differentiating circuit*, if the time constant *CR* is small compared with the duration of the input signal pulse $v_i(t)$. The output voltage, $v_o(t)$, generated across *R* is equal to iR, where i is the transient current flowing through *R*. At any point in time, i is equal to dq/dt; that is, the

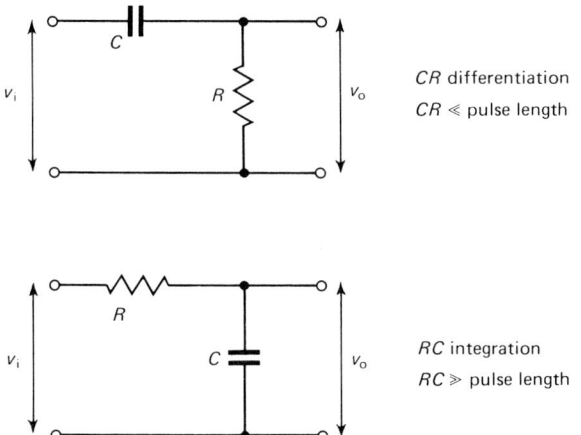

Figure 12.7 Resistance–capacitance differentiating and integrating circuits

rate of change of q with time, where q is the instantaneous charge induced on the capacitance, C. By the definition of capacitance, q is equal to Cv_C, where v_C is the instantaneous voltage across C. Hence

$$i = \frac{dq}{dt} = \frac{d(Cv_C)}{dt} = C\frac{dv_C}{dt} \qquad (12.4)$$

Because CR is very small, the impedance of C, which is proportional to $1/C$, is much greater than the resistance, R, so the voltage across C, v_C, is approximately equal to v_i, and

$$i = C\frac{dv_i}{dt}$$

so that

$$v_o(t) = CR\frac{dv_i(t)}{dt} \qquad (12.5)$$

Thus, the output pulse shape is the derivative of the input pulse shape with respect to time and the voltage at any point on its profile is proportional to the gradient at the corresponding point on the input pulse. In practice, because of the approximation made, the process is not exact differentiation but, as shown in *Figure 12.8*, it is very close to it. A differentiated pulse is *bipolar*, being positive while the input pulse is increasing in size, zero when it reaches its peak and negative while the signal pulse decays. The point at which it changes polarity is known as the *crossover point* and is a very accurate time reference point. Differentiated pulses are used for timing measurements in a technique known as *crossover pickoff*. If a longer time constant CR is used, the pulse is longer than that shown in *Figure 12.8* and has less negative overshoot, but it is still referred to as a differentiated pulse.

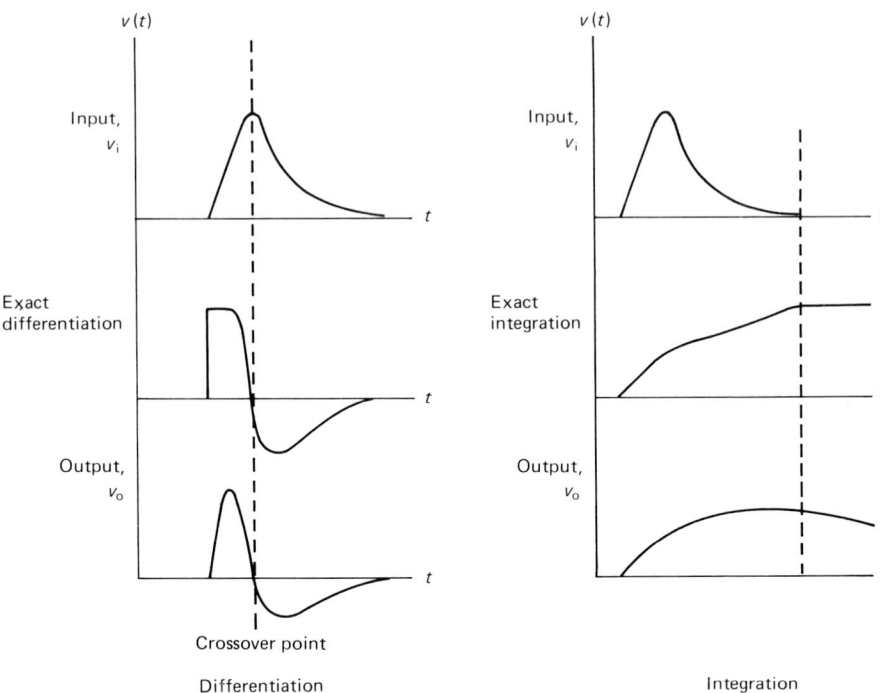

Figure 12.8 Differentiation and integration of a detector signal pulse

In the *integrating circuit* shown in *Figure 12.7*, the time constant, RC, is much longer than the pulse duration. At any point in time, the output voltage across C is given by the ratio q/C, where q is the instantaneous charge accumulated on C; that is, the integral, with respect to time, of the current, i_C, that has flowed through C. Thus

$$q = \int i_C \cdot dt \tag{12.6}$$

Because RC is large, the impedance of C is small compared with R so it can be assumed that the instantaneous voltage across R is v_i. The current through C is the same as that through R which, by Ohm's Law, is equal to v_i/R. The output voltage is equal to q/C, so that

$$v_o(t) = \frac{1}{RC} \int v_i(t) \cdot dt \tag{12.7}$$

Hence the output pulse shape is proportional to the integral of the input pulse shape and the output voltage, v_o, at any point in time is equal to the area under the graph of v_i, up to that time, divided by RC. Again, as shown in *Figure 12.8*, exact integration is not achieved in practice. C is never large enough to hold its charge indefinitely, so $v_o(t)$ decreases gradually, as leakage takes place, instead of retaining its maximum, integrated value. Integration lengthens and flattens a pulse to an extent that increases with the magnitude of the circuit time constant, RC.

There are other methods of shaping pulses. They produce shapes different from those obtained by *CR* differentiation and *RC* integration but basically, like these processes, they shorten or lengthen signal pulses over a range of selectable time constants.

Pulse shaping circuits can also be used as *filters* to transmit selected ranges of pulse shape and suppress others. A differentiating circuit responds well to high-frequency and fast pulses, and is known as a *high pass circuit*. It attenuates pulses that are longer than the differentiating time constant by allowing the voltage to discharge across the resistance as fast as it builds up across the capacitance so that the full pulse height is never realised. An integrating circuit, on the other hand, responds well to low-frequency and slow pulses, and is known as a *low pass circuit*. It attenuates pulses that are shorter than the integrating time constant by flattening and lengthening them.

In nuclear electronics, a high pass filter is ideal for timing and high count rate measurements. For example, it differentiates a double-peaked pulse, corresponding to two poorly resolved events, into two clearly separated, narrow pulses. It does not suppress the relatively fast, high-frequency noise pulses and, if its time constant is very small, it tends to reduce analog pulse heights by amounts that vary with the pulse width, so it is unsuitable for pulse height spectrometry. A low pass filter smoothes out noise pulses and reduces the effect of electronic noise but it tends to produce signal pulse overlap which is bad for timing and for measuring high count rates as well as pulse resolution and reduces dead-time counting losses, while a low pass circuit minimises the effects of electronic noise. The best time constant for pulse height spectrometry and energy resolution is one that is long enough to integrate the charge pulse without introducing pulse overlap. It is often obtained by a combination of differentiation and integration and its ideal value depends on the count rate. For example, if this is 10^4 counts s^{-1}, then a pulse duration of 10 μs will result in a 10% overlap (as given by equation 5.10 for dead-time losses).

Unplanned pulse shaping has numerous adverse effects. One of these occurs in a coaxial cable used to transmit signal pulses from one unit to another. Such a cable has a *characteristic impedance* per unit length, typically of the order of 50 or 100 Ω. If pulses are fed to it from a detector or nucleonic unit with an output impedance (that is, an impedance measured across its output terminals) which is much different from the characteristic impedance, then reflection and refraction processes take place as they do, for example, when light pulses pass between media of different refractive indices. The result is pulse shape distortion and attenuation. It is avoided by *impedance matching*. The output impedance of the detector or unit must be the same as the cable impedance. The same effect is observed at the other end of the cable if it connects to a high input impedance. Nucleonic units are designed to avoid such a mismatch by having suitable input impedances.

A second problem arises when attempts are made to monitor pulse shape at intermediate points in a processing system. This is a standard method of determining whether or not a system is functioning properly. The pulses are observed by means of a cathode ray oscilloscope (CRO) which must be fast enough to trigger at the required count rate and to reproduce fast pulse shapes. To have a good transient response, the CRO time base should go down to about 0.1 μs cm^{-1}, but even then the displayed pulse is likely to be slightly longer than the actual circuit signal pulse. Impedance mismatching at the CRO input socket can be eliminated by shorting across the socket with an impedance equal to that of the cable. Special T-shaped connectors are available for this purpose.

12.2 NUCLEONIC UNITS

Amplifiers

Nucleonic units are designed to accept a relatively narrow, standard range of voltage pulses, with heights typically in the range of 1 to 10 V. Radiation detectors supply a wide variety of pulse sizes, ranging from about 0.1 mV from an ionisation chamber to about 1 V from a scintillation counter or a GM counter. In at least one case, that of the semiconductor detector, a charge pulse rather than a voltage pulse is produced. Consequently, the first unit in a nucleonic system is invariably an amplifier, which not only amplifies pulses (that is, increases their heights by a *gain* or *amplification factor*), but may also have to integrate them into voltage pulses. In addition, because it is the first processing unit, it has to perform a number of other functions, some of which depend on the type of detector being used while others depend on the nucleonic system being addressed, so the amplifier is an interface unit between the detector and the nucleonics.

In general, the detector has to be remote from the nucleonics, even if only by a metre or so. It has to be in the radiation environment while the nucleonic system and the experimenter should be further from the source of radiations. There are several reasons for locating the amplifier close to the detector, not least of which is the fact that a long cable between the two will attenuate pulses and provide a serious impedance mismatch between the detector and the cable. On the other hand, the experimenter must have easy access to the amplifier controls in order to vary the gain and pulse shaping parameters to suit the experimental requirements. It is possible, in certain specific cases, to do without the latter facility, in which case a single *head amplifier* is mounted on the detector, but in general it is necessary to distribute the amplifier functions between two separate units. One of these is known as a *preamplifier* (PA) and located at the detector. The other is known as a *main amplifier*, or a *linear amplifier* or, simply, an *amplifier*. It is located at the nucleonic system.

The *preamplifier* is detector-orientated. It is designed specifically for one type of detector and its major functions are as follows:

(1) *Amplification*: it incorporates a *fixed gain* to bring the signal pulse heights up to a suitable range for further processing. For pulse height spectrometry, the gain has to be very *stable* and for this reason preamplifiers operate on the principle of *negative feedback*.
(2) *Noise reduction*: any noise pulses generated at the input to the preamplifier will be amplified along with the signal pulses and contribute more noise effects than will pulses generated after amplification. For this reason, pre-amplifiers are based on *field effect transistors* (FET) or valves, rather than the noisier bipolar transistors used in most electronic circuits, and they may be cooled. As a further measure, an integrating, filter circuit is included to smooth out noise pulses. The effect of electronic noise may be expressed in terms of a *signal-to-noise ratio*, which is the ratio of the mean signal pulse height to the mean noise pulse height. Since noise fluctuations are bipolar,

their mathematical value is zero, so in practice the root mean square (r.m.s.) of their pulse heights is used. Equipment manufacturers often state this value in their literature. Alternatively, they may quote the noise specification of a preamplifier in r.m.s.-equivalent energy units, such as keV, or as the r.m.s. value of the number of ion pairs required to generate an equivalent signal pulse. A low-noise preamplifier typically has a noise level of about 500 ion pairs. It must be noted, however, that the effectiveness of the filter circuit depends on the capacitance, C, of the detector, which is located across the input of the preamplifier. As C increases, its impedance decreases and the filter becomes less effective, so noise increases with detector capacitance and is quoted as a function of C by the manufacturer.

(3) *Impedance matching*: in order to avoid pulse distortion and attenuation, the preamplifier's input impedance must match the very large impedance of a radiation detector while its output impedance must match the relatively small impedance of coaxial cable connecting it with the main amplifier. If the detector has a large, stable capacitance and is therefore able to integrate its charge pulse into a voltage pulse, it requires a *voltage-sensitive* preamplifier with a resistive input impedance. If it has a small capacitance that varies with operating conditions it requires a *charge-sensitive* preamplifier. This has a large, stable input capacitance which integrates the signal from a charge pulse to a voltage pulse. In general, a scintillation counter operates best with a voltage-sensitive preamplifier while a semiconductor counter requires a charge-sensitive preamplifier. A gas counter can usually be satisfied with either type, but nowadays there is a preference for charge-sensitive preamplifiers in spectroscopic work.

(4) *Pulse shaping*: if the detector is to be used for timing measurements, its preamplifier should include a differentiating circuit with a short time constant. If it is used for pulse height spectrometry, it should have a longer time constant, usually produced by differentiation, to avoid overlap, then integration, to improve signal-to-noise ratio and preserve the proportionality of the pulse height to charge deposited in the chamber. Frequently, a preamplifier incorporates both circuits, with separate output sockets, one for timing and the other for spectrometry. A scintillation counter with a preamplifier and amplifier is shown in *Figure 12.9*.

The *main amplifier* is essentially nucleonics-orientated and performs the following functions:

(1) *Amplification*: it must have a *variable* gain which can be adjusted by the experimenter to meet the needs of the particular nucleonic units being used and to observe, for example, different regions of a pulse height spectrum. In addition, for pulse height spectrometry, the gain must be *linear* over its full range of values, and as a function of count rate. For this latter reason, it should be a wide-band amplifier.

(2) *Pulse shaping*: a wide-band amplifier with variable, linear gain is sufficient for many purposes but, for a wider range of applications, it should have provision for *variable* pulse shaping. This may be obtained by means of CR and RC circuits or with more sophisticated circuits which eliminate the long pulse tails. For spectrometry, the optimum time constant is deduced by experiment and is the value that provides the best energy resolution.

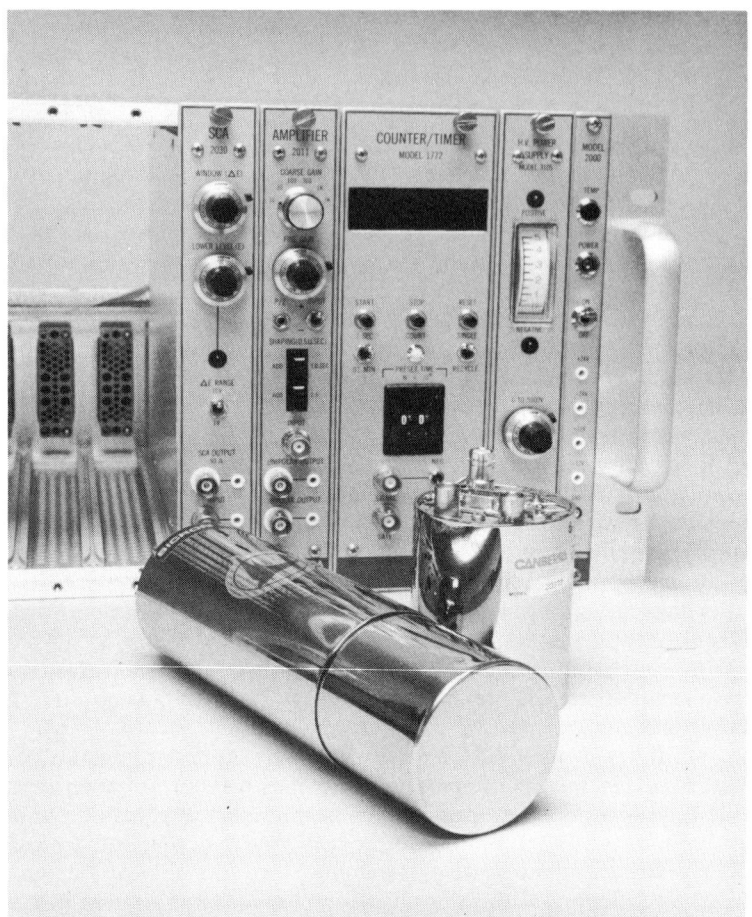

Figure 12.9 A scintillation counter with a preamplifier and amplifier. (Courtesy of Canberra Instruments Ltd)

Normally, it is also the value that gives the largest pulses for a given amplification factor since short time constants tend to attenuate pulses by clipping them while long time constants may attenuate them by integration. For high count rates, of course, shorter time constants are required to avoid pulse overlap. These may be obtained by a process known as *baseline restoration* in which pulses are clipped after they have peaked. For timing measurements, short and, usually, differentiated pulses are required and the circuit should be stable against jitter and other time fluctuations.

(3) *Preamplifier servicing*: the amplifier is usually mounted in a rack along with the other nucleonic units and draws its power supplies, to operate its transistors or valves, from the rack. The preamplifier is remote from this source of power so its supplies are usually drawn, by a multicore cable, from a socket in the amplifier. In modern equipment, a single voltage, such as +24 V, may be sufficient, in which case a coaxial cable may be used.

Scalers

A scaler, or counter, is similar in many respects to a small, unintelligent computer of limited function. It reads in data, in the form of signal pulses, and adds them to form a running total which it stores in a memory and reads out, usually continuously. It records the number of signal pulses read into it over a given period of time and is an essential component of any counting system.

Unless the scaler is fully or partly automated, it has a minimum of three manual controls, in the form of pushbuttons or switches. One of these is a START button, which initiates the accumulation of data, the second is a STOP button, which terminates the process, and the third is a RESET button, which clears the memory in readiness for another measurement. Of these, it is the STOP control that is most frequently automated and replaced by a PRESET TIME control. This is a clock mechanism, based on an oscillator circuit, which allows the experimenter to preselect a counting time, after which the count stops automatically. It may be calibrated in *real time* — that is, actual, chronological time — or in *live time*, which excludes pulse dead time so that no correction has to be made for dead-time counting losses. Time control may be built into the scaler or it may be derived from a separate *timing unit*. Another unit that may be built into the scaler, giving rise to another manual control, is a *pulse height discriminator*, which eliminates pulses smaller than its voltage setting and is mainly used to reject noise pulses. Some of these controls are shown in *Figure 12.10*.

Although most scalers will accept a range of pulse heights and size, they do not require analog pulses and in fact perform better with digital pulses. These should be narrow, square pulses, such as those produced by a monostable multivibrator, which may be included in the scaler, in order to provide the maximum voltage for the shortest possible time. In this way a maximum count rate of about 20 MHz can safely be accommodated, with a corresponding time resolution of about 50 μs. The pulse height is usually a few volts and depends on the type of memory and circuit logic employed.

The memory of a modern scaler is similar to that used for main storage in a digital computer. It may be an array of *bistable multivibrator*, or flip-flop, circuits in the form of integrated circuits (IC) or as large-scale integrated circuits (LSI), in which several circuits are incorporated in one solid-state element. Alternatively, it may consist of an array of *ferrite cores*. Each of these is a small annulus of ferromagnetic material with a square magnetic hysteresis loop and two strongly defined magnetic polarities. When a small current, derived from a signal pulse, is passed along a wire through the annulus, the core switches its polarity from one state to the other. Thus, both the bistable multivibrator and the ferrite core are bistable switches, with two stable states which conveniently form the basis of a *binary counting* system. The bistable multivibrator is slightly less expensive than the ferrite core but erases when the apparatus is switched off.

A single scaler, or counting channel, consists of several of these binary switches connected in series so that the output of one is passed on to the next in line, as illustrated in *Figure 12.11*. Each switch has two stable states, designated as 0 and 1, and reads out one complete pulse for every two input pulses. Before counting commences, all the switches are RESET to their zero states. On the START command, pulses enter the array. The first one changes the first switch to its 1 state. The second pulse returns this switch to its 0 condition and causes one pulse to be transferred to the second switch, which flips to its 1 state. Thus

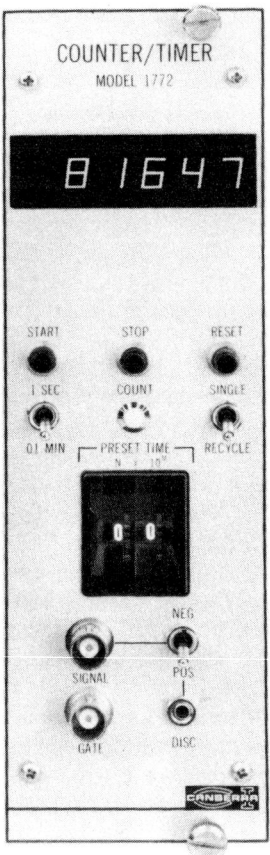
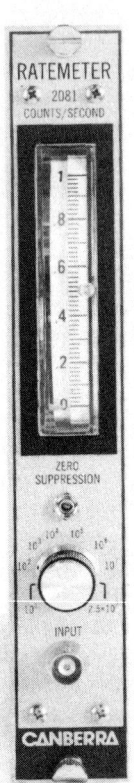

Figure 12.10 A scaler (counter/timer) and a ratemeter. (Courtesy of Canberra Instruments Ltd)

one pulse is recorded as 1, two pulses are recorded as 10, three pulses would be recorded as 11, four as 100, and so on. In other words, the array counts in *binary code*. Each switch represents one binary digit, or *bit*, and the whole array may be described as a *word*. In *Figure 12.11*, six switches are shown, representing a six-bit word. A 1 in the first bit denotes 1 count, a 1 in the second bit denotes 2^1 (= 2) counts, a 1 in the third bit denotes 2^2 (= 4) counts, and so on to the sixth bit, where a 1 denotes 2^5 (= 32) counts. The whole word can contain up to $2^6 - 1$ (= 63) counts.

There are several alternatives to the straightforward binary code. In nucleonics, the most widely used of these is the *binary-coded decimal* (BCD) notation. This uses a 4-bit binary array to represent each decade of a decimal number, as illustrated in *Figure 12.12*. Each 4-bit circuit counts up to 9 (that is, 1001 in binary code) then, with the next input pulse, resets all these bits to zero and passes one count on to the next decade array. The BCD system is less efficient than the pure binary system, in that it requires more bits to generate a given number, but it is more easily converted to decimal notation in the readout process.

Figure 12.11 Binary code counting in a six-bit word

Count number	2^5	2^4	2^3	2^2	2^1	2^0
Bit number	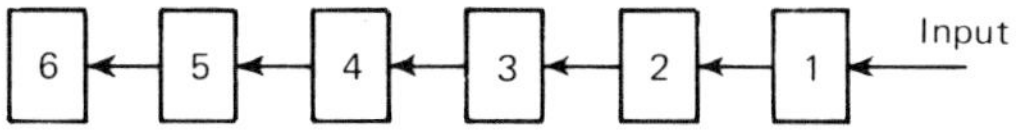 6 ←	5 ←	4 ←	3 ←	2 ←	1 ← Input

Decimal equivalent						
0						0
1						1
2					1	0
3					1	1
4				1	0	0
5				1	0	1
6				1	1	0
7				1	1	1
8			1	0	0	0
9			1	0	0	1
10			1	0	1	0

Figure 12.11 Binary code counting in a six-bit word

The simplest readout mechanism is one in which each switch operates a light bulb in the front panel of the scaler, so that the 1 state corresponds to the light being switched on. Such a display is difficult to read in binary notation but it has been very popular in the BCD system. A more recent development has been the introduction of the light-emitting diode (LED) display. A typical example is shown in *Figure 12.12*, in which each numeral is coded onto a seven-segment LED display.

Some older scalers use *decade switches*, with direct decimal readout. Mainly, they are ring counters, like the *dekatron*, with 10 analog positions reading in clockwise rotation. Essentially they are gas-filled valves with a central disc anode surrounded by 10 point cathodes. An input pulse transfers a gas discharge

Decade	4	3	2	1	Bit
		1	0	0	4
		0	1	1	3
	1	0	1	0	2
	0	1	0	1	1
Decimal number	2	9	6	5	

Seven-segment LED display

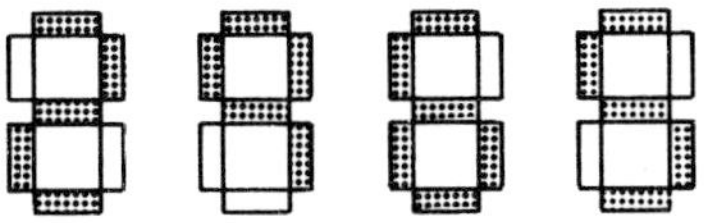

Figure 12.12 Binary-coded decimal (BCD) system

from one cathode to the next so that a visible glow is seen to move round the perimeter. When this reaches the top, zero position, a pulse is passed to the next decade switch, and so on.

One of the features which is common to binary and decimal counters is the transfer of an overflow pulse from one full binary or decimal digit to the next. The same can be done from one scaler to another. If, for example, a scaler has a capacity of six decades, then it can count up to 999 999. If a further pulse arrives, it resets all the nines to zero but generates an *overflow* pulse which can be read out to a second scaler. Thus, two six-decade scalers, connected in series, have a total capacity of $10^{12} - 1$.

Ratemeters

A ratemeter is a unit which measures the mean pulse *count rate*, in counts per second, averaged over an adjustable time interval known as the *time constant*. Alternatively, the count rate may be averaged over an adjustable number of counts and related to the *standard deviation* of the counts instead of the time interval during which they accumulate. In either case, selection is by means of a switch on the front panel of the unit, calibrated in time intervals, such as 0.5, 1, 2, 3, 5 and 10 seconds, or in terms of standard deviations, such as 1%, 2%, 5% and 10%. For a given count rate, the time constant is, of course, inversely proportional to the square of the standard deviation. When high accuracy is required, long time constants or small standard deviations should be selected, but when the need is to observe fluctuations in count rate, small time constants or large deviations may have to be tolerated.

The count rate is usually read from an analog meter, as illustrated in *Figure 12.10*. This can be calibrated on a *linear* scale or on a *logarithmic* scale, like that of a slide rule. A linear ratemeter must provide several, selectable count rate ranges to satisfy different experimental conditions. It is easier to read, but may require a considerable amount of switching from one range to another in the course of an experiment. Unless each range is accurately zeroed, there may be some relative error between one range and another. A logarithmic scale is continuous from very low to very high count rates but may be intrinsically less accurate and is more difficult to read because of the need for non-linear interpolation between the scale markings.

The simplest circuit for a linear ratemeter is one in which the incoming pulses are digitised then allowed to charge up a capacitance C through a resistance R. Basically, it is an integrating circuit with a leakage time constant RC. Since each digital pulse gives the same increment to the voltage across the capacitance, the total voltage developed across it in a given time is proportional to the count rate and is measured by means of an ordinary analog voltmeter. The integrating time constant of the ratemeter can be varied by altering the value of R or C and, therefore, changing the time constant RC. This process alters the voltage produced by a given count rate, so the different range scales have to be calibrated accordingly. Alternatively, range selection can be based on the absolute magnitude of this voltage and, therefore, on the number of counts accumulated and the standard deviation. A logarithmic ratemeter can operate on the same principle, but the voltage across the capacitance is read with a logarithmic voltmeter or amplified by a logarithmic amplifier.

Compared with a scaler, a ratemeter offers an immediate measure of count rate and variation in source strength with time. The scaler has to be operated over discrete time intervals but it provides the better accuracy of a digital device relative to an analog one.

Discriminators

There are two types of discriminator: *pulse height discriminators* and *timing discriminators*. The former determine whether or not the height of an incoming analog pulse exceeds an adjustable *bias* or *threshold* voltage and accordingly transmit an output, digital pulse suitable for counting. The latter measure the time at which an incoming analog pulse exceeds the bias, or threshold voltage and generate an output analog pulse whose leading edge defines that time.

The most common version of the pulse height discriminator is the *low-level discriminator*, or lower discriminator, which digitises and transmits pulses whose heights exceed the bias voltage and stops the smaller pulses. It is mostly used to suppress noise pulses. The alternative is the *high-level discriminator*, or upper discriminator, which transmits pulses whose heights are smaller than the bias voltage and stops larger pulses. It may be used to eliminate summed pulses, or pulses due to high-energy events that are not being investigated. *Figure 12.13* illustrates the effects of both types and their effect on a pulse height spectrum. The units themselves may be separate or they may be incorporated into other units, such as amplifiers or scalers. In either case, a multi-turn potentiometer is provided on the front panel to allow the bias voltage to be adjusted to suit experimental requirements.

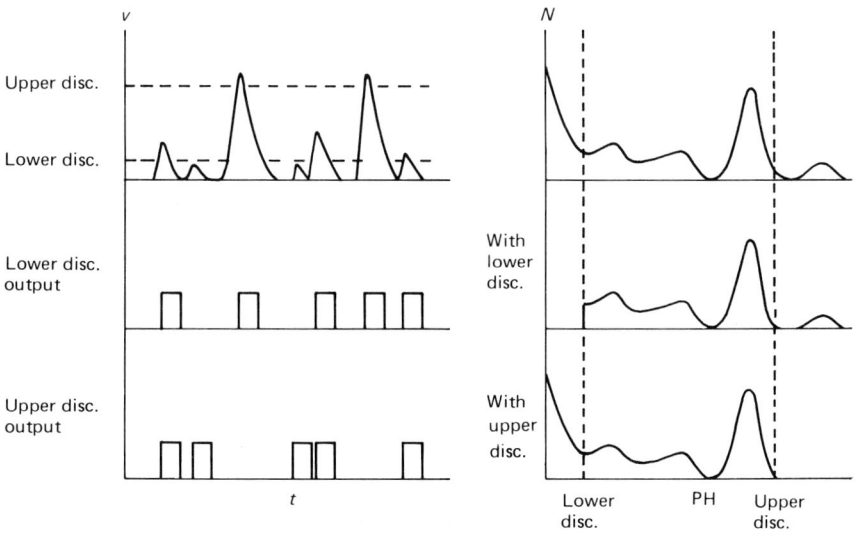

Figure 12.13 The effects of pulse height discrimination on analog pulses and on a pulse height spectrum

A number of circuits have been devised to provide pulse height discrimination. These include tunnel diode circuits, Schmitt trigger circuits and monostable multivibrators. Basically, they have a similar effect. When the leading edge of the incoming pulse exceeds a reverse bias, produced by a potentiometer, the circuit is triggered into an unstable mode from which it returns, after a preset time, to generate a square, digital pulse. This process results in low-level discrimination. High-level discrimination can be obtained by splitting the incoming pulse so that one part goes through a low-level discriminator to an anticoincidence unit and the other goes directly to the same unit. The anticoincidence unit produces an output pulse only when it receives one input pulse, namely the direct one.

For timing purposes, the discriminating voltage is usually set to about 30% of the expected analog pulse height. This is high enough to exclude noise pulse, but low enough to ensure that the circuit will trigger in the steep, leading edges of pulses of various sizes. The result is a narrow, digital pulse with an almost vertical leading edge which is excellent for timing measurements. A discriminator used in this mode is known as a *timing discriminator* or a *constant-fraction discriminator*.

Single-channel analysers (SCA)

If a low-level and a high-level discriminator are used together, the arrangement forms a single-channel analyser which passes only those pulses whose heights fall in the *channel*, or *window*, between the two bias voltages. There are therefore three variables, namely the lower, or *threshold voltage*, the *upper voltage*, and their separation, or *channel width*, of which only two need be adjusted to select a particular pulse height channel. In most cases, the channel width is much smaller than the other two voltages and, for accuracy, it must be one of the controllable variables. Thus, the threshold voltage, V, and the channel width, ΔV, are the variables for which potentiometer adjustment is provided.

There are two ways of using a single-channel analyser: for pulse height analysis (PHA) and for pulse height selection. In the first mode, the channel width, ΔV, is fixed at a suitable small value, while the threshold voltage, V, is varied from a minimum value, such as zero, to a maximum value which exceeds the maximum pulse height. At each point, the count rate is measured and plotted against the prevailing threshold voltage to obtain a graph of count rate against pulse height; that is, a *pulse height spectrum*. If, as is usually the case, the pulse height intervals, ΔV, form a contiguous set, then the result is a histogram rather than a continuous graph, since it is plotted over a series of discrete increments. If ΔV is relatively large — say, about 10% of the maximum pulse height — a very coarse histogram is produced, with poor resolution of structural features. If ΔV is smaller, the process takes longer to complete, but a more detailed histogram is obtained, with better resolution. It can be seen, therefore, that channel width may contribute to pulse height and energy resolution.

A smoothly varying pulse height spectrum is generated, as illustrated in *Figure 12.14*, if the threshold voltage is increased continuously. The process is known as *scanning* and it may be automated by employing a motor-driven potentiometer. Again, the count rate recorded at any point is the average value over the pulse height interval, ΔV. Spectral resolution improves as ΔV decreases and counting statistics improve as the scanning speed decreases, so the most

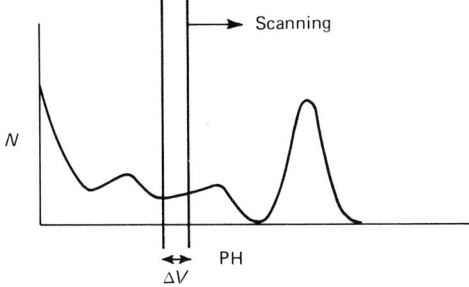

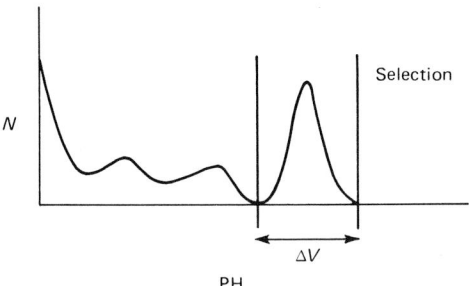

Figure 12.14 The function of a single-channel analyser (SCA)

accurate pulse height spectra are obtained with narrow channel widths and slow scanning speeds. A typical scan might take ten or twenty minutes.

In practice, a multichannel analyser is much faster and more accurate than a single-channel analyser. The latter is used only when a multichannel analyser is not available or when the purpose is to calibrate a single-channel analyser for a subsequent application involving *pulse height selection*. This is the major function of a single-channel analyser, in which the values of V and ΔV are fixed in order to count the pulses in only one specific region of a pulse height spectrum.

A typical use of this form of pulse height analysis is the selective assay of one radionuclide in the presence of others. The analyser is calibrated by recording a pulse height spectrum for a standard preparation of the radionuclide to be investigated. Further spectra are obtained for other radionuclides which may be present in the mixed source to ensure that their spectral peaks are resolved from that of the first nuclide. The mixed source is then assayed with the V and ΔV discriminators adjusted to transmit only those pulses due to the radionuclide of interest. The process is illustrated in *Figure 12.14* for a ^{137}Cs source whose photopeak is to be measured separately. The technique is widely used in liquid scintillation counting, when two or three separate counting channels may be set up to assay different nuclides, such as ^{3}H, ^{14}C and ^{35}S.

Multichannel analysers (MCA)

A multichannel analyser is really a composite of several nucleonic units but it is usually purchased as a single package and it performs a single function, so it is conveniently regarded as one unit. Its basic function is *multichannel analysis*. It provides an array of *counting channels*, or scalers, and systematically stores incoming pulses in these channels according to some prescribed criterion. The criterion defines a mode of operation. In *pulse height analysis* (PHA), pulses are allocated to counting channels on the basis of their heights. In *multichannel scaling* (MCS) they are allocated on the basis of their arrival times. These are the most common operational modes.

The structure of a multichannel analyser is illustrated in the block diagram of *Figure 12.15*. It comprises an *analog to digital converter* (ADC), a *memory unit*,

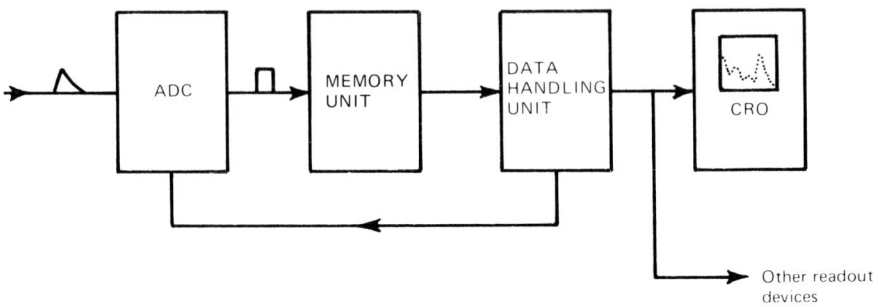

Figure 12.15 Block diagram of a basic multichannel analyser (MCA)

a *data handling unit* and a number of data readout units which must include a *cathode ray oscilloscope* (CRO) or *visual display unit* (VDU). The memory unit consists of a large number of counting channels, typically 128, 256, 512, 1024 or 4096 channels. Each of these is a scaler, made up of a group of ferrite cores or integrated circuits counting in binary code or BCD, as described above. The ADC allocates incoming pulses to one channel or another and the CRO displays the number of counts accumulated in each channel. The data handling unit controls the operation of the ADC and the readout process.

In the *PHA mode*, each incoming analog pulse is recorded in a channel on the basis of its height, so that channel number is proportional to pulse height. The ADC converts the analog pulse to a digital one but, in this case, unlike the digitisation process described in the above sections, it is the *pulse height* that is digitised. In a 512-channel analyser, for example, the digital pulse has one of 512 discrete values of height and is unambiguously allocated to the appropriate channel. When the memory contents are read out, they produce a graph, or histogram, of counts against channel number, which is a *pulse height spectrum*. Since all the incoming pulses, apart from dead-time losses, are recorded, the process is much more efficient and, therefore, faster than the action of a scanning SCA. A 512-channel analyser is effectively equivalent to 512 single-channel analysers operating simultaneously. In fact, the pulses are not counted simultaneously but are processed sequentially as they arrive at the ADC. This unit contains a clock oscillator generating high-frequency pulses. The clock pulses address each counting channel in turn and, in effect, increase the size of a digital pulse. When this exceeds the height of the analog input pulse, a count is recorded in the channel that is currently being addressed. Thus the data storage mechanism involves a form of *random-access memory* (RAM) and the time taken to address a memory location defines the counting dead time of the MCA. Consequently,

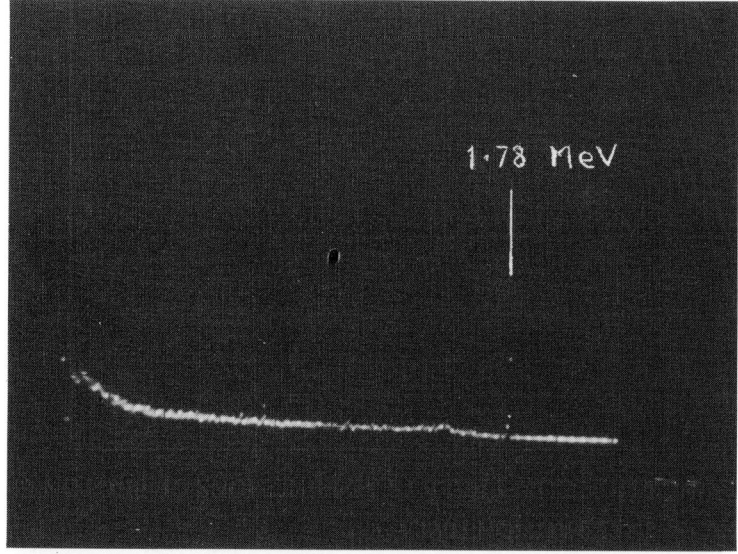
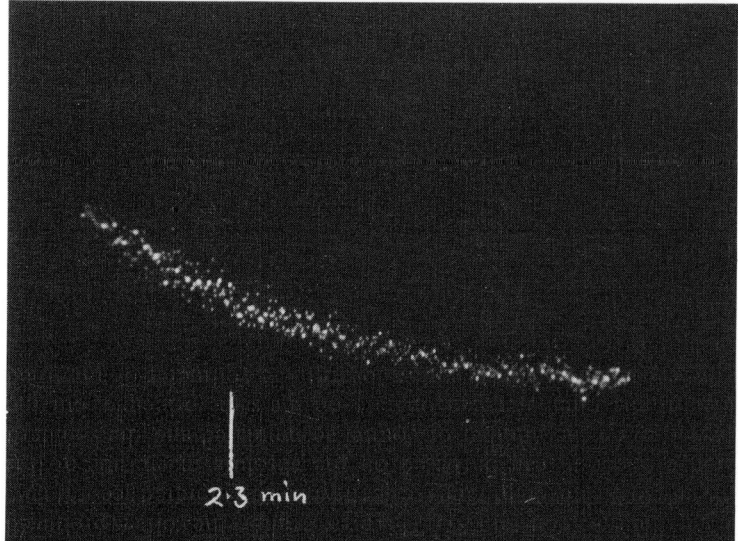

Figure 12.16 Pulse height spectrum and a decay curve for ^{28}Al recorded with a Ge(Li) detector

high clock frequencies, of the order of 20 MHz, are required to minimise dead-time losses.

In the *MCS mode*, the ADC clock addresses each channel in turn for a selected *dwell time*, which may be as large as about 1 s. During this time, *all* incoming pulses are recorded in the channel being addressed. Thus, channel number is proportional to the time of arrival of the pulses and, when the memory contents are read out, they form a graph of counts against time.

Figure 12.16 shows the results obtained in a study of the radioactive emissions of ^{28}Al (produced by neutron irradiation of ^{27}Al) using a Ge(Li) detector and an MCA operated in the PHA mode and the MCS mode. In the PHA mode, a gamma ray spectrum was obtained, showing the 1.78-MeV peak. In the MCS mode, the exponential decay of the aluminium, with a half-life of about 2.3 min, was observed.

There are many different MCA systems currently available and they vary considerably in design and capability. One model is shown in *Figure 12.17* to illustrate some of the controls that may be provided. These are distributed between the ADC and the data handling unit. If no separate *amplifier* is included in the MCA package then such a unit must be incorporated in the ADC in order

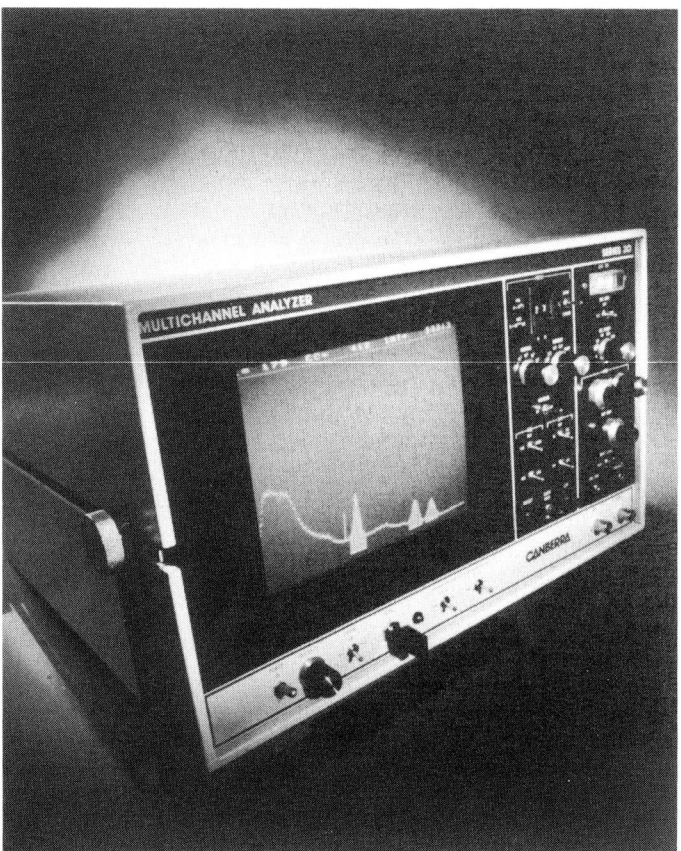

Figure 12.17 Multichannel analyser. (Courtesy of Canberra Instruments Ltd)

to expand or contract the range of pulse heights being recorded. It may be calibrated in terms of *conversion gain*, as the number of channels (e.g. 256, 512 or 1024) over which the pulse height spectrum is spread. If the analyser has only 256 channels, this means that the full spectrum, the lower half or the lower quarter of the spectrum, respectively, is displayed. Amplification alters the *dispersion* of a pulse height spectrum; that is, the pulse height or energy interval per channel. In addition, the ADC should have a *low-level discriminator* to eliminate noise pulses, and it may have a *zero shift* control. This moves the entire spectrum to higher or lower channels without changing the dispersion.

In setting up the ADC, the zero shift should be adjusted to bring the electronic noise spectrum into the first few channels, then the low-level discriminator should be wound up to exclude this distribution. If the pulse height spectrum then occupies only a small fraction of the total channel range, the gain should be increased and the above process repeated. If the pulse height spectrum is not contained within the available channels, the gain should be reduced and the zeroing process repeated.

The controls of the data handling unit are usually more easily interpreted, although the terminology used may vary from one system to another. Normally, a user's manual or handbook is provided with the equipment. There should be the following:

(1) a PHA–MCS switch to select the analysis mode;
(2) a COUNT–STOP–READOUT switch to acquire data and read them out;
(3) a READOUT MODE control to select the readout device from, for example, a CRO, X–Y plotter, teletype and a tape punch;
(4) a CLEAR control which erases the memory and usually takes the form of two pushbuttons which must be pressed simultaneously;
(5) a PRESET LIVE TIME control to select the live counting time;
(6) a PRESET DWELL TIME control to select the dwell time for MCS mode counting; and
(7) a y-scale amplifier to adjust the size of the CRO display in the direction of the counts, or N, axis. This can be adjusted after the data have been collected to optimise the vertical spread of the output graph.

A multichannel analyser can be regarded as a small, dedicated computer. It reads data into a formatted array, performs some data processing and reads out the results in the form of a graph or a list of channel contents. In the system described so far, these functions are hardware-controlled and of limited range. Greater flexibility can be obtained by adding more computing capability. Two further stages of development can be identified, leading to what may be described as *intelligent MCA* and *computer-based MCA*.

An *intelligent MCA* has a number of data analysis functions built in. They are *hardware-controlled*; that is, executed by specific-purpose electronic circuits and selected by front panel controls. The options available include alphanumeric character generation in the visual display, isolation of selected spectral areas, mean peak channel location, spectrum subtraction, automatic energy calibration, integration of peak counts, logarithmic readout, and a data library for isotope identification.

A computer-based MCA is an MCA interfaced, by means of suitable nucleonic units, with an on-line microcomputer. This consists of a processor unit, with its own random-access memory, and, possibly, further bulk storage memory devices and additional peripheral units for data readout. The system may be *software-controlled*. If so, it has an *operating system* capable of compiling a high-level language, such as FORTRAN, so that experimental data can be analysed according to programs and data files provided by the user. The data acquisition process can also be controlled by software, so the system is much more flexible than the hardware-controlled MCA.

Timing units

A timing unit is any unit whose primary function relates to the alteration, selection or measurement of pulse arrival times. A number of devices fall within this category, including *linear gates, timers, delay units, coincidence* and *anti-coincidence units, time to amplitude converters*, and *time to digital converters*.

As illustrated in *Figure 12.18*, the basic function of a *linear gate* is to supply a second unit with a gating pulse which effectively switches the other unit on or

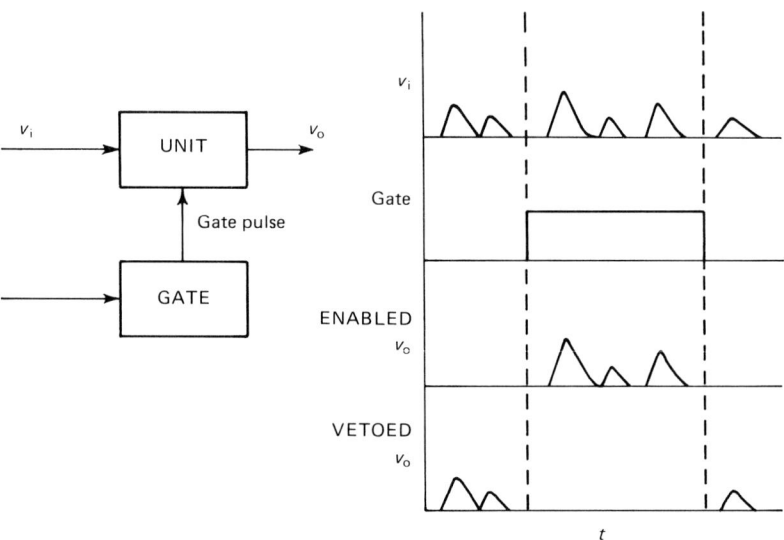

Figure 12.18 Functions of a linear gate

off for the duration of the pulse. Typically, the gating pulse is a long, digital pulse such as may be generated by a monostable multivibrator with a long recovery time constant. It can be applied in the ENABLE mode, to switch the other unit on, or in the INHIBIT or VETO mode, to switch the unit off. It is used, for example, as a fixed dead time generator. Some nucleonic units, such as MCA, have fairly long processing times and, if a second pulse arrives while a previous one is still being processed, the two may be summed. To avoid this possibility, the first pulse triggers a VETO gate pulse which inhibits the response of the input circuit for the dead-time interval required for processing.

A *timer* is a clock pulse generator connected to a scaler which counts the clock pulses and, therefore, time intervals. An *automatic timer* causes the equipment being controlled by it to count for a preset time, then stop. It does so by applying some form of ENABLING gate pulse to the circuit. The counting is terminated by an overflow pulse from some preselected point in the binary array of the timing scaler. This results in a preset real time. Preset live time is obtained by means of gating circuits which, triggered by the signal pulses, switch the timing circuit off while the signals are being processed.

A *delay unit* delays the passage of a signal pulse for an adjustable time; that is, it imposes an adjustable delay time between the arrival of an input pulse and the production of the output pulse. The simplest delay unit is a length of coaxial cable, which delays a pulse by about 3 ns m^{-1}. Longer delays are obtained with other forms of inductive–capacitative transmission lines, typically of the order of a few ns mm^{-1}.

A *coincidence unit* accepts two (or more) input pulses and generates a single output pulse only if the two inputs are received in coincidence, or, more precisely, within an adjustable resolving time of each other. In effect, the input circuit of one pulse is gated, in the ENABLE mode, by the input circuit of the other and the resolving time is determined by the length of the gating pulse. In fact, coincidence units employ logic gates rather than linear gates, and these require digital pulses, so coincidence units invariably convert input analog pulses to digital ones. The analog property may be retained by using a linear gate instead of the coincidence unit.

It frequently happens that two coincident events do not produce coincident pulses at the coincidence unit because they are delayed by different amounts in preceding electronic units and coaxial cables. The delay can be measured with a CRO by observing one pulse on a time base externally triggered by the other, earlier pulse. It can be nullified by inserting a variable delay unit in each circuit, or in the faster circuit, and adjusting these until true coincidence is observed as such. Simultaneous pulses from a pulse generator are useful for this purpose.

An *anticoincidence unit* also has at least two input points. If it receives two coincident inputs it generates no output pulse. If it receives only one input pulse within its preset resolving time, it does generate an output pulse. In effect, one circuit is gated, in the VETO mode, by the other. Again, a gating circuit can be used instead of an anticoincidence unit to preserve analog properties.

A *time to amplitude converter* (TAC), or time to height converter, accepts two input pulses and measures the time interval between them. The earlier pulse is designated as a START pulse and the later arrival as a STOP pulse. The START pulse starts a ramp voltage generator and the STOP pulse stops it, so that a single analog pulse is generated whose height is proportional to the time interval between the two inputs.

The output pulse can be passed on to an SCA to select and count a particular time interval. This provides a time of flight method of discriminating against one type of radiation, as described in an earlier chapter. Alternatively, a complete *pulse time spectrum*, that is, a graph of counts against time interval, can be obtained by reading the TAC pulses into an MCA. This method is used, for example, in neutron energy spectrometry in which neutron transit times are measured, and in position measurement in the plane of a spark chamber, as described above. In the latter mode, the TAC is followed by the ADC of the analyser. In a specific-purpose system, these two units can be collapsed into one *time to digital converter* (TDC) in which the START pulse starts a clock oscillator addressing the memory and a STOP pulse stores a count in the location being addressed.

If a TAC unit is not available, there are, of course, two other ways to measure the time interval between two pulses, provided it is approximately constant. One is to observe the STOP pulses on a CRO triggered by the START pulses. The other is to use a coincidence unit and a variable delay, in the START pulse line, which is adjusted until coincidence counting is observed.

Multiple-input units

There are several units which accept input pulses from two or more detector circuits, perform some operation other than searching for coincidences, and output signals in a predetermined format. They include the following:

(1) A *mixer* transfers the pulses from two or more input circuits to a common output unit, such as a scaler or an MCA. In whole-body counting, for example, several detectors, each with its own preamplifier and amplifier, may be used to examine a patient. The outputs of these circuits are fed into a mixer and counted in a single scaler, improving the overall detection efficiency and sampling a larger area of the subject.
(2) A *fan-out* unit accepts several inputs and, for each one, produces two or more outputs for parallel processing. One output, for example, may lead to a scaler and another to an MCA.
(3) A *subtraction ratemeter* measures the difference between two input pulse count rates. It is used to compare two activities, in renography, for example, where one detector examines each kidney and differential activity is to be determined.
(4) A *multiplexer* scans a number of input channels and routes output signals according to preselected criteria. In a typical arrangement, up to four channels are interrogated and their contents are transferred, in a selected mode, to the memory of an MCA so that any permutation of pulse height spectrum displays are obtained ranging from one integrated spectrum to four separate spectra.

12.3 NUCLEONIC SYSTEMS

A system consists of a number of units arranged to perform some specific function or range of functions. In addition to the units described above, it must include a *power supply* to provide all the units with appropriate operating voltages and current, and a *high-voltage* (HT) unit to provide the detector with its applied potential. Both of these must be *stabilised* against random fluctuations, a.c. ripple and steady drift in the values of their output voltages. The whole package can be assembled as an *integral*, or *hard-wired, system*, or as a *modular system*.

An *integral system* contains all of its units in one case — with the possible exception of preamplifiers, which may have to change with detector type. It is treated as a single piece of equipment and its component units are not immediately recognisable as such. They are circuits, or circuits boards, or integrated circuits or large-scale-integrated circuits (LSI) mounted inside the case. The system has some advantages over the modular system. It is more compact and simpler to use because it has no external connections to be made and fewer, clearer controls. Standard systems, for which there is a large market, are usually less expensive than modular alternatives. One such system, a scaler–timer–single-channel analyser, is illustrated in *Figure 12.19*. This particular system includes its own built-in preamplifiers with different, rear-mounted input sockets for Geiger–Müller counters and scintillation counters. It is shown as a block diagram here and as a physical representation in *Figure 6.1*.

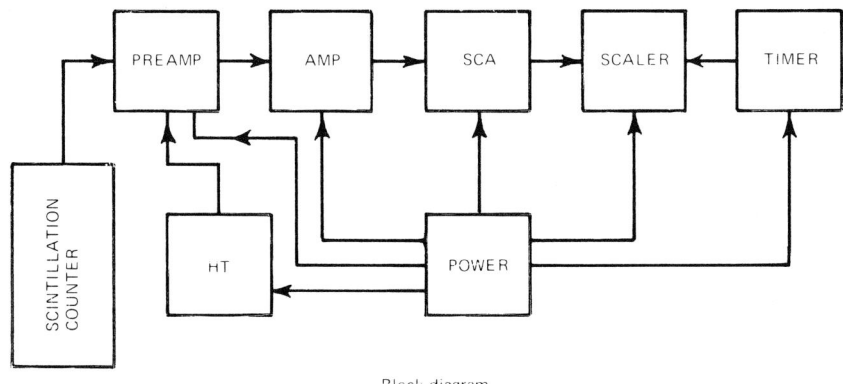

Figure 12.19 Scintillation counting system with automatic timing and single-channel analysis. (Courtesy of Nuclear Enterprises Ltd.) A modular system to perform the same function is shown in Figure 12.9

In a *modular system*, each unit is a separate *module* which plugs into a multipin socket in a *bin* or *crate*. The bin contains a fixed number of spaces, each provided with a multipin socket, and a module occupies one, two or more spaces depending on whether it is a single-width, double-width or wider unit. The sockets are wired up to convey power to the modules from a supply unit which may be plugged in as a module or suspended on the rear of the bin to leave more spaces free for other modules. Two or more bins may be connected together and stacked on top of each other, so that one power supply can serve all the modules, provided their total current requirement does not exceed the maximum output of the supply unit.

In general, the multipin sockets have more connections than are required for power supplies and the extra ones can be used to transfer signal pulses from one unit to another. This is especially likely when the system is provided as a complete assembly. In a multichannel analyser, for example, signal transfers between the ADC and the memory, and between the memory and the readout devices, are routed through the rear connections, as are the control signals from the data handling unit. Additional intermodule connections are provided on the rear panels if the existing pathways are insufficient. In other cases, however, modular systems are made up by the user from individually selected units and signal pulses are transferred via coaxial cables which are usually connected to the front panels. A typical system, designed to perform the same function as the integral system given above as an example, is described in *Figure 12.9*.

A modular system is more complicated than an integral one and requires more technical expertise to set up and operate properly, but it has some advantages. First, it is relatively easy for a user with no knowledge of circuit theory to locate a malfunctioning unit, by observing the input and output pulses at each module, in turn. The faulty module can be replaced or dispatched for repair, on its own, so maintenance is simple. Secondly, a given set of modules can be arranged in a number of ways to construct different systems, so the modular system has greater flexibility and is the least expensive way of establishing a range of functions.

334 Nuclear electronics

The earliest modular systems varied considerably from one manufacturer to another in dimensions, signal pulse properties and power supply requirements, so that units were not interchangeable between different brands. This product incompatibility partly defeated the aims of modularisation. In 1964, the USAEC Committee on Nuclear Instrument Modules drew up specifications for a standard system, known as the Nuclear Instrumentation Module Standard (NIMS). This is now widely adopted by all manufacturers so that modules are interchangeable. The NIMS (or, to use the popular abbreviation, the NIM) bin has 12 single-width

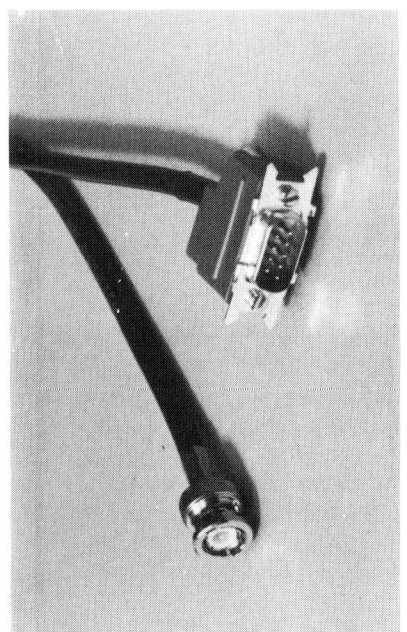 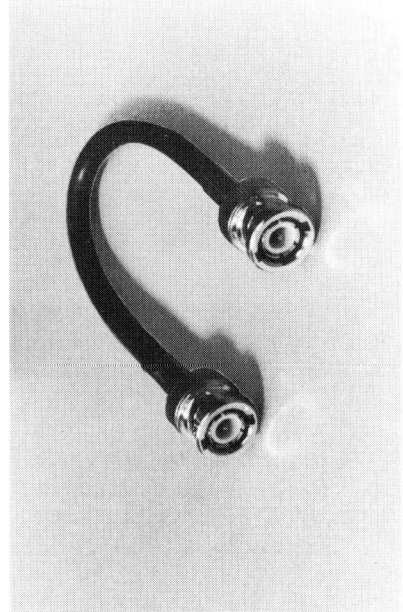

Figure 12.20 Coaxial cables and connectors. (Courtesy of Canberra Instruments Ltd)

modules in a length of 19 in (4826 mm). Shorter bins are available, for fewer modules, and two different module heights can be obtained but otherwise the dimensions are standard. Power supplies are ±24 V, ±12 V and ±6 V, although the last two are not always required or provided.

A remaining problem is that cable connectors and sockets are not standardised. Many types are used, including BNC, amphenol and lemo, for signal transfer, special high-voltage versions for detector–preamplifier connections, and a wide range of multipin connections to provide supplies from the bin to a free-standing preamplifier. Some connectors are shown in *Figure 12.20*.

Block diagrams

From the user's point of view, the most convenient method of designing and discussing a nucleonic system is to refer to a block diagram, and the quickest way of learning how to do so is by considering some examples, as follows:

(1) *Figure 12.19* shows a block diagram for a *scintillation counting system* with provision for automatic timing and single-channel pulse height analysis. *Figure 6.1* shows what the system might look like in integral and *Figure 12.9* in modular form. The analog signal pulses are processed, first of all, in a free-standing voltage-sensitive preamplifier. Power supplies to this unit are provided by a multicore cable from the rear panel of the amplifier, and the HT supply is routed direct from the high-voltage unit. The pulses pass on to the amplifier, then to the SCA, which can be operated in the differential mode, as a single-channel analyser, or in the integral mode, as a low-level discriminator, to eliminate only the noise pulses. The SCA output pulses are counted by a scaler, which is controlled by the timer. The power supply in this case is a modular unit.

(2) The second example, described in *Figure 12.21*, is of a *gamma ray spectrometer* based on a Ge(Li) detector and an MCA. The charge-sensitive preamplifier has a TEST input which accepts pulses from a signal generator as a means of testing the system.

The amplifier should be a spectroscopic amplifier with provision for pulse shaping, to optimise signal-to-noise ratio, and baseline restoration, to cope with high count rates. The MCA is shown as a single unit, although it comprises an ADC, memory unit, data handling unit and a VDU. Alternative readout systems are provided. The X–Y plotter produces a permanent copy of the pulse height spectrum, the teletype provides a digital printout of the channel contents, and the tape punch reproduces the same data on paper tape for subsequent computer analysis. In this example, the bin and power supply are added as an unconnected block to simplify the diagram. In many cases, they are omitted entirely from the diagram.

(3) *Figure 12.22* shows a *coincidence/anticoincidence counting system*. In the *coincidence mode*, true coincident events are observed simultaneously in detectors A and B whose output signals are processed in fast, timing preamplifiers and timing amplifiers. The amplifier outputs are passed to a coincidence unit so that the scaler records only true coincidences. Additional features include a timing unit, to select a preset counting time, a single-channel analyser, to select a particular energy and, therefore, type of event and, in the other detector line, a delay unit. The delay unit can be adjusted to match the pulse transit times in the two lines so that coincident detector pulses arrive at the coincidence unit within its resolving time, but it has another function. When a large delay is inserted in one line, the detector signal pulses are moved out of coincidence and the scaler then records random coincidences between noise pulses. The random-coincident count, which is not affected by the delay time, should be subtracted from the apparently true coincidence count to obtain the exact value of coincidence count.

This type of system has a number of applications. One of these is in *liquid scintillation counting*, where detectors A and B are photomultiplier tubes examining the same sample bottle. It provides a means of separating the low-energy signals typical of the ^{3}H and ^{14}C sources employed in liquid scintillation counting from the substantial photomultiplier noise. Another application is in the counting of annihilation photons from *positron emitters* and, in this form, the circuit can be included in the processing system of a positron camera.

In the *anticoincidence mode*, the system can be used for *low-level counting* as, for example, in carbon dating measurements and environmental radiation

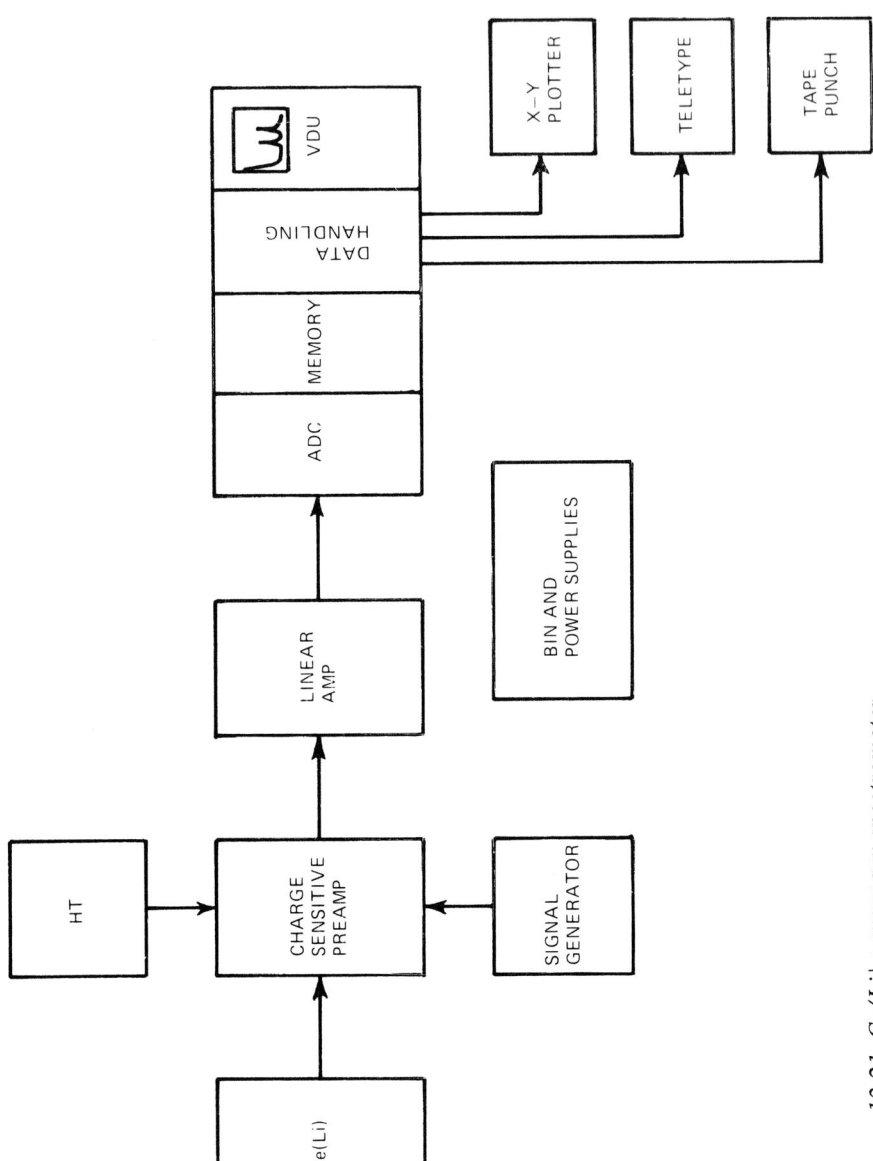

Figure 12.21 Ge(Li) gamma ray spectrometer

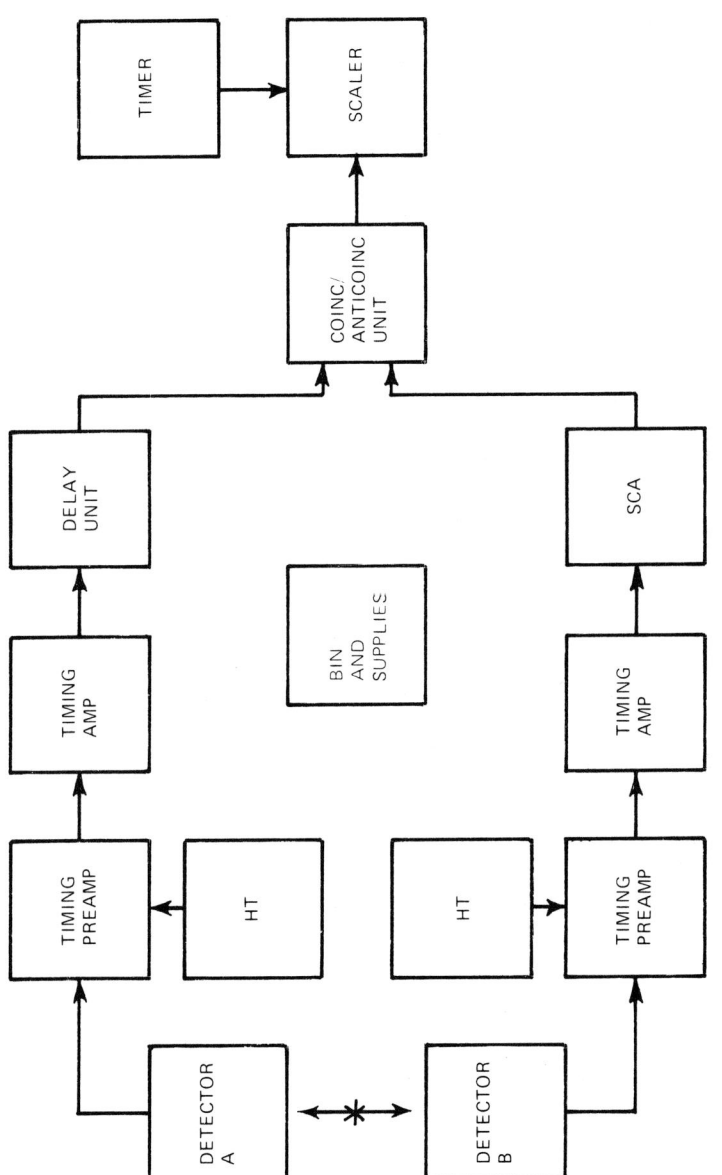

Figure 12.22 Coincidence/anticoincidence counting system

studies. In this case, detector B is the primary detector, such as a Ge(Li) detector, which records the source radiations, and detector A is an anti-coincidence shield, such as an annular scintillator. Incident cosmic rays are detected, in coincidence, by both detectors and eliminated by the anti-coincidence unit so that only the source radiations are counted by the scaler.

(4) A *time of flight spectrometry* system is illustrated in *Figure 12.23*. An incident particle registers in detector A then in detector B. These generate a START pulse and a STOP pulse which are fed into a TAC unit. The latter produces an analog pulse whose height is proportional to the time interval between the two input pulses and, therefore, to the time taken for the incident particle to travel between the two detectors. The TAC pulses are recorded in an MCA as a pulse height spectrum which effectively shows a graph of counts against particle velocity. For timing accuracy the preamplifiers, amplifiers and low-level discriminators are all fast, timing units. An added feature is the SCA and gating unit, which ENABLES the MCA to respond only to particles of a specified energy. This simultaneous measurement of velocity and energy identifies the mass of the particle.

(5) A *delay line readout system* for bidimensional imaging is described in *Figure 12.24*. As pointed out in a previous chapter, it is widely used with imaging multiwire proportional counters and is shown in the figure with such a detector. An event in the chamber produces three output pulses, namely, a PROMPT signal from the central anode and delayed pulses from the cathode arrays on each side of it. The anode signal acts as a START pulse for two TAC units. The cathode pulses are delayed by the times taken to reach the ends of their respective delay lines and provide a STOP pulse for each TAC. Since the delay times are proportional to the x and y coordinates, so too are the heights of the TAC output pulses.

The image is reconstructed, in this example, in a *display oscilloscope*. The Y pulse is fed into the SIGNAL input socket to deflect the beam in the y direction and the X pulse is entered via the EXTERNAL TIME BASE socket to deflect the beam, simultaneously, in the x direction. A third pulse, known as the Z unblanking pulse, is input via a third socket. This unblanks the beam when it has been deflected to its target coordinates so that a single point is displayed rather than a continuous line from the origin. An ordinary oscilloscope shows each event as a transient light flash. These have to be integrated into an image. This can be done by taking a long-exposure photograph of the screen or by using a *storage oscilloscope* which holds each image point for a time that is long enough to collect many such points in the same display.

(6) *Figure 12.25* shows another method of reading out position coordinates and an alternative means of displaying the image. The system is that used with the Anger *scintillation camera*. The hexagonal array of, in this example, 19 photomultiplier tubes produces five output pulses per event through an impedance network. X^+ and X^- pulses are obtained from opposite ends of the network in such a way that the pulse height difference, $X^+ - X^-$, is proportional to the x coordinate and, of course, to the total energy deposited in the NaI(Tl) crystal. This difference is measured then divided by the Z pulse, which is proportional to the energy deposited, to yield a normalised X signal whose height is proportional to the x coordinate. The Y^+ and Y^- pulses are similarly processed to form a Y pulse.

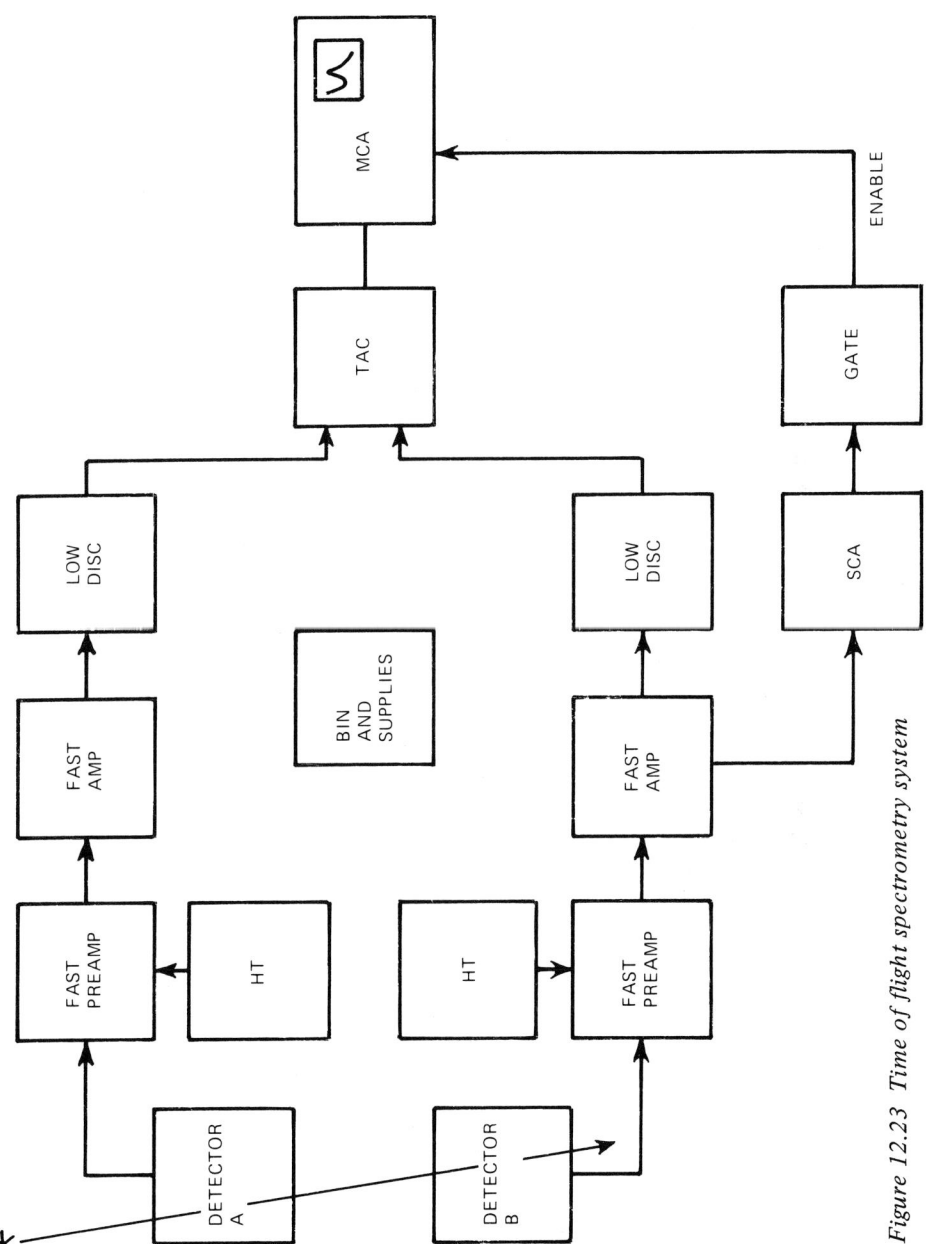

Figure 12.23 Time of flight spectrometry system

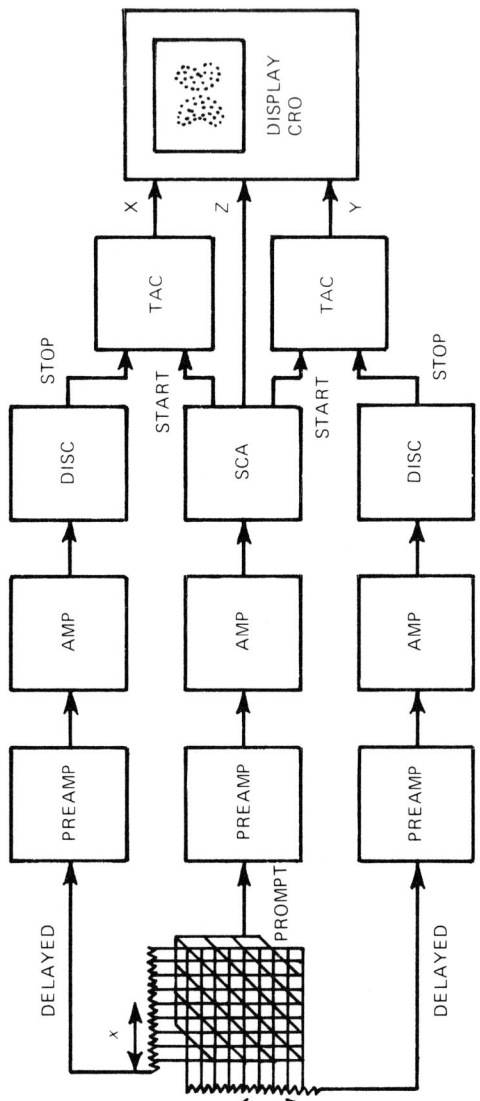

Figure 12.24 Delay line imaging system with a display oscilloscope for a multiwire proportional counter

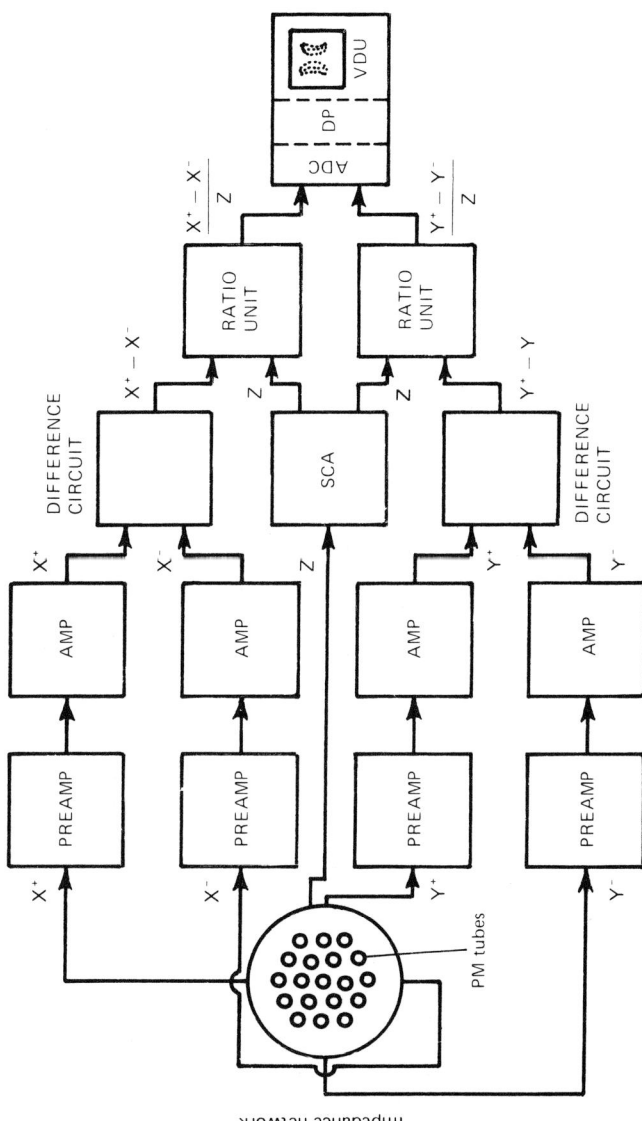

Figure 12.25 Scintillation camera system with digital display

Events may be displayed as image points by feeding the X and Y pulses into a display oscilloscope, as described in the previous example, but *Figure 12.25* illustrates another method of producing an integrated image, by means of an MCA. This incorporates a *dual-parameter input and display controller* which stores data in a two-dimensional array in the memory unit and reads out in the *contour* mode to form an integrated image. The number of elements in the image matrix is limited by the size of the memory and, for this reason, image quality is usually poorer than that obtained with a display oscilloscope, but the image is a fully quantitative one, in that the number of counts in each element is recorded and stored in digital form. The advantage of this feature is that it facilitates image processing. In addition, it leads naturally to the concept of digital data storage with on-line computer processing.

12.4 COMPUTERISED SYSTEMS

A radiation detection system has two functions: *data acquisition* and *data processing*. It acquires data in the form of counts, times of events, pulse heights, and so on, and processes them into more directly useful forms, such as count rates, coincidence count rates, pulse height spectra and spatial distributions. The investigator, using the equipment, also performs two functions, namely, further *data processing* and *data analysis*. Both tend to require the injection of externally available information in the form of published data, correlations between two or more measurements, or established criteria. Data processing may involve the calibration of a pulse height spectrum and its conversion to an energy spectrum, the comparison of many count rates in a liquid scintillation counting experiment and their conversion to source activities, the organisation of an image display to enhance certain features, the reconstruction of an image in computed tomography, and so on. Data analysis involves the use of experimental, processed data to draw conclusions regarding the system under investigation. The results of a liquid scintillation counting experiment may be used to study a biochemical process, a a bidimensional image provides the basis for medical diagnosis and the results of a high-energy physics experiment are applied to the study of fundamental particles and nuclear systems.

Data analysis is usually performed away from the detection system. It may require considerable interpretation and computation, and involve the use of an off-line digital computer. It is not part of the radiation detection system. Data processing, however, is part of that system and can also be computerised, but in this case there is some advantage in using an on-line computer that is an integral part of the experimental equipment. Such a computer can provide processed results in real time, during or soon after the measurement stage, and it can control the experiment. It is programmed beforehand to carry out a specific data processing function that would otherwise have to be performed by the user. The major benefit is a considerable saving in time, as well as some improved reliability, especially when the process involves *data logging*, that is, the organisation of substantial quantities of data, or *computation* of a complex or lengthy nature.

Electronic data processing can be achieved by *analog* or *digital* methods. Analog processors, such as pulse shapers, amplifiers, ratemeters and display oscilloscopes perform specific and direct mathematical operations. They deal in continuous variables such as voltages or currents, and tend to produce approximate results. Digital processors, such as scalers, clock oscillators and MCA, count in terms of discrete variables and store data as absolute numbers. In this respect, at least, they are more accurate than analog equivalents. A combination of scaler and timer, for example, yields a more accurate estimate of count rates than does a ratemeter, while a digital image is more quantitative than that of a display oscilloscope because it shows the absolute number of counts at each image point rather than the relative intensity.

These examples illustrate two other points of comparison. First, the digital process involves more operations and is less direct than the analog one. For simple operations, where absolute accuracy is not required, analog circuits are preferred. Secondly, because it does store intermediate data rather than a single complete measurement, the digital process allows further calculations to be made, as required. For example, the scaler—timer combination provides sufficient data for the retrospective estimation of statistical counting errors and dead-time losses, while the digital image facilitates similar forms of quantitative assessment as well as further processing to highlight specific features. These operations can be performed by an on-line digital computer.

Such a computer extends the data processing capability of a nucleonic system in two major respects. First of all, it has a *processor*, or central processor unit (CPU), which is able to perform a wide range of arithmetic and logical operations by means of digital circuitry. Secondly, it has a *memory* unit, and this can store a set of instructions, known as an *object program*, which controls the function of the processor, so that the user can program the computer to carry out any of a wide range of specific operations.

The memory consists of an array of ferrite cores or metal-oxide-silicon (MOS) circuits organised into *words* or *bytes*. Typically, there are two bytes per word and each byte may comprise 8 bits of binary storage space. The total memory size is expressed in K words or K bytes, where one K is equal to 1024. A typical minicomputer has about 24 K words of memory. Numbers are stored in binary code and instructions, such as those of an object program, are stored in a similar code known as *machine code*. All data are stored in a formatted array which can be accessed randomly by the processor, so the memory is described as a *random-access memory* (RAM). Machine code is inconvenient for the user because it is orientated towards the needs of the processor rather than the user's problem. For this reason, the memory may contain another set of instructions, known as a compiler, which translates the user's instructions, or *source program*, from a high-level, problem-orientated language, such as FORTRAN, to the low-level machine code. The compiler is usually part of a larger set of instructions, referred to as an *operating system*, which automatically controls all the functions of the computer. Thus, a computer consists of physical components, or *hardware*, and various stored data and operating instructions, or *software*.

The basic structure of a computer is described in *Figure 12.26*. Besides the processor and its memory unit, there is a *control console*, through which the operator can interrogate the processor and override the conditions prevailing in that unit, a number of *peripherals* and, usually, a *bulk-storage device*. These are connected to the processor through an arrangement of many parallel wires

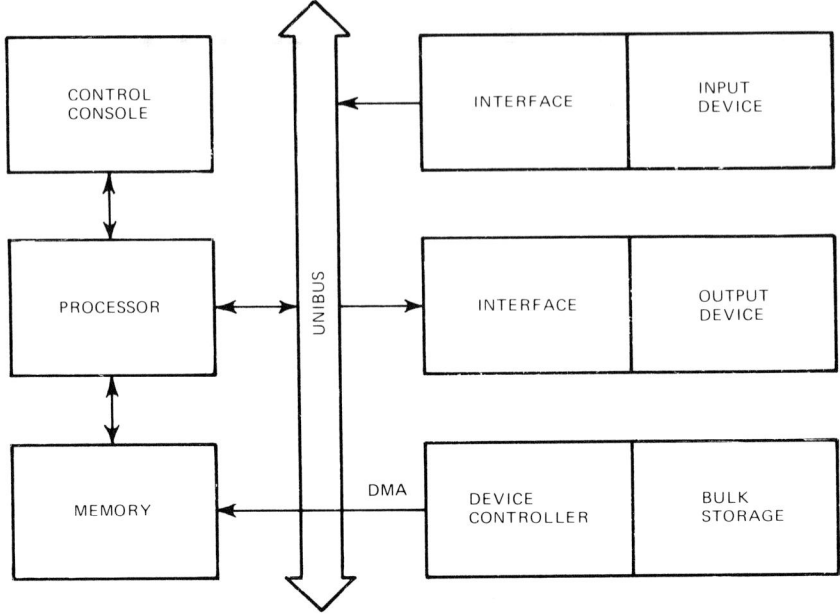

Figure 12.26 Physical structure of a computer

known as a *unibus* or, sometimes, a bus, a bus-bar or a data highway. Each peripheral unit and bulk storage device is connected to the unibus through an *interface* unit which matches any type of unit to the particular processor in use. Peripherals are *input devices*, such as card readers, paper tape readers, analog to digital converters (ADC), and teletype keyboards, as well as *output devices*, such as teletype printers, line printers, card and tape punches, and digital to analog converters for use with X—Y plotters and visual display units (VDU). The bulk storage devices include magnetic tapes, hard and floppy magnetic discs, and magnetic drums. They are used for intermediate storage of large quantities of data or as backing stores for programs and other software, which can be transferred through a device controller interface, by direct memory access (DMA) to the main memory unit.

With on-line computers, the input devices are part of the experimental equipment and, if experimental control is required, so too are some of the output devices. If very large quantities of data are generated, they are often read into a bulk storage device and transferred later, in data blocks, to the main memory. In order to meet the interfacing requirements of the computer and to provide suitable digitised data, a special nucleonic system has been developed. It is known as a computer-aided measurement and control system (CAMAC), and is a modified version of the NIMS modular system.

The CAMAC crate is the same size as a NIM bin but its modules are half the width of NIM modules. The crate contains 25 module locations, of which the last two are reserved for a *crate controller*. This acts as an interface and control unit between the CAMAC system and the particular computer being used. The crate has a unibus, or data highway, like that of a computer, and each module

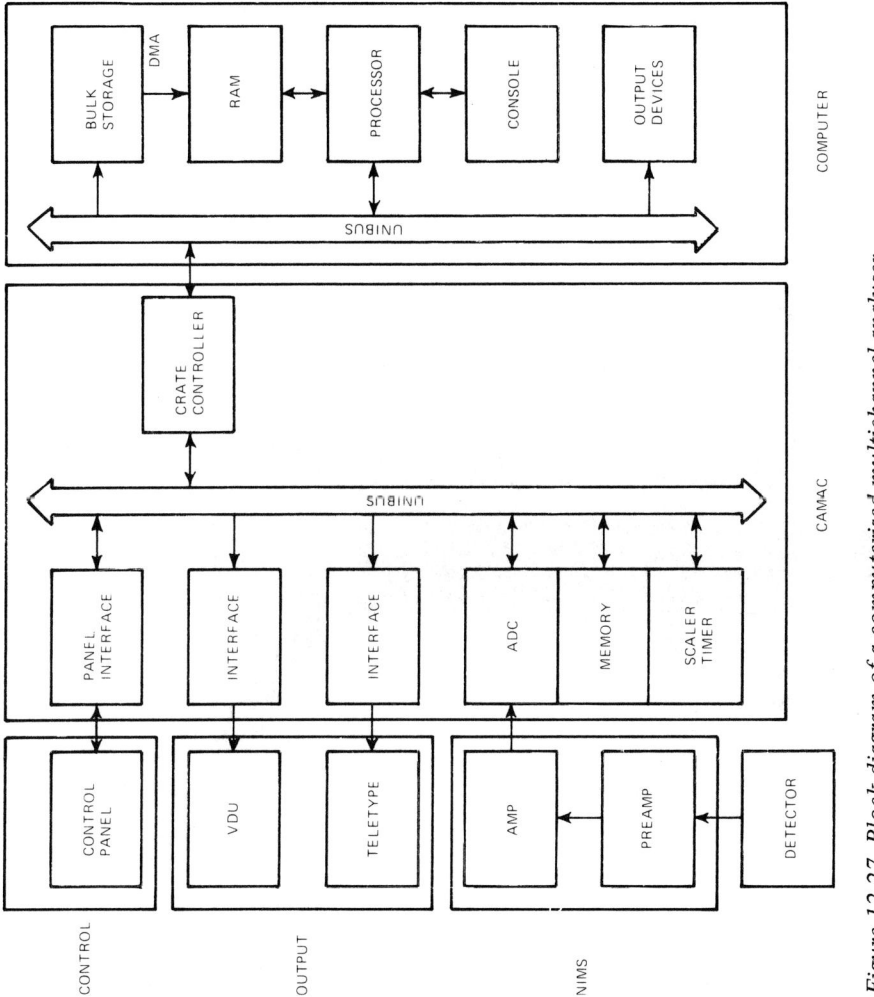

Figure 12.27 Block diagram of a computerised multichannel analyser

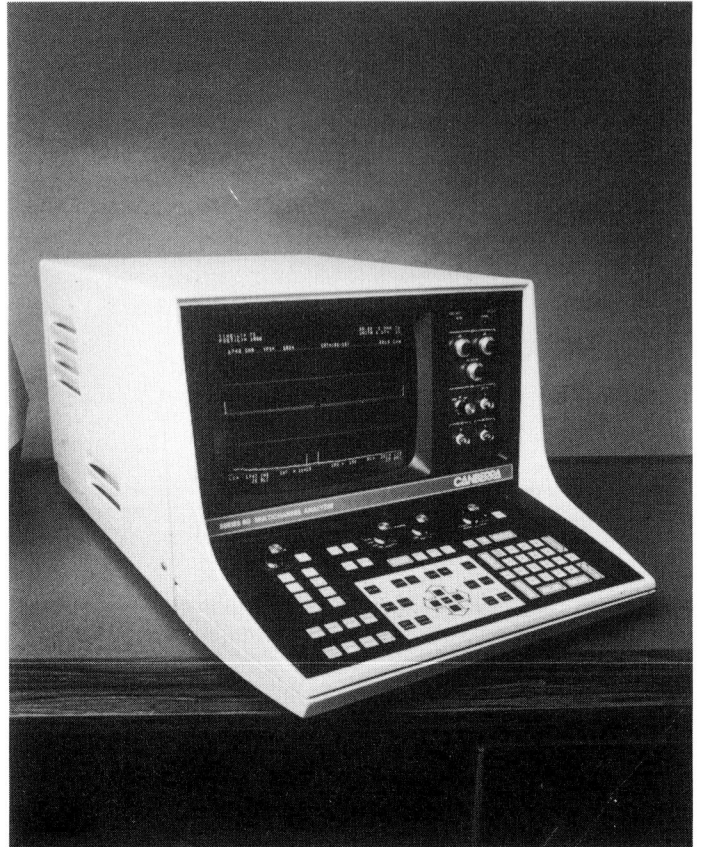

*Figure 12.28 Computerised MCA system.
(Picture by courtesy of Canberra Instruments Ltd)*

plugs into it via an 86-way printed edge card connector. Signals transferred through the data highway are coded to address specific units and sub-addresses, such as memory areas, and to define functions, such as READ or WRITE. They are controlled by the computer software, through the crate controller, or directly by the operator's console which can be interfaced to the CAMAC crate.

The CAMAC system is itself a peripheral of the computer and its modules can be regarded mainly as interface units. Basically, there are three types of module, concerned with input, output and control functions. The *input units* include ADC, scalers, timers, multiplexers, and so on, which interface with analog measurement systems, such as NIMS units, and convert experimental data into digitised, computer-compatible forms. The *output units* interface with readout equipment such as VDU, teletypes and line printers, while the *control units* interface with the operator's control console. A typical arrangement, for a computerised MCA system, is illustrated in *Figure 12.27*. This omits some detail but does describe the main features of a basic system. A photograph of such a system is shown in *Figure 12.28*.

In concluding this chapter, it may be noted that nuclear electronics is a technology-dependent and fast-changing subject. Textbooks tend to become outdated in a short time. Fortunately, the equipment manufacturers produce comprehensive catalogues and it is recommended that these should be consulted for further information.

BIBLIOGRAPHY

CHIANG, HAI HUNG, *Basic Nuclear Electronics*, Wiley–Interscience, New York (1968)
HERBST, L. J. (Ed.), *Electronics for Nuclear Particle Analysis*, Oxford University Press, London (1970)
KOWALSKI, E., *Nuclear Electronics*, Springer-Verlag, Heidelberg (1970)
NICHOLSON, P. W., *Nuclear Electronics*, John Wiley, New York (1974)

Appendix 1

PERIODIC TABLE OF THE ELEMENTS

Legend:
- I — Group
- Z number
- Symbol
- Outer electron configuration
- Ionisation potential of a single atom (eV)

I	II								
1 H 1s 13.6									
3 Li 2s 5.39	4 Be $2s^2$ 9.32								

		IIIa	IVa	Va	VIa	VIIa	VIII	VIII	
11 Na 3s 5.14	12 Mg $3s^2$ 7.64								
19 K 4s 4.34	20 Ca $4s^2$ 6.11	21 Sc $3d\,4s^2$ 6.56	22 Ti $3d^2\,4s^2$ 6.83	23 V $3d^3\,4s^2$ 6.74	24 Cr $3d^5\,4s$ 6.76	25 Mn $3d^5\,4s^2$ 7.43	26 Fe $3d^6\,4s^2$ 7.90	27 Co $3d^7\,4s^2$ 7.86	
37 Rb 5s 4.18	38 Sr $5s^2$ 5.69	39 Y $4d\,5s^2$ 6.5	40 Zr $4d^2\,5s^2$ 6.95	41 Nb $4d^4\,5s$ 6.77	42 Mo $4d^5\,5s$ 7.18	43 Tc $4d^6\,5s$ 7.28	44 Ru $4d^7\,5s$ 7.36	45 Rh $4d^8\,5s$ 7.46	
55 Cs 6s 3.89	56 Ba $6s^2$ 5.21	57 La $5d\,6s^2$ 5.61	72 Hf $4f^{14}\,5d^2\,6s^2$ 7.0	73 Ta $5d^3\,6s^2$ 7.88	74 W $5d^4\,6s^2$ 7.98	75 Re $5d^5\,6s^2$ 7.87	76 Os $5d^6\,6s^2$ 8.7	77 Ir $5d^9$ 9.0	

Lanthanide series — IIIa

58 Ce $4f^2\,6s^2$ 6.91	59 Pr $4f^3\,6s^2$ 5.76	60 Nd $4f^4\,6s^2$ 6.31	61 Pm $4f^5\,6s^2$	62 Sm $4f^6\,6s^2$ 5.6	63 Eu $4f^7\,6s^2$ 5.67

87 Fr 7s	88 Ra $7s^2$ 5.28	89 Ac $6d\,7s^2$ 6.9

Actinide series — IIIa

90 Th $6d^2\,7s^2$	91 Pa $5f^2\,6d\,7s^2$	92 U $5f^3\,6d\,7s^2$ 4.0	93 Np $5f^5\,7s^2$	94 Pu $5f^6\,7s^2$	95 Am $5f^7\,7s^2$

			III	IV	V	VI	VII	NOBLE GAS
								2 He $1s^2$ 24.58
			5 B $2s^2\,2p$ 8.3	6 C $2s^2\,2p^2$ 11.26	7 N $2s^2\,2p^3$ 14.54	8 O $2s^2\,2p^4$ 13.61	9 F $2s^2\,2p^5$ 17.42	10 Ne $2s^2\,2p^6$ 21.56
			13 Al $3s^2\,3p$ 5.98	14 Si $3s^2\,3p^2$ 8.15	15 P $3s^2\,3p^3$ 10.55	16 S $3s^2\,3p^4$ 10.36	17 Cl $3s^2\,3p^5$ 13.01	18 Ar $3s^2\,3p^6$ 15.76
VIII	Ia	IIa						
28 Ni $3d^8\,4s^2$ 7.63	29 Cu $3d^{10}\,4s$ 7.72	30 Zn $3d^{10}\,4s^2$ 9.39	31 Ga $4s^2\,4p$ 6.00	32 Ge $4s^2\,4p^2$ 7.88	33 As $4s^2\,4p^3$ 9.81	34 Se $4s^2\,4p^4$ 9.75	35 Br $4s^2\,4p^5$ 11.84	36 Kr $4s^2\,4p^6$ 14.00
46 Pd $4d^{10}$ 8.33	47 Ag $4d^{10}\,5s$ 7.57	48 Cd $4d^{10}\,5s^2$ 9.00	49 In $5s^2\,5p$ 5.78	50 Sn $5s^2\,5p^2$ 7.34	51 Sb $5s^2\,5p^3$ 8.64	52 Te $5s^2\,5p^4$ 9.01	53 I $5s^2\,5p^5$ 10.45	54 Xe $5s^2\,5p^6$ 12.13
78 Pt $5d^9\,6s$ 8.96	79 Au $5d^{10}\,6s$ 9.22	80 Hg $5d^{10}\,6s^2$ 10.43	81 Tl $6s^2\,6p$ 6.11	82 Pb $6s^2\,6p^2$ 7.41	83 Bi $6s^2\,6p^3$ 7.29	84 Po $6s^2\,6p^4$ 8.43	85 At $6s^2\,6p^5$	86 Rn $6s^2\,6p^6$ 10.74

64 Gd $4f^7\,5d\,6s^2$ 6.16	65 Tb $4f^8\,5d\,6s^2$ 6.74	66 Dy $4f^{10}\,6s^2$ 6.82	67 Ho $4f^{11}\,6s^2$	68 Er $4f^{12}\,6s^2$	69 Tm $4f^{13}\,6s^2$	70 Yb $4f^{14}\,6s^2$ 6.2	71 Lu $4f^{14}\,5d\,6s^2$ 5.0

96 Cm $5f^7\,6d\,7s^2$	97 Bk	98 Cf	99 Es	100 Fm	101 Md	102 No	103 Lw

Appendix 2 PROPERTIES OF THE ELEMENTS (Courtesy of Nuclear Enterprises Ltd)

Element	Atomic Number, Z	Atomic Weight, A	Density at N.T.P., g/cm³	K Absorption Edge Energy E_K keV	Mass Absorption Coefficients, cm²/g $\mu 1$	Mass Absorption Coefficients, cm²/g $\mu 2$	Principle K X-ray Energies (keV) K_{α_1}	Principle K X-ray Energies (keV) K_{β_1}	L Absorption Edge L_I	L Absorption Edge L_{II}	L Absorption Edge L_{III}	Principle L X-ray Energies (keV) L_{α_1}	Principle L X-ray Energies (keV) L_{β_1}	Principle L X-ray Energies (keV) L_γ
Hydrogen	1	1.008	0.0898×10^{-3}	0.0136										
Helium	2	4.003	0.178×10^{-3}	0.0246										
Lithium	3	6.940	0.53	0.055			0.052							
Beryllium	4	9.02	1.84	0.116			0.110							
Boron	5	10.82	2.34	0.192			0.185							
Carbon	6	12.010	2.25	0.283	1000		0.282							
Nitrogen	7	14.008	1.25×10^{-3}	0.399	840		0.392							
Oxygen	8	16.000	1.43×10^{-3}	0.531	720	11000	0.523							
Fluorine	9	19.000	1.70×10^{-3}	0.687	600	8600	0.677							
Neon	10	20.183	0.90×10^{-3}	0.874	500	6800	0.851				0.022			
Sodium	11	22.997	0.97	1.08	420	5400	1.041	1.067	0.048					
Magnesium	12	24.32	1.74	1.303	350	4500	1.254	1.297	0.055	0.034	0.034			
Aluminium	13	26.97	2.70	1.559	300	3700	1.487	1.553	0.063	0.050	0.049			
Silicon	14	28.06	2.35	1.838	250	3000	1.740	1.832	0.087	0.073	0.072			
Phosphorus	15	30.98	2.2	2.142	215	2500	2.015	2.136	0.118	0.099	0.098			
Sulphur	16	32.066	2.0	2.470	185	2100	2.308	2.464	0.153	0.129	0.128			
Chlorine	17	35.457	3.21×10^{-3}	2.819	160	1800	2.622	2.815	0.193	0.164	0.163			
Argon	18	39.944	1.78×10^{-3}	3.203	140	1500	2.957	3.192	0.238	0.203	0.202			
Potassium	19	39.096	0.85	3.607	120	1250	3.313	3.589	0.287	0.247	0.245			
Calcium	20	40.08	1.54	4.038	104	1050	3.691	4.012	0.341	0.297	0.294	0.341	0.344	
									0.399	0.352	0.349			

Element	Z	at. wt.	density										
Scandium	21	45.10			91	900	4.090	4.460	0.462	0.411	0.406	0.395	0.399
Titanium	22	47.90	4.5	4.496	80	760	4.510	4.931	0.530	0.460	0.454	0.452	0.458
Vanadium	23	50.95	5.69	4.964	72	660	4.952	5.427	0.604	0.519	0.512	0.510	0.519
Chromium	24	52.01	6.9	5.463	64	580	5.414	5.946	0.679	0.583	0.574	0.571	0.581
Manganese	25	54.93	7.42	5.988	57	500	5.898	6.490	0.762	0.650	0.639	0.636	0.647
Iron	26	55.85	7.9	6.537	51	450	6.403	7.057	0.849	0.721	0.708	0.704	0.717
Cobalt	27	58.94	8.71	7.111	45	390	6.930	7.649	0.929	0.794	0.779	0.775	0.790
Nickel	28	58.69	8.8	7.709	42	345	7.477	8.264	1.015	0.871	0.853	0.849	0.866
Copper	29	63.54	8.9	8.331	37	310	8.047	8.904	1.100	0.953	0.933	0.928	0.948
Zinc	30	65.38	7.1	8.980	33.5	275	8.638	9.571	1.200	1.045	1.022	1.009	1.032
Gallium	31	69.72	5.90	9.660	30.5	245	9.251	10.263	1.30	1.134	1.117	1.096	1.122
Germanium	32	72.60	5.46	10.368	27.5	220	9.885	10.981	1.42	1.248	1.217	1.186	1.216
Arsenic	33	74.91	5.7	11.103	25	200	10.543	11.725	1.529	1.359	1.323	1.282	1.317
Selenium	34	78.96	4.5	11.863	23	180	11.221	12.495	1.652	1.473	1.434	1.379	1.419
Bromine	35	79.92	3.12	12.652	21.4	162	11.923	13.290	1.794	1.599	1.552	1.480	1.526
Krypton	36	83.7	3.71×10^{-3}	13.475	19.6	150	12.648	14.112	1.931	1.727	1.675	1.587	1.638
Rubidium	37	85.48	1.53	14.323	18.2	134	13.394	14.960	2.067	1.866	1.806	1.694	1.752
Strontium	38	87.63	2.55	15.201	16.9	121	14.164	15.834	2.221	2.008	1.941	1.806	1.872
Yttrium	39	88.92	5.51	16.106	15.5	111	14.957	16.736	2.369	2.154	2.079	1.922	1.996
Zirconium	40	91.22	6.44	17.037	14.4	102	15.774	17.666	2.547	2.305	2.220	2.042	2.124	2.302
Niobium	41	92.91		17.998	13.4	94	16.614	18.621	2.706	2.467	2.374	2.166	2.257	2.462
Molybdenum	42	95.95	9.0	18.987	12.5	86	17.478	19.607	2.884	2.627	2.523	2.293	2.395	2.623
Technetium	43			20.002	11.7	79	18.410	20.585	3.054	2.795	2.677	2.424	2.538	2.792
Ruthenium	44	101.7	12.1	21.054	11.0	73	19.278	21.655	3.236	2.966	2.837	2.558	2.683	2.964
Rhodium	45	102.91	12.4	22.118	10.2	67	20.214	22.721	3.419	3.145	3.002	2.696	2.834	3.144
Palladium	46	106.7	12.2	23.224	9.8	62	21.175	23.816	3.617	3.329	3.172	2.838	2.990	3.328
Silver	47	107.88	10.5	24.347	9.2	58	22.162	24.942	3.810	3.528	3.352	2.984	3.151	3.519
Cadmium	48	112.41	8.6	25.517	8.6	53	23.172	26.093	4.019	3.727	3.538	3.133	3.316	3.716
Indium	49	114.76	7.28	26.712	8.2	49	24.207	27.274	4.237	3.939	3.729	3.287	3.487	3.920
Tin	50	118.70	7.3	27.928	7.7	46	25.270	28.483	4.464	4.157	3.928	3.444	3.662	4.131
Antimony	51	121.76	6.7	29.190	7.2	43	26.357	29.723	4.697	4.381	4.132	3.605	3.843	4.347
Tellurium	52	127.61	6.0	30.486	6.8	39.5	27.471	30.993	4.938	4.613	4.341	3.769	4.029	4.570
Iodine	53	126.92	4.94	31.809	6.5	37.0	28.610	32.292	5.190	4.856	4.559	3.937	4.220	4.800
Xenon	54	131.3	5.85×10^{-3}	33.164	6.2	34.5	29.802	33.644	5.452	5.104	4.782	4.111	4.422	5.036

Element	Atomic Number, Z	Atomic Weight, A	Density at N.T.P., g/cm³	K Absorption Edge Energy, E_K keV	Mass Absorption Coefficients cm²/g μ1	μ2	Principle K X-ray Energies (keV) K_{α_1}	K_{β_1}	L Absorption Edge L_I	L_{II}	L_{III}	Principle L X-ray Energies (keV) L_{α_1}	L_{β_1}	L_{γ_1}
Cesium	55	132.91	1.87	35.959	5.8	32.0	30.970	34.984	5.720	5.358	5.011	4.286	4.620	5.280
Barium	56	137.36	3.78	37.410	5.5	30.0	32.191	36.376	5.995	5.623	5.247	4.467	4.828	5.531
Lanthanum	57	138.92	6.15	38.931	5.2	28.5	33.440	37.799	6.283	5.894	5.489	4.651	5.043	5.789
Cerium	58	140.13	6.8	40.449	5.0	26.5	34.717	39.255	6.561	6.165	5.729	4.840	5.262	6.052
Praseodymium	59	140.92	6.48	41.998	4.75	25.0	36.023	40.746	6.846	6.443	5.968	5.034	5.489	6.322
Neodymium	60	144.27	6.96	43.571	4.5	23.5	37.359	42.269	7.144	6.727	6.215	5.230	5.722	6.602
Promethium	61	(146)		45.207	4.35	22.5	38.649	43.945	7.448	7.018	6.466	5.431	5.956	6.891
Samarium	62	150.43	7.8	46.846	4.15	21.0	40.124	45.400	7.754	7.281	6.721	5.636	6.206	7.180
Europium	63	152.0		48.515	4.0	19.5	41.529	47.027	8.069	7.624	6.983	5.846	6.456	7.478
Gadolinium	64	156.9		50.229	3.8	18.5	42.983	48.718	8.393	7.940	7.252	6.059	6.714	7.788
Terbium	65	159.2		51.998	3.7	17.5	44.470	50.391	8.724	8.258	7.519	6.275	6.979	8.104
Dysprosium	66	162.46		53.789	3.55	16.6	45.985	52.178	9.083	8.621	7.850	6.495	7.249	8.418
Holmium	67	164.94		55.615	3.4	15.7	47.528	53.934	9.411	8.920	8.074	6.720	7.528	8.748
Erbium	68	167.2	4.77(?)	57.483	3.25	14.8	49.099	55.690	9.776	9.263	8.364	6.948	7.810	9.089
Thulium	69	169.4		59.335	3.15	14.0	50.730	57.576	10.144	9.628	8.652	7.181	8.101	9.424
Ytterbium	70	173.04		61.303	3.0	13.3	52.360	59.352	10.486	9.977	8.943	7.414	8.401	9.779
Lutetium	71	174.99		63.304	2.9	12.7	54.063	61.282	10.867	10.345	9.241	7.654	8.708	10.142

Hafnium	72	178.6				55.757	63.209	11.264	10.734	9.556	7.898	9.021	10.514
Tantalum	73	180.88	16.6	65.313	2.85	12 1	57.524	65.210	11.676	11.130	9.876	8.145	9.341	10.892
Tungsten	74	183.92	18.9	67.400	2.75	11 8	59.310	67.233	12.090	11.535	10.198	8.396	9.670	11.283
Rhenium	75	186.31		69.508	2.7	11 3	61.131	69.298	12.522	11.955	10.531	8.651	10.008	11.684
Osmium	76	190.2	22.5	71.662	2.6	10 5	62.991	71.404	12.965	12.383	10.869	8.910	10.354	12.094
Iridium	77	193.1	22.4	73.860	2.5	10 2	64.886	73.549	13.413	12.819	11.211	9.173	10.706	12.509
Platinum	78	195.23	21.4	76.097	2.4	9 7	66.820	75.736	13.873	13.268	11.559	9.441	11.069	12.939
Gold	79	197.2	19.3	78.379	2.35	9 3	68.794	77.968	14.353	13.733	11.919	9.711	11.439	13.379
Mercury	80	200.61	13.6	80.713	2.3	8 8	70.821	80.258	14.841	14.212	12.285	9.987	11.823	13.828
Thallium	81	204.39	11.9	83.106	2.2	8 4	72.860	82.558	15.346	14.697	12.657	10.266	12.210	14.288
Lead	82	207.21	11.3	85.517	2.15	8 0	74.957	84.922	15.870	15.207	13.044	10.549	12.611	14.762
Bismuth	83	209.00	9.8	88.001	2.1	7.7	77.097	87.335	16.393	15.716	13.424	10.836	13.021	15.244
Polonium	84	(210)		90.521	2.04	7.3	79.296	89.809	16.935	16.244	13.817	11.128	13.441	15.740
Astatine	85	(221)		93.112	2.0	7.0	81.525	92.319	17.490	16.784	14.215	11.424	13.873	16.248
Radon	86	222	9.73×10^{-3}	95.740	1.93	6.6	83.800	94.877	18.058	17.337	14.618	11.724	14.316	16.768
				98.418	1.90	6.3								
Francium	87	(224)		101.147	1.83	6.0	86.119	97.483	18.638	17.904	15.028	12.029	14.770	17.301
Radium	88	226.05		103.927	1.76	5.75	88.485	100.136	19.233	18.481	15.442	12.338	15.233	17.845
Actinium	89	(227)		106.759	1.72	5.5	90.894	102.846	19.842	19.078	15.865	12.650	15.712	18.405
Thorium	90	232.12	11.5	109.630	1.67	5.2	93.334	105.592	20.460	19.688	16.296	12.966	16.200	18.977
Protactinium	91	231		112.581	1.64	4.95	95.851	108.408	21.102	20.311	16.731	13.291	16.700	19.559
Uranium	92	238.07	18.7	115.591	1.62	4.7	98.428	111.289	21.753	20.943	17.163	13.613	17.218	20.163
Neptunium	93			118.619	1.57	4.55	101.005	114.181	22.417	21.596	17.614	13.945	17.740	20.774
Plutonium	94			121.720	1.53	4.35	103.653	117.146	23.097	22.262	18.066	14.279	18.278	21.401
Americium	95			124.876	1.50	4.15	106.351	120.163	23.793	22.944	18.525	14.618	18.829	22.042
Curium	96			128.088	1.47	4.0	109.098	123.235	24.503	23.640	18.990	14.961	19.393	22.699
Berkelium	97			131.357			111.896	126.362	25.230	24.352	19.461	15.309	19.971	23.370
Californium	98			134.683			114.745	129.544	25.971	25.080	19.938	15.661	20.562	24.056

Notes – Relative intensities in K and L spectra are: $K\alpha : K\beta = 6:1$; $L\alpha : L\beta : L\gamma = 9:6:1$.

Appendix 3 NUCLEAR DATA (Courtesy of the Radiochemical Centre, Amersham, Bucks., England)

nuclide	half life	type of decay	particle energies and transition probabilities		electromagnetic transitions				
			energy MeV	transition probability	photon energy MeV	photons emitted	transitions internally converted	photon energy MeV	photons emitted	transitions internally converted

nuclide	half life	type of decay	energy MeV	transition probability	photon energy MeV	photons emitted	transitions internally converted	photon energy MeV	photons emitted	transitions internally converted
Actinium-227	21·8y		see Uranium/Actinium (4n+3) Radioactive series, p. 384							
Americium-241	433y	α	5·387	1·6%	0·026	2·5%	~10%	0·103	0·02%	
			5·442	12·5%	0·033	0·1%	~20%	0·125	0·004%	
			5·484	85·2%	0·043	0·1%	~10%	others	low intensity	
			5·511	0·20%	0·0595	35·3%	~40%	0·012–		
			5·543	0·34%	0·099	0·02%		0·022	~40% (Np L X-rays)	
			others	low						
Antimony-122	2·70d	β⁻	0·72	4·5%	0·564	70%		others	low intensity	
			1·41	67%	0·692	3·7%				
			1·97	26%	1·140	0·7%				
		e.c.		2·5%	1·255	0·76%				
Antimony-124	60·2d	β⁻	0·21	9%	0·603	98·0%	0·4%	1·045	1·9%	
			0·61	52%	0·646	7·2%		1·325	1·5%	
			0·86	4%	0·709	1·4%		1·355	0·9%	
			0·94	2%	0·714	2·3%		1·368	2·5%	
			1·57	5%	0·723	11·2%		1·437	1·1%	
			1·65	3%	0·791	0·7%		1·691	50·4%	
			2·30	23%	0·968	1·9%		2·091	6·1%	
			others	low				others	<0·5% each	
Antimony-125	2·77y	β⁻	0·094	13·5%	0·035	4·5%	60%	0·636	11·4%	
			0·124	5·7%	0·176	6·8%	1%	0·671	1·7%	
			0·130	18·1%	0·321	0·5%		others	<0·5% each	
			0·241	1·5%	0·381	1·5%		0·027–		
			0·302	40·2%	0·428	29·8%	0·4%	0·031	~50% (Te K X-rays)	
			0·332	0·3%	0·463	10·4%				
			0·445	7·2%	0·601	17·8%				
			0·621	13·5%	0·607	4·9%				

23% of ¹²⁵Sb decays to 58d ¹²⁵ᵐTe (see p. 373)

Isotope	Half-life	Decay	Energy (MeV)	Intensity	Energy (MeV)	Intensity	γ Energy (MeV)	Intensity
Argon-41	1·83h	β−	0·82	0·05%	1·293	99·15%		
			1·20	99·15%	1·67	0·05%		
			2·49	0·8%				
Arsenic-71	65h	β+	0·81	~33%			0·175	83·4%	7·7%	0·527	0·7%
		e.c.		~67%			0·327	2·7%		1·096	4·0%
							0·391	0·6%		1·139	0·8%
							0·500	3·0%		others	low intensity
							0·511	from β+			
		Daughter	⁷¹Ge								
Arsenic-72	26·0h	β+	1·874	6%			0·511	from β+		1·476	0·5%
			2·504	62%			0·630	7·6%		2·106	0·7%
			3·338	20%			0·834	75·5%		2·202	0·5%
			others	2%			0·894	0·7%		2·621	0·4%
		e.c.		10%			1·051	0·9%		others	<0·5% each
							1·464	1·1%			
Arsenic-73	80·3d	e.c.		100%			0·0135	0·1%	99·9%	0·053	9·9%	90·1%
Arsenic-74	17·7d	β−	0·72	15·1%			0·511	from β+		0·994	0·02%
			1·36	18·2%			0·596	59·4%	0·1%	1·204	0·3%
		β+	0·91	25·3%			0·609	0·6%		2·198	0·02%
			1·51	3·9%			0·635	15·0%		others	<0·01% each
		e.c.		37·5%			0·887	0·03%			
Arsenic-76	26·3h	β−	0·542	2%			0·559	46%		1·229	1·5%
			1·184	2%			0·624	0·4%		1·438	0·3%
			1·756	8%			0·657	6·4%		1·788	0·3%
			1·854	4%			0·666	0·3%		2·097	0·6%
			2·413	29%			0·740	0·2%		2·112	0·24%
			2·972	54%			1·213	2%		others	low intensity
			others	low			1·216	4%			
Barium-131	12d	e.c.		100%			0·124	28·6%	25·0%	0·496	43·1%	0·5%
							0·134	1·9%	0·9%	0·585	1·1%
							0·216	22·2%	2·0%	0·620	1·3%
							0·240	2·7%	0·2%	1·048	1·3%
							0·249	3·1%	0·2%	others	<1% each
							0·373	13·4%	0·3%	0·031−	
							0·404	1·3%		0·036	~96% (Cs K X-rays)
							0·487	1·9%			

355

nuclide	half life	type of decay	particle energies and transition probabilities		electromagnetic transitions					
			energy MeV	transition probability	photon energy MeV	photons emitted	transitions internally converted	photon energy MeV	photons emitted	transitions internally converted
Barium-133	10·8y	e.c.		100%	0·053	2·2%	11·9%	0·303	18·7%	0·8%
					0·080	2·4%	4·2%	0·356	61·9%	1·5%
					0·081	33·8%	55·6%	0·384	8·9%	0·2%
					0·160	0·7%	0·2%	0·030–		
					0·223	0·5%		0·036	~123% (Cs K X-rays)	
					0·276	7·1%	0·4%			
Barium-135m	28·7h	i.t.		100%	0·268	16%	84%	0·032–0·037	~50% (Ba K X-rays)	0·2%
Barium-140	12·80d	β−	0·468	24%	0·014	1·3%	72%	0·537	23·8%	
			0·582	10%	0·030	14%	79%	0·602	0·6%	
			0·886	2·6%	0·163	6·2%	0·7%	0·661	0·7%	
			1·005	46%	0·305	4·5%	0·3%	others	low intensity	
			1·019	17%	0·424	3·2%				
			others	0·4%	0·438	2·1%				
		Daughter ¹⁴⁰La								
Beryllium-7	53·3d	e.c.		100%	0·478	10·4%	~0%			
Bismuth-206	6·24d	e.c.		100%	0·184	16%		0·803	99%	
					0·344	24%		0·881	67%	
					0·497	15%		0·895	16%	
					0·516	41%		1·098	14%	
					0·537	31%		1·718	32%	
								others	low intensity	
Bismuth-207	38y	e.c.		100%	0·570	97·8%	2%	1·44	0·15%	
					0·897	0·16%		1·770	7·3%	
					1·064	74·3%	10%	0·072–0·087	~78% (Pb K X-rays)	
								(including transitions via 0·8s ²⁰⁷mPb)		

Isotope	Half-life	Decay	Energy (MeV)	Intensity	Energy (MeV)	Intensity
Bismuth-210	5.01d	α	~4.7	~1.3×10^{-4}%		
		β−	1.161	~100%		
		Daughter	^{210}Po			
Bromine-82	35.4h	β−	0.263	1.7%	1.008	1.7%
			0.444	98.3%	1.044	29%
					1.317	28%
					1.475	17%
					1.651	0.9%
					others up to 1.96	<1% each
Cadmium-109	453d	e.c.		100%	0.022–0.026	67.7% (Ag K X-rays)
			via 40s ^{109m}Ag			
			0.088	3.6%	0.022–0.026	34.6% (Ag K X-rays)
Cadmium-115	53.5h	β−	0.58	35%	0.0356	0.4%
			0.62	5%	0.231	0.8%
			0.85	1%	0.261	2.1%
			1.11	59%	0.267	0.06%
					0.492	9.5%
					0.528	30.3%
		Daughter	^{115m}In	4%		
				96.4%		
Cadmium-115m	44d	β−	0.33	0.8%	0.158	0.016%
			0.69	1.8%	0.485	0.25%
			1.62	97%	0.934	1.81%
			others	low	1.133	0.076%
					1.291	0.83%
					1.450	0.015%
					others	low intensity
Caesium-131	9.7d	e.c.		100%	0.029–0.035	~74% (Xe K X-rays)
Caesium-132	6.47d	β−	0.2	0.4%	0.464	1.8%
		β+	0.7	1.6%	0.506	0.8%
		e.c.	0.4	0.4%	0.511	from β+
				97.6%	0.630	1.0%
					0.668	98%
					1.136	0.5%
					1.318	0.6%
					others	low intensity
Caesium-134	2.06y	β−	0.09	26%	0.475	1.5%
			0.42	2.5%	0.563	8.1%
			0.66	71.5%	0.569	14.0%
					0.605	97.5%
					0.796	85.4%
					0.802	8.6%
					1.038	1.0%
					1.168	2.0%
					1.365	3.3%
						0.5%
						0.2%

nuclide	half life	type of decay	particle energies and transition probabilities		electromagnetic transitions					
			energy MeV	transition probability	photon energy MeV	photons emitted	transitions internally converted	photon energy MeV	photons emitted	transitions internally converted
Caesium-137	30·1y	β−	0·512 1·174	94·6% 5·4%	via 2·6m ^{137m}Ba 0·662	85·1%	9·5%	0·032–0·038	8% (Ba K X-rays)	
Calcium-45	164d	β−	0·257	100%						
Calcium-47	4·54d	β−	0·69 1·22 1·99	82% 0·1% 17·9%	0·489 0·530 0·767	6·8% 0·1% 0·2%		0·808 1·297 others	6·8% 75·1% low intensity	
	Daughter	^{47}Sc								
Californium-252	2·65y	α s.f.	5·974 6·075 6·118 others	0·3% 15% 81·6% low 3·1%	γ-radiation associated with the α-transitions is of low intensity. The spontaneous fission events and the subsequent decay of fission products produce approximately 20 γ-rays per fission. This γ-radiation covers an energy range to ~9MeV. Each fission also produces on average ~4 fast neutrons					
Carbon-14	5730y	β−	0·156	100%						
Cerium-139	137d	e.c.			0·166	80%	20%	0·033–0·039	~90% (La K X-rays)	
Cerium-141	32·5d	β−	0·436 0·581	70% 30%	0·145	48%	22%	0·035–0·042	~17% (Pr K X-rays)	
Cerium-143	33·0h	β−	0·71 1·09 1·39 others	~15% ~50% ~30% low	0·057 0·293 0·351	~11% ~46% ~3·5%	~80% ~3%	0·664 0·722 0·811 others	~6·5% ~6·5% ~1·2% <4% each	
	Daughter	^{143}Pr								

Isotope	Half-life	Decay	Energy (MeV)	Intensity	Energy (MeV)	Intensity	Energy (MeV)	Intensity
Cerium-144	284d	β−	0·182	19·1%	0·034	0·1%	0·100	0·03%
			0·216	0·2%	0·040	0·4%	0·134	10·8%
			0·236	4·4%	0·053	0·1%		0·07%
			0·316	76·3%	0·080	1·5%		6·2%
			via 7·2m ^{144m}Pr					
		β−	1·534	0·05%	0·059	0%	0·814	0·05%
		i.t.		1·15%	0·696	0·05%		
			via 17m ^{144}Pr					
		β−	0·808	1·0%	0·696	1·53%		
			2·298	1·2%	1·489	0·28%		
			2·994	97·75%	2·186	0·72%		
Chlorine-36	3·01 × 10⁵y	β−	0·709	98·1%				
		e.c.		1·9%				
Chromium-51	27·7d	e.c.		100%	0·320	9·83%	0·005–0·006	~22% (V K X-rays)
Cobalt-56	78·8d	β+	0·4	1%	0·511	from β+	2·015	3·1%
			1·5	18%	0·847	99·97%	2·035	7·9%
		e.c.		81%	0·977	1·4%	2·599	16·9%
					1·038	14·0%	3·010	1·0%
					1·175	2·3%	3·202	3·0%
					1·238	67·6%	3·254	7·4%
					1·360	4·3%	3·273	1·8%
					1·771	15·7%	3·452	0·9%
						0·03%	others	<1% each
Cobalt-57	270·5d	e.c.		100%	0·014	9·4%	0·692	0·16%
					0·122	85·2%	others	low intensity
					0·136	11·1%	0·006–0·007	~55% (Fe K X-rays)
					0·570	0·02%		
Cobalt-58	70·8d	β+	0·475	15·0%	0·511	from β+	1·675	0·5%
		e.c.		85·0%	0·811	99·4%	0·006–0·007	~26% (Fe K X-rays)
					0·864	0·7%		
Cobalt-60	5·27y	β−	0·318	99·9%	1·173	99·86%		0·02%
			1·491	0·1%	1·333	99·98%		0·01%
					others	<0·01%		

nuclide	half life	type of decay	particle energies and transition probabilities		electromagnetic transitions					
			energy MeV	transition probability	photon energy MeV	photons emitted	transitions internally converted	photon energy MeV	photons emitted	transitions internally converted
Copper-64	12.8h	β− β+ e.c.	0.573 0.654	38% 18.4% 43.6%	0.511 1.347	from β+ 0.55%		0.007− 0.009	~14% (Ni K X-rays)	
Copper-67	61.8h	β−	0.182 0.392 0.483 0.576	1% 56% 23% 20%	0.091 0.185 0.209 0.300	7.3% 47.0% 0.1% 0.7%	0.6% 0.9%	0.394 0.008− 0.010	0.2% 0.6% (Zn K X-rays)	
					via 9.2μs 67mZn 0.093	16.9%	14.6%	0.008− 0.010	6% (Zn K X-rays)	
Curium-242	163d	α	6.068 6.112 others	26% 74% low	0.044 0.102	3% 0.004%	23%	0.158 others up to ~1.0	0.002% <0.0002% each	
Curium-244	17.8y	α	5.763 5.806 others	23.6% 76.4% low	0.043 0.099 0.152	~0.02% ~0.0013% ~0.0014%	23.6%	others up to ~0.8 0.012− 0.023	low intensity ~8% (Pu L X-rays)	
Erbium-169	9.4d	β−	0.332 0.340	42% 58%	0.008	~0.15%	~41.8%			
Europium-152	13.0y	β− β+ e.c.	0.185 0.394 0.705 1.073 1.484 others	1.6% 2.1% 13.5% 1.0% 10.2% 0.6% 0.02% 71%	0.122 0.245 0.344 0.411 0.444 0.779 0.867	28.9% 7.4% 26.2% 2.1% 3.0% 12.3% 3.9%	~33%	0.964 1.086 1.090 1.112 1.213 1.299 1.408 others	13.8% 9.7% 1.6% 13.0% 1.3% 1.6% 20.1% <1% each	
Gadolinium-153	241.5d	e.c.		100%	0.070 0.075 0.083 0.089	2.6% 0.07% 0.23% 0.12%	11% 0.5% 0.2%	0.097 0.103 0.173 0.041− 0.048	30% 20% 0.04% ~110% (Eu K X-rays)	8% 30%

Nuclide	Half-life	Decay mode	Energy (MeV)	Intensity	Energy (MeV)	Intensity	Notes
Gallium-66	9.4h	β+	0.362	1.3%	0.511	from β+	
			0.772	0.7%	0.834	5.9%	
			0.924	4.1%	1.039	37.3%	
			1.781	0.5%	1.333	1.2%	
			4.153	51.8%	1.919	2.1%	
			others	0.4%	2.190	5.6%	
		e.c.		41.2%	2.422	1.9%	
					2.752	22.8%	
					3.229		1.5%
					3.381		1.4%
					3.422		0.8%
					3.791		1.0%
					4.086		1.1%
					4.295		3.4%
					4.462		0.7%
					4.807		1.4%
					others		<0.7% each
Gallium-67	78.26h	e.c.		100%	0.091	3.6%	0.494 0.1%
					0.185	23.5%	0.704 0.02%
					0.209	2.6%	0.795 0.06%
					0.300	16.7%	0.888 0.17%
					0.394	4.4%	0.008–0.010 43% (Zn K X-rays)
					via 9.2μs ^{67m}Zn		
					0.093	37.6%	32.4%
					0.008–0.010		13% (Zn K X-rays)
Gallium-72	14.1h	β−	0.658	14.9%	0.601	5.5%	1.597 4.3%
			0.675	22.0%	0.630	24.9%	1.861 5.3%
			0.964	27.4%	0.786	3.2%	2.202 26.0%
			1.485	9.0%	0.834	95.9%	2.491 7.7%
			2.536	9.2%	0.894	9.9%	2.508 12.8%
			3.166	9.7%	1.051	6.9%	others <2% each
			others	7.8%	1.464	3.6%	
Germanium-71	11.8d	e.c.		100%	0.009–0.011	~45% (Ga K X-rays)	
Gold-195	183d	e.c.		100%	0.031	1.1%	~37%
					0.099	10.4%	75%
					0.130	0.7%	1.3%
					others		low intensity
					0.065–0.078		~90% (Pt K X-rays)
Gold-198	2.696d	β−	0.285	1.32%	0.412	95.45%	4.3%
			0.961	98.66%	0.676	1.06%	0.03%
			1.373	0.02%	1.088	0.23%	
Gold-199	3.13d	β−	0.25	21%	0.050	0.3%	3.5%
			0.29	72%	0.158	39.6%	36.4%
			0.45	7%	0.208	8.8%	8.3%
					0.069–0.083		~18% (Hg K X-rays)

nuclide	half life	type of decay	particle energies and transition probabilities		electromagnetic transitions					
			energy MeV	transition probability	photon energy MeV	photons emitted	transitions internally converted	photon energy MeV	photons emitted	transitions internally converted
Hafnium-175	70d	e.c.		100%	0.089	2.3%	10.7%	0.343	88%	9%
					0.114	0.32%	0.8%	0.354	0.24%	
					0.161	0.02%		0.433	1.5%	0.1%
					0.230	0.78%	0.1%	0.053–	~93% (Lu K X-rays)	
					0.319	0.18%		0.063		
Hafnium-181	42.4d	β–	0.405	7%	0.133	35.8%	62.2%	0.346	16.4%	0.9%
			0.409	93%	0.136	4.5%	8.0%	0.476	0.4%	0.1%
					0.137	0.6%	1.2%	0.482	77.9%	4.2%
								others	low intensity	
Holmium-166	27h	β–	0.19	0.3%	0.081	6.7%	46.5%	1.750	0.03%	
			0.39	0.9%	0.674	0.02%		others	<0.02% each	
			1.77	52%	1.379	0.93%		0.048–	~11% (Er K X-rays)	
			1.85	46.8%	1.582	0.18%		0.058		
			others	low	1.662	0.12%				
Hydrogen-3 (Tritium)	12.35y	β–	0.0186	100%						
Indium-111	2.83d	e.c.		100%	0.171	90.9%	9.1%	0.023–	~84% (Cd K X-rays)	
					0.245	94.2%	5.8%	0.027		
Indium-113m	99.5m	i.t.		100%	0.392	64.9%	35.1%	0.024–	24% (In K X-rays)	
								0.028		
Indium-114m	49.5 d	e.c.		3.7%	0.192	16.6%	79.7%	0.024–	~37% (In K X-rays)	
		i.t.		96.3%	0.558	3.7%		0.028		
					0.725	3.7%				
		via 72s ^{114}In								
		β–	0.686	0.14%	0.511	from β+				
			1.986	94.4%	0.576	~4 × 10^{-3}%				
		β+	0.42	~4 × 10^{-3}%	1.300	0.14%				
		e.c.		1.8%						
Indium-115m	4.5h	β–	0.84	5.5%	0.336	44.9%	49.6%	0.023–	~1% (Cd K X-rays)	
		i.t.		94.5%				0.027		

Isotope	Half-life	Decay	Energy (MeV)	%	Energy (MeV)	%	Energy (MeV)	%
Iodine-123	13.2h	e.c.		100%	0.159	83.0%	0.529	1.05%
					0.347	0.10%	0.539	0.27%
					0.440	0.35%	others	<0.1% each
					0.506	0.26%	0.027–0.032	~86% (Te K X-rays)
Iodine-125	60.0d	e.c.		100%	0.035	7%	0.027–0.032	138% (Te K X-rays)
						93%		
Iodine-126	13d	β−	0.38	3%	0.389	32%	0.880	0.8%
			0.88	30%	0.491	2%	1.420	0.3%
			1.27	15%	0.511	from β+	others	<0.1% each
		β+	0.46	~0.1%	0.666	30%	0.027–0.032	~38% (Te K X-rays)
			1.1	~0.4%	0.754	4%		
		e.c.		51.5%				
Iodine-129	1.57 × 10⁷y	β−	0.150	100%	0.040	7.5%	0.030–0.035	~69% (Xe K X-rays)
						92.5%		
Iodine-131	8.06d	β−	0.247	1.8%	0.080	2.4%		
			0.304	0.6%	0.284	5.9%		3.8%
			0.334	7.2%	0.364	81.8%		0.3%
			0.606	89.7%	0.637	7.2%		1.7%
			0.806	0.7%	0.723	1.8%		

1.3% of ¹³¹I decays via 12d ¹³¹mXe

| (Xenon-131m) | | i.t. | | 100% | 0.164 | 2% | | 98% |

(percentages relate to disintegrations of ¹³¹ᵐXe)

Iodine-132	2.29h	β−	0.84	16.0%	0.506	5.0%	0.810	2.9%
			1.01	3.5%	0.523	16.1%	0.812	5.6%
			1.07	6.5%	0.621	2.0%	0.955	18.1%
			1.09	3.0%	0.630	13.7%	1.136	3.0%
			1.10	2.6%	0.651	2.7%	1.295	2.0%
			1.26	2.9%	0.668	98.7%	1.372	2.5%
			1.29	18.4%	0.670	4.9%	1.399	7.1%
			1.57	10.8%	0.672	5.2%	1.433	1.4%
			1.72	12.7%	0.727	6.5%	1.921	1.2%
			2.24	20.2%	0.773	76.2%	2.002	1.1%
			others	3.4%			others	<1.5%

nuclide	half life	type of decay	particle energies and transition probabilities		electromagnetic transitions					
			energy MeV	transition probability	photon energy MeV	photons emitted	transitions internally converted	photon energy MeV	photons emitted	transitions internally converted
Iridium-192	74.0d	β−	0.075	0.1%	0.201	0.4%		0.468	47.0%	1.3%
			0.250	5.4%	0.206	3.4%	0.2%	0.484	2.9%	0.1%
			0.530	42.6%	0.283	0.3%	1.0%	0.489	0.3%	
			0.670	47.2%	0.296	29.6%		0.589	4.4%	0.1%
		e.c.		4.7%	0.308	30.7%	3.3%	0.604	8.2%	0.2%
					0.316	82.7%	3.0%	0.612	5.3%	0.1%
					0.374	0.7%	7.2%	0.884	0.3%	
					0.416	0.6%		1.062	0.05%	
Iron-55	2.7y	e.c.		100%	0.006	~28% (Mn K X-rays)				
Iron-59	44.6d	β−	0.084	0.1%	0.143	0.8%		1.292	43.8%	
			0.132	1.1%	0.192	2.8%		1.482	0.06%	
			0.274	45.8%	0.335	0.3%				
			0.467	52.7%	0.383	0.02%				
			1.566	0.3%	1.099	55.8%				
Krypton-85	10.73y	β−	0.158	0.43%	via 0.96μs 85mRb					
			0.672	99.57%	0.514	0.43%				
Lanthanum-140	40.27h	β−	1.247	11%	0.131	0.8%		0.920	2.5%	
			1.253	6%	0.242	0.6%		0.925	6.9%	
			1.288	1%	0.266	0.7%		0.950	0.6%	
			1.305	5%	0.329	21%		1.597	95.6%	
			1.357	45%	0.432	3.3%		2.348	0.9%	
			1.421	5%	0.487	45%		2.522	3.3%	
			1.685	18%	0.752	4.4%		others	<0.5% each	
			2.172	7%	0.816	23%				
			others	low	0.868	5.5%				
Lead-210	22.3y	α		2 × 10⁻⁶%	0.046	~4%	~76%			
		β−	0.015	~80%	0.009–					
			0.061	~20%	0.017	~21% (Bi L X-rays)				
			Daughter	²¹⁰Bi						

Isotope	Half-life	Decay	Energy (MeV)	Intensity	Energy (MeV)	Intensity	
Lutetium-177	6·71d	β⁻	0·176	12%	0·208	0·1%	11·0%
			0·384	7·5%	0·250	13·3%	0·2%
			0·497	80·5%	0·321	0·1%	0·2% 0·7%
Magnesium-28	21·1h	β⁻	0·459	100%	0·031	96%	30%
					0·400	30%	70%
		via 2·24m ²⁸Al in equilibrium					
		β⁻	2·855	100·2%	1·779	100·2%	
Manganese-52	5·67d	β⁺	0·574	28·3%	0·503	1·0%	93%
		e.c.		71·7%	0·511	from β⁺	4·7%
					0·744	85%	5·1%
					0·848	3·2%	100%
					others		<1% each
Manganese-54	312·5d	e.c.		100%	0·835	100%	0·0055 ~25% (Cr K X-rays)
Manganese-56	2·58h	β⁻	0·32	1·4%	0·847	98·9%	2·523 1·2%
			0·73	16·0%	1·038	0·06%	2·599 0·03%
			1·03	29·0%	1·238	0·14%	2·658 0·6%
			2·84	53·5%	1·811	28·4%	2·960 0·3%
			others	0·1%	2·113	15·7%	3·370 0·2%
Mercury-197	64·4h	e.c.		100%	0·077	19·2%	0·067–
					0·192	~1·1%	0·080 ~73% (Au K X-rays)
					0·268	~0·1%	
Mercury-197m	24h	e.c.		6·5%	0·134	31·8%	0·067–
		i.t.		93·5%	0·165	0·3%	0·083 36% (Au/Hg K X-rays)
					via 7·8s ¹⁹⁷ᵐAu		
					0·130	0·5%	0·067–
					0·279	5·0%	0·080 ~2% (Au K X-rays)
					0·409	<0·005%	
		Daughter ¹⁹⁷Hg					
Mercury-203	46·6d	β⁻	0·212	100%	0·279	81·5%	18·5% 0·071–
							0·085 12·8% (Tl K X-rays)

nuclide	half life	type of decay	particle energies and transition probabilities		electromagnetic transitions				
			energy MeV	transition probability	photon energy MeV	photons emitted	transitions internally converted	photon energy MeV	photons emitted	transitions internally converted
Molybdenum-99	66·2h	β−	0·454	18·3%	0·041	1·2%	4·8%	0·621	0·02%	
			0·866	1·4%	0·141	5·4%	0·7%	0·740	13·6%	
			1·232	80%	0·181	6·6%	1·0%	0·778	4·7%	
			others	0·3%	0·366	1·4%		0·823	0·13%	
					0·412	0·02%		0·961	0·1%	
					0·529	0·05%				
					via 6·02h 99mTc in equilibrium					
					0·002	~0%	93·9%	0·143	0·03%	0·8%
					0·141	83·9%	10·0%			
Neodymium-147	11·0d	β−	0·208	2%	0·091	27·9%	57·4%	0·398	0·9%	
			0·363	15%	0·121	0·5%		0·440	1·2%	
			0·803	81%	0·275	0·9%		0·531	13·3%	0·2%
			others	2%	0·319	2·1%	0·1%	0·686	0·8%	
								others	<0·5% each	
Daughter 147Pm										
Neptunium-237	2·14 × 10⁶y	α	4·638	6%	0·029	12%	~11%	0·155	0·10%	
			4·663	3·3%	0·087	13%	18%	0·169	0·08%	
			4·765	8%	0·106	0·08%		0·193	0·06%	
			4·770	25%	0·118	0·18%		0·195	0·21%	
			4·787	47%	0·131	0·09%		0·202	0·05%	
			4·802	~3%	0·134	0·07%		0·212	0·17%	
			4·816	2·5%	0·143	0·44%		0·214	0·05%	
			4·872	2·6%	0·151	0·25%		0·238	0·07%	
			others	<2%				others	low intensity	
Daughter 233Pa										
Nickel-63	100y	β−	0·066	100%						
Niobium-95	35·1d	β−	0·160	>99·9%	0·766	99·8%	0·1%			
			0·925	low						

Nuclide	Half-life	Decay mode	β energy	%	γ energy	%	Other energy	%
Niobium-97	74m	β−	0.906 1.271 others	1.1% 98.3% 0.6%	0.658 0.719 1.024	98.2% 0.11% 1.1%		0.2%
Osmium-185	94d	e.c.		100%	0.072 0.125 0.163 0.233 0.592 0.646	0% 0.6% 0.6% 1.2% 1.2% 80.1%	1.269 1.516 others 0.718 0.871 0.879 0.059– 0.071	1.5% 1.8% 0.5% 0.2% 0.9% 0.17% 0.12% <0.1% each 4.1% 6.4% 6.4% 0.1% 0.1% 0.1% ~66% (Re K X-rays)
Osmium-191	15.4d	β−	0.143	100%	via 4.9s ¹⁹¹ᵐIr 0.042 0.129 others	0% 24% <1% each	0.063– 0.076	100% 76% ~51% (Ir K X-rays)
Palladium-103	18.4d	e.c.		100%	0.062 0.295 0.357	0.004% 0.01% 0.06%	0.497 0.020– 0.023	0.005% 0.01% ~67% (Rh K X-rays)
Palladium-109	13.5h	β−	1.028	100%	via 57m ¹⁰³ᵐRh 0.040	0.1%	0.020– 0.023	100.1% ~8% (Rh K X-rays)
Phosphorus-32	14.3d	β−	1.709	100%	via 40s ¹⁰⁹ᵐAg 0.088	3.6%	0.022– 0.026	96.4% ~35% (Ag K X-rays)
Phosphorus-33	25.5d	β−	0.248	100%				
Platinum-195m	4.0d	i.t.		100%	0.031 0.099 0.129 0.130 0.140	2.0% 11.4% 0.08% 2.8% 0.03%	0.211 0.239 0.065– 0.078	90.8% 81.4% 99.7% 4.3% 0.04% 0.06% ~91% (Pt K X-rays)

nuclide	half life	type of decay	particle energies and transition probabilities		electromagnetic transitions					
			energy MeV	transition probability	photon energy MeV	photons emitted	transitions internally converted	photon energy MeV	photons emitted	transitions internally converted
Platinum-197	20·0h	β⁻	0·450 0·642 0·719	10% 79% 11%	0·077 0·191 0·269	17% 4·4% 0·3%	72% 5·2% 0·1%			
Plutonium-238	87·75y	see Uranium/Radium (4n+2) Radioactive series, p. 382								
Plutonium-239	2·44 ×10⁴y	α	5·103 5·142 5·155 others	11% 15% 73% <0·05% each	0·039 0·052 0·129	0·008% 0·023% 0·006%	2·7% 5·6%	0·375 0·414 others	0·0016% 0·0015% <0·001% each	
Plutonium-240	6537y	see Uranium/Thorium (4n) Radioactive series, p. 378			via 26·1m ²³⁵mU 0·00007	~0%	~100%			
Plutonium-241	15y	see Americium/Neptunium (4n+1) Radioactive series, p. 380								
Polonium-208	2·90y	α e.c.	5·114	~100% 1·8 × 10⁻³%	0·292 0·539 0·571	~1·3 × 10⁻³% ~0·3 × 10⁻³% ~0·8 × 10⁻³%		0·603 0·862	~0·6 × 10⁻³% ~0·4 × 10⁻³%	
Polonium-210	138·38d	α	4·5 5·305	0·001% ~100%	0·802	0·00012%				
Potassium-40	1·27 ×10⁹y	β⁻ e.c.	1·314	89·5% 10·5%	1·461	10·3%				
Potassium-42	12·36h	β⁻	1·683 1·995 3·520 others	0·3% 17·6% 82% 0·1%	0·312 0·900 1·021	0·3% 0·05% 0·02%		1·525 1·921 2·424 others	17·9% 0·04% 0·02% <0·01% each	
Potassium-43	22·2h	β⁻	0·44 0·84 1·24 1·46 1·83	2·4% 92·2% 3·5% 0·5% 1·4%	0·221 0·373 0·397 0·593	4·3% 87·2% 11·6% 11·0%		0·617 0·989 1·015 1·022 others	80·4% 0·3% 0·2% 2·0% <0·2% each	

Nuclide	Half-life	Decay	β energies (MeV)	%	γ energies (MeV)	%	(extra)
Praseodymium-142	19.1h	β−	0.592 2.164	3.7% 96.3%	1.572	3.7%	
Praseodymium-143	13.6d	β−	0.931	100%			
Promethium-147	2.623y	β−	0.103 0.225	low ~100%	0.121	4 × 10⁻³%	
Promethium-149	53.08h	β−	0.784 1.071 others	3% 97% low	0.286 0.582 0.850	~2.7% ~0.15% ~0.18%	
Promethium-151	28h	β−	0.85 1.03 1.20 others	39% 10% 11% 40%	0.065 0.066 0.100 0.105	~2% ~1% ~4% ~3%	0.168 ~11% 0.178 ~5% ~1% 0.275 ~6% 0.340 ~20% ~1%
Protactinium-233	27.0d	see Americium/Neptunium (4n+1) Radioactive series, p. 380					
Radium-226	1600y	see Uranium/Radium (4n+2) Radioactive series, p. 382					
Rhenium-186	90h	β− e.c.	0.934 1.071	21% 74% 5%	0.122 0.137 0.630	0.5% 9.5% ~0.04%	0.767 ~0.03% 1.0% 11.5%
Rhenium-188	16.9h	β−	1.50 1.98 2.13 others	2.5% 26% 70% 1.5%	0.155 0.478 0.633	16.2% 1.2% 1.6%	0.672 0.2% 0.828 0.6% 0.931 0.8% others <0.2% each 12.8%
Rhodium-105	35.4h	β−	0.246 0.259 0.565 others	19.5% 5.3% 75% 0.2%	0.281 0.306 0.319 0.443	0.2% 5.2% 19.2% 0.04%	0.1% 0.3%
Rubidium-86	18.7d	β−	0.69 1.77	8.8% 91.2%	1.077	8.8%	

nuclide	half life	type of decay	particle energies and transition probabilities		electromagnetic transitions					
			energy MeV	transition probability	photon energy MeV	photons emitted	transitions internally converted	photon energy MeV	photons emitted	transitions internally converted
Ruthenium-97	2·9d	e.c.		100%	0·216	85·4%	3·4%	0·569	0·9%	
					0·324	10·8%	0·2%	others	<0·1% each	
		Daughter ⁹⁷Tc								
Ruthenium-103	39·5d	β⁻	0·101	6·3%	0·053	0·4%	0·8%	0·497	88·2%	0·4%
			0·214	89·0%	0·113	~0·01%		0·557	0·8%	
			0·456	0·3%	0·242	~0·01%		0·610	5·5%	
			0·711	4·4%	0·295	0·3%		0·020–	~0·9% (Rh K X-rays)	
					0·444	0·4%		0·023		
					via 56m ¹⁰³ᵐRh					
					0·040	0·1%	99·7%	0·020–0·023	~8% (Rh K X-rays)	
Ruthenium-106	369d	β⁻	0·039	100%						
		via 30·4s ¹⁰⁶Rh								
		β⁻	1·98	1·7%	0·512	20·6%		1·050	1·5%	
			2·41	10·5%	0·616	0·7%		1·128	0·4%	
			3·03	8·4%	0·622	9·9%		1·562	0·2%	
			3·54	78·9%	0·874	0·4%		others	<0·1% each	
			others	0·5%						
Samarium-153	46·7h	β⁻	0·632	33%	0·070	4·04%		0·097	0·71%	
			0·702	46%	0·075	0·60%		0·103	28·2%	
			0·805	20%	0·083	0·14%		0·173	0·08%	
			others	1%	0·089	0·13%		0·531	0·06%	
								others	<0·05% each	
Scandium-46	83·8d	β⁻	0·357	~100%	0·889	100%				
			1·48	0·004%	1·121	100%				
Scandium-47	3·40d	β⁻	0·44	70%	0·159	69·7%	0·3%			
			0·60	30%						

Isotope	Half-life	Decay	Energy (MeV)	Intensity	Energy (MeV)	Intensity
Selenium-75	120d	e.c.		100%	0.066	1.1%
					0.097	2.9%	0.3%
					0.121	15.7%	3.0%
					0.136	54.0%	0.7%
					0.199	1.5%	1.6%
					0.265	56.9%	0.4%	0.280	18.5%
							0.401	11.7%
							others	<0.05% each
							0.010– 0.012	~50% (As K X-rays) 0.2%
					via 16.4ms ^{75m}As			
					0.024	0.03%	0.010– 0.012	~2.6% (As K X-rays)
					0.280	5.4%	5.5%
					0.304	1.2%	0.1%
Silver-105	40d	e.c.		100%	0.064	10.8%	14.7%	0.345	42.3%
					0.280	30.6%	0.7%	0.443	11.8%
					0.319	4.5%		0.645	11.8%
					0.332	4.5%		1.088	4.1%	0.8%
Silver-110m	253d	β⁻	0.084	67.6%	0.116	0%	1.4%	0.764	22.5%
			0.531	31%	0.447	3.4%		0.818	7.2%
			others	low	0.620	2.7%		0.885	71.7%
		i.t.		1.4%	0.658	94.2%	0.3%	0.937	34.4%
					0.678	11.1%		1.384	25.7%
					0.687	6.9%		1.476	4.1%
					0.707	16.3%		1.505	13.7%
					0.744	4.5%		1.562	1.2%	0.1%
		via 24.5s ^{110}Ag							
		β⁻	2.23	0.1%	0.658	0.1%			
			2.89	1.3%	others	lower intensity			
Silver-111	7.47d	β⁻	0.16	0.4%	0.096	0.28%		0.374	0.003%
			0.69	6.8%	0.245	1.35%	0.08%	0.524	0.003%
			0.79	0.8%	0.278	<0.001%		0.621	0.018%
			1.03	92%	0.342	6.55%		0.867	0.005%
Sodium-22	2.60y	β⁺	0.546	90.49%	0.511	from β⁺			
			1.820	0.05%	1.275	99.95%			
		e.c.		9.46%					
Sodium-24	15.02h	β⁻	0.284	0.08%	1.369	100%		3.861	0.08%
			1.392	99.92%	2.754	99.85%			

nuclide	half life	type of decay	particle energies and transition probabilities		electromagnetic transitions					
			energy MeV	transition probability	photon energy MeV	photons emitted	transitions internally converted	photon energy MeV	photons emitted	transitions internally converted
Strontium-85	65d	e.c.		100%	0.36 0.88	0.002% 0.01%		0.013– 0.015	~60% (Rb K X-rays)	
Strontium-87m	2.805h	e.c. i.t.			via 0.96µs ^{85m}Rb 0.514	99.2%	0.8%			
Strontium-89	50.5d	β⁻	0.554 1.463	0.3% 99.7% ~0.01% ~100%	0.388	82.3%	17.4%	0.014– 0.016	9.4% (Sr K X-rays)	
Strontium-90	28.5y	β⁻	0.546	100%	via 16s ^{89m}Y 0.909	~0.01%				
		Daughter ^{90}Y								
Sulphur-35	87.4d	β⁻	0.167	100%						
Tantalum-182	115d	β⁻	0.254 0.320 0.433 0.518 0.549 0.585 others	29% 2% 20% 40% 3% 1% 5%	0.068 0.100 0.152 0.179 0.222 0.229 0.264	41.8% 14.3% 7.2% 3.2% 7.7% 3.8% 3.7%		1.113 1.121 1.189 1.221 1.231 1.257 1.289 others	0.4% 35.0% 16.5% 27.5% 11.6% 1.5% 1.4% <3% each	
Technetium-97	2.6 ×10⁶y	e.c.		100%	0.017– 0.020	~60% (Mo K X-rays)				

Nuclide	Half-life	Decay mode	Energy (MeV)	%	Energy (MeV)	%	Energy (MeV)	%
Technetium-99	2.13 × 10⁵y	β⁻	0.204 0.293	low ~100%	0.089	6 × 10⁻⁴%		
Technetium-99m	6.02h	i.t.		100%	0.002 0.141	~0% 88.5%	99.1% 10.6%	0.03% 0.87%
		Daughter ⁹⁹Tc					0.143	
Tellurium-123m	119.7d	i.t.		100%	0.088 0.159	0.1% 83.5%	99.9% 16.5%	0.027–0.032 ~50% (Te K X-rays)
Tellurium-125m	58d	i.t.		100%	0.035 0.109	7% 0.3%	93% 99.7%	0.027–0.032 ~110% (Te K X-rays)
Tellurium-127m	109d	β⁻ i.t.	0.723	2.4% 97.6%	0.058 0.088 0.659	~0.5% ~0.1% ~0.012%	~1.9% 97.5%	others 0.027–0.032 ~37% (Te K X-rays) others
		via 9.35h ¹²⁷Te in equilibrium β⁻	0.277 0.695	~1.2% 96.8%	0.058 0.203 0.215	0.03% 0.06% 0.04%	0.12%	0.360 0.418 others 0.13% 0.96% lower intensity
Tellurium-132	78h	β⁻	0.215	100%	0.050 0.112	13.9% 1.8%	86.1% 1.2%	0.116 0.228 1.9% 89% 1.1% 8%
		Daughter ¹³²I						
Terbium-160	72.1d	β⁻	0.441 0.481 0.553 0.575 0.791 0.874 others	4.4% 10.0% 3.3% 46.4% 6.8% 26.8% 2.3%	0.087 0.197 0.216 0.299 0.765 0.879	13.8% 5.2% 3.9% 26.9% 2.0% 29.8%	61.2% 1.3% 0.1% 0.5% 0.1%	0.962 0.966 1.178 1.200 1.272 1.312 others 9.9% 24.9% 15.1% 2.3% 7.6% 2.9% <2% each 0.1%
Thallium-204	3.78y	β⁻ e.c.	0.763	97.4% 2.6%	0.069–0.083	~1.5% (Hg K X-rays)		
Thorium-228	1.913y	see Uranium/Thorium (4n) Radioactive series, p. 378						
Thorium-232	1.405 × 10¹⁰y	see Uranium/Thorium (4n) Radioactive series, p. 378						

nuclide	half life	type of decay	particle energies and transition probabilities		electromagnetic transitions					
			energy MeV	transition probability	photon energy MeV	photons emitted	transitions internally converted	photon energy MeV	photons emitted	transitions internally converted
Thulium-170	128d	β−	0·884	22·8%	0·084	3·4%	19·4%	0·051– 0·061	~5% (Yb K X-rays)	
			0·968	77%						
		e.c.		0·2%						
Tin-113	115d	e.c.		100%	0·255	2·1%	0·1%	0·024– 0·028	73% (In K X-rays)	
		Daughter ¹¹³ᵐIn								
Tin-119m	245d	i.t.		100%	0·024	16·3%	83·7%	0·025– 0·029	~28% (Sn K X-rays)	
					0·065	~0%	100%			
Tin-121	27·0h	β−	0·383	100%						
Titanium-44	47·3y	e.c.		100%	0·068	74%	26%	0·078	91%	7%
		via 3·92h ⁴⁴Sc in equilibrium			0·511 from β+			1·501	0·9%		
		β+	1·467	95%	1·156	99·9%		2·657	0·1%	
		e.c.		5%						
Tritium	see Hydrogen-3, p. 362									
Tungsten-181	121·2d	e.c.		100%	0·0062	1%	33%	0·056– 0·067	~65% (Ta K X-rays)	
					0·136	0·04%	0·07%			
					0·152	0·10%	0·13%			
Tungsten-185	75d	β−	0·304	low	0·125	~0·005%				
			0·429	>99·9%						
Tungsten-187	23·9h	β−	0·538	4·4%	0·072	12·3%	9·3%	0·625	1·7%	0·1%
			0·625	56·1%	0·134	9·3%	18·9%	0·686	28·4%	
			0·686	5·5%	0·479	21·9%	0·4%	0·746	0·4%	
			0·693	5·5%	0·512	0·8%		0·773	4·3%	0·1%
			1·311	26·5%	0·552	5·3%		0·865	0·4%	
			others	2%	0·618	8·4%	0·2%	0·880	0·16%	
								others	low intensity	

Isotope	Half-life	Decay	Energy (MeV)	Intensity	Notes
Uranium-232	72y	see Uranium/Thorium (4n) Radioactive series, p. 378			
Uranium-233	1.585×10^5y	see Americium/Neptunium (4n+1) Radioactive series, p. 380			
Uranium-235	7.1×10^8y	see Uranium/Actinium (4n+3) Radioactive series, p. 384			
Uranium-238	4.49×10^9y	see Uranium/Radium (4n+2) Radioactive series, p. 382			
Vanadium-48	16.1d	β+	0.695	50%	
		e.c.		50%	
			0.511	from β+	
			0.929	1.2%	1.312 98.5%
			0.945	8%	1.438 0.1%
			0.983	100%	2.240 2.4%
					others <0.1% each
Vanadium-49	330d	e.c.	0.0045	100%	~20% (Ti K X-rays)
Xenon-133	5.29d	β−	0.080	0.4%	0.030– 0.5%
			0.081	36.6%	0.036 63.3% ~46% (Cs K X-rays)
			0.160	0.05%	99.1%
Xenon-133m	2.26d	i.t.	0.233	8%	0.029– 92%
					0.035 ~59% (Xe K X-rays)
	Daughter ^{133}Xe			100%	
Ytterbium-169	30.7d	e.c.	0.008	0.4%	0.177 94.7% 21.2% 12.6%
			0.021	0.2%	0.198 11.9% 34.8% 15.3%
			0.063	43.4%	0.261 48.9% 1.7% 0.1%
			0.094	2.6%	0.308 10.0% 10.7% 0.7%
			0.110	17.4%	others 42.2% low intensity
			0.118	1.9%	0.049– 3.1%
			0.131	11.1%	0.059 12.9% 179% (Tm K X-rays)
Ytterbium-175	4.2d	β−	0.114	1.7%	0.251 4.3% 0.2%
			0.138	0.1%	0.283 0.1% 3.4%
			0.145	0.4%	0.396 5.7%
				10%	0.2%
				3%	0.3%
				87%	
Yttrium-87	80.3h	β+	0.485	92.2%	0.014– 0.3%
		e.c.	0.511	from β+	0.016 62% (Sr K X-rays)
				0.2%	
				99.8%	
			via 2.805h ^{87m}Sr		
			C.388	85.3%	0.014– 18.0%
					0.016 9.7% (Sr K X-rays)
					0.76
					0.071
					0.352
					0.467
					0.266
					0.346

nuclide	half life	type of decay	particle energies and transition probabilities		electromagnetic transitions			electromagnetic transitions		
			energy MeV	transition probability	photon energy MeV	photons emitted	transitions internally converted	photon energy MeV	photons emitted	transitions internally converted
Yttrium-87m	12·7h	β+ e.c. i.t.	1·2	0·7% ~0·8% 98·5%	0·381 0·511	77% from β+	21·5%	0·014– 0·017	~10·5% (Sr K X-rays)	
		Daughter	⁸⁷Y							
Yttrium-88	106·6d	β+ e.c.	0·763	~0·2% 99·8%	0·511 0·898 1·383 1·836 2·734 3·219 3·52 0·014– 0·016	from β+ 93·2% 0·04% 99·4% 0·6% 0·009% 0·007% ~60% (Sr K X-rays)				
Yttrium-90	64·1h	β–	0·513 2·274	~0·02% ~99·98%	1·761	0%	~0·02%			
Yttrium-91	58·5d	β–	0·340 1·545	0·3% 99·7%	1·205	0·3%				
Zinc-65	243·8d	β+ e.c.	0·325	1·46% 98·54%	0·345 0·511 0·770 1·115 0·008– 0·009	~0·003% from β+ ~0·003% 50·7% ~38% (Cu K X-rays)				
Zinc-69	57m	β–	0·925	100%						
Zinc-69m	13·8h	i.t.		100%	0·439	95%	5%			
		Daughter	⁶⁹Zn							

Zirconium-95	64.0d	β–	0.365	54.7%		0.724	44.5%	0.1%
			0.398	44.6%		0.757	54.6%	0.1%
			0.887	0.7%				
			1.12	low				
						via 86.6h ^{95m}Nb in equilibrium		
		Daughter	^{95}Nb			0.235	0.2%	0.5%
Zirconium-97	16.8h	β–	0.52	5%		0.254	1.2%	1.021	1.3%
			0.86	2%		0.355	2.2%	1.148	2.6%
			1.37	4%		0.508	5.1%	1.276	1.0%
			1.88	86%		0.602	1.4%	1.363	1.3%
			others	3%		0.704	1.0%	1.750	1.3%
								others lower intensity
						via 53s ^{97m}Nb		
		Daughter	^{97}Nb			0.743	93.0%	1.6%

URANIUM/THORIUM (4n) SERIES

nuclide	half life	type of decay	particle energies and transition probabilities		electromagnetic transitions					
			energy MeV	transition probability	photon energy MeV	photons emitted	transitions internally converted	photon energy MeV	photons emitted	transitions internally converted
Plutonium-240	6537y	α	5·014 5·123 5·168	~0·09% ~24% ~76%	0·045	0%	24%			
Uranium-236	2·40 × 10⁷y	α	4·331 4·443 4·493	~0·3% ~26% ~74%	0·050 0·113	0% 0%	26% 0·3%			
Thorium-232	1·405 × 10¹⁰y	α	3·830 3·953 4·012	~0·2% ~23% ~77%	0·059 0·124	0% 0%	22% 0·2%			
Radium-228	5·75y	β⁻	0·048	100%	0·007	0%	100%			
Actinium-228	6·13h	β⁻	0·45 0·49 0·62 0·99 1·02 1·12 1·17 1·76 2·10	4·9% 5·6% 5·7% 7·3% 4·0% 6·5% 33% 20% 13%	0·058 0·099 0·129 0·184 0·209 0·270 0·328 0·338 0·410 0·463	0·6% 0% 2·8% 1·7% 4·3% 4·1% 5·3% 15% 2·5% 3·6%	79·4% 5% 7·2% 5·6% 0·2% 0·2% 0·2% 1% 0·5% 0·6%	0·783 0·796 0·836 0·912 0·966 0·970 1·464 1·503 1·593 1·642	1·4% 4% 2·5% 23% 7% 13% 1·1% 2·5% 4·5% 2·4%	
Uranium-232	72y	α	5·137 5·263 5·320	0·3% 31·7% 68%	0·058 0·129	0·2% 0·08%	31·8% 0·22%			
Thorium-228	1·913y	α	5·140 5·176 5·211 5·341 5·424 others	0·03% 0·2% 0·4% 28% 71% low	0·085 0·132 0·167 0·216	1·6% 0·19% 0·12% 0·29%	27·4% 0·01% 0·18% 0·01%			

Nuclide	Half-life	Decay	Energy	%	Energy	%	Energy	%
Radium-224	3.64 d	α	5.447	5.2%	0.241	4.2%		1.0%
			5.684	94.8%				
Radon-220	55.3 s	α	5.747	0.07%	0.542	0.07%		
			6.288	99.93%				
Polonium-216	0.15 s	α	5.984	~0.002%	0.808	0.002%		
			6.777	~100%				
Lead-212	10.6 h	β⁻	0.155	5%	0.115	0.6%		4.1%
			0.332	82%	0.239	44.8%		37.5%
			0.571	13%	0.300	3.4%		1.5%
Bismuth-212	60.6 m	α	5.607	0.4%	0.040	1%		28%
			5.768	0.6%	0.288	0.3%		0.1%
			6.051	25.2%	0.328	0.1%		
			6.090	9.6%	0.453	0.4%		
			others	low	0.727	6.6%		0.1%
					0.785	1.1%		0.1%
		β⁻	0.445	0.7%	0.893	0.4%		
			0.572	0.3%	0.952	0.2%		
			0.630	1.9%	1.079	0.5%		
			0.738	1.5%	1.513	0.3%		
			1.524	4.5%	1.621	1.5%		
			2.251	55.2%	1.680	0.1%		
			others	low	1.806	0.1%		
Polonium-212	3.05 × 10⁻⁷ s	α	8.785	100%				
Thallium-208	3.07 m	β⁻	1.032	3.4%	0.253	0.8%	0.860	12.3%
			1.073	0.6%	0.277	6.9%	1.093	0.4%
			1.285	24%	0.511	23.0%	2.615	99.8%
			1.518	22%	0.583	85.8%		
			1.795	50%	0.763	1.8%		
Lead-208	stable							

Percentages relate to the disintegrations of the individual nuclides.

AMERICIUM/NEPTUNIUM (4n + 1) SERIES

nuclide	half life	type of decay	particle energies and transition probabilities		electromagnetic transitions					
			energy MeV	transition probability	photon energy MeV	photons emitted	transitions internally converted	photon energy MeV	photons emitted	transitions internally converted
Plutonium-241	15y	α β−	~4·9 0·0208	2·45 × 10⁻³% 100%						
Americium-241	433y	α	5·387 5·442 5·484 others	1·6% 12·5% 85·2% 0·7%	0·026 0·033 0·043 0·060	2·5% 0·1% 0·1% 35·3%	~10% ~20% ~10% ~40%			
Uranium-237	6·75d	β−	present in the series only in very low abundance							
Neptunium-237	2·14 × 10⁶y	α	4·638 4·663 4·765 4·770 4·787 4·802 4·816 4·872 others	6% 3·3% 8% 25% 47% ~3% 2·5% 2·6% ~2·6%	0·029 0·087	12% 13%	~11% 18%			
Protactinium-233	27·0d	β−	0·15 0·17 0·23 0·26 0·53 0·57	27% 15% 36% 17% 2% 3%	0·075 0·087 0·104 0·300 0·312 0·340	1·3% 2·0% 0·7% 6·5% 28% 4·3%	~10% ~10% ~2·5% 5% 27% 2·3%	0·375 0·298 0·416	0·7% 1·3% 1·7%	
Uranium-233	1·585 × 10⁵y	α	4·729 4·783 4·824 others	1·6% 13·2% 84·4% 0·8%	0·042	0·16%	14·3%			

381

Nuclide	Half-life	Decay	Energy (MeV)	Intensity			γ Energy (MeV)	Intensity
Thorium-229	7340y	α	4·806	11·4%	0·017	~0%	0·137	~2% ~8%
			4·837	58·2%	0·025	~0%	0·155	~2% ~7%
			4·894	10·7%	0·031	~4%	0·194	~5% ~10%
			4·961	6·0%	0·075	~0% ~19%	0·211	~3% ~7%
			5·048	6·7%	0·086	~0% ~13%		
			others	7·0%	0·125	~4% ~3%		
Radium-225	14·8d	β⁻	0·320	~63%	0·040	~31% ~32%		
			0·392	~37%				
Actinium-225	10·0d	α	5·732	9·5%	0·063	0·5% 0·7%	0·100	3·0% 0·8%
			5·793	29%	0·074	1·2% 0·2%	0·111	0·9% 1·0%
			5·825	52%	0·086	3·0%	0·150	1·1% 0·1%
			others	9·5%	0·096	1·0% 0·3%		
Francium-221	4·8m	α	6·120	15·3%	0·099	~0·2% ~1·4%		
			6·340	82%	0·218	~13% ~5%		
			others	2·7%				
Astatine-217	3 x 10⁻²s	α	7·066	100%				
Bismuth-213	46m	α	5·549	0·2%	0·440	32% 6%		
			5·869	2·0%				
		β⁻	0·32	1%				
			1·02	31%				
			1·42	65·8%				
Polonium-213	4·2 x 10⁻⁶s	α	8·377	100%				
Thallium-209	2·20m	β⁻	1·83	100%	0·117	82% 18%	1·566	100%
					0·465	82% 18%		
Lead-209	3·31h	β⁻	0·635	100%				
Bismuth-209	stable							

Percentages relate to disintegrations of the individual nuclides.

URANIUM/RADIUM (4n + 2) SERIES

nuclide	half life	type of decay	particle energies and transition probabilities		electromagnetic transitions					
			energy MeV	transition probability	photon energy MeV	photons emitted	transitions internally converted	photon energy MeV	photons emitted	transitions internally converted
Plutonium-242	3.87 x 10⁵y	α	4.856 4.900	21.1% 78.9%	0.045	0%	21.1%			
Uranium-238	4.49 x 10⁹y	α	4.145 4.195	23% 77%	0.048	0%	23%			
Thorium-234	24.1d	β⁻	0.100 0.101 0.193	12% 21% 67%	0.030 0.063 0.092	0% 5.7% 3.2%	8.2% 2.3% 16.5%	0.093	3.6%	
Protactinium-234m	1.17m	β⁻ i.t.	2.29 others	98% low 0.13%	0.043 0.767	0% 0.2%	1%	0.810 1.001	0.5% 0.6%	
Protactinium-234	6.70h	β⁻	present in the series only in very low abundance							
Plutonium-238	87.75y	α	5.445 5.499 others	28.7% 71.1% 0.2%	0.043– 0.094– 0.115	~0% ~2.1 x 10⁻⁴% (U K X-rays)	~28.5%	0.011– 0.022	~13% (U L X-rays)	
Uranium-234	2.48 x 10⁵y	α	4.723 4.773	27.5% 72.5%	0.053	0.1%	27.4%			
Thorium-230	7.7 x 10⁴y	α	4.618 4.684 others	23.4% 76.3% 0.3%	0.068	0.4%	22.6%			
Radium-226	1600y	α	4.598 4.781	5.5% 94.5%	0.186	3.4%	2.1%			
Radon-222	3.824d	α	5.486	100%						
Polonium-218	3.05m	α β⁻	6.000 0.277	~100% ~0.02%						

Nuclide	Half-life	Decay	Energies (MeV)		Intensities	
Astatine-218	~2s	α, β⁻	present in the series only in very low abundance			
Radon-218	3.0×10^{-2}s	α	present in the series only in very low abundance			
Lead-214	26.8m	β⁻	0.21		0.5%	
			0.51		15.5%	
			0.69		42%	
			0.74		36%	
			1.03		6%	
			0.053	~0%		~11%
			0.242	6.7%		5.3%
			0.295	16.9%		8.1%
			0.352	32.0%		10.0%
Bismuth-214	19.8m	α	4.9–5.5		0.02%	
		β⁻	0.42		11%	
			1.02		23%	
			1.51		18%	
			1.55		15%	
			1.88		9%	
			2.6		4%	
			3.27		20%	
			0.273		5.3%	0.3%
			0.609		41.7%	
			0.769		5.3%	
			1.120		14.3%	
			1.238		5.0%	
			1.378		4.8%	
			1.764		15.9%	
			2.204		5.3%	
Polonium-214	1.62×10^{-4}s	α	7.688		100%	
Thallium-210	1.30m	β⁻	present in the series only in very low abundance			
Lead-210	22.3y	β⁻	0.015		~80%	
			0.061		~20%	
			0.046	~4%		~76%
			0.009–			~21% (Bi L X-rays)
			0.017			
Bismuth-210	5.01d	α	~4.67		~1.3 × 10⁻⁴%	
		β⁻	1.161		~100%	
Polonium-210	138.38d	α	5.305		100%	
Thallium-206	4.20m	β⁻	present in the series only in very low abundance			
Lead-206	stable					

Percentages relate to disintegrations of the individual nuclides.

URANIUM/ACTINIUM (4n + 3) SERIES

nuclide	half life	type of decay	particle energies and transition probabilities		electromagnetic transitions					
			energy MeV	transition probability	photon energy MeV	photons emitted	transitions internally converted	photon energy MeV	photons emitted	transitions internally converted
Plutonium-239	2.44×10^4y	α	5.103	11%	0.039	0.008%	2.7%			
			5.142	15%	0.052	0.023%	5.6%			
			5.155	73%						
			others	1%						
Uranium-235	7.1×10^8y	α			via 26·1m ^{235m}U					
					0·00007 ~0%	~0%	~100%			
			4.216	6%	0.109	2.5%	0.2%			
			4.325	3%	0.144	11.0%	2.0%			
			4.368	12%	0.163	5.0%	0.8%			
			4.374	6%	0.183	0.5%	1.5%			
			4.400	56%	0.185	54.0%	26.0%			
			4.556	3%	0.201	0.8%	2.2%			
			4.598	4%	0.202	5.02	0.5%			
			others	10%						
Thorium-231	25·5h	β⁻	0.140	2.0%	0.027	~12%	~58%			
			0.203	11.0%	0.059	~0%	~80%			
			0.213	1.6%	0.081	~1%	~9%			
			0.303	85.0%	0.082	~0.3%	~4%			
			others	0.4%	0.084	~5%	~23%			
Protactinium-231	3.25×10^4y	α	4.734	8%	0.027	9.5%	~30%			
			4.933	3%	0.030	0.1%	~23%			
			4.950	23%	0.034	~0%	~10%			
			5.013	25%	0.038	0.12%	~10%			
			5.028	20%	0.284	1.6%	0.1%			
			5.030	3%	0.300	2.3%	~1%			
			5.058	11%	0.303	2.3%				
			others	7%	0.330	1.3%	~0.5%			
Actinium-227	21·8y	α	4.938	0.5%	0.009	0%	~44%			
			4.951	0.7%	0.015	0%	~9%			
			others	low	0.025	0%	~1%			
		β⁻	0.019	~10%						
			0.034	~35%						
			0.044	~54%						

Nuclide	Half-life	Decay	Energies (MeV)	%			
Thorium-227	18.5 d	α	5.707 5.712 5.755 5.976 6.037 others	8.2% 4.9% 20.3% 23.4% 24.5% 18.7%	0.030 0.032 0.045 0.050 0.062 0.236	~0% ~0% ~0% ~10% ~0% 12.5%	~27% ~12% ~6% ~6% ~10% 0.5%	0.256 others	6.3% low intensity	1.2%
Francium-223	22 m	β⁻	0.78 0.914 1.069 1.099 others	2.7% 15.5% 12.5% 68% 1.3%	0.050 0.080 0.205 0.235	43% 10.8% 1.4% 4.3%	29% 2.2% 1.8% 6.0%			
Radium-223	11.4 d	α	5.534 5.603 5.712 others	9.2% 24.2% 52.5% 14.1%	0.122 0.144 0.154 0.270	~1% ~3% ~5% ~14%	~10% ~15% ~23% ~11%	0.324 0.338	~3% ~2.5%	~3% ~1.5%
Radon-219	4.0 s	α	6.423 6.551 6.817	7.5% 11.5% 81.0%	0.131 0.271 0.402	0.1% 9.9% 6.5%	0.5% 2.2% 0.4%			
Polonium-215	1.78×10^{-3} s	α β⁻	7.384	100% very low						
Astatine-215	1.0×10^{-4} s	α	present in the series only in very low abundance							
Lead-211	36.1 m	β⁻	0.544 0.971 1.376 others	4.8% 1.4% 93% low	0.405 0.427 0.832	3.4% 1.7% 2.9%				
Bismuth-211	2.13 m	α β⁻	6.278 6.622 0.590	16.0% 83.7% 0.3%	0.351	13.3%	2.72			
Polonium-211	0.56 s	α	present in the series only in very low abundance							
Thallium-207	4.77 m	β⁻	0.534 1.436	0.24% 99.76%	0.896	0.24%				
Lead-207	stable									

Percentages relate to disintegrations of the individual nuclides.

Appendix 4 INTERACTION DATA
(A) MASS ATTENUATION COEFFICIENTS AS FUNCTIONS OF PHOTON ENERGY

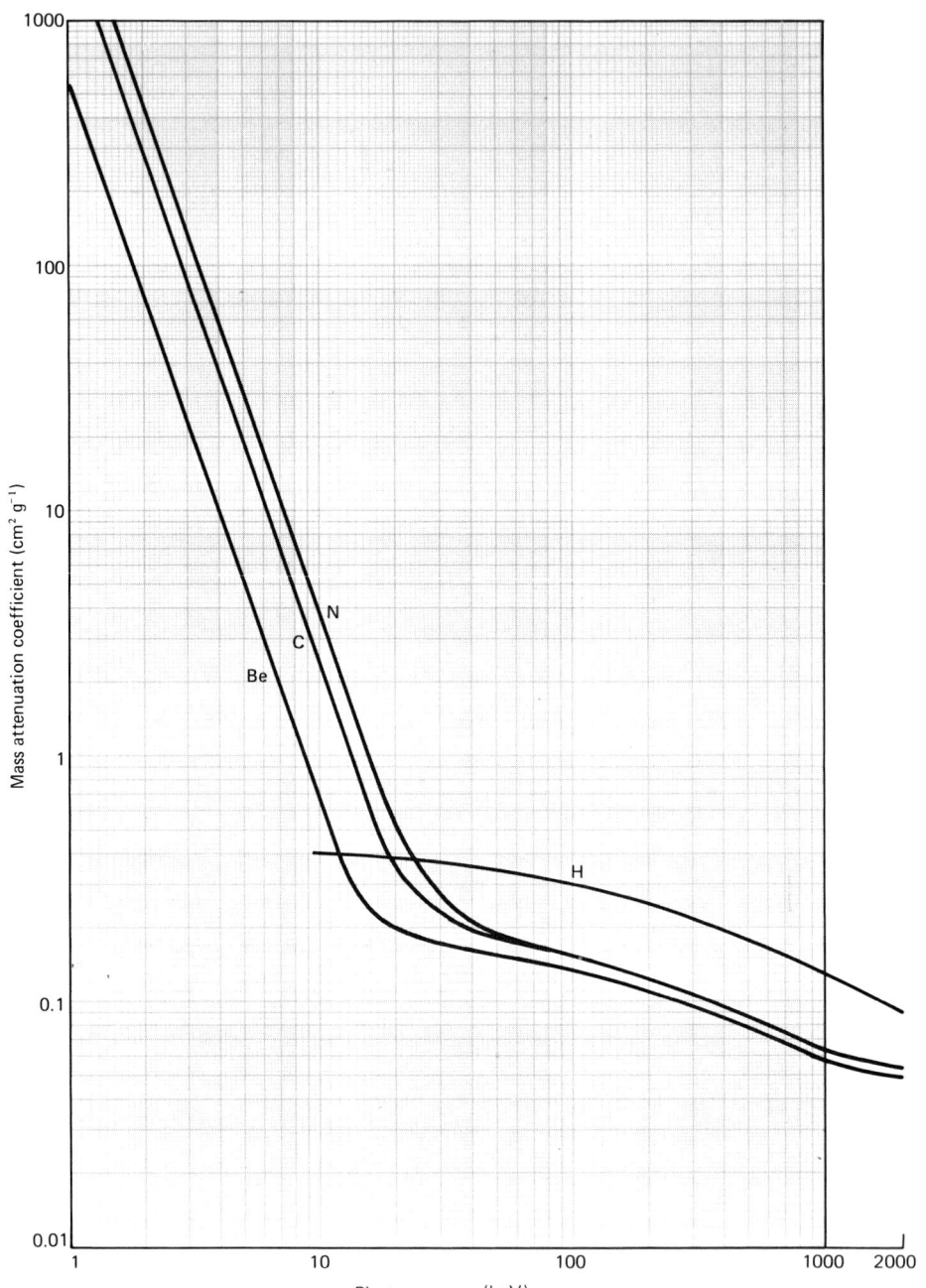

From Thiesen, R. and Vollath, D., *Tables of X-ray Mass Attenuation Coefficients*, Verlag Stahleisen, Düsseldorf (1967) and White Grodstein, G., 'X-ray Attenuation Coefficients from 10 keV to 100 MeV', Nat. Bur. Standards Circ. 583 (1957)

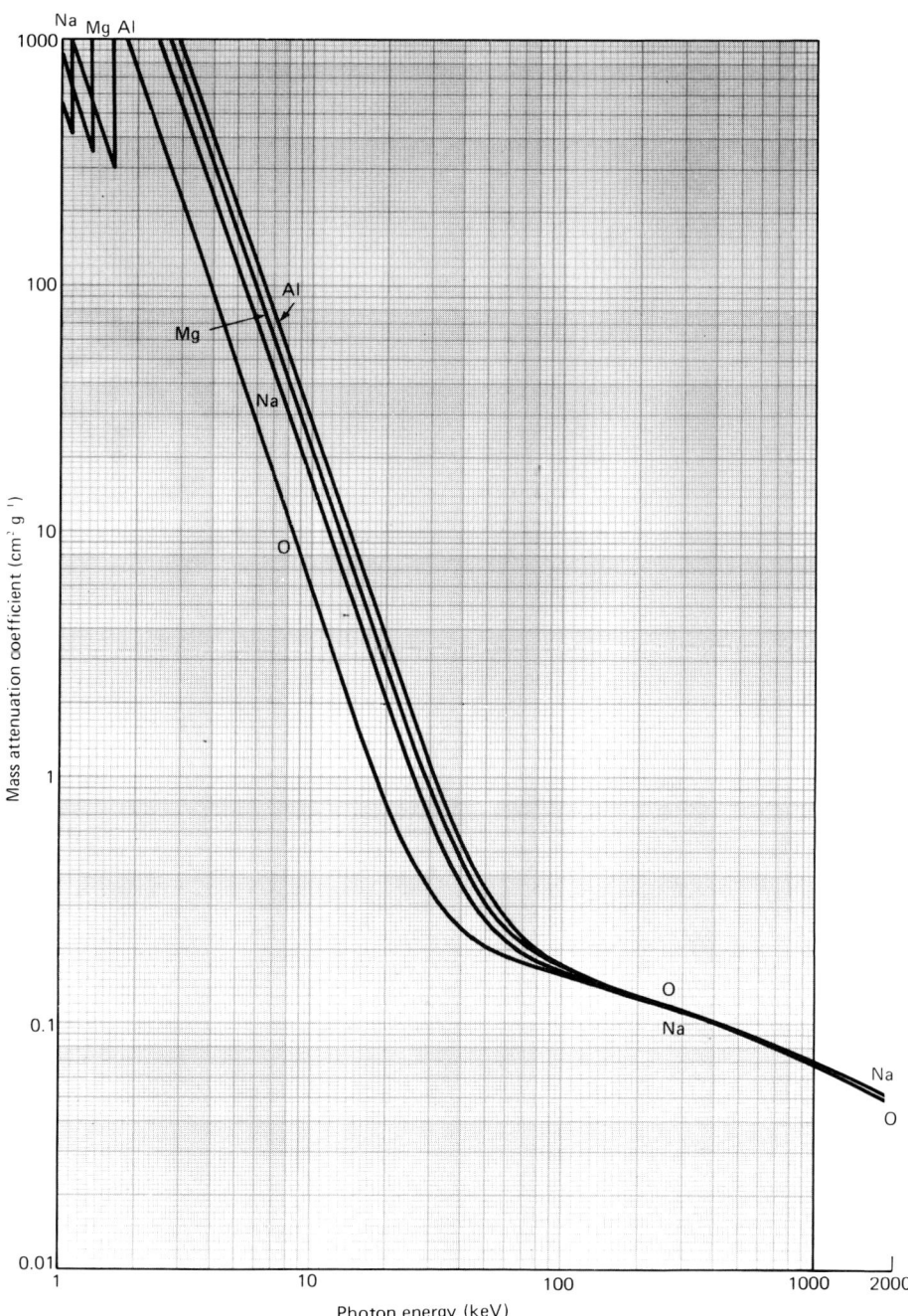

Appendix 4(A) (continued)

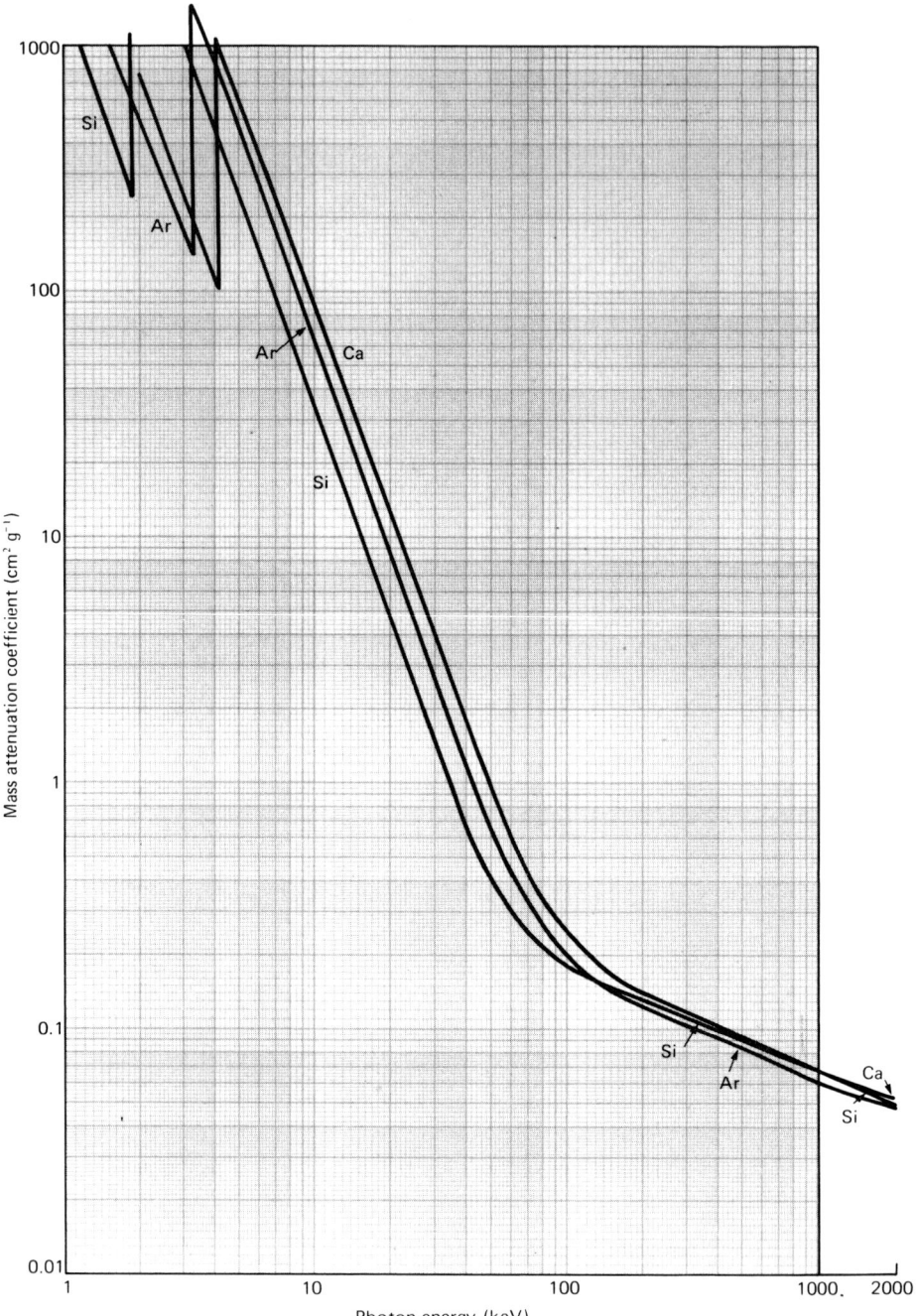

Appendix 4(A) (continued)

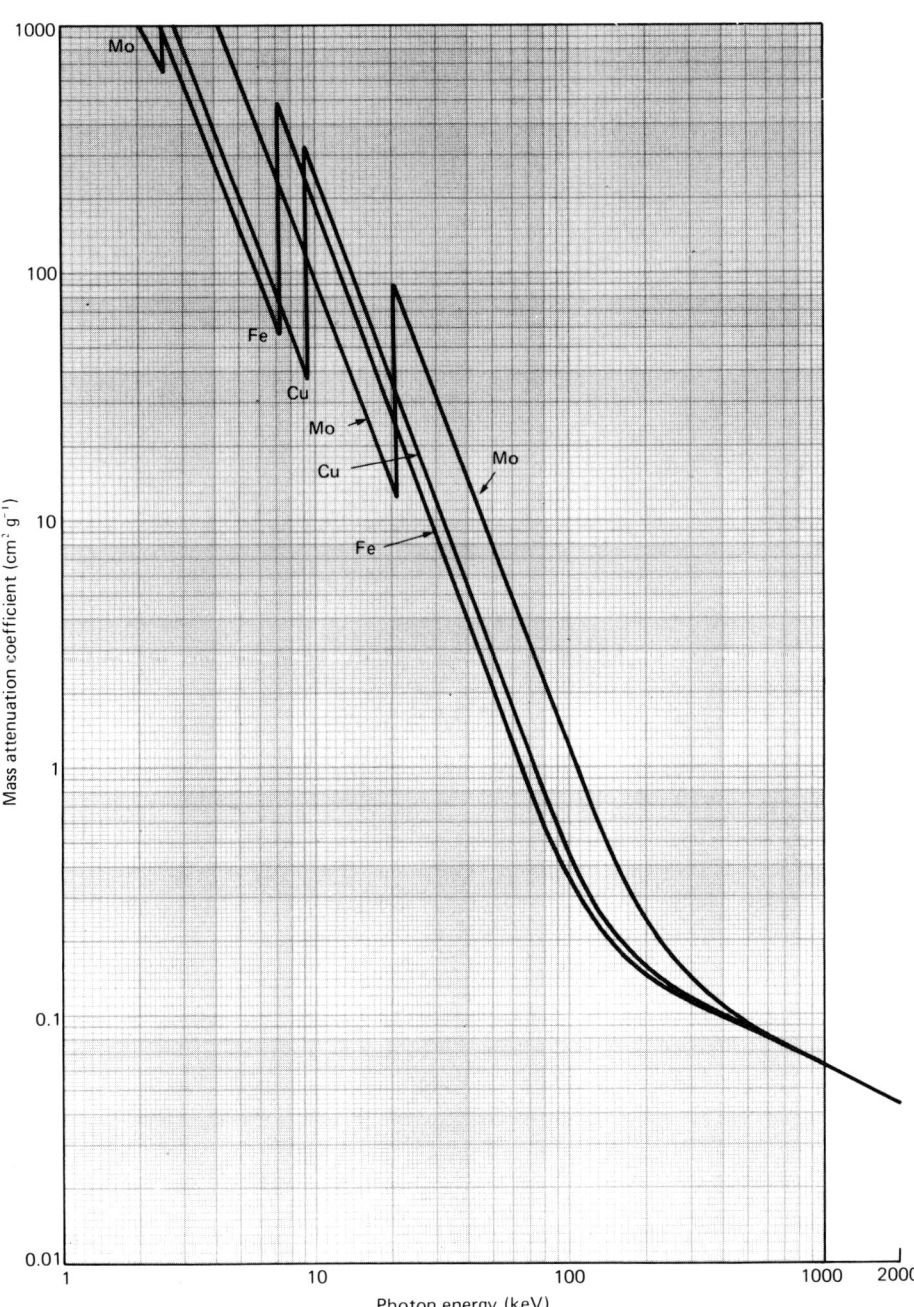

Appendix 4(A) (continued)

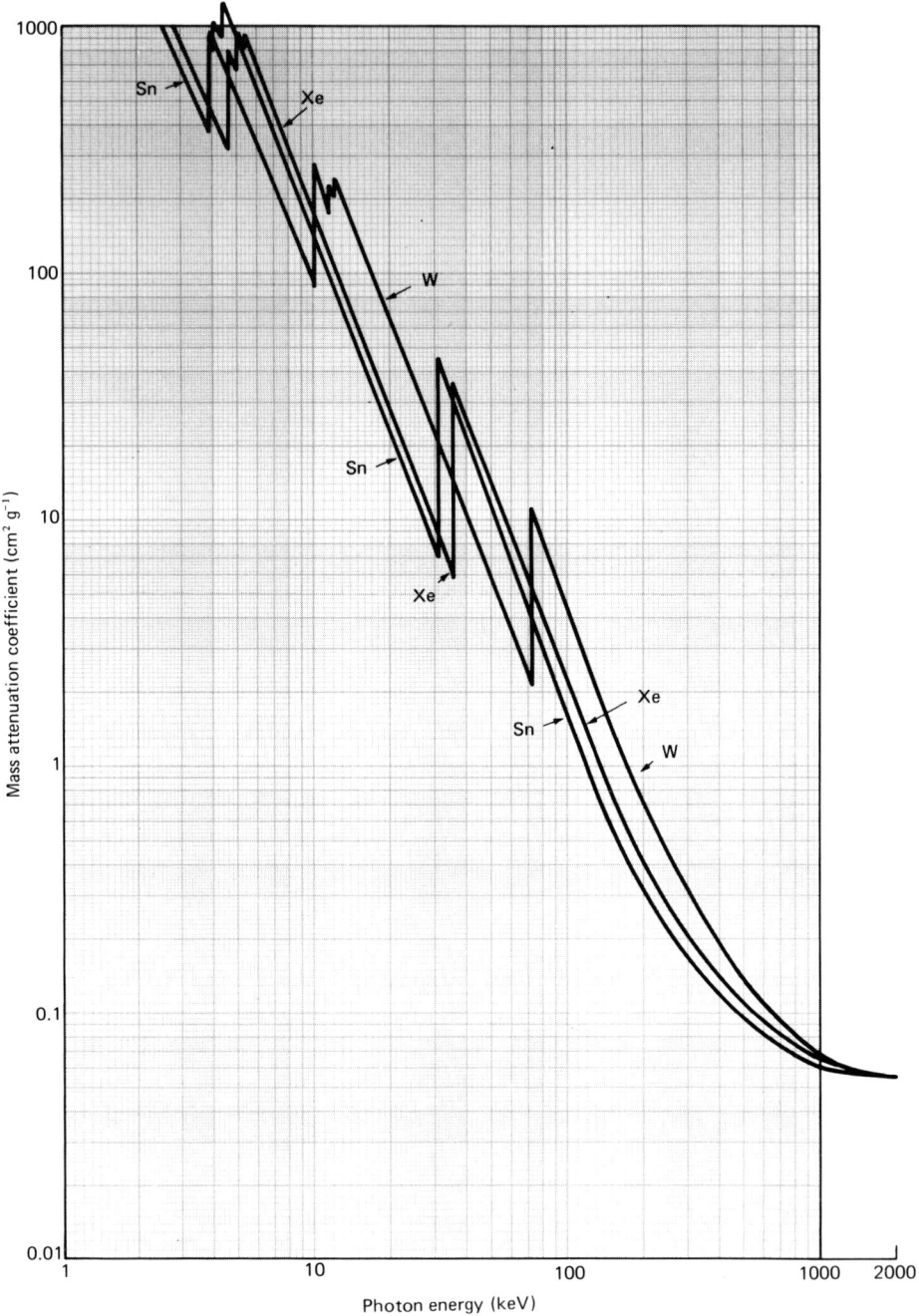

Appendix 4(A) (continued)

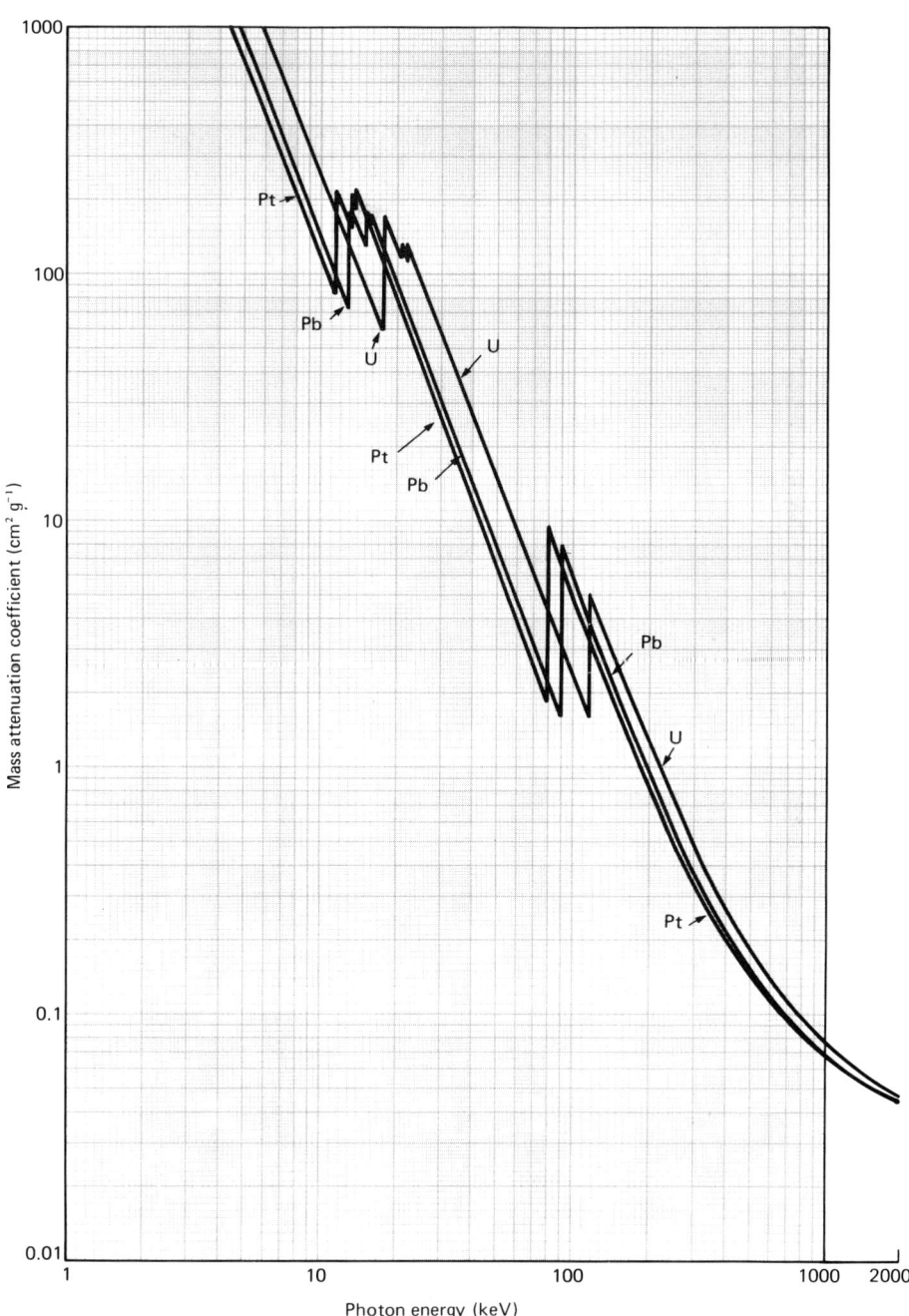

Appendix 4(A) (continued)

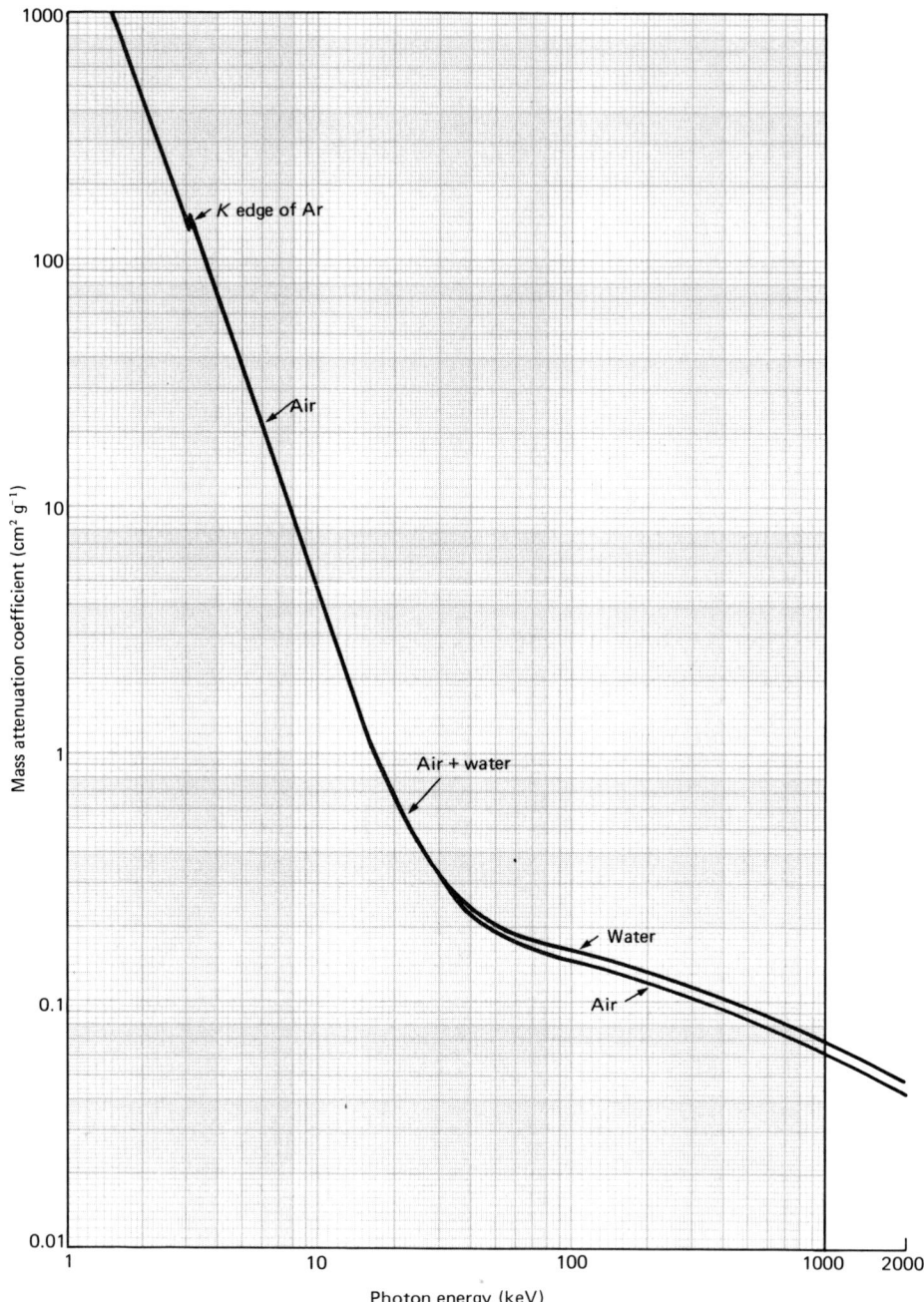

Appendix 4(A) (continued)

(B) ELECTRON/BETA RANGE AS A FUNCTION OF ENERGY FOR ALL MATERIALS EXCEPT HYDROGEN AND THE HEAVY ELEMENTS

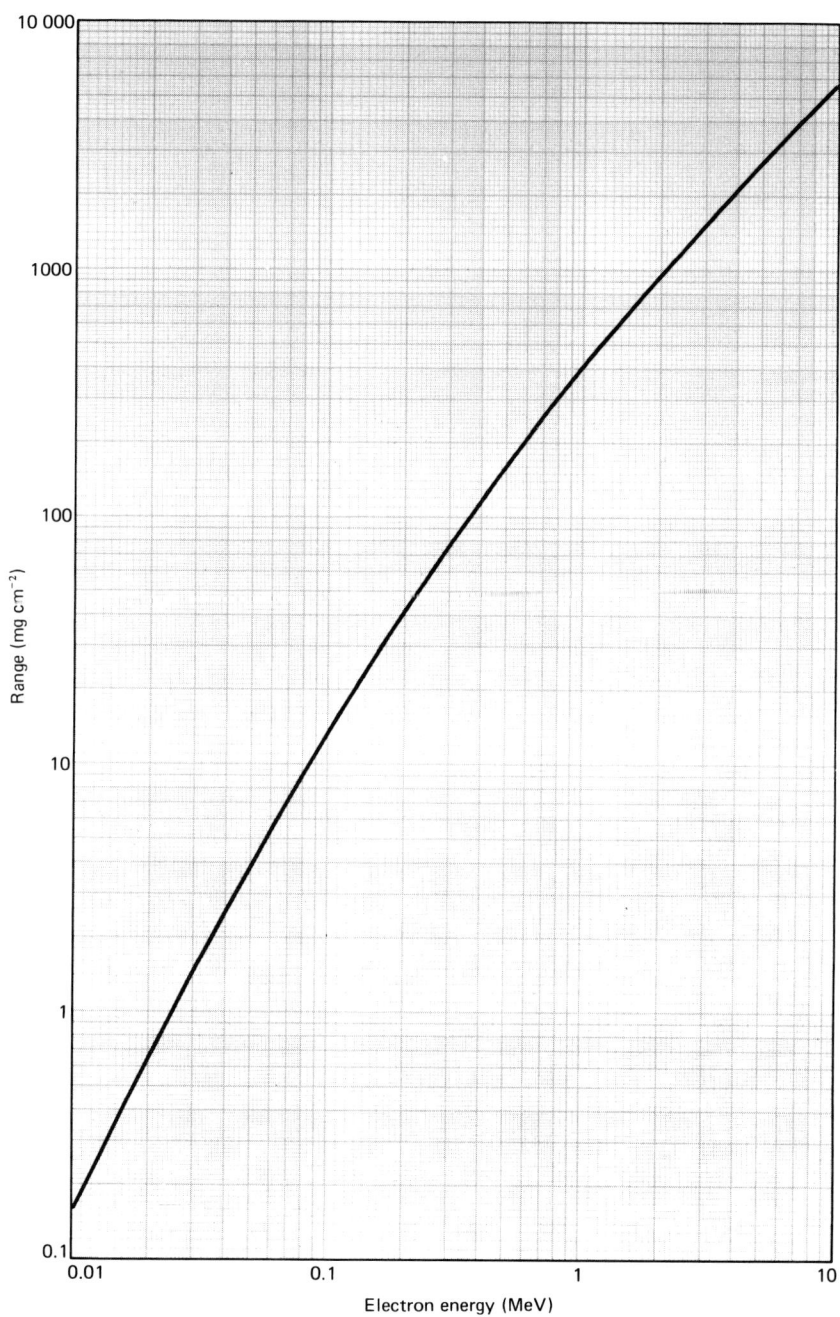

From Katz, L. and Penfold, A.S., Rev. Mod. Phys., 24, 28 (1952)

(C) ELECTRON/BETA dE/dx AS A FUNCTION OF ENERGY

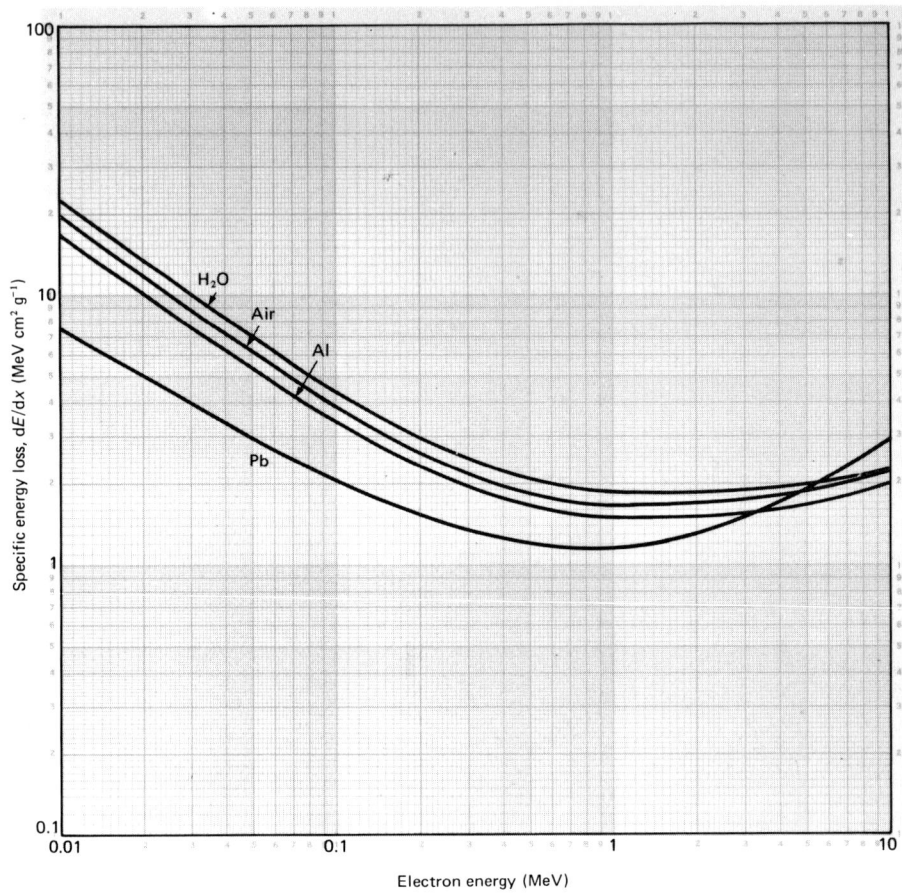

From Heitler, W., *The Quantum Theory of Radiation*, Oxford University Press, London (1954)

(D) ALPHA PARTICLE RANGE AS A FUNCTION OF ENERGY

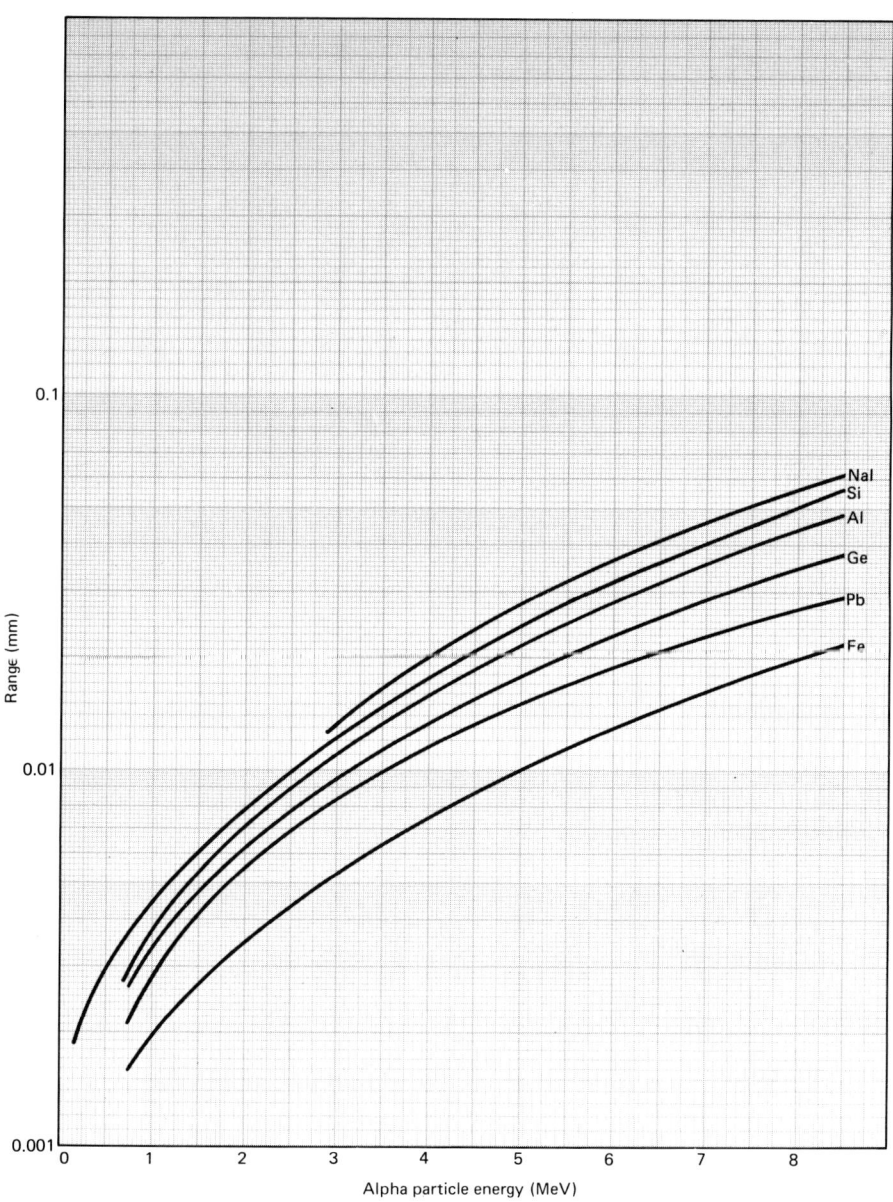

(E) NEUTRON INTERACTION COEFFICIENTS AT THERMAL ENERGY

Element	At. No.	(μ/ρ) true	(μ/ρ) scattering	(μ/ρ) total	μ total
H	1	0.11	48.4	48.5	
Li	3	3.5	0.17	3.7	2.0
Be	4	0.0003	0.50	0.50	0.92
B	5	24		24	60
C	6	0.00015	0.26	0.26	0.60
N	7	0.048	0.43	0.48	
O	8	<0.00002	0.15	0.15	
F	9	<0.0003	0.11	0.11	
Ne	10	0.006		0.006	
Na	11	0.007	0.092	0.099	0.097
Mg	12	0.001	0.092	0.093	0.16
Al	13	0.003	0.033	0.036	0.97
Si	14	0.001	0.043	0.044	0.10
P	15	0.002	0.060	0.062	0.12
S	16	0.0055	0.023	0.029	0.058
Cl	17	0.33	0.255	0.59	
Ar	18	0.0060		0.006	
K	19	0.018	0.031	0.049	0.042
Ca	20	0.0037	0.053	0.057	0.088
Sc	21	0.09	0.175	0.27	0.68
Ti	22	0.044	0.075	0.119	0.54
V	23	0.033	0.060	0.093	0.56
Cr	24	0.021	0.044	0.065	0.46
Mn	25	0.083	0.024	0.107	0.79
Fe	26	0.015	0.126	0.141	1.1
Co	27	0.21	0.051	0.26	2.2
Ni	28	0.028	0.185	0.213	1.9
Cu	29	0.021	0.074	0.095	0.85
Zn	30	0.0055	0.039	0.045	0.32
Ga	31	0.015		0.015	0.089
Ge	32	0.011	0.071	0.082	0.45
As	33	0.020	0.056	0.076	0.44
Se	34	0.056	0.076	0.132	0.59
Br	35	0.029	0.045	0.074	
Kr	36	0.0002		0.0002	
Rb	37	0.0029	0.039	0.042	0.064
Sr	38	0.0048	0.065	0.070	0.18
Y	39	0.0056		0.0056	0.021
Zr	40	0.0006	0.046	0.047	0.31
Nb	41	0.0041	0.040	0.044	0.37
Mo	42	0.009	0.046	0.055	0.55
Ru	44	0.009		0.009	0.11
Rh	45	0.53		0.53	6.6
Pd	46	0.023	0.027	0.050	5.7
Ag	47	0.20	0.039	0.24	2.5
Cd	48	11.2		11.2	97
In	49	0.60		0.60	4.4
Sn	50	0.002	0.025	0.027	0.20
Sb	51	0.016	0.021	0.037	0.25

Element	At. No.	(μ/ρ) true	(μ/ρ) scattering	(μ/ρ) total	μ total
Te	52	0.013	0.018	0.031	0.19
I	53	0.018	0.018	0.036	0.18
Xe	54	0.083		0.083	
Cs	55	0.077	0.032	0.109	0.20
Ba	56	0.0027	0.015†	0.018	0.068
La	57	0.023	0.040	0.063	0.39
Ce	58	0.0021	0.012	0.014	0.097
Pr	59	0.029	0.017	0.046	0.30
Nd	60	0.11	0.10	0.21	1.5
Sm	62	25		25	195
Eu	63	10		10	52
Gd	64	84		84	497
Tb	65	0.09		0.09‡	0.75‡
Dy	66	2.0		2.0‡	17.2‡
Ho	67	0.015		0.015‡	1.3‡
Er	68	0.36	0.054	0.41	2.0
Tm	69	0.25		0.25‡	2.3‡
Yb	70	0.076		0.076‡	0.42‡
Lu	71	0.22		0.22‡	2.1‡
Hf	72	0.20		0.20‡	2.3‡
Ta	73	0.044	0.023	0.067	1.1
W	74	0.036	0.022	0.058	1.1
Re	75	0.16		0.16‡	3.4
Os	76	0.028		0.028‡	0.63‡
Ir	77	0.80		0.80‡	18‡
Pt	78	0.015	0.035	0.050	11
Au	79	0.17	0.027	0.20	3.9
Hg	80	0.63	0.080	0.71	9.6
Tl	81	0.006	0.021†	0.027†	0.32†
Pb	82	0.0003	0.034	0.034	0.39
Bi	83	<0.00003	0.029	0.029	0.28
Th	90		0.033	0.033*	0.37*
U	92	0.005	0.023†	0.028†	0.52

From Thewlis, J., *Brit. J. Appl. Phys.*, **7**, 345 (1956) and *Prog. Nondestructive Testing*, **1**, 111 (1958)

Index

Abrasion-resistant source, 64
Absorbed dose, 117, 150
Absorption coefficient, 152
Absorption edge, 41, 91, 350
Absorption interaction, 66
Accelerator, 59
Activation, 63, 69, 104
Activator, 248
Activity, 46
Air equivalent, 156
Air-equivalent chamber, 184
Air monitor, 157
Air wall chamber, 183
Alpha decay, 47
Alpha particle, 47
    energy spectrum, 49
    range, 86, 395
Aluminium-28, 327
Americium-241, 58, 232
Amplification, 316
Amplifier, 138, 316
Amplitude
    pulse, 305
    wave, 4
Analog processing, 342
Analog pulse, 118, 125, 307
Analog-to-digital converter (ADC), 326
Anger, 165, 338
Ångstrom (Å), 5
Angular frequency, 4
Angular momentum, 3
Annihilation, 51, 88
Anticoincidence counting, 173, 257, 331, 335
Antiparticle, 51
Area monitor, 157
Artefact, 270
Artificial radionuclide, 63
ASA film speed, 275
Astable multivibrator, 310
Atom, 18
Atomic mass unit, 33
Atomic number, 18, 30, 350
Atomic weight, 350
Attenuation, 73

Attenuation coefficient
    neutron, 102, 396
    photon, 76, 90, 386
    photon in air, 153
    photon in germanium, 229
    photon in lead, 92
    photon in silicon, 229
    photon in sodium iodide, 256
Auger electron, 44
Automatic sample changer, 260
Autoradiography, 166, 281
Avalanche, 186

Background radiation, 64, 129
Backscatter peak, 96, 148
Band structure, 27
    inorganic scintillator, 247
    insulator, 210
    semiconductor, 213, 216
    silicon, 212
    silver bromide, 269
    solid-state devices, 225
Bandwidth, 307
Barium-133, 233
Barn, 72
Baryon, 61
Baseline restoration, 318
Beam chopper, 174
Becquerel, 46
Beta decay, 50
Beta particle, 50
    energy spectrum, 52
    range, 86, 393
Bethe–Bloch formula, 83
Beva-electronvolt (BeV), 3
Bias, 305, 323
Bin, 333
Binary code, 319
Binary coded decimal (BCD), 320
Binding energy
    atom, 17
    nucleus, 33
Bipolar pulse, 310, 313
Bismuth-207, 149

399

400   Index

Bismuth-212, 49
Bistable multivibrator, 311
Bit, 320, 343
Block diagram, 305
Bohr model, 18
Bond, 26
Boron trifluoride counter, 204
Bound system, 6, 10, 14
Bragg curve, 85
Bragg–Gray principle, 185
Bremsstrahlung, 40, 87
Broad beam geometry, 79
BSI film speed, 275
Bubble chamber, 296
Built-in potential, 216
Bus, 343
Byte, 343

Cadmium sulphide, 283
Caesium-137, 56, 149, 233, 260
Calibration
  energy, 146
  source strength, 115
Californium-252, 58
CAMAC, 344
Camera
  gamma, 164, 263, 338
  television, 284
Cancellation, 6
Capture interaction, 66
Carbon, 31
Castle, lead, 140
Cathode ray oscilloscope (CRO), 326
Cavity chamber, 184
Central processor unit (CPU), 343
Čerenkov counter, 299
Čerenkov effect, 88, 299
Chain, radioactive, 62
Channel, pulse height, 145, 324
Channel plate, 283
Characteristic curve
  film, 171, 273
  G–M counter, 192
  scintillation counter, 243
Characteristic impedance, 315
Characteristic X-rays, 39, 350
Charge carrier, 214
Charge-sensitive preamplifier, 221, 317
Charpak counter, 197
Chart of the Nuclides, 35
Circuit diagram, 305
Classical model, 2
Cloud chamber, 294
Coaxial cable, 334
Cobalt-60, 56, 233, 260
Coherent scattering, 66
Coincidence counting, 116, 135, 141, 173, 257, 331, 335
Coincidence counting losses, 114
Collective Model of the nucleus, 31, 33

Collimator, 136, 164
  multihole, 165
  pinhole, 165
Coloration, 225
Colour quenching, 253
Compensated diode, 222
Compiler, 343
Compound nucleus, 69
Compton edge, 97, 148
Compton scattering, 93
Computed tomography, 163, 208, 264
Computer, 342
Computer-based multichannel analyser, 329, 346
Conduction band, 27
Conduction counter, 209
Conductor, 29
Conservation laws, 2
Constant fraction discriminator, 324
Contrast, 168, 171, 275
Conversion efficiency, 239
Converter, 268
Copper-64, 54
Cosmic rays, 136, 141
Cosmic ray telescope, 136
Counter, 138, 319
Counting
  gas counter, 202
  scintillation counter, 254
  semiconductor, 226
Counting errors, 110
Counting statistics, 142
Counting system, 139
Coupling capacitance, 309
Covalent bond, 25, 27
Cow system, 63
Crate, 333
Critical radius, 189
Crossover, 127, 313
Cross section, 63, 70, 90
Crystal detector, 209
Curie, 46
Current ion chamber, 185

Dark current, 244
Data processing, 198
Data recording efficiency, 110
Dead time, 113, 141
de Broglie, 11
Decay, radioactive, 45
Decay constant, radioactivity, 45
Decay scheme
  ^{212}Bi, 50
  ^{131}I, ^{64}Cu, 54
  ^{137}Cs, ^{60}Co, 56
  ^{203}Hg, ^{133}Xe, 57
Decay time constant, 125, 178, 309
De-excitation, 19
Dekatron, 321
Delay line, 198

Delay line readout, 338
Delay unit, 331
Delta rays, 82
Density of the elements, 350
Depletion layer, 216
Detection efficiency, 109
    film, 272
    gas counter, 202
    semiconductor, 230
    sodium iodide, 111
Detector types, 108
Deuterium, deuteron, 45
Development, 270
Diamond, 28
Differential attenuation, 74
Differential Čerenkov detector, 303
Differential cross section, 72
Differentiation (pulse), 127, 312
Diffused junction, 222
Diffusion chamber, 296
Digital processing, 342
Digital pulse, 118, 307
DIN film speed, 275
Discriminator, 323
Disintegration constant, 45
Dispersion, 328
Display oscilloscope, 338
Dose, absorbed, 117, 150
Dose equivalent, 117, 151
Dose measurement, 156
Dosemeter, 156
Dosimetry, 150
    film, 279
    gas counter, 206
    scintillator, 261
    solid state, 234
Drift chamber, 198, 207
Drift velocity, 177
D–T reaction, 59
Dwell time, 172, 327
Dynode, 241

Edge contrast, 286
Edge enhancement, 286
Eigenfunction, 7, 14
Eigenstate, 7, 14
Einstein, 8
Elastic scattering, 103
Electrical deflection, 131
Electrical interference, 312
Electrode, 188
Electromagnetic interaction, 69
Electromagnetic radiation, 5
Electron, 18
Electron affinity, 186
Electron capture (EC), 43, 52
Electron configuration, 22
Electron equilibrium, 183
Electron microscopy, 167, 281
Electron radiography (ERG), 287

Electron range, 395
Electronegative element, 25
Electronic noise, 121, 128, 244
Electronvolt (eV), 3
Electropositive element, 25
Electrostatic force, 69
Element, 18, 348, 350
Emulsion, 267
Endoergic reaction, 69
Energy, 2
Energy band structure, 27
Energy calibration, 146
Energy fluence, 152
Energy level diagram, 16, 19, 23, 26, 34, 36
Energy resolution, 117, 119, 148
Energy spectrum
    alpha, 49, 119
    beta, 52
    gamma, 56
Entry window, 107
Equation of motion, 2
Equivalent thickness, 77
Error bar, 143
Escape peak, 148
Excitation, 19
Excitation potential, 24
Excited state, 21
Exciton, 28, 214, 246
Exclusion Principle (Pauli), 14
Exoergic reaction, 69
Expansion chamber, 294
Exponential attenuation, 76
Exponential decay law, 46
Extrapolated range, 82
Extrinsic semiconductor, 215

Fano factor, 121
Fan-out, 332
Fast neutron, 98, 103
Feedback, 316
Fermi, 31
Fermion, 60
Ferrite core, 319
Field, 3
Field effect transistor (FET), 232
Figure of merit, 172
Film, 267
Film badge, 279
Film speed, 171, 272, 275
Filter circuit, 315
Fission reaction, 58
Fixing, 270
Flip-flop circuit, 311, 319
Fluence
    energy, 152
    particle, 117, 152, 273
Fluorescence, 42, 81
Fluorescent screen, 282
Fluorescent yield, 44
Fluoroscopy, 283

Focused dynode, 242
Fogging, 270, 275
Forbidden energy gap, 28
Forbidden transition, 54
Forces, 68
Forward bias, 217
Free space, 3
Frequency (wave), 4
Full absorption peak, 148
Full width half maximum (FWHM), 119, 124, 148
Fundamental particle, 60
Fusion reaction, 59

Gain, 316
Gamma camera, 164, 263, 338
Gamma radiography, 160
Gamma ray/radiation, 38, 55
Gamma ray spectrometry, 56, 147, 335
Gamma scanner, 166, 264
Gas amplification factor, 186
Gas counters, 176
Gas mixture
  ion chamber, 185
  proportional counter, 189
Gas properties, 179
Gas scintillator, 249
Gaussian attenuation, 74
Gaussian distribution, 142
Geiger–Müller counter, 191
Geiger–Nuttall law, 49
Ge(Li) detector, 223
Geometrical attenuation factor, 110
Germanium, 211, 227
Giga-electronvolt (GeV), 3
Glass scintillator, 249
Glendenin formula, 85
Gravitational force, 68
Gray, 150
Grid ion chamber, 181
Ground state, 21
Guard ring chamber, 182

Hadron, 60
Half-life, 30, 47, 354
Half-life measurement, 172
Half-value thickness, 78
Halogen quenching, 192
Hardware, 343
Head amplifier, 316
Heisenberg Uncertainty Principle, 14
Hertz (Hz), 5
High-level discriminator, 323
High pass circuit, 315
High tension (HT), 305
Hornyak button, 257
Hydrogen atom, 15
Hygroscopic scintillators, 249

Image formation, film, 269
Image intensifier, 283
Image quality, 168
Image unsharpness, 168
Imaging, 157
  film, 280
  gas counter, 193, 206, 304
  scintillator, 262
  semiconductor, 234
Impedance matching, 315
Impurity quenching, 253
Incoherent scattering, 66
Inelastic scattering, 103
Inertia, film, 275
Information content, 137
Inorganic scintillator, 245
Input device, 344
Integral nucleonic system, 332
Integrating circuit, 314
Integrating ion chamber, 182
Integrating solid-state devices, 224
Integration, pulse, 314
Interaction cross section, 70
Interaction efficiency, 110, 116, 200
Interaction rate, 73
Interface, 343
Interference, electrical, 312
Internal conversion (IC), 43, 44, 57
Intrinsic layer, 216
Intrinsic scintillator, 246
Intrinsic semiconductor, 213
Iodine-131, 54
Ion implantation, 222
Ion pair, 19
Ion quenching, 190
Ionic bond, 26
Ionisation, 19
Ionisation chamber, 177
Ionisation potential, 21, 37, 348
Ionising radiation, 37
Ionography, 287
Iron-55, 52, 205, 255
Isobar, 31
Isobaric process, 52
Isomeric transition, 55
Isotone, 31
Isotope, 30
Isotope imaging, 163

Joule, 3
Junction diode, 216

K capture, 52
Kilo-electronvolt (keV), 3
Kinetic energy, 2
K shell, 21, 39, 350
Kurie plot, 55

Labelled compound, 163
Latitude, 171, 275
Lead castle, 140
Leakage current, 220
Lepton, 60
Light-emitting diode (LED), 321
Linear attenuation coefficient, 77
Linear energy attenuation coefficient, 152
Linear energy transfer (LET), 83
Linear gate, 330
Linear pulse, 118, 307
Linear stopping power, 82
Line broadening, 118
Line spectrum, 38
Liquid Drop Model, 31
Liquid scintillation counting, 257, 335
Liquid scintillator, 251
Lithium drifted diode, 222
Lithium drifted germanium, 223
Lithium drifted silicon, 223
Live time, 319
Load, 308
Logic pulse, 118, 307
Lorentz contraction, 8
Low-level counting, 141, 257
Low-level discriminator, 128, 323
Low pass circuit, 315
L shell, 21, 39, 350

Machine code, 343
Macroscopic cross section, 99
Magic mixture, 199
Magic number, 33
Magnetic analyser, 136, 293
Magnetic deflection, 132
Magnetic quantum number, 16
Mammography, 287
Mass absorption coefficient, 152
Mass attenuation coefficient, 77
Mass-energy attenuation coefficient, 152
Mass number, 30
Mass stopping power, 83
Material attenuation factor, 110
Maximum permissible dose, 155
Mean energy per ion pair, 38, 82, 120
Mean free path
   particle, 75
   photon, 78
Mean life, 47
Mega-electronvolt (MeV), 3
Memory, 343
Mercury-203, 57
Metal, 29
Metastable state, 55
Microcomputer, 329
Mixer, 332
Model, 1
   atom, 18
   nucleus, 31
Moderator, 103

Modular system, 332
Modulation transfer function (MTF), 169
Module, 333
Molecule, 25
Momentum, 3
Monitor, 156
Monostable multivibrator, 310
Multichannel analyser (MCA), 145, 325
Multichannel scaling (MCS), 172, 326
Multihole collimator, 165
Multiplexer, 332
Multivibrator, 310
Multiwire proportional counter (MWPC), 197
Muon, 60

Narrow beam geometry, 78
Natural radionuclide, 61
Negative feedback, 316
Negatron, 51
Neutrino, 50, 60
Neutron, 29, 57
Neutron activation, 104
Neutron current, 98
Neutron detection, 202, 254, 261
Neutron flux, 98
Neutron generator, 59
Neutron half-life, 30
Neutron interaction coefficient, 102, 396
Neutron interactions, 98
Neutron number, 30
Neutron radiography, 101, 161
Neutron source, 58
NIMS, 334
Noble gas, 24
Noise
   electronic, 121, 169, 311
   image, 169
n-type semiconductor, 215
Nuclear data, 354
Nuclear emulsion, 267, 292
Nuclear energy level diagram, 34
Nuclear reaction, 58, 63, 69
Nucleon, 30
Nucleus, 18, 29, 31
Nuclide, 30

Open source, 64
Operating system, 329
Optical coupling, 240
Optical density, 272
Orbital angular momentum quantum
   number, 15
Orbital model of the atom, 19
Organic scintillator, 250
Oscillator, 310
Output device, 344

Pair production, 97
Paralysis time, 113

Particle, 2
Particle current, 117
Particle fluence, 117, 152, 273
Particle flux density, 117
Particle identification system, 134
Pauli Exclusion Principle, 14
Pauli neutrino hypothesis, 54
Peak-to-Compton ratio, 149
Peak-to-total ratio, 114, 148
Period, wave, 4
Periodic Table, 24, 348
Peripheral, 343
Personal monitor, 157
Phonon, 28, 214, 247
Photocathode, 241
Photoelectric effect, 43, 90
Photoelectron, 43, 90
Photographic emulsion, 267
Photoionisation, 90, 190
Photomultiplier, 241
Photon, 11, 18
Photon attenuation coefficient, 90, 386
Photon detection
    gas counter, 203
    scintillator, 254
    semiconductor, 230
Photon quenching, 190
Photopeak, 148
Pile-up, 125
Pinhole collimator, 165
Pion, 61
Pion therapy, 291
Planchet, 64
Planck's constant, 11
Plastic scintillator, 254
Point spread function (PSF), 168
Poisson distribution, 142
Polaroid-Land, 270
POPOP, 251
Positive hole, 29, 214
Positron, 51, 88
Positron camera, 165, 263
Positronium, 88
Potential energy, 2
Potential well, 16
PPO, 251
Preamplifier, 138, 306, 316
Primary solute, 251
Principal quantum number, 15
Probability amplitude, 13
Probe, 140
Processor, 343
Program, 343
Proportional counter, 186
Proton, 29, 45
p-type semiconductor, 215
Pulse, 305
Pulse generator, 310
Pulse height/amplitude, 178, 305
Pulse height analysis (PHA), 145, 324, 326
Pulse height discriminator, 323

Pulse height selection, 325
Pulse height spectrometry, 119, 132, 144, 324
    gas counter, 205
    scintillator, 260
    semiconductor, 231
Pulse shape discrimination (PSD), 134, 254
Pulse shaping, 312
Pulse time spectrum, 331

Quality factor, 117, 151
Quantum, 11
Quantum efficiency, 116, 272
Quantum number, 7, 15
Quantum theory, 2, 10
Quenching
    gas counter, 189
    scintillator, 249, 253
Q-value, 70

Rad, 150
Radiation damage, 231
Radiation quenching, 253
Radioactive series, 62
Radioactive source, 45
Radioactivity, radioactive decay, 30, 45
Radioassay, 141
Radiochromatography, 167, 208
Radiography
    X-, 159, 280
    neutron, 101, 161
Radionuclide, 30, 45, 61, 354
Radionuclide imaging, 163
Radiopharmaceutical, 30
Radiophotoluminescence (RPL), 226
Radiothermoluminescence, 226
Ramp voltage, 310
Random access memory (RAM), 326, 343
Random coincidence, 114, 125
Range, 75, 82
    alpha, beta, 86
Ratemeter, 322
Reactor, 60
Real time, 319
Recovery time, 127
Recycling, 190
Reduced wavenumber, 4
Reinforcement, 6
Relative biological effectiveness (RBE), 151
Relativity, 2, 8
Rem, 151
Remote handling, 140
Removal coefficient, 99
Resistivity, 214
Resolution
    energy, 119, 148
    spatial, 124
    time, 114, 124

Index 405

Resolving time, 114, 124
Resonance capture, 100
Rest mass, 8
 electron, 9
Rest mass energy, 9
Restrahl, 28
Reverse bias, 217
Rise time, 125, 178
Röntgen, 39
Röntgen unit, 151
Rutherford Model of the atom, 18

Sawtooth voltage, 310
Scaler, 319
Scanner, 166, 264
Scanning, pulse height, 324
Scanning electron microscope, 167
Scatter grid, 160
Scattering, 67
Schrödinger equation, 14, 20
Scintillating proportional counter, 250
Scintillation camera, 338
Scintillation counter, 237, 243, 318, 335
Scintillation efficiency, 239
Scintillation process, 237, 247, 251
Scintillator, 238
Screen factor, 276
Screened film, 267
Sealed source, 64
Secondary solute, 253
Secular equilibrium, 62
Semiconductor, 28, 212, 215
Semiconductor counter, 210
Sensitivity, 109, 272, 275
Shell Model of the atom, 21
Shell Model of the nucleus, 31, 33
Sievert, 151
Signal, 126, 137, 308
 gas counter, 192
Signal-to-noise ratio (SNR)
 electronic, 121, 169, 317
 image, 169
Si(Li) detector, 223
Silicon, 211, 227
Silver bromide, 267, 277
Single-channel analyser (SCA), 145, 324
Sodium chloride, 26, 28
Sodium energy levels, 22
Sodium iodide, 255
Software, 329, 343
Solarisation, 275
Solid, 26
Solid-state devices, 209
Source, radioactive, 209
Source program, 343
Source strength, 46
Space quantisation, 16
Spark chamber, 193, 207
Spark discharge, 194
Spatial frequency, 124, 169

Specific activity, 46
Specific energy loss, 82, 394
Specific gamma ray constant, 154
Specific ionisation, 83
Spectral quality, 239
Spectral response, 243
Spectrometry, 132
 pulse height, 119, 132, 144, 324
Spectroscopic notation, 22
Spin, 3
Spin quantum number, 16
Standard deviation, 143
State function, 14
Stationary wave, 6
Statistical uncertainty, 143
Stereophotography, 295
Stereoradiography, 161
Storage oscilloscope, 338
Straggling, 76
Streamer chamber, 207
Strong interaction, 69
Subtraction ratemeter, 332
Summing, 125
Surface barrier, 222
Surface monitor, 157

Technetium-99m, 55
Telescope, 136, 159
Television camera, 284
Thermal neutron, 58, 98, 100
Thermoelectron, 43
Thermoluminescence, 226
Thermoluminescent dosimetry (TLD), 226
Thimble chamber, 184
Thorium-232, 49
Threshold Čerenkov detector, 303
Time constant, 308
Time dilation, 9
Time-of-flight, 135, 173, 338
Time-to-amplitude converter (TAC), 136, 174, 198, 331
Time-to-digital converter (TDC), 331
Timing, 172, 235, 266
Timing discriminator, 323
Timing unit, 330
Tissue-equivalent, 156
Tissue-equivalent chamber, 184
Tomography, 162
Townsend avalanche, 186
Townsend coefficient, 187
Tracer, 30
Track resolution, 124
Tracking, 159
Transient, 308
Transmission electron microscope, 167
Transmutation, 69
Trigger system, 134, 196, 295
Triode, semiconductor, 235
Tritium, 30, 45, 257
Triton, 45

Index

True coincidence, 116, 135
Tunnel effect, 48

Unibus, 343
Unsharpness, 168

Valence bond, 27
Valency, 24
Veall counter, 204
Venetian blind dynode, 242
Visual display unit (VDU), 326
Volt, 3
Voltage-sensitive preamplifier, 317
Voltage supplies, 305

Wave, 3
Wave function, 4
Wave motion, 3
Wave packet, 13
Wave–particle duality, 10
Wavelength, 4
    electron, neutron, 12
Wavelength shifter, 248, 253
Wavenumber, 5

Weak interaction, 69
Weizsacker, 32
Whole body monitoring, 262
Wipe test, 157
Word, 320, 343

Xenon-133, 57
Xerography, 284
Xeroradiography, 285
Xerox, 284
X-radiation, 38
X-radiography, 160, 281
X-ray, 38
X-rays of the elements, 350
X-ray crystallography, 281
X-ray detector
    gas counter, 203
    semiconductor, 231
X-ray fluorescence, 42
X-ray machine, 41
X-ray spectrometry
    gas counter, 205
    semiconductor, 231

Zinc-64, 36

/539.77T135R>C1/